AAPG TREATISE OF PETROLEUM GEOLOGY
REPRINT SERIES

The American Association of Petroleum Geologists
gratefully acknowledges and appreciates the leadership and support
of the AAPG Foundation in the development of the
Treatise of Petroleum Geology.

TREATISE OF PETROLEUM GEOLOGY REPRINT SERIES

Compiled by Edward A. Beaumont and Norman H. Foster

1. Geologic Basins I: Classification, Modeling, and Predictive Stratigraphy

2. Geologic Basins II: Evaluation, Resource Appraisal, and World Occurrence of Oil and Gas

3. Reservoirs I: Properties

4. Reservoirs II: Sandstones

5. Reservoirs III: Carbonates

6. Traps and Seals I: Structural/Fault-Seal and Hydrodynamic Traps

7. Traps and Seals II: Stratigraphic/Capillary Traps

8. Geochemistry

9. Structural Concepts and Techniques I: Basic Concepts, Folding, and Structural Techniques

10. Structural Concepts and Techniques II: Basement-Involved Deformation

11. Structural Concepts and Techniques III: Detached Deformation

12. Geophysics I: Seismic Methods

13. Geophysics II: Tools for Seismic Interpretation

14. Geophysics III: Geologic Interpretation of Seismic Data

15. Geophysic IV: Gravity, Magnetic, and Magnetotelluric Methods

16. Formation Evaluation I: Log Evaluation

17. Formation Evaluation II: Log Interpretation

18. Photogeology and Photogeomorphology

19. Remote Sensing

REMOTE SENSING

COMPILED BY
EDWARD A. BEAUMONT
AND
NORMAN H. FOSTER

TREATISE OF PETROLEUM GEOLOGY
REPRINT SERIES, NO. 19

PUBLISHED BY
THE AMERICAN ASSOCIATION OF PETROLEUM GEOLOGISTS
TULSA, OKLAHOMA 74101-0979, U.S.A.

ISBN: 0-89181-418-3
ISSN: 1043-6103

Available from:
The AAPG Bookstore
P. O. Box 979
Tulsa, OK 74101-0979
U.S.A.

Phone: 918-584-2555
Telex: 49-9432
Fax: 918-584-0469

Association Editor: Susan Longacre
Science Director: Gary D. Howell
Publications Manager: Cathleen P. Williams
Special Projects Editor: Anne H. Thomas
Project Editor: Michael H. Blechner

Library of Congress Cataloging-in-Publication Data

Remote sensing / compiled by Edward A. Beaumont and Norman H. Foster.
     p.   cm. — (Treatise of petroleum geology reprint series, ISSN 1043-6103 ; no. 19)
     includes bibliographical references.
     ISBN 0-89181-418-3
     1. Petroleum—prospecting—Remote sensing. 2.Petroleum—Prospecting—Remote
sensing—Case studies. I. Beaumont, E.A. (Edward A.) II. Foster, Norman H.
III. American Association of Petroleum Geologists. IV. Series.
TN271. P4R45   1992
622' . 1828—dc20
                                              92-9077
                                               CIP

AMERICAN ASSOCIATION OF PETROLEUM GEOLOGISTS FOUNDATION
TREATISE OF PETROLEUM GEOLOGY FUND*

Major Corporate Contributors
($25,000 or more)

Amoco Production Company
BP Exploration Company Limited
Chevron Corporation
Exxon Company, U.S.A.
Mobil Oil Corporation
Oryx Energy Company
Pennzoil Exploration and Production Company
Shell Oil Company
Texaco Foundation
Union Pacific Foundation
Unocal Corporation

Other Corporate Contributors
($5,000 to $25,000)

ARCO Oil & Gas Company
Ashland Oil, Inc.
Cabot Oil & Gas Corporation
Canadian Hunter Exploration Ltd.
Conoco Inc.
Marathon Oil Company
The McGee Foundation, Inc.
Phillips Petroleum Company
Transco Energy Company
Union Texas Petroleum Corporation

Major Individual Contributors
($1,000 or more)

John J. Amoruso
Thornton E. Anderson
C. Hayden Atchison
Richard A. Baile
Richard R. Bloomer
A. S. Bonner, Jr.
David G. Campbell
Herbert G. Davis
George A. Donnelly, Jr.
Paul H. Dudley, Jr.
Lewis G. Fearing
Lawrence W. Funkhouser
James A. Gibbs
George R. Gibson
William E. Gipson
Mrs. Vito A. (Mary Jane) Gotautas
Robert D. Gunn
Merrill W. Haas
Cecil V. Hagen
Frank W. Harrison
William A. Heck

Roy M. Huffington
J. R. Jackson, Jr.
Harrison C. Jamison
Thomas N. Jordan, Jr.
Hugh M. Looney
Jack P. Martin
John W. Mason
George B. McBride
Dean A. McGee
John R. McMillan
Lee Wayne Moore
Grover E. Murray
Rudolf B. Siegert
Robert M. Sneider
Estate of Mrs. John (Elizabeth) Teagle
Jack C. Threet
Charles Weiner
Harry Westmoreland
James E. Wilson, Jr.
P. W. J. Wood

The Foundation also gratefully acknowledges the many who have supported this endeavor with additional contributions.

*Based on contributions received as of December 31, 1991.

TREATISE OF PETROLEUM GEOLOGY
ADVISORY BOARD

INTRODUCTION

The *Treatise of Petroleum Geology* is AAPG's Diamond Jubilee project, commemorating the 75th anniversary of the organization of the Association in 1916. *The Treatise of Petroleum Geology* is made up of three series: the Handbook of Petroleum Geology, the Atlas of Oil and Gas Fields, and the Reprint Series. With input from more than 260 geologists, geophysicists, geochemists, and engineers from around the world, we designed the *Treatise* to represent the cutting edge in petroleum exploration knowledge and application. The Handbook of Petroleum Geology series is a professional explorationist's guide to the methodology and technology used to find oil and gas. The Atlas of Oil and Gas Fields series collects detailed field studies to illustrate the myriad ways oil and gas are trapped. The third part of the *Treatise*, the Reprint Series, provides landmark papers from diverse worldwide geological, geophysical, geochemical, and engineering publications. In some cases, articles are reprinted from very obscure sources. *Remote Sensing* is part of the Reprint Series.

The papers in the various volumes of the Reprint Series complement the subjects covered in the Handbook series. Papers were selected on the basis of their current usefulness in petroleum exploration and development. Many "classic papers" that led to our present state of knowledge have not been included because of space limitation. In some cases, it has been difficult to decide in which Reprint Series volume a particular paper should be published because that paper covers several topics. We suggest, therefore, that interested readers become familiar with all the Reprint Series volumes if they are looking for a particular paper.

Remote sensing is defined by the American Geological Institute's *Glossary of Geology* as "The collection of information about an object by a recording device that is not in physical contact with it. The term is usually restricted to mean methods that record reflected or radiated electromagnetic energy, rather than methods that involve significant penetration into the Earth. The technique employs such devices as the camera, infrared detectors, microwave frequency receivers, and radar systems." Aerial photographs were the first remotely sensed data, sensing the visible portion of the electromagnetic spectrum. Because photogeology and photogeomorphology have been widely used for many years with great success in petroleum exploration, we included papers on those subjects in a separate volume in the Reprint Series.

Other portions of the electromagnetic spectrum have been sensed using devices and instruments carried by aircraft or satellites. Sensing in the near infrared and thermal infrared range and by radar have proven to be very effective in petroleum exploration, as have multispectral scanner images and photographs.

Mapping surface geology and interpreting subsurface geology from aerial photographs or other remote sensing data are two of the most effective and least expensive methods of exploration. The surface of the earth represents the most extensive, easily accessible, and best-exposed cross section available: a horizontal cross section. In petroleum exploration, one should never ignore the surface but, sadly, often it is not studied. Sometimes the surface will provide few or no clues to the subsurface geology but, in many cases, it may add a critical element necessary for discovery of oil and gas. One must always study aerial photographs and other available remote sensing data from the prospect area. A close correlation should be made between the known subsurface geology and the surface in order to better interpret the sometimes subtle clues on aerial photographs or other remote sensing data. This is especially true over known producing fields, where features such as drainage patterns, single stream morphology, topographic features and patterns, tonal patterns, and lineations may be used as critical elements in finding similar patterns in frontier areas. Explorationists sometimes dismiss the possibility of the surface expression of subsurface structure because of the depth involved or the presence of unconformities in the stratigraphic section. Frequently, they do not even look at the aerial photographs or other remote sensing data to determine if anomalous surface features are present in association with a subsurface anomaly. It is important to remember that once an anomaly is established in the geologic column, that anomaly will tend to continue to affect sedimentation and even structural development, thus propagating itself through time. Anomalies get even stronger with time, in some cases. Unconformities do not necessarily wipe out anomalies—they may enhance them if drainage and topographic patterns form during that erosional cycle. It is therefore very important to always make at least a preliminary photogeologic, photogeomorphic, and remote sensing study of an area of interest. If anomalous features are found, then a more complete study can be undertaken as indicated.

This reprint volume begins with a series of general papers chosen to cover most aspects of remote sensing. These are followed by papers on thermal infrared imagery and on radar. Finally, several papers present case histories illustrating the use of remote sensing in geologic interpretation and petroleum exploration. We appreciate the help of the Treatise of Petroleum Geology Advisory Board, especially the contribution of Floyd F. Sabins, Jr., in choosing papers for this reprint volume

We hope that you will find these landmark papers to be of great value to you in your search for the elusive trap.

Norman H. Foster
Denver, Colorado

Edward A. Beaumont
Tulsa, Oklahoma

TABLE OF CONTENTS

REMOTE SENSING

GENERAL METHODS

THERMAL INFRARED IMAGERY

RADAR

Case Histories

GENERAL METHODS

CHAPTER

8

Resource Exploration

This chapter deals with nonrenewable resources, specifically minerals and fossil fuels. Other resources such as vegetation, water, and soil are discussed in Chapter 9 and elsewhere. Remote sensing methods have great promise as techniques for both reconnaissance and detailed exploration of nonrenewable resources. The following sections give theory, techniques, and examples of these applications.

MINERAL EXPLORATION

Remote sensing has proven valuable for mineral exploration in at least four ways:

1. Mapping regional lineaments along which groups of mining districts may occur

2. Mapping local fracture patterns that may control individual ore deposits

3. Detecting hydrothermally altered rocks associated with ore deposits

4. Providing basic geologic data

Regional Lineaments and Ore Deposits of Nevada

Prospectors and mining geologists have long realized that, in many mineral provinces, mining districts occur along linear trends that range from tens to hundreds of kilometers in length. These trends are referred to as *mineralized belts* or *zones,* and many deposits have been found by exploring along the projections of such trends. The state of Nevada is rich in historic and active mining districts of great wealth. In the late 1800s, rich gold and silver deposits (Virginia City and Goldfield) were discovered in the western part of the state. Later, porphyry copper deposits such as the Ruth, Ely, Eureka, and Yerington deposits were discovered. Exploration has continued, resulting in the more recent discovery of large deposits of gold at Carlin, Alligator Ridge, Cortez, and elsewhere.

Lineament Interpretation It was long recognized that Nevada mining districts were not randomly distributed but tended to occur in linear zones or belts. The availability of Landsat images has enabled geologists to evaluate the relationship between mineral deposits and linear structural features. Rowan and Wetlaufer (1975) of the U.S. Geological Survey interpreted a mosaic of Landsat MSS images of Nevada (Figure 8.1); they recognized 367 lineaments, 80 percent of which correlated with previously mapped faults. Fifty-seven percent of the lineaments are formed by the linear contact between the bedrock of mountain ranges and the alluvium of adjacent valleys. These lineaments are the expression of basin-and-range faults that dominate the structure of Nevada. Other lineaments are formed by linear ridges, aligned ridges, and tonal boundaries.

Seven lineaments of regional extent and importance are shown in Figure 8.2A. These lineaments transect the topography of the Basin and Range province and are several hundred kilometers in length. The Walker Lane, Las Vegas, Midas Trench, and Oregon-Nevada lineaments have previously been documented as major crustal features. The following description of the major lineaments is summarized from Rowan and Wetlaufer (1975).

Walker Lane lineament This zone of right-lateral transcurrent faulting extends southeast from Pyramid Lake to south-central Nevada, where it merges with the lineament called the Las Vegas shear zone. The Walker Lane lineament is expressed by a distinct northwest-trending discontinuity on the mosaic (Figure 8.1). On either side of the lineament, mountain ranges have the regional northeast trend; within the zone, however, the orientation is abruptly changed to northwest trends. Right-lateral, strike-slip movement is thought to be the dominant sense of displacement. A number of mining districts occur along the lineament, from Virginia City in the northwest to Tonopah and Goldfield in the southeast.

Las Vegas shear zone The extension of the Walker Lane lineament into southern Nevada has long been known as the Las Vegas shear zone. Prior to Landsat it was recognized from regional topographic maps that the mountain ranges on either side of the zone are bent in a pattern suggesting drag along a right-lateral strike-slip fault. In the Landsat mosaic, this drag effect is clearly seen in the mountain ranges northwest of Las Vegas. Few ore deposits occur along the Las Vegas shear zone, possibly because the sedimentary bedrock in southern Nevada is less suitable for ore formation than the volcanic and plutonic rocks elsewhere in the state.

Midas Trench lineament This feature extends northeast from Lake Tahoe for 460 km to the northern border of Nevada. The lineament is named for the old mining camp of Midas, where it forms a prominent linear depression. Tuscarora, Golconda, Winnemucca, and the Eagle-Picher mining districts also occur along the lineament. Elsewhere the Midas Trench is expressed by linear escarpments and aligned stream segments. Recent lateral movements along faults of the lineament zone are indicated by offset stream channels.

Oregon-Nevada lineament This lineament extends 750 km from central Nevada to central Oregon and consists of aligned tonal and textural boundaries caused by closely spaced faults.

A linear belt of lava flows and flow domes of late Miocene age mark the Nevada portion of the lineament. An aeromagnetic map shows a prominent linear belt of high values in this portion of the lineament that is attributed to dikes that provided magma for the volcanic features (Stewart, Walker, and Kleinhampl, 1975).

Rye Patch lineament This 250-km-long feature trends northwest and is located midway between the Walker Lane and Oregon-Nevada lineaments. The Rye Patch lineament, named for the Rye Patch reservoir, has a diffuse appearance on the Landsat mosaic where it is mapped by tonal and textural changes. Mountain ranges are terminated or offset by the lineament, which coincides with the existence of regional deep-seated fracture zones proposed by earlier workers.

East-northeast lineament This feature is actually a pair of parallel lineaments that terminate or disrupt the north-trending ranges that they intersect. The lineaments are marked by aligned streams, canyons and tonal boundaries. Both lineaments are associated with mapped faults with right-lateral strike-slip displacement.

Ruby Mountains lineament This relatively poorly documented lineament has a length of 230 km. The southern part coincides with the normal fault that forms the east boundary of the Ruby Mountains. The northern part of the lineament is marked by a series of smaller faults.

Analysis of Lineaments The maps in Figure 8.2 were prepared by Rowan and Wetlaufer (1975) to evaluate the relationship between the lineaments and ore deposits. Nevada mining districts, ranked by dollar value of production, are plotted in Figure 8.2B. Dollar values are based on prices at time of production. During late 1800s and early 1900s, when much of the ore was mined, the price of gold was $20 per ounce and silver was less than $1 per ounce. One linear belt of mining districts coincides with the northeast-trending Midas Trench lineament. The districts in the southwest part of the state occur in a broad belt parallel with the northwest-trending Walker Lane lineament.

To aid in evaluating the relationship of mining districts to lineaments, Rowan and Wetlaufer laid a grid over Figure 8.2B and counted the number of districts in each grid square. These data were contoured to produce the map of Figure 8.2C. This map emphasizes the concentration of districts along the Midas Trench lineament. The high concentration is interrupted by an area of low mining density at the intersection with the Oregon-Nevada lineament. The lack of mining districts along the Oregon-Nevada lineament may be caused by the extensive cover of Tertiary volcanic rocks that masks any underlying deposits. The greatest density of ore deposits along the Midas Trench occurs at the intersection with the Rye Patch lineament. In south-central Nevada the East-Northeast lineament system coincides with two of the three east-trending belts of high mining density on Figure 8.2C. Ore deposits along the Walker Lane

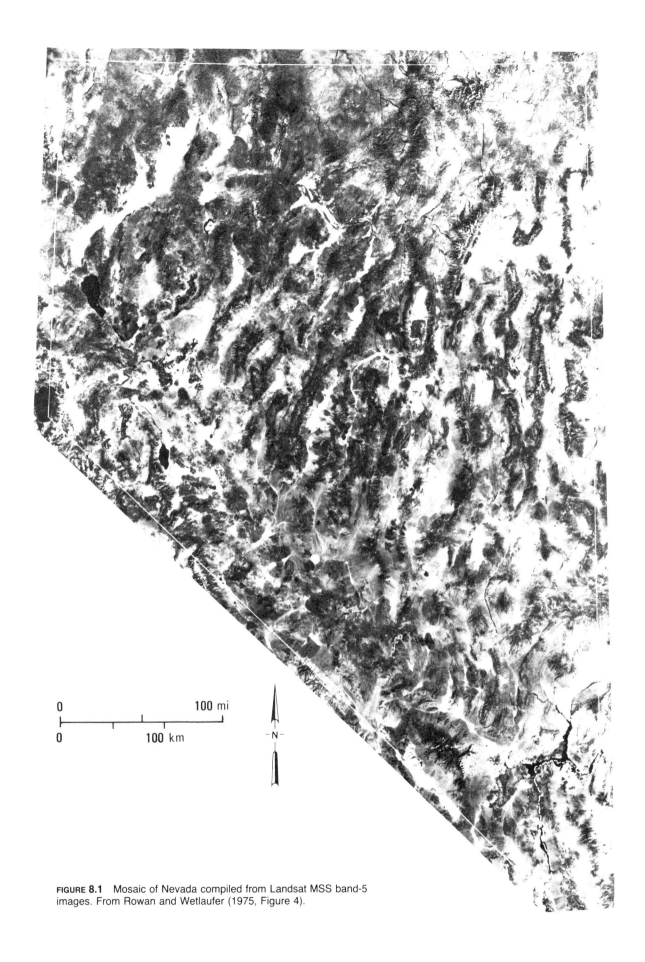

0   100 mi

0   100 km

– N –

FIGURE 8.1   Mosaic of Nevada compiled from Landsat MSS band-5 images. From Rowan and Wetlaufer (1975, Figure 4).

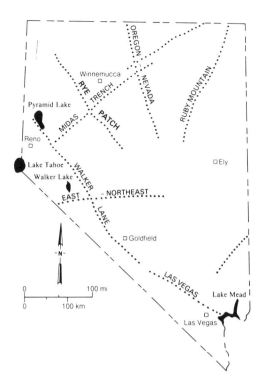

A. MAJOR LINEAMENTS INTERPRETED FROM A LANDSAT MOSAIC.

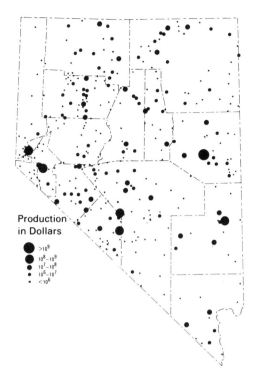

B. MINING DISTRICTS.

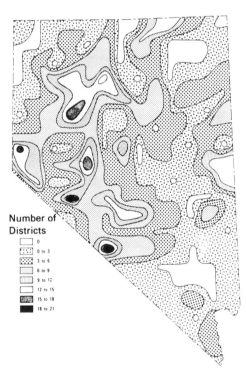

C. CONTOUR MAP OF THE NUMBER OF MINING DISTRICTS.

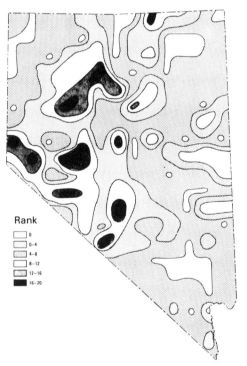

D. CONTOUR MAP OF THE DISTRIBUTION OF MINING DISTRICTS WEIGHTED ACCORDING TO DOLLAR VALUE.

FIGURE 8.2 Landsat lineaments and mining districts of Nevada. Maps from Rowan and Wetlaufer (1975); mining data from Horton (1964).

lineament are generally concentrated at the intersections with east-trending lineaments.

Figure 8.2D was prepared by gridding and contouring the weighted dollar value of production for the districts. These value trends closely resemble the trends of mining density. The Midas Trench and Walker Lane lineaments are marked by aligned concentrations of mining districts.

Local Fractures and Ore Deposits of Central Colorado

Within a mineral province, areas with numerous fracture intersections are good prospecting targets because fractures are conduits for ore-forming solutions. Local fracture patterns are mappable on enlarged Landsat images, especially those acquired at low sun angles and those that have been digitally filtered to enhance fractures.

Relationships between Landsat fracture patterns and ore deposits are illustrated in the example from central Colorado, which is summarized from the work of Nicolais (1974). A winter image (Figure 8.3) was used for interpretation because the snow cover and low sun elevation enhance the expression of fractures. On the interpretation map (Figure 8.4), fractures and circular features are plotted together with location of major mining districts. The Landsat interpretation reduced the original 33,500 km² image to 10 target areas, each 165 km² in area. These areas were selected because they show concentrations of fracture intersections or intersections of fractures and circular features. Five of these target areas coincide with, or are directly adjacent to, major mining districts. The other five target areas may be sites of undiscovered ore deposits.

Mapping Hydrothermally Altered Rocks

Many ore bodies are deposited by hot aqueous fluids, called *hydrothermal solutions,* that invade the host rock, or *country rock.* During formation of the ore minerals, these solutions also interact chemically with the country rock to alter the mineral composition for considerable distances beyond the site of ore deposition. The hydrothermally altered country rocks contain distinctive assemblages of secondary, or alteration, minerals that replace the original rock constituents. Alteration minerals commonly occur in distinct sequences, or *zones of hydrothermal alteration,* relative to the ore body. These zones are caused by changes in temperature, pressure, and chemistry of the hydrothermal solution at progressively greater distances from the ore body. At the time of ore deposition, the zones of altered country rock may not extend to the surface of the ground. Later uplift and erosion expose successively deeper alteration zones and eventually the ore body itself.

Not all alteration is associated with ore bodies, and not all ore bodies are marked by alteration zones, but these zones are valuable indicators of possible deposits. Fieldwork, laboratory analysis of rock samples, and interpretation of aerial photographs have long been used to explore for hydrothermal alteration zones.

In regions where bedrock is exposed, multispectral remote sensing is useful for recognizing altered rocks because their reflectance spectra differ from those of the country rock. An instructive example of remote sensing of alteration zones is provided by the gold and silver vein deposits of Goldfield, Nevada.

Goldfield, Nevada

The Goldfield district in southwest Nevada (Figure 8.2A) was noted for the richness of its ore. Over 4 million troy ounces (130,000 kg) of gold with silver and copper were produced, largely in the boom period between 1903 and 1910. During peak production the town had a population of 15,000 but today is largely a ghost town.

Geology and Hydrothermal Alteration The geology and hydrothermal alteration have been thoroughly mapped and analyzed by the U.S. Geological Survey (Ashley, 1974, 1979). Volcanism began in the Oligocene epoch with eruption of rhyolite and quartz latite flows and the formation of a small caldera and ring-fracture system. Hydrothermal alteration and ore deposition occurred during a second period of volcanism in the early Miocene epoch when the dacite and andesite flows that host the ore deposits were extruded. Heating associated with volcanic activity at depth caused convective circulation of hot, acidic, hydrothermal solutions through the rocks. Fluid movement was concentrated in the fractures and faults of the ring-fracture system. Following ore deposition, the area was covered by younger volcanic flows. Later doming and erosion have exposed the older volcanic center with altered rocks and ore deposits.

In the generalized map (Figure 8.5), the hydrothermally altered rocks are cross-hatched and the unaltered country rocks are blank. The map also identifies alluvial deposits and post-ore (formed after ore deposition) volcanic rocks. Approximately 40 km² of the area are underlain by altered rocks, but less than 2 km² of the altered area are underlain by mineral deposits, shown in black. The irregular oval band of alteration was controlled by the zone of ring fractures, which have a linear extension toward the east. The central patch of alteration shown in Figure 8.5 was controlled by closely spaced faults and fractures.

Figure 8.6 is a vertical section through an ore-bearing vein and the associated altered rocks. Alteration is most intense where hydrothermal solution entered the dacite

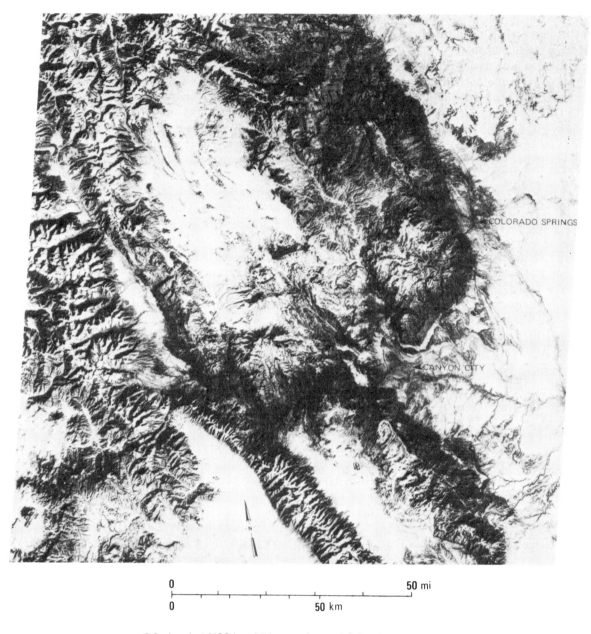

```
0                                    50 mi
|----|----|----|----|----|----|----|----|
0                           50 km
```

FIGURE 8.3 Landsat MSS band-7 image of central Colorado acquired January 11, 1973, at a sun elevation of 23°.

and andesite country rocks through a fault or fracture, and gold was deposited to form a vein. Intensity of alteration decreases laterally away from the vein. Characteristics of the various alteration zones are summarized below.

Silicic zone Predominantly quartz replaces the ground mass of host rock; subordinate amounts of alunite and kaolinite replace feldspar phenocrysts. Fresh rock of this zone, which is gray and resembles chert, is resistant to erosion and weathers to ridges with conspicuous dark coatings of desert varnish. Contact with adjacent argillic zones is sharp. All ore deposits occur in veins of the silicic zone, but not all veins contain ore.

Argillic zone Alteration minerals are predominantly clay. The argillic zone is divided into three subzones (Figure 8.6) based on the predominant clay species. Disseminated grains of pyrite (iron sulfide) are present that weather to iron oxides. The argillic rocks generally have a bleached appearance, but

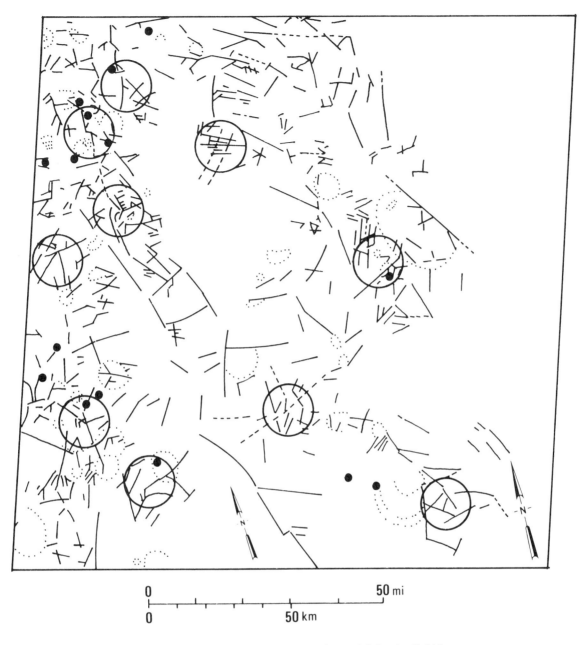

FIGURE 8.4 Interpretation of Landsat image of central Colorado. Solid lines are distinct lineaments, dashed lines are possible lineaments, and dotted lines are curvilinear features. Large circles are target areas selected for exploration. Solid dots indicate major mining districts. From Nicolais (1974, Figure 3).

the secondary iron oxide minerals (limonite and goethite) impart local patches of red, yellow, and brown to the outcrops. No ore deposits occur in the argillic zone, but the presence of these altered rocks may be a clue to the occurrence of veins.

Alunite-kaolinite subzone Relatively narrow and locally absent. In addition to alunite and kaolinite, some quartz is present.

Illite-kaolinite subzone Marked by the occurrence of illite.

Montmorillorite subzone Montmorillonite is the dominant clay

mineral in this subzone, which has a pale yellow color due to jarosite, an iron sulfate mineral.

Propylitic zone These rocks represent regional alteration of lower intensity than the argillic and silicic zones. Chlorite, calcite, and antigorite are typical minerals in this zone and impart a green or purple color to the rocks. Propylitic alteration is absent at numerous localities in Goldfield, where the argillic zone is in sharp contact with unaltered rocks.

Country rock Dacite and andesite. These gray volcanic rocks are hard and resistant to erosion. As shown by the blank

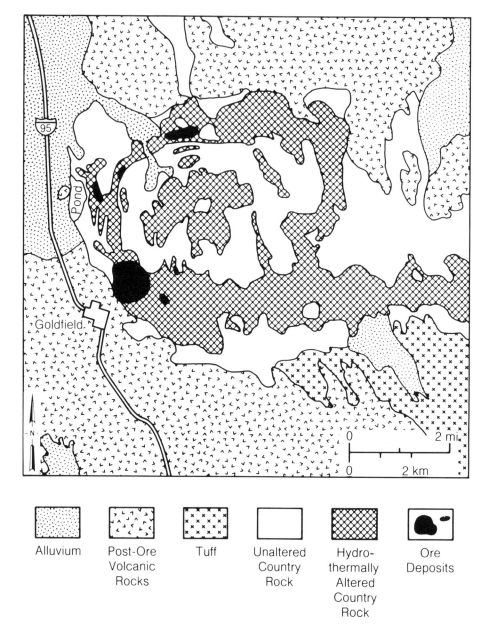

Alluvium

Post-Ore
Volcanic
Rocks

Tuff

Unaltered
Country
Rock

Hydro-
thermally
Altered
Country
Rock

Ore
Deposits

FIGURE 8.5 Map showing hydrothermal alteration and geology of the Goldfield district, Nevada. After Ashley (1979, Figures 1 and 8).

area in the map (Figure 8.5), the unaltered rocks surround the inner and outer margins of the circular belt of altered rocks.

In the field the orderly sequence of subzones shown in Figure 8.6 rarely occurs because the veins are so closely spaced that the subzones coalesce and overlap to form the altered outcrops shown in the map (Figure 8.5).

Clay and iron minerals of the altered rocks have distinctive spectral characteristics that are recognizable in multispectral images such as Landsat thematic mapper. Reflectance spectra of quartz, alunite, and the clay minerals are shown in Figure 8.7A. Spectral bands of Landsat MSS and TM are shown in Figure 8.7C. The spectrum of quartz has no distinctive feature, but the spectra of the clay minerals (kaolinite, illite, and montmorillonite) and alunite have distinctive absorption features at wavelengths of approximately 2.2 μm that coincide with band 7 of TM. Spectra of the iron oxide minerals limonite, hematite, and goethite are shown in Figure 8.7B.

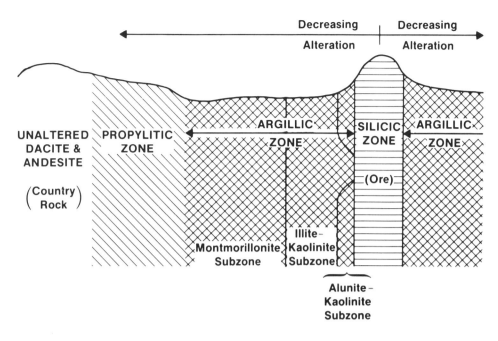

FIGURE **8.6** Model cross section of zones of hydrothermal alteration at Goldfield, Nevada. Not to scale. Silicic zone has maximum width of a few meters. Argillic zone is several tens of meters wide. After Ashley (1974) and Harvey and Vitaliano (1964).

Reflectance spectra of altered and unaltered rocks were measured in the field at Goldfield using a portable spectrometer (Rowan, Goetz, and Ashley, 1979). In the average rock spectra (Figure 8.7C), the two gaps at 1.4 μm and 1.9 μm are due to absorption by water vapor in the atmosphere, as explained in Chapter 1.

This background information on geology and spectral properties of rocks at Goldfield sets the stage for using remote sensing methods to recognize the hydrothermally altered rocks as a guide for ore exploration.

Earlier Remote Sensing Investigations An early investigation by Rowan and others (1974) employed Landsat MSS data. As shown in Figure 8.7C these detectors are restricted to wavelengths of less than 1.1 μm and do not record the diagnostic absorption features of clay and alunite at 2.2 μm. The iron oxide staining commonly associated with altered rocks is recognizable in the visible region, however, and may be detected in MSS data. The red color of iron oxide causes high reflectance in MSS band 5 and low reflectance in band 4. The ratio of band 4 divided by band 5 (Chapter 7) has a low value for reddish rocks because the higher digital numbers of band 5 are in the denominator. Rowan and others (1974, Figure 17) prepared a color ratio image by projecting the ratio 4/5 image with blue light, the ratio 5/6 image with yellow light, and the ratio 6/7 image with magenta light. In the resulting color image, green tones correlate with limonitic areas of hydrothermally altered rocks at Goldfield and the other mining districts in the area. The green tones also correlate with outcrops of ferruginous sandstone and siltstone and with some plutonic rocks.

Digitally Processed TM Images Investigations at Goldfield and elsewhere in the mid-1970s pointed out some shortcomings of Landsat MSS spectral bands for mineral exploration. At that time, NASA had designed the Landsat TM scanner with six spectral bands: bands 1, 2, and 3 in the visible region, band 4 centered at 0.85 μm, band 5 centered at 1.6 μm, and band 6 in the thermal region. No coverage was planned for the critical band at 2.2 μm. Representatives from the remote sensing user community for exploration (Geosat Committee, U.S. Geological Survey, NASA, and JPL) argued successfully for adding band 7, centered at 2.2 μm, to the TM scanner.

Landsat 4 acquired an excellent TM image of Goldfield on October 4, 1984. The data were digitally processed at Chevron Oil Field Research Company, using methods described in Chapter 7. The normal color image of the Goldfield subscene (Plate 12A) was enhanced with a linear contrast stretch and an IHS transformation. The subscene covers the area shown in Figure 8.5. A yellow patch directly northeast of the town of Goldfield is caused by the mine dumps and disturbed ground of the main mineralized area. A white patch 3 km north of Goldfield

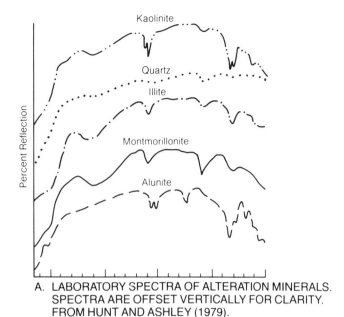

A. LABORATORY SPECTRA OF ALTERATION MINERALS. SPECTRA ARE OFFSET VERTICALLY FOR CLARITY. FROM HUNT AND ASHLEY (1979).

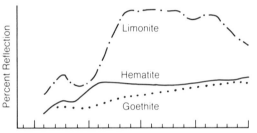

B. LABORATORY SPECTRA OF IRON OXIDE MINERALS. SPECTRA ARE OFFSET VERTICALLY. FROM HUNT, SALISBURY, AND LENHOF (1971).

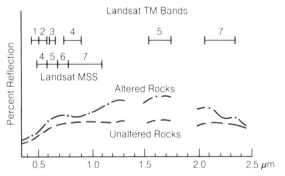

C. FIELD SPECTRA AVERAGED FOR ALTERED AND UNALTERED ROCKS AT GOLDFIELD, NEVADA. FROM ROWAN, GOETZ, AND ASHLEY (1977).

FIGURE 8.7 Reflectance spectra of hydrothermally altered rocks and minerals.

is the dry tailings pond of the abandoned Columbia Mill, where gold was separated from the altered host rock. The tailings pond is a useful reference standard since it contains a concentration of alteration minerals. The dark rocks in the margins of the image are volcanic flows that are younger than the ore deposits and altered rocks. Distinctive light blue signatures in the southeast portion are outcrops of volcanic tuff. Some of the hydrothermally altered rocks have a tan signature in the normal color image (Plate 12A), but other rocks have similar colors.

Neither the normal color TM image nor the IR color image (not illustrated) are diagnostic for recognizing the hydrothermally altered rocks. The spectra of altered and unaltered rocks at Goldfield (Figure 8.7C) suggest a means of distinguishing between these rocks. Absorption caused by alunite and clay minerals results in low reflectance at 2.2 μm, which corresponds to TM band 7. Altered rocks have high reflectance at 1.6 μm, which corresponds to TM band 5. Enhanced images of these bands are shown in Figure 8.8A,B where the altered rocks have brighter signatures in band 5 than in band 7. A ratio 5/7 image (Figure 8.8C) has bright signatures for altered rocks because the lower reflectance values of band 7 are in the denominator, which results in higher ratio values of approximately 1.5.

The unaltered rocks, however, have nearly equal reflectance values in bands 5 and 7 (Figure 8.7C) resulting in a ratio 5/7 of approximately 1.0. In the ratio 5/7 image (Figure 8.8C), the high values of altered rocks have brighter signatures than unaltered rocks. A color density slice was applied to the ratio image to convert gray-scale variations into color (Plate 12C). Red and yellow colors were assigned to the highest ratio values; green and blue were assigned to the lowest values. Red and yellow hues closely match the distribution of altered rocks in the map of Figure 8.5.

Spectra of the weathered iron minerals (Figure 8.7B) have weak reflectance in the blue region (TM band 1) and strong reflectance in the red region (TM band 3). The ratio 3/1 has high values for iron-stained areas, resulting in bright tones on the image (Figure 8.8D). A color density slice of this ratio (Plate 12D) assigns red and yellow colors to the highest ratio values. These colors delineate the iron-stained portions of the altered rocks.

Another way to use ratio images is to combine three selected ratios in color as shown in Plate 12B, where ratios 5/7, 3/1, and 3/5 are combined in red, green, and blue respectively. The orange and yellow tones delineate the outer and inner areas of altered rocks in a fashion similar to that of the density slice of the ratio 5/7 image (Plate 12C). An advantage of the color ratio image is

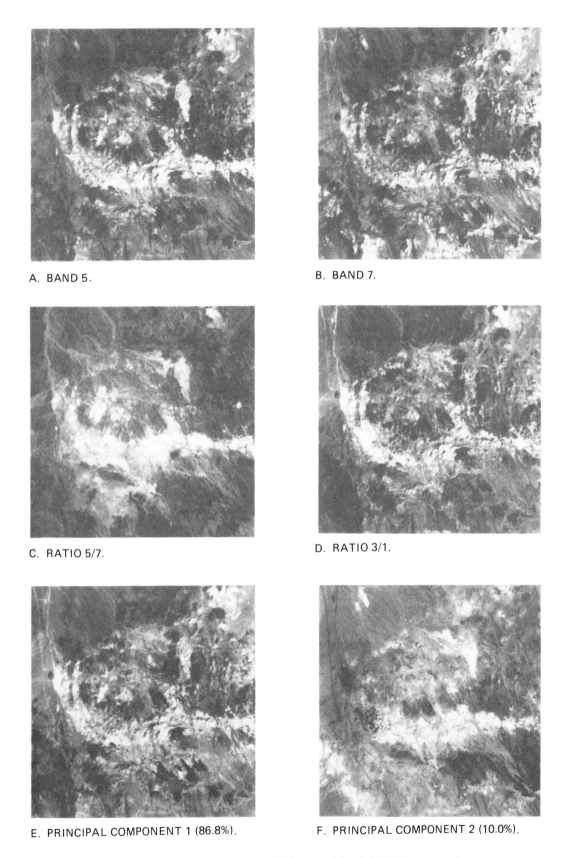

A. BAND 5.

B. BAND 7.

C. RATIO 5/7.

D. RATIO 3/1.

E. PRINCIPAL COMPONENT 1 (86.8%).

F. PRINCIPAL COMPONENT 2 (10.0%).

FIGURE **8.8** Digitally processed Landsat TM images of the Goldfield, Nevada, subscene.

11

that it displays distribution patterns of both iron-staining and hydrothermal clays. A disadvantage is that the individual patterns of iron-staining and clay minerals may overlap and obscure each other in the color ratio image.

Principal component (PC) images were prepared for the six TM visible and reflected IR bands of the Goldfield subscene. In PC 1 (Figure 8.8E) the hydrothermally altered rocks and the tailings pond have bright signatures that distinguish them from the other materials. In PC 2 (Figure 8.8F) the altered rocks are bright, but so are outcrops of unaltered volcanic rocks southwest of Goldfield town site. The tailings pond and the mine dumps north of Goldfield are dark in PC 2. A statistical analysis of the PC images show that TM band 5 is the most heavily weighted component of PC 1 and PC 2. Altered rocks have their maximum reflectance in TM band 5 (Figure 8.7C), which explains their bright signatures in PC 1 and PC 2.

An unsupervised multispectral classification applied to the six visible and reflected IR bands of the subscene resulted in 12 classes. These classes were aggregated into the 6 classes shown in the map and explanation of Plate 12E,F. Two types of altered rocks were classified. The class shown in red ("altered rocks, A") is confined to the altered rocks and to the tailings pond but does not indicate the full extent of alteration. The class shown in orange ("altered rocks, B") includes all of the remaining altered rocks as well as some rocks outside the alteration zone. Basalt (blue), volcanic tuff (purple), and unaltered rocks (green) are reasonably portrayed. Alluvium (yellow) is considerably more extensive in the classification map (Plate 12E) than in the geologic map (Figure 8.5). Field checking and comparison with the normal color image (Plate 12A) shows that much of the bedrock is thinly covered with detritus and is correctly classed as alluvium by the computer. The field geologist, however, is able to infer and map the lithology of the underlying bedrock.

In summary, the hydrothermally altered rocks at Goldfield are successfully distinguished by three types of digitally processed TM data: ratio images, principal-component images, and unsupervised multispectral classification maps. Much of this success is due to the availability of TM band 7, but the high spatial resolution and spectral sensitivity of TM also contribute.

Porphyry Copper Deposits

Much of the world's copper is mined from porphyry deposits, which occur in a different geologic environment from the gold deposits of the Goldfield type. Porphyry deposits are named for the *porphyritic* texture of the granitic host rock, in which large feldspar crystals are surrounded by a fine-grained matrix of quartz and other minerals. Granite porphyry occurs as plugs (or *stocks*) up to several kilometers in diameter that intruded the older country rock and reached to within several kilometers of the surface. Intensive fracturing of the porphyry and country rock occurred during the emplacement and cooling of the stock. The heat of the magma body caused convective circulation of hydrothermal fluids through the fracture system that resulted in alteration of the porphyry stock and adjacent country rock.

Hydrothermal Alteration The model of alteration zones shown in Figure 8.9 was developed from the study of many deposits in southwest United States (Lowell and Guilbert, 1970) and is applicable elsewhere.

The most intense alteration occurs in the core of the stock and diminishes radially outward in a series of zones described below.

Potassic zone Most intensely altered rocks in the core of the stock. Characteristic minerals are quartz, sericite, biotite, and potassium feldspar.

Phyllic zone Quartz, sericite, and pyrite are common.

Ore zone · Disseminated grains of chalcopyrite, molybdenite, pyrite, and other metal sulfides. Much of the ore occurs in a cylindrical shell near the gradational boundary between the potassic and phyllic zones. Copper typically constitutes only a few tenths of a percent of the ore, but the large volume of ore is suitable for open pit mining. Where the ore zone is exposed by erosion, pyrite may oxidize to form a red to brown limonitic crust called a *gossan*. Gossans can be useful indicators of underlying mineral deposits, although not all gossans are associated with ore deposits.

Argillic zone Quartz, kaolinite, and montmorillonite are characteristic minerals of the argillic zone in porphyry deposits, just as they are associated with the argillic zone at Goldfield and elsewhere.

Propylitic zone Epidote, calcite, and chlorite occur in these weakly altered rocks. Propylitic alteration may be of broad extent and have little significance for ore exploration.

Few, if any, porphyry ore deposits and alteration patterns have the symmetry and completeness of the model in Figure 8.9. Structural deformation, erosion, and deposition commonly conceal large portions of the system. Nevertheless, recognition of small patches of altered rock on remote sensing images may be a valuable clue to a potential ore deposit. Porphyry copper deposits are one category of nonrenewable resource that was investigated by the NASA/Geosat test case project described in the following section.

NASA/Geosat Test Case Project—Copper Mines
The Geosat Committee is a nonprofit organization that promotes the use and development of remote sensing

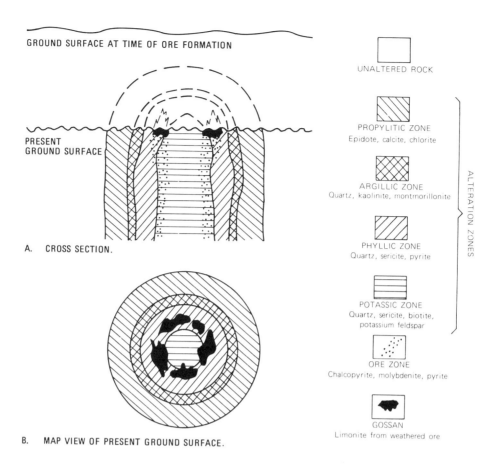

GROUND SURFACE AT TIME OF ORE FORMATION

PRESENT
GROUND SURFACE

A.   CROSS SECTION.

B.   MAP VIEW OF PRESENT GROUND SURFACE.

UNALTERED ROCK

PROPYLITIC ZONE
Epidote, calcite, chlorite

ARGILLIC ZONE
Quartz, kaolinite, montmorillonite

PHYLLIC ZONE
Quartz, sericite, pyrite

POTASSIC ZONE
Quartz, sericite, biotite,
potassium feldspar

ORE ZONE
Chalcopyrite, molybdenite, pyrite

GOSSAN
Limonite from weathered ore

ALTERATION ZONES

FIGURE 8.9   Model of hydrothermal alteration zones associated with porphyry copper deposits. From Lowell and Guilbert (1970, Figure 3).

by industry. The committee consists of, and is supported by, approximately 100 companies involved in various aspects of remote sensing. In 1977, NASA and Geosat jointly agreed to evaluate the use of remote sensing data in exploration for porphyry copper, uranium, and oil and gas. Test sites were selected for these three categories. NASA then acquired a variety of remote sensing data for the sites, which were digitally processed at JPL. Representatives of 38 Geosat member companies plus JPL personnel formed teams to interpret the image data, collect and analyze samples, and prepare a report (Abrams, Conel, and Lang, 1985). Settle (1985) prepared an executive summary of the project.

The image-processing and interpretation phases of the project were essentially completed before Landsat TM images became available in late 1982. To compensate for the lack of TM images, NASA acquired images with an airborne *thematic mapper simulator* (TMS). The eight TMS bands compare with Landsat TM bands as follows:

TMS band	TM band	Spectral range, μm
1	1	0.45 to 0.52
2	2	0.52 to 0.60
3	3	0.63 to 0.69
4	4	0.76 to 0.90
5	No band	1.00 to 1.30
6	5	1.55 to 1.75
7	7	2.08 to 2.35
8	6	10.40 to 12.50

Aircraft images were also acquired with a *modular multispectral scanner* (M^2S) in the 0.3-to-1.1-μm range. Spatial resolution of TMS and M^2S images ranges from 10 to 20 m.

The Geosat Committee chose three porphyry copper sites: the Silver Bell, Helvetia, and Safford deposits, all located in southern Arizona. The alteration zones shown in Figure 8.9 crop out in the Silver Bell district, where TMS data were digitally processed to produce a color ratio composite image (TMS ratio 3/2 = green, ratio 4/5 = blue, ratio 6/7 = red). The phyllic and potassic alteration zones have a distinct yellow-orange hue; the adjacent argillic and propylitic zones have yellowish-green and yellowish-brown hues (Abrams and Brown, 1985, Figure 4-32). A supervised classification map of TMS data also defined the outcrops of altered rocks (Abrams and Brown, 1985, Figure 4-41).

These results at Silver Bell using TMS data are similar to those described earlier for Goldfield using TM data. This is surprising because the geology and ore deposits of the two districts are completely different. The comparable remote sensing results are explained by the following similarities:

1. Hydrothermal alteration produced similar suites of secondary minerals (quartz, clays, and iron oxide) at both districts. Alunite occurs at Goldfield, but is absent at Silver Bell.

2. Spectral ranges of the TMS bands used at Silver Bell are comparable to the TM bands used at Goldfield.

3. The digital-processing methods were similar, although Silver Bell data were processed at JPL and Goldfield data were processed at Chevron Oil Field Research Company.

The similarity in results for the two mining districts indicates the potential value of remote sensing and digital image processing for mineral exploration.

Basic Data for Geologic Mapping

In addition to locating specific mineral target areas of fracture intersections or rock alteration, remote sensing provides data for preparing and improving geologic maps, which are the fundamental tool for exploration. Geologic maps, even at reconnaissance scales, are not available for large areas of the earth. For example, approximately two-thirds of southern Africa lacks published geologic maps at scales of 1:500,000 or larger; this can be improved by use of Landsat images. A previously unknown major fault was discovered on a Landsat image of southern South-west Africa and the Cape Province of South Africa by Viljoen and others (1975, Figure 3), who named it the Tantalite Valley fault zone. The fault zone appears to have right-lateral strike-slip displacement and has been mapped for 450 km along the strike. A number of large mafic intrusives have been emplaced along the Tantalite Valley fault zone and are recognized on Landsat images. On a Landsat color mosaic of the

northwest Cape Province of South Africa, Viljoen and others (1975, Figures 11 and 12) mapped a pronounced structural discontinuity, called the Brakbos fault zone, that separates the Kaapvaal craton on the east from the Bushmanland Metamorphic Complex on the west. The contact between these structural provinces is obscure in the field and had previously been drawn approximately 30 km to the east of the Brakbos fault zone, which is also defined on gravity maps. The use of seasonal Landsat images for mapping rock types was illustrated for the Transvaal Province in Chapter 4.

In the Nabesna quadrangle of east-central Alaska, Albert (1975) combined lineament analysis and digital processing of Landsat MSS data to evaluate known and potential mineral deposits. A preliminary analysis indicates that 56 percent of the known mineral deposits occur within 1.6 km of Landsat lineaments. Color anomalies on the enhanced images coincide with 72 percent of the known mineral occurrences. Of the remaining color anomalies, some coincide with areas of known rock alteration and others constitute potential exploration targets.

URANIUM EXPLORATION

Most uranium deposits in the United States occur in nonmarine fluvial sandstone and conglomerate beds of the Colorado Plateau, Rocky Mountain basins, and southern Texas. The solutions that deposited the uranium also altered the host rocks and caused local color anomalies on the outcrops. Aerial photographs have been used extensively in exploring for these areas of altered rock. Digital processing of Landsat CCTs has the potential to recognize subtle alteration effects that may not be obvious to the eye or on aerial photographs. A brief description of typical sedimentary uranium deposits will aid in understanding the following examples of Landsat applications.

Origin of Sedimentary Uranium Deposits

The model shown in Figure 8.10 for the formation of sedimentary uranium deposits is widely accepted, although there is debate about the origin of the uranium and the chemistry of the transporting solutions. Granite, granitic detritus, and silicic volcanic ash and flows contain disseminated uranium in concentrations up to 10 parts per million. Some of this uranium is leached from the source rocks by oxygen-rich groundwater that then migrates into porous sandstone and conglomerate beds carrying the uranium in solution. Within these sedimentary rocks, the migrating water encounters reducing conditions caused by the presence of organic material, nat-

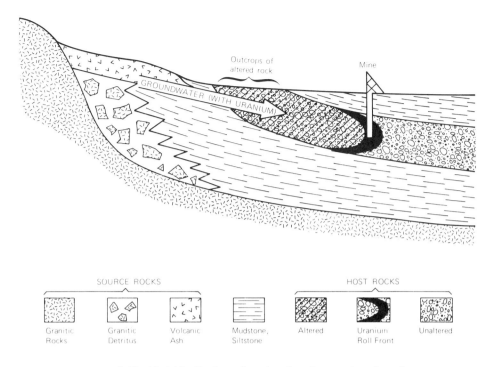

SOURCE ROCKS

SOURCE ROCKS

Granitic Rocks

Granitic Detritus

Volcanic Ash

Mudstone, Siltstone

HOST ROCKS

Altered

Uranium Roll Front

Unaltered

FIGURE 8.10   Model for the formation of sedimentary uranium deposits.

ural gas, or hydrogen sulfide and pyrite. The change from oxidizing to reducing conditions causes the uranium to precipitate as oxide minerals, primarily uraninite, that coat sand grains and fill pore spaces in the host rock. The ore deposits typically contain from 0.1 to 0.5 percent U_3O_8 and occur as tabular layers or as arcuate bodies called *roll fronts* (Figure 8.10). Later uplift and erosion may expose the ore to secondary oxidation.

Outcrops of altered host rock are clues to possible ore deposits below the surface. The unaltered host rocks are typically drab in color and contain organic carbon and pyrite. Oxidation by the migrating solutions destroys the carbon and converts the dark pyrite to iron oxide minerals that impart characteristic yellow, red, and brown colors to the altered rocks.

The following sections describe remote sensing applications to uranium exploration in Arizona and Texas.

Cameron Uranium District, Arizona

In the Cameron district of north-central Arizona, uranium host rocks are conglomerates, sandstones, and siltstones interbedded with mudstone and limestone of the Chinle Formation of Upper Triassic age (Figure 8.11). The original pyrite, calcite, and aluminous mineral components of the host rocks were altered by the mineralizing solutions to limonite, alunite, gypsum, and jarosite. The resulting light brownish yellow color contrasts with the typical purple to gray color of unmineralized parts of the Chinle Formation. Areas of altered rock form elongate halos up to 400 m long surrounding the ore deposits. The alteration colors are valuable guides to the ore deposits but are not uniquely related to ore deposits for the following reasons: (1) the normal color of some unmineralized parts of the Chinle Formation resembles that of the altered zones; (2) uranium may have been remobilized and removed after the alteration occurred; and (3) alteration may have occurred without any deposition of ore. This geologic description is summarized from the work of Spirakis and Condit (1975) of the U.S. Geological Survey, who also reported the following Landsat interpretation.

Plate 13A shows a digitally processed subscene of a Landsat MSS image of the Cameron district. Grabens, faults, collapse structures, volcanic cones, basalt flows, and sedimentary rock formations are recognizable. On this color image the light gray and light brown altered rocks cannot be distinguished from the surrounding unaltered rocks of the Chinle Formation, which have similar colors. To enhance the appearance of altered rocks, a color ratio image (Plate 13B) was prepared by combining the ratio images 4/7, 6/4, and 7/4 in blue, green, and red, respectively. The color ratio image is dominated by yellow and brown tones, but there are a few conspicuous blue patches located along the outcrop of the Chinle Formation (Figure 8.11). These blue patches

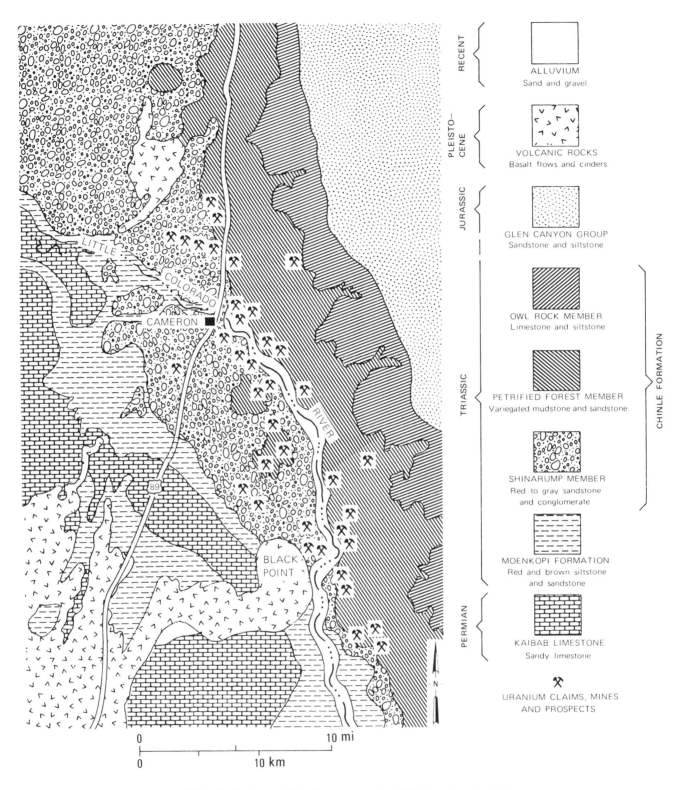

RECENT

ALLUVIUM
Sand and gravel

PLEISTO—
CENE

VOLCANIC ROCKS
Basalt flows and cinders

JURASSIC

GLEN CANYON GROUP
Sandstone and siltstone

OWL ROCK MEMBER
Limestone and siltstone

PETRIFIED FOREST MEMBER
Variegated mudstone and sandstone

CHINLE FORMATION

SHINARUMP MEMBER
Red to gray sandstone
and conglomerate

TRIASSIC

MOENKOPI FORMATION
Red and brown siltstone
and sandstone

PERMIAN

KAIBAB LIMESTONE
Sandy limestone

⚒ URANIUM CLAIMS, MINES
AND PROSPECTS

0   10 mi

0   10 km

FIGURE **8.11**   Geologic map of the Cameron uranium district, north-central Arizona. From Chenoweth and Magleby (1971).

correlate with areas of altered rock as shown by an aerial reconnaissance of the region (Spirakis and Condit, 1975). The uranium mines and claims shown in the geologic map occur within or adjacent to the altered outcrops indicated by the blue signatures in the color ratio image.

The color ratio image is not a perfect exploration method because some unaltered rocks may have reflectance properties similar to those of altered host rocks. In the southeast portion of the Cameron subscene, for example, some outcrops of unaltered Moenkopi Formation west of the Colorado River have blue signatures on the color ratio image. These areas can be eliminated as exploration targets because they are not associated with known host rocks.

Freer–Three Rivers Uranium District, Texas

In this typical south Texas uranium district, the host rocks are channels filled with sandstone or conglomerate in the Catahoula Tuff (Miocene age). One important exploration method is to map the occurrence of outcrops of the channels filled with potential host rocks. Geologic mapping in this area of low relief is hampered by lack of outcrops, nondistinctive rock types, a partial cover of younger gravel, and restricted land access. Digital processing of Landsat images of the district has not been successful. The U.S. Geological Survey conducted airborne scanner surveys in an attempt to map the sandstone channel deposits (Offield, 1976).

Figure 8.12A shows a daytime cross-track scanner image acquired in the visible band. The road network and boundaries of agricultural fields are shown by reflectance differences. Some known occurrences of sandstone channels are indicated by arrows, but these rocks are not recognizable in the daytime visible image. The nighttime thermal IR image (Figure 8.12B) was acquired following a week of heavy rain. The high moisture content greatly reduced thermal contrasts between different rock types. Despite these poor conditions for image acquisition, the sandstone channels have distinct bright (warm) signatures that contrast with the dark (cool) signatures of the clay-rich tuff units. These nighttime thermal IR signatures are consistent with the densities and thermal characteristics described in Chapter 5. Sandstone has a higher density, which results in higher thermal inertia and a warmer nighttime radiant temperature relative to tuff, which has lower density, lower thermal inertia, and cooler nighttime temperature. The warm signature of the sandstone host rock is caused by the thermal properties of the rock, not by radiogenic heat (heat from decay of radioactive elements). Calculations show that radiogenic heat produced by typical sedimentary uranium deposits is insufficient to produce a

detectable thermal anomaly (Kappelmeyer and Haenel, 1974, p. 170).

NASA/Geosat Test Case Project— Uranium Mines

The Lisbon Valley, Utah, and Copper Mountain, Wyoming, areas were selected as uranium test sites for the NASA/Geosat project. At Lisbon Valley, uranium occurs in the Chinle Formation (Triassic age), which is poorly exposed at the surface. The overlying Wingate Sandstone (Triassic age) is widely exposed at Lisbon Valley and has a characteristic red color caused by iron-oxide minerals. Where it overlies uranium deposits in the Chinle Formation, however, the Wingate Sandstone is white because the iron oxides are removed (bleached). Mining geologists have attributed the bleaching of the Wingate Sandstone to the reducing conditions that precipitated uranium in the Chinle Formation. Under reducing conditions, ferric iron is converted to ferrous iron, which is soluble and may be removed by groundwater. Thus the present-day white outcrops of Wingate Sandstone record the subsurface geochemical conditions that precipitated uranium in the Chinle Formation. Conel and Alley (1985) prepared and interpreted principal-component images from TMS data. These images distinguished the white from the red portion of the Wingate Sandstone and separated a number of geologic formations. This ability to recognize anomalous color patterns may be useful in uranium exploration.

FUTURE MINERAL EXPLORATION METHODS

Landsat MSS was a new tool for mineral exploration; Landsat TM was a major improvement with its higher spatial resolution and additional spectral bands. The Goldfield example demonstrated the importance of TM band 5 centered at 1.6 μm and band 7 centered at 2.2 μm for mapping areas of hydrothermal alteration. TM bands have relatively broad spectral ranges, however, and are not capable of discriminating the various zones and subzones of the alteration models. The subzones of the Goldfield model (Figure 8.6) are characterized by the presence of specific alteration minerals. Within the atmospheric window from 2.0 to 2.5 μm, kaolinite, montmorillonite, illite, and alunite have distinctive absorption features in their reflectance spectra as shown in Figure 8.13. All four minerals have an absorption minimum near 2.2 μm, but the exact location of the minimum is different for each mineral. There are additional spectral features that can be used to identify specific minerals. For example, kaolinite has a ''shoulder''

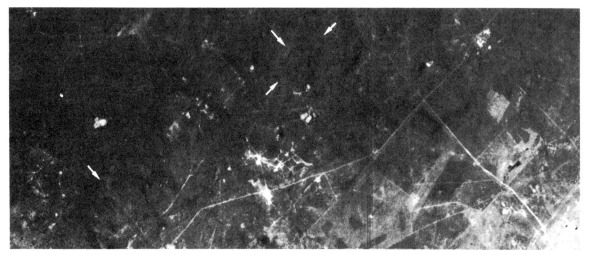

A. IMAGE ACQUIRED AT MIDDAY IN THE VISIBLE
   SPECTRAL REGION. ARROWS MARK CHANNEL–FILL
   CONGLOMERATES WITH TOPOGRAPHIC EXPRESSION.

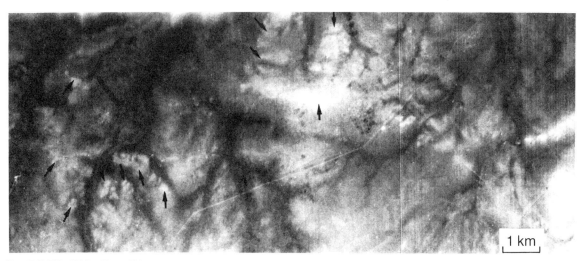

1 km

B. PREDAWN THERMAL IR IMAGE (8 TO 14 μm).
   ARROWS MARK CHANNEL–FILL CONGLOMERATES THAT
   ARE WARMER (BRIGHTER TONE) THAN THE
   SURROUNDING CATAHOULA TUFF.

FIGURE **8.12** Freer–Three Rivers uranium district, southern Texas. Airborne
scanner images acquired November 1974. From Offield (1976, Figure 4). Courtesy
T. W. Offield, U.S. Geological Survey.

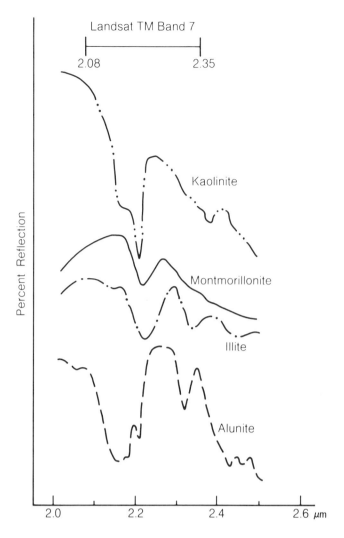

FIGURE 8.13 Laboratory spectra of alteration minerals in the 2.0-to-2.5-μm atmospheric window. Spectra are offset vertically for clarity.

tered. Figure 8.14 shows one of the AIS image bands of the Cuprite district with several ground resolution cells indicated. For each of these cells, a reflectance spectrum was plotted and displayed in Figure 8.14 together with laboratory reference spectra of alunite and kaolinite. The AIS spectra from the northern part of the district are similar to the alunite reference spectrum; those from southern part of the district are similar to the kaolinite reference. The distribution of alteration minerals determined remotely by AIS agree with the mineral patterns mapped at Cuprite. This example demonstrates the exploration potential of multispectral scanner images with high spectral resolution. AIS is currently an experimental system, but this technology will soon become operational.

MINERAL EXPLORATION IN COVERED TERRAIN

The examples of mineral exploration in this book, except for the Freer–Three Rivers deposit, are in arid to semiarid terrain with extensive exposures of bedrock and little vegetation cover. Remote sensing images from these and similar areas can be analyzed for evidence of alteration zones that are surface expressions of mineral deposits. In most of the temperate and humid climate zones of the world, however, mineral deposits are commonly

in the absorption feature at 2.18 μm, and alunite has a secondary absorption minimum at 2.21 μm. The broad spectral range of Landsat TM band 7 (Figure 8.13) is incapable of distinguishing these spectral features that characterize the different alteration minerals.

The airborne imaging spectrometer (AIS), described in Chapter 2, has the capability of recording detailed spectral features and thereby the potential for identifying specific alteration minerals. AIS acquires multiple images at wavelength intervals of 0.01 μm. In the important region from 2.00 to 2.30 μm, AIS acquires 30 spectral readings for each ground resolution cell that can be displayed as a reflectance spectrum.

Jet Propulsion Laboratory has acquired AIS image data over the Cuprite district, 20 km south of Goldfield, where the volcanic rocks have been hydrothermally al-

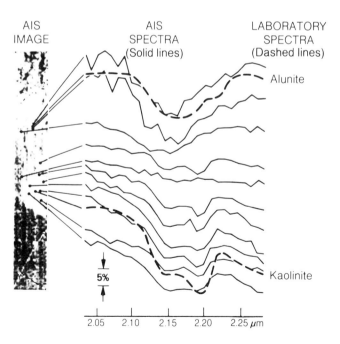

FIGURE 8.14 Reflectance spectra derived from AIS data of the Cuprite district, Nevada. From Goetz (1984, Figure 14). Courtesy A. F. H. Goetz, Jet Propulsion Laboratory.

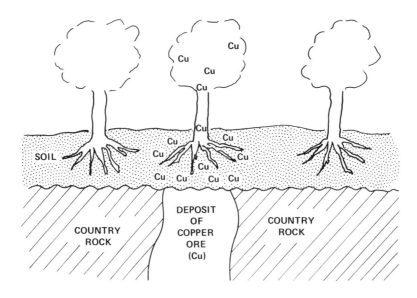

FIGURE 8.15 Copper enrichment of vegetation and soil overlying a concealed copper deposit.

concealed beneath a cover of soil and vegetation. The composition of residual soil reflects the composition of the underlying bedrock from which the soil was derived by weathering processes.

Geochemical exploration techniques involve collecting water and soil samples and analyzing their metal content. Areas with high metal concentrations are then tested by core drilling. Figure 8.15 illustrates the copper enrichment of soil overlying a copper deposit in the bedrock. Vegetation growing in mineralized soil may have higher metal content in its tissue than vegetation in normal, or background, soil, as the figure shows. This concentration of metals in vegetation is the basis for biogeochemical and geobotanical prospecting methods. However, these techniques for testing soil and analyzing vegetation are not applicable in areas where the soil has been transported rather than formed in place. Alluvial and glacial soils are examples of transported soils.

Biogeochemical and Geobotanical Exploration

Biogeochemical exploration consists of collecting vegetation samples that are analyzed chemically for high metal concentrations that may indicate a concealed ore deposit.

Geobotanical exploration searches for unusual vegetation conditions that may be caused by high metal concentrations in the soil. Sampling and chemical analyses of vegetation are not required, but skill and experience are needed to recognize the more subtle vegetation anomalies. Remote sensing techniques are being investigated as possible geobotanical exploration methods. The major geobotanical criteria for recognizing concealed ore deposits are:

1. *Lack of vegetation.* This may be caused by concentrations of metals in the soil that are toxic to plants. These areas are sometimes called *copper barrens* where they are caused by high concentrations of that metal. Areas that lack vegetation may be seen on aerial photographs. These barren areas may result from causes other than mineralization, however.

2. *Indicator plants.* These are species that grow preferentially on outcrops and soils enriched in certain elements. Cannon (1971) prepared an extensive list of indicator plants. For example, in the Katanga region of southern Zaire, a small blue-flowered mint, *Acrocephalus robertii,* is restricted entirely to copper-bearing rock outcrops.

3. *Physiological changes.* High metal concentrations in the soil may cause abnormal size, shape, and spectral reflectance characteristics of leaves, flowers, fruit, or entire plants. A relationship between spectral reflectance properties of plants and the metal content of their soils could form the basis for remote sensing of mineral deposits in vegetated terrain.

Remote Sensing for Minerals in Vegetated Terrain

Chlorosis, or yellowing of leaves, is an example of a spectral change visible to the eye. Chlorosis results from an upset of the iron metabolism of plants that may be caused by an excess concentration of copper, zinc, manganese, or other elements. Relatively high metal concentrations are required to produce chlorosis, which has been observed in plants growing near mineral deposits. Chlorosis is not a reliable prospecting criterion, however, for these reasons: (1) Many areas of known mineralized soil support apparently healthy plants with no visible toxic symptoms. (2) Chlorosis may result from conditions unrelated to mineral deposits, such as soil salinity.

The large, low-grade, copper-molybdenum deposit at Catheart Mountain, Maine, has been used as a geobotanical remote sensing test site (Yost and Wenderoth, 1971). Field spectrometers measured reflectance of trees growing in normal soil and in mineralized soil overlying the deposit (Figure 8.16). Red spruce and balsam fir growing in the mineralized soil had higher metal concentrations than trees of the same species in unmineralized soil. In the reflected IR spectral region, the mineralized balsam firs have a higher reflectance than the normal trees, whereas mineralized red spruce have a lower reflectance than the normal trees (Figure 8.16). In the green spectral region, the mineralized trees of both species have a higher reflectance.

In the years following the 1970 study, a number of additional investigations have been made and were summarized by Labovitz and others (1983, Figure 1). With some exceptions, vegetation reflectance in the green and red bands generally increased with increasing metal concentration in the soil. In the reflected IR region, however, there is less agreement; some studies show an increase in vegetation reflectance while others show a decrease. Labovitz and others (1983, p. 759) also noted that the geobotanical model of Figure 8.15 is not universally true. In Virginia they found that the leaves of oak trees growing in metal-rich soil may have a lower metal content than leaves from trees in normal soil.

Geophysical Environmental Research has taken a different approach to geobotanical remote sensing. The company operates a nonimaging airborne system that acquires detailed reflectance spectra. The spectra in Figure 8.17 were acquired from conifers growing over a mineralized area and in an adjacent nonmineralized area. In the green band (0.5 to 0.6 μm) reflectance is higher for trees in the mineralized area, which is consistent with other studies. Beginning at a wavelength of about 0.7 μm, vegetation spectra have a steep upward slope

to the high reflectance values in the IR region. In Figure 8.17, this steep slope is shifted slightly toward shorter wavelengths for the conifers growing in the mineralized area. This shift, called the *blue shift,* has been noted in vegetation over several mineralized areas (Collins and others, 1983) and has exploration potential.

In summary, remote sensing of mineral deposits in vegetated terrain is in the research and development stage. The relationships between mineral deposits, soil chemistry, reflectance properties of vegetation, and remote sensing systems are complex, but they are being investigated at several research centers.

OIL EXPLORATION

The search for oil in unexplored onshore areas, such as portions of Africa and China, normally follows a pattern.

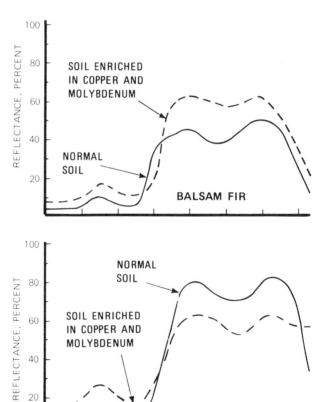

FIGURE 8.16 Reflectance spectra of balsam fir and red spruce growing in normal soil and in soil enriched in copper and molybdenum. Spectra were recorded with a field spectrometer by Yost and Wenderoth (1971, Figures 5 and 6).

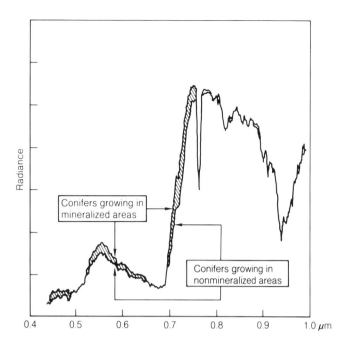

FIGURE 8.17 Airborne reflectance spectra of conifers in the Cotter Basin, Lewis and Clark County, Montana. Note the blue shift for conifers in the mineralized area. From Collins and others (1983, Figure 4B).

It begins with regional reconnaissance and is succeeded by progressively more detailed (and expensive) steps that culminate in drilling a wildcat well. *Wildcat wells* are exploratory tests in previously undrilled areas, whereas *development wells* are drilled to produce oil from a previously discovered field. A typical exploration program proceeds as follows:

1. *Regional remote sensing reconnaissance.* Small-scale Landsat mosaics covering hundreds of thousands of square kilometers are especially useful in this phase. The objective is to locate *sedimentary basins*, which are areas underlain by thick sequences of sedimentary rocks. These basins are essential for the formation of oil fields.

2. *Reconnaissance geophysical surveys. Aerial magnetic* surveys are made to produce maps that record the intensity of the earth's magnetic field. Sedimentary basins have lower magnetic intensities than do areas underlain by nonsedimentary rocks such as granite and volcanic rocks. The aerial magnetic maps thus can confirm the presence of sedimentary basins. Surface *gravity surveys* are made that record the intensity of the earth's gravity field. Sedimentary rocks have a lower specific gravity than nonsedimentary rocks; hence sedimentary basins are shown by lower values on the gravity maps. Gravity and magnetic maps may also show regional structural features.

3. *Detailed remote sensing interpretation.* Individual digitally processed Landsat images are studied to identify and map geologic structures, such as anticlines and faults, that may form oil traps. Promising structures may be mapped in detail using stereo pairs of aerial photographs. Radar images are used in regions of poor weather where it is difficult to acquire good photographs and Landsat images. At this stage, geologists go into the field to check the interpretation and collect samples of the exposed rocks.

4. *Seismic surveys.* Explosives or mechanical devices are used to transmit waves of sonic energy into the subsurface, where they are reflected by geologic structures. The reflected waves are recorded at the surface and processed to produce *seismic maps and cross sections* that show details of subsurface geologic structure.

5. *Drilling.* Wildcat wells are drilled to test the subsurface targets defined by the preceding steps. Because of the inevitable uncertainties of predicting geologic conditions thousands of meters below the surface, the success rate in 1983 of wildcat wells drilled in the United States was only 17.5 percent.

Each of the three following examples illustrates a particular aspect of remote sensing for oil exploration. The Kenya project was Chevron's first major utilization of Landsat images. No oil was discovered, but Landsat was shown to be a reliable source of regional geologic information. As a result, Chevron geologists then interpreted Landsat images of southern Sudan to locate a major sedimentary basin. The Sudan project, a classic example of the systematic exploration process described above, resulted in the discovery of several oil fields. Kenya and the Sudan are frontier exploration areas where relatively few wells have been drilled. The final example in northwest Colorado is a mature area that has been thoroughly explored and drilled. Thus it is a good training site for learning how to recognize oil-producing structures on Landsat images.

Kenya

Chevron Overseas Petroleum acquired an exploration license in eastern Kenya and completed a photogeologic and field study in 1972. Landsat MSS images of the area became available in 1973 and were compiled into the mosaic shown in Figure 8.18 that was interpreted by Miller (1975). The drainage patterns on the interpretation map (Figure 8.19) provide valuable geographic reference in this region of generalized base maps. Much geologic information is present on the Landsat images, despite the relatively featureless nature of the terrain.

Lineaments are a major feature in Kenya, as shown in the interpretation map and in the image covering the northwest part of the Chevron license area (Figure 8.20A). The lineaments and other features are more apparent on color composite images (not illustrated) than on the black-and-white versions shown here. The Lagh Bogal lineament that trends northwest across Figure 8.20A is particularly significant because it marks the northeast boundary of a sedimentary basin that was confirmed by later geophysical surveys. The major lineaments extend beyond the limits of the mosaic. Large volcanoes occur along the extensions of several lineaments beyond the Chevron license area.

Young volcanic flows form the black pattern in the west-central part of Figure 8.20A. In the southeast part, large arcuate tonal bands may represent depositional patterns in clastic beds. A dark lobe extends southeast from the northwest part of the image and marks the outcrop of crystalline basement rocks with discernible foliation trends.

The image in Figure 8.20B covers the southwest part of the license area, where dark basement rocks, with north-trending foliation patterns, crop out in the southwest part of the image. East of the basement outcrops, and probably in fault contact, are sands and clays of Pliocene age that form a triangular light-colored outcrop through which flows the Tana River with its associated vegetation. North of the river, the Pliocene strata are capped by a dark layer of duricrust. In the southeast part of Figure 8.20B, a thin wedge of strata with a gray tone occurs between the duricrust and the underlying light-toned strata. The eastward expansion of this wedge represents basinward thickening of the sedimentary section that was first observed on the Landsat image.

Figure 8.21 summarizes the results of gravity, magnetic, and seismic surveys. The Lagh Bogal lineament coincides in part with a subsurface fault, independently interpreted from the geophysical surveys, that forms the northeast boundary of the basin. Geophysical data indicate that the west border of the basin is not a single major fault zone but a combination of faulting and tilting. The northeast and northwest trends of the image lineaments in the northeast part of the Chevron license area are parallel with the gravity and magnetic positive trends shown in Figure 8.21. Based largely on the Landsat interpretations, Chevron acquired a second exploration license area adjoining the original area on the northwest (Figure 8.21). Several wildcat wells were drilled to test geophysical prospects but were dry because subsurface conditions were not conducive to hydrocarbon generation. However, the geophysical surveys and drilling confirmed the accuracy of the geologic interpretations of Landsat images, which encouraged Chevron to use remote sensing to explore farther north in the Sudan.

The Sudan

The Sudan, the largest nation in Africa, is about equal in area to the United States east of the Mississippi River. The Sudan is incompletely mapped, both geographically and geologically. Little oil exploration had been done prior to 1975. Using the experience gained in Kenya, Chevron compiled a mosaic of Landsat MSS images at a scale of 1:1,000,000 for the southern part of the Sudan. Figure 8.22 shows a reduced-scale version of part of the mosaic. The White Nile River flows northward from the highlands of Uganda into the vast Sudd Swamp, which occupies much of the eastern portion of the mosaic (Figure 8.22). The river emerges from the swamp and flows northward toward Khartoum.

The mosaic was interpreted by J. B. Miller of Chevron, who recognized the presence of a previously unknown sedimentary basin in the vicinity of the present-day Sudd Swamp. Miller also analyzed the drainage patterns and noted that while there were numerous local bends and meanders, the major streams were relatively straight at the regional scale of the Landsat mosaic. These stream lineaments were interpreted as the expression of faults that formed the boundaries of the major basin and smaller subbasins. Based on this regional Landsat interpretation, Chevron negotiated with the Sudanese government to obtain exploration rights to the area outlined in the mosaic (Figure 8.22), which included the inferred sedimentary basin. Figure 8.23A shows the location of the original concession, which covered an area of 5.1×10^5 km². For comparison, note that the state of California covers an area of less than 4.1×10^5 km².

The next step in the campaign was to have the Landsat MSS images covering the concession area digitally processed, enlarged, and registered to ground-control points to produce a base map for the area that was used throughout the campaign. Additional geologic information was interpreted from the individual enhanced color images to guide subsequent exploration.

Chevron then made aerial magnetic surveys that confirmed the existence of the sedimentary basin. A gravity survey added details to the evolving regional picture. The next step was to conduct seismic surveys, which were difficult and expensive in this area of few roads and towns. Equipment and supplies were moved by aircraft to airstrips built for that purpose. The seismic surveys defined a number of potential oil traps, and wildcat drilling began in 1977. The first five wells were dry, but in 1979 the first oil was discovered in Sudan at Abu Gabra (Figure 8.23B). Subsequent drilling discovered the Heglig, Unity, and Melut fields. Full extent of the fields is not known at this early stage of development, but reserves exceed several hundred million barrels. A

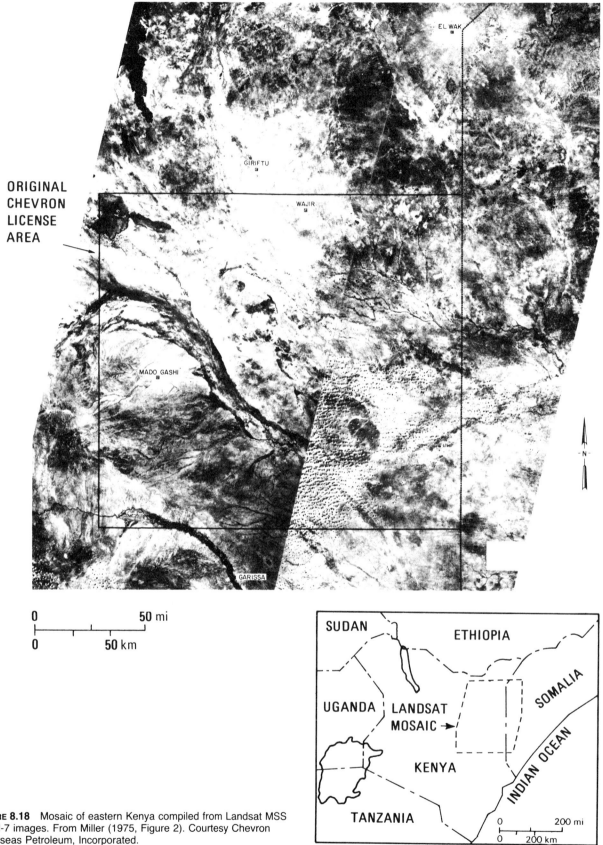

ORIGINAL
CHEVRON
LICENSE
AREA

EL WAK

GIRIFTU

WAJIR

MADO GASHI

GARISSA

| 0 | | 50 mi |
| 0 | | 50 km |

SUDAN   ETHIOPIA

UGANDA   LANDSAT   SOMALIA
          MOSAIC →

          KENYA

TANZANIA

INDIAN OCEAN

| 0 | | 200 mi |
| 0 | | 200 km |

FIGURE 8.18   Mosaic of eastern Kenya compiled from Landsat MSS
band-7 images. From Miller (1975, Figure 2). Courtesy Chevron
Overseas Petroleum, Incorporated.

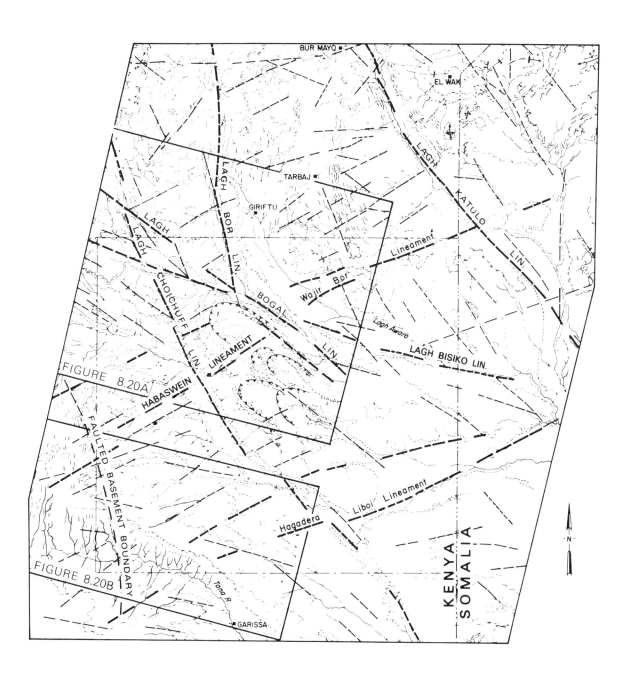

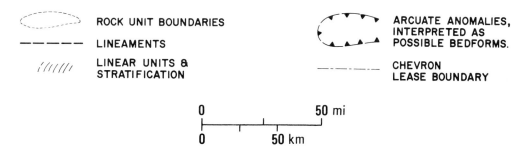

ROCK UNIT BOUNDARIES

LINEAMENTS

LINEAR UNITS & STRATIFICATION

ARCUATE ANOMALIES, INTERPRETED AS POSSIBLE BEDFORMS.

CHEVRON LEASE BOUNDARY

0   50 mi

0   50 km

FIGURE 8.19 Interpretation of Landsat mosaic of eastern Kenya. From Miller (1975, Figure 4). Courtesy Chevron Overseas Petroleum, Incorporated.

A. NORTHWEST PART OF THE CHEVRON LICENSE AREA.

B. SOUTHWEST PART OF THE CHEVRON
   LICENSE AREA.

FIGURE 8.20 Landsat MSS band-5 images that cover parts of the Chevron
exploration license area in Kenya. See Figure 8.19 for locations. From Miller
(1975). Courtesy Chevron Overseas Petroleum, Incorporated.

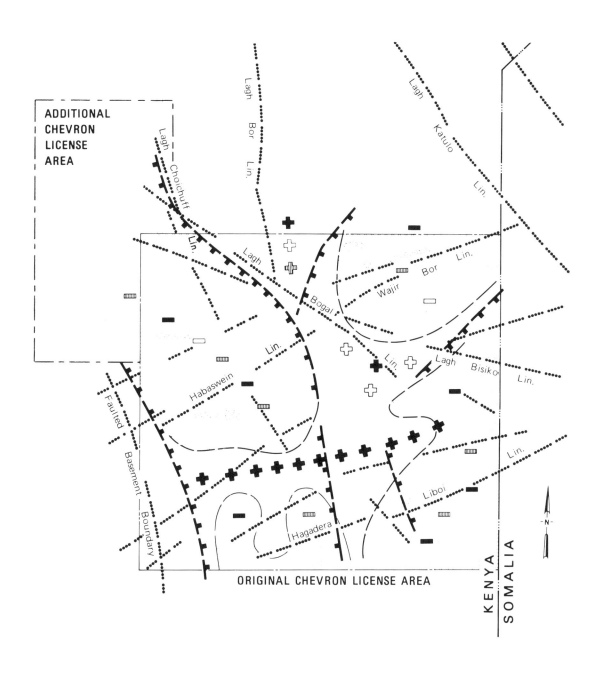

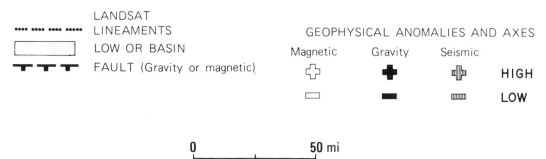

FIGURE 8.21 Comparison of Landsat features with geophysical trends. From Miller (1975, Figure 6). Courtesy Chevron Overseas Petroleum, Incorporated.

FIGURE 8.22 Mosaic of Landsat MSS band-5 images of southern Sudan showing the outline of the original Chevron exploration concession area.

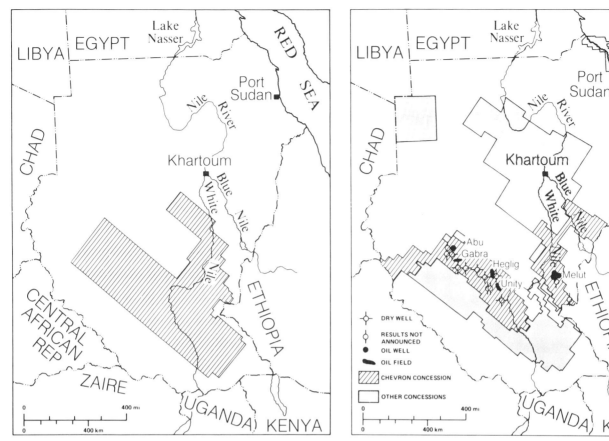

A. ORIGINAL CHEVRON EXPLORATION CONCESSION
   GRANTED IN 1974.

B. STATUS OF EXPLORATION AND CONCESSIONS
   IN 1982.

FIGURE 8.23 Maps showing status of oil exploration in the Sudan.

pipeline is planned to Port Sudan on the Red Sea, where a shipping terminal and refinery will be constructed.

The maps in Figure 8.23 show major changes between 1974 and 1982. The original Chevron exploration area has diminished in size because the agreement requires the company to relinquish portions of the concession at stated times. Rights to much of the relinquished areas were acquired by competing companies, which also acquired other large concession areas. The progress of the Sudan project has been reported by Miller and Vandenakker (1977), Vandenakker and Ryan (1983), and Schull (1984).

Landsat data were also employed in the operational phase of the project. Southern Sudan has a dry season and a wet season with drastically different terrain con-

ditions. Plate 13C is a subscene of an MSS image in the Sudd Swamp acquired during the dry season. The sinuous red band is a stream channel with vegetation, and associated small lakes which have dark signatures. The dark terrain on either side of the channel is grassland that was burned by the local people to produce better forage during the ensuing wet season. Plate 13D is the same subscene imaged during the wet season and digitally processed in the following manner. Field crews in the area noted localities of major terrain categories and communicated this information to Chevron staff personnel in San Francisco. The field localities were used as training sites to produce the supervised classification map of Plate 13D. The map colors and corresponding terrain categories are listed below.

Color	Terrain category
Dark blue	Open water
Light blue	Shallow water with vegetation
Red	Papyrus and water hyacinth
Orange	Wet grass with standing water
Green	Bullrushes
Black	Bullrushes with standing water
Dark yellow	Dry grass
Light yellow	Upland areas, driest areas

These terrain classification maps were valuable to the seismic crews for planning operations. For example, swamp buggies were used in marsh areas, but where papyrus plants occurred, the stalks jammed the drive mechanisms. Fieldworkers were able to avoid driving through areas with papyrus plants by consulting the classification map (Plate 13D), which shows these areas in red.

Northwest Colorado

In contrast to Kenya and the Sudan, northwest Colorado is a mature exploration area where the surface structures have been drilled and one can directly evaluate the relationship of Landsat features to oil and gas fields. The winter MSS image of Figure 8.24 demonstrates the advantages of low sun elevation and light snow cover for structural mapping. The major geologic features and the local structures associated with oil and gas fields are shown in the map of Figure 8.25. The White River and Uinta Mountain uplifts and the Piceance Creek and Sand Wash basins are well expressed. The Piceance Creek basin contains major oil shale reserves of the Green River Formation (Eocene age) that crops out around the basin margins. The Grand Hogback monocline separating the White River uplift from the Piceance Creek basin is especially prominent on the image.

The Rangely anticline, in the southwest part of the map (Figure 8.25), is a major Chevron oil field that is outlined by strike ridges of resistant Cretaceous sandstones surrounding the eroded Mancos Shale outcrops in the core. The asymmetry of the anticline is indicated on the image by the gentle dip slopes on the north flank and the steeper slopes on the south. The Blue Mountain anticline to the north is equally well expressed, but is nonproductive. Moffat, Iles, and Thornburgh are small oil fields trapped at anticlinal closures that are outlined by resistant sandstone ridges on the image. The Danforth Hills and Wilson Creek oil fields are also anticlinal

structures but are marked on the image by a change to very fine texture. The Piceance Creek gas field is formed by an anticline that is clearly indicated by the radial drainage pattern and by the streams that "wrap around" the structure.

Radar Images for Oil Exploration

The essentially all-weather capability and the ability to enhance geologic structure in forested terrain have made side-looking airborne radar useful for oil exploration, especially in tropical regions. Wing and Mueller (1975) of Continental Oil Company described their structural reconnaissance mapping in Irian Jaya, Indonesia, using SLAR images. Magnier, Oki, and Kaartidiputra (1975) published a SLAR mosaic of the Mahakam delta on the east coast of Kalimantan, Indonesia. The anticlinal trends of the onshore oil fields are clearly visible on the mosaic despite the dense vegetation cover. In the 1980s the Space Shuttle acquired radar images of Indonesia and other oil-producing tropical areas (Chapter 6). One can readily interpret lithologic terrains and geologic structures from these images. When more coverage is available, satellite radar images will be a valuable exploration resource.

Oil Exploration Research

Landsat and radar images may be analyzed for surface evidence of geologic structures such as folds and faults that may form petroleum traps at depth. Some oil fields are marked at the surface by direct indications of the underlying hydrocarbons. Surface seeps of oil and gas that have leaked from subsurface traps are a well-known example. The original Drake well in Pennsylvania was located on the basis of oil seeps. As recently as the early 1900s, oil fields were located in California on the basis of oil seeps. The recognition of oil and gas seeps is called *direct detection*, and much research has been done on this subject.

Escaping oil and gas may interact with surface rocks, soil, and vegetation to produce anomalous conditions that may be clues to the underlying hydrocarbon deposit. Classic examples of rock alteration occur at the Cement and Velma oil fields in southern Oklahoma. Sandstone outcrops in the area are typically red, but over the fields they are tan and gray. Gypsum ($CaSO_4 \cdot nH_2O$) is locally replaced by calcite ($CaCO_3$) over the fields. The color change from red to gray has long been attributed to escaping hydrocarbons that have chemically reduced the red iron oxide in the sandstone to a nonred iron compound. Donovan (1974) recognized another surface alteration effect: the secondary calcite and dolomite ($Ca_{1/2}Mg_{1/2}CO_3$) in the surface rocks have

A. TIMS COLOR COMPOSITE IMAGE. FROM
   KAHLE (1983, COVER). COURTESY A. B. KAHLE,
   JET PROPULSION LABORATORY.

B. IR COLOR IMAGE FROM LANDSAT TM.

PLATE 9   Thermal IR multispectral scanner (TIMS) image and Landsat image in the
Panamint Mountains and Death Valley, California. Each image covers a width of 12
km.

A. NORMAL COLOR IMAGE BEFORE IHS
   TRANSFORMATION.

B. NORMAL COLOR IMAGE AFTER IHS
   TRANSFORMATION.

C. PRINCIPAL–COMPONENT COLOR IMAGE.
   PC IMAGE 2 = RED, PC IMAGE 3 = GREEN,
   AND PC IMAGE 4 = BLUE.

D. RATIO COLOR IMAGE.
   3/1 = RED, 5/7 = GREEN, 3/5 = BLUE.

PLATE 10   Digital enhancement and information extraction of Landsat TM data for
the Thermopolis, Wyoming, subscene. Each image covers a width of 20 km.

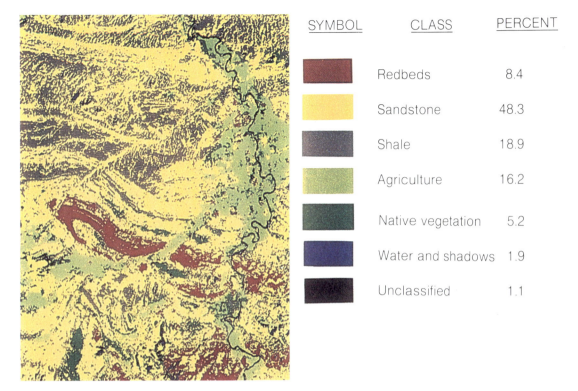

SYMBOL	CLASS	PERCENT
	Redbeds	8.4
	Sandstone	48.3
	Shale	18.9
	Agriculture	16.2
	Native vegetation	5.2
	Water and shadows	1.9
	Unclassified	1.1

A. SUPERVISED–CLASSIFICATION MAP AND EXPLANATION.

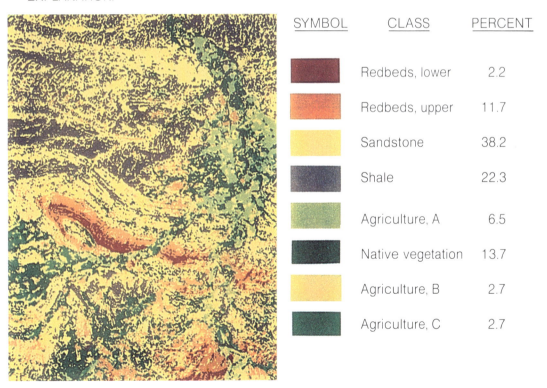

SYMBOL	CLASS	PERCENT
	Redbeds, lower	2.2
	Redbeds, upper	11.7
	Sandstone	38.2
	Shale	22.3
	Agriculture, A	6.5
	Native vegetation	13.7
	Agriculture, B	2.7
	Agriculture, C	2.7

B. UNSUPERVISED–CLASSIFICATION MAP AND EXPLANATION.

PLATE 11 Multispectral classification maps of Landsat TM data for the Thermopolis, Wyoming, subscene. Each image covers a width of 20 km.

A. NORMAL COLOR IMAGE, ENHANCED.

B. COLOR RATIO IMAGE. 5/7 = RED,
3/1 = GREEN, 3/5 = BLUE.

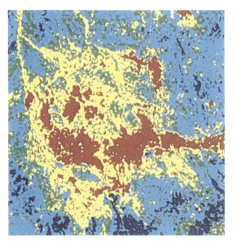

C. RATIO 5/7 IMAGE WITH DENSITY SLICE.

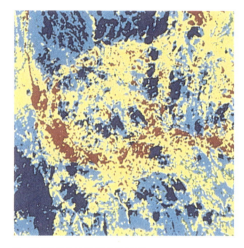

D. RATIO 3/1 IMAGE WITH DENSITY SLICE.

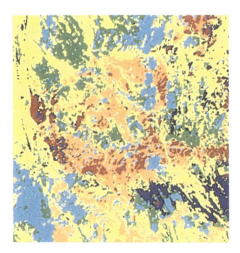

E. UNSUPERVISED–CLASSIFICATION MAP.

F. CLASSIFICATION EXPLANATION.

SYMBOL	CLASS	PERCENT
	Alluvium	39.2
	Basalt	14.0
	Tuff	6.6
	Altered rocks, A	5.3
	Altered rocks, B	18.3
	Unaltered rocks	16.6

PLATE 12 Digitally processed Landsat TM images of the Goldfield, Nevada,
subscene. Each image covers a width of 15 km.

A. COLOR COMPOSITE IMAGE, CAMERON
   URANIUM DISTRICT, ARIZONA. AREA
   COVERS A WIDTH OF 3.5 km.

B. COLOR RATIO IMAGE, CAMERON
   URANIUM DISTRICT, ARIZONA.

C. COLOR COMPOSITE IMAGE, SUDD
   SWAMP, THE SUDAN. AREA COVERS A
   WIDTH OF 45 km.

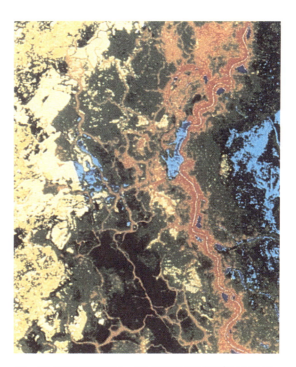

D. SUPERVISED–CLASSIFICATION MAP,
   SUDD SWAMP, THE SUDAN.

PLATE 13   Digitally processed Landsat MSS images for resource exploration.
Plates A and B are from Spirakis and Condit (1975, Figures 4, 6). Plates C and D
are from Vandenakker and Ryan (1983).

35

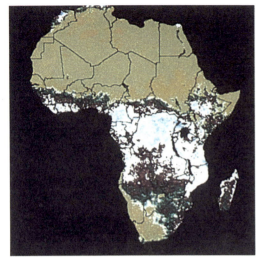

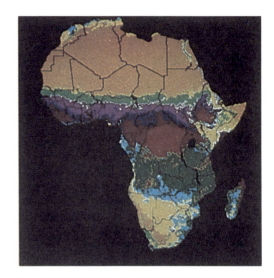

A. SPECTRAL–VEGETATION–INDEX
   MAP OF AFRICA, APRIL 12
   TO MAY 2, 1982.

B. VEGETATION CLASSIFICATION
   MAP OF AFRICA. COLORS ARE
   EXPLAINED IN CHAPTER 9.

C. CZCS RATIO 1/3 IMAGE,
   GULF OF MEXICO AND FLORIDA.
   AREA COVERS A WIDTH OF 600 km.

D. CZCS RATIO 2/3 IMAGE,
   GULF OF MEXICO AND FLORIDA.

PLATE 14 Digitally processed images from NOAA environmental satellites. Plates
A and B are from Tucker, Townshend, and Goff (1985, Figures 1 and 5). Plates C
and D are from Gordon and others (1980, Figure 2).

A. MSS IR COLOR IMAGE, THE OXNARD
PLAIN, CALIFORNIA. AREA COVERS
A WIDTH OF 14 km.

B. SUPERVISED–CLASSIFICATION MAP.
COLORS ARE EXPLAINED IN CHAPTER 10.

C. TM IR COLOR IMAGE, LAS VEGAS,
NEVADA. AREA COVERS A WIDTH OF 10 km.

D. UNSUPERVISED–CLASSIFICATION MAP.
COLORS ARE EXPLAINED IN CHAPTER 10.

PLATE 15 Land use and land cover interpreted from digitally classified Landsat
images. Plates A and B are from Estes and others (1979, Figures 2 and 3). Plates
C and D were digitally processed at Chevron Oil Field Research Company.

A. IR COLOR IMAGE. B. RATIO 5/1, DENSITY SLICED.

PLATE 16 Landsat TM images of Death Valley, California. Each image covers a width of 30 km.

38

unusual carbon isotopic values. These values indicate that hydrocarbons leaking from the reservoir were oxidized and the carbon incorporated into the secondary carbonate minerals of the surface rocks. Similar carbon isotopic values occur over the Davenport oil field in central Oklahoma (Donovan, Friedman, and Gleason, 1974).

Several attempts have been made to identify on Landsat MSS and TM images any spectral signatures of the color and mineralogic alteration patterns at the Cement oil field. Such signatures could then be used to explore for other fields. Various digital processes have been applied to Landsat data of the Cement field, but no successes have been reported. A field investigation found that the color changes at the Cement and Velma fields are only visible at limited exposures in road cuts and stream beds. Most of the area is covered with soil and agriculture that obscures the alteration effects.

Everett and Petzel (1973) interpreted Landsat MSS images of the Anadarko Basin in the Texas Panhandle and western Oklahoma. They reported "hazy" anomalies over a number of oil and gas fields, that were said to appear as if image detail had been smudged or erased. They are recognizable only on certain Landsat images and are not visible on aerial photographs. However, the anomalies are not artifacts of the image reproduction process. No conclusive explanation has been given for them, but it has been suggested that they are caused by human activities. These anomalies were of interest to other investigators, but no one has reported similar features. The Anadarko Basin investigation was summarized by Short (1975), who noted the lack of agreement about causes of the hazy anomalies.

NASA/Geosat Test Case Project—Oil and Gas Fields

The NASA/Geosat project selected three oil and gas fields for study: the Coyanosa field in west Texas, the Lost River field in West Virginia, and the Patrick Draw field in southwest Wyoming. Lang, Nicolais, and Hopkins (1985) interpreted TMS images and Landsat MSS images of the Coyanosa test site. They found no evidence of surface alteration caused by hydrocarbon seepage, such as the bleached rocks that occur at the Cement and Velma fields in Oklahoma. Some significant structural features were expressed in the images, however.

The Lost River field is particularly interesting because it is located in forested terrain of the Appalachian Mountains and provided the opportunity to investigate possible vegetation anomalies associated with a gas field. Lang, Curtis, and Kovacs (1985) prepared a supervised classification map from TMS data that shows distribution of plant types. Concentrations of maple trees oc-

curred at localities normally occupied by oak and hickory trees. A soil gas survey identified unusually high concentrations of methane and ethane that coincide with the maple anomaly. It is postulated that the gas concentration causes anaerobic soil conditions that inhibit growth of oak and hickory trees but does not inhibit maples.

At the Patrick Draw field, the vegetation cover is predominantly sagebrush. Lang, Alderman, and Sabins (1985) described a ratio color image of TMS data that indicated an anomalous area of sagebrush at the west margin of the field. Ground investigation found a few square kilometers of blighted sagebrush with stunted growth and small leaves. A soil gas survey showed high concentrations of gas that could be responsible for this blighted sagebrush. The blighted vegetation, however, does not extend over a significant proportion of the field, and soil gas concentrations can occur with no associated vegetation anomaly.

GEOTHERMAL ENERGY

Geothermal energy is obtained from subsurface reservoirs of steam or hot water that are shallow enough to be drilled and exploited economically. Most of the steam and hot water is used to generate electricity, but some is used directly for heating, as in Iceland. The following conditions are necessary for a *geothermal reservoir:*

1. A large, high-temperature heat source must be present at relatively shallow depth. Intrusive masses of young igneous rock are the usual heat source, and most geothermal areas are associated with surface or subsurface igneous rocks of Cenozoic age.

2. Porous and permeable reservoir rocks filled with steam or hot water must occur near the heat source. A variety of rocks can serve as geothermal reservoirs. In the Imperial Valley, California, poorly consolidated sandstones of late Tertiary age are the reservoir rocks. At the Geysers in northern California, the reservoir occurs in highly fractured sedimentary and volcanic rocks.

3. A natural recharge system must replenish the reservoir with water as steam or hot water is produced.

4. An impermeable zone above the reservoir is necessary to prevent the escape of steam and hot water. Convective flow to the surface would dissipate the heat of an unconfined reservoir. Heat losses due to thermal conduction through the rocks are relatively minor because of the low thermal conductivity of rocks.

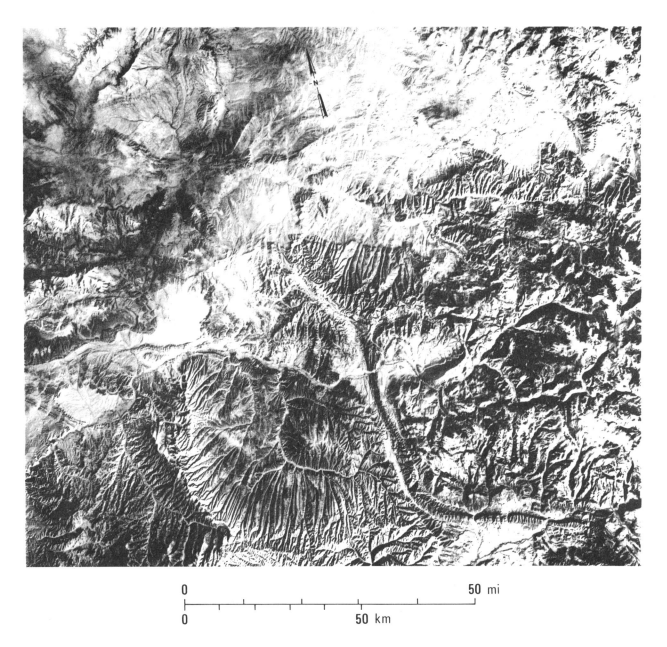

0   50 mi

0   50 km

FIGURE 8.24   Landsat MSS band-5 winter image of northwest Colorado.

Some geothermal reservoirs have no visible or thermal expression at the surface and are not detectable by remote sensing methods. Many geothermal reservoirs, however, have surface thermal expressions ranging in intensity from a minor increase in ground temperature to the presence of hot springs and geysers. Hot springs and geysers commonly occur along faults and fractures that allow hot water to escape from the reservoir. Thermal springs and zones of hydrothermal alteration associated with a geothermal area in Mexico have been interpreted from thermal IR images (Valle and others, 1970). At the Geysers area, hot springs and fumaroles were also detected on thermal IR images and there was some local evidence of higher ground temperatures (Moxham, 1969). However, there was little evidence of a regional surface temperature anomaly on the images of the Geysers area.

Iceland Geothermal Reconnaissance

Iceland is the site of frequent volcanic eruptions, including six since 1946, and associated geothermal activ-

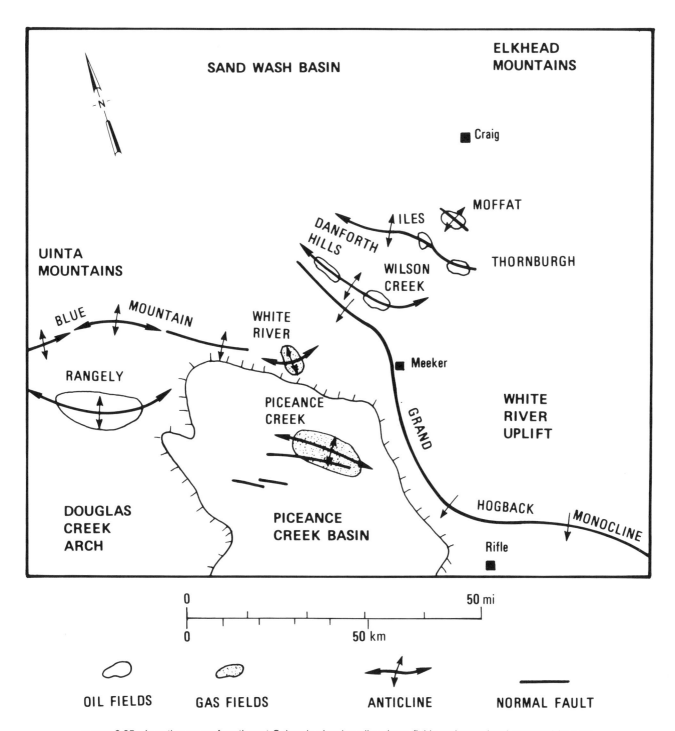

FIGURE 8.25 Location map of northwest Colorado showing oil and gas fields and associated structural features.

ity. Space heating and hot water for the capital city of Reykjavik have long been supplied from geothermal sources. The 17 high-temperature geothermal areas are concentrated along the zones of active rifting and volcanism. Vatnajökull is an ice cap approximately 100 km in diameter that overlies two known high-temperature geothermal areas.

The aerial photograph and thermal IR image of Figure 8.26 show the Kverkfjöll geothermal area, which is located on the northern edge of the Vatnajökull ice cap. As shown on the interpretation map of Figure 8.27, the geothermal area is located between the Kverkjökull outlet glacier on the east and a re-entrant of bedrock on the west. The thermal IR image (Figure 8.26B) is not

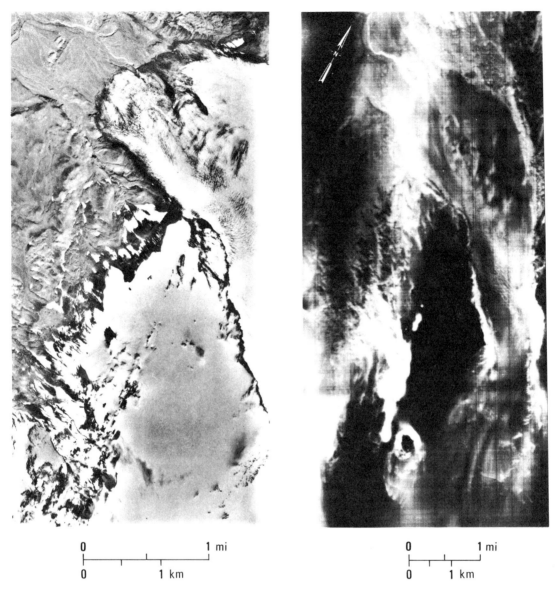

0 |———————|——————| 1 mi
0 |————|————| 1 km

A. AERIAL PHOTOGRAPH ACQUIRED
   AUGUST 24, 1960, BY THE U.S. AIR
   FORCE.

0 |———————|——————| 1 mi
0 |————|————| 1 km

B. NIGHTTIME THERMAL IR (1.0 TO 5.5 μm)
   IMAGE (NOT RECTILINEARIZED).
   ACQUIRED AUGUST 22,1966, BY THE
   U.S. AIR FORCE, CAMBRIDGE
   RESEARCH LABORATORIES.

FIGURE 8.26 Kverkfjöll geothermal area and Kverkjökull outlet glacier, Iceland. From Friedman and others (1969, Figures 10 and 11). Courtesy R. S. Williams, U.S. Geological Survey.

rectilinearized, which accounts for the geometric compression at the east and west margins. The ice and bedrock have relatively cool signatures (dark tones). The geothermal features, the meltwater, and the bedrock ridges that confine the outlet glacier have warm signatures. The north-trending geothermal feature is at least 2 km long and includes two separate hot areas at

the northeast end. Near the south margin of the image, warm signatures mark concentric crevasses and an ice cauldron subsidence feature that are caused by subsurface melting of the glacier (Figure 8.27). The warm stream emerging from the snout of the outlet glacier is meltwater that has flowed along the base of the glacier from a subglacial geothermal source (R. S. Williams, Jr., per-

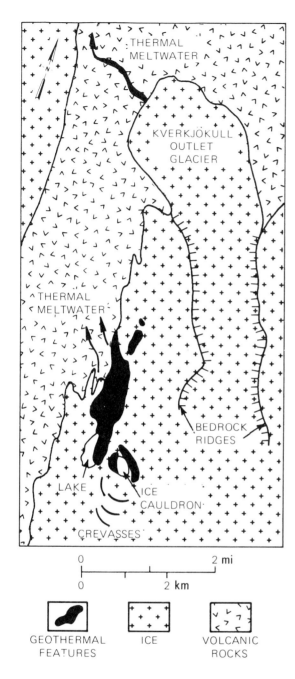

FIGURE 8.27 Interpretation map of the thermal IR image of the Kverkfjöll geothermal area. From Friedman and others (1969, Figure 11).

sonal communication). The topographic expressions of some geothermal features are detectable on the aerial photograph of Figure 8.26A, but are less pronounced than on the IR image.

Geothermal vents and hot springs have been detected on thermal IR images in such diverse localities as Japan, Italy, Ethiopia, and the United States. Remote sensing is especially useful in remote areas where the surface expression of geothermal activity has not been located by conventional means.

Low-Intensity Geothermal Anomalies

Thermal IR images are ideal for detecting fumaroles, steaming ground, and hot springs associated with very active geothermal areas such as Iceland and Yellowstone National Park. Elsewhere, however, there are so-called *blind geothermal areas* that lack surface thermal or alteration expression. Intermediate in character between the active areas and the blind areas is a broad category of geothermal areas where the surface temperature is only slightly higher than the surrounding areas. These areas with low-intensity surface temperature anomalies are difficult to detect on IR images or on airborne radiometer profiles. Watson (1975) pointed out that natural variations of geology and topography can readily overwhelm surface geothermal anomalies of several hundred *heat-flow units* (HFU) (1 HFU = 1×10^{-6} cal $\cdot$ cm^{-1} $\cdot$ sec^{-1}). Mathematical modeling was used to evaluate the relative effect of various factors on the surface radiant temperature. This theoretical analysis suggests that both thermal and reflectance images should be acquired at least three times during the diurnal cycle (Watson, 1975, p. 136). Comparison of the images may reveal subtle anomalies. An 8-to-14-μm image in the Raft River area of Idaho revealed a weak thermal anomaly that was confirmed by ground measurements.

OTHER ENERGY SOURCES

The location and distribution of large reserves of coal, oil shale, and tar sands are already known in the United States and Canada. Therefore exploration is relatively inactive and there is little application of remote sensing. Landsat images are potentially useful during mining of coal. The status of strip mining and land reclamation may be monitored by digitally processing data acquired during the repetitive cycles of Landsat. Roof falls are hazards in underground coal mines. In Indiana it was demonstrated that areas of intense fracturing on Landsat images coincided with areas of roof falls in the mines and could be used to predict the hazards (Wier and others, 1973).

COMMENTS

Remote sensing has proven a valuable aid in exploring for mineral and energy resources. Many ore deposits are localized along regional and local fracture patterns that provided conduits along which ore-forming solutions penetrated host rocks. Landsat images are used to map these fracture patterns. Ore-forming hydrothermal solutions alter the host rocks to distinctive assemblages of secondary minerals. Iron sulfide minerals weather to iron oxide minerals with reddish colors that are recognizable on Landsat MSS ratio 4/5 images and Landsat TM ratio 3/1 images. Clay minerals and alunite formed in hydrothermally altered rocks have distinctive absorption features at wavelengths between 2.08 and 2.35 μm that are recorded by band 7 of Landsat TM. The TM ratio 5/7 image has distinctive bright signatures associated with hydrothermally altered rocks. In the future, multispectral scanners with high spectral resolution may identify specific alteration minerals.

Detection of hydrothermally altered rocks is not possible in vegetated areas, so this environment requires other remote sensing methods. Reflectance spectra of foliage growing over mineralized areas may differ from spectra of foliage in adjacent nonmineralized areas. The spectral differences, however, are variable for different plant species. More research and development is needed for remote detection of mineral deposits in vegetated terrain.

For oil exploration, remote sensing is especially valuable in poorly mapped regions. Interpretation of Landsat images of southern Sudan revealed the presence of a previously unrecognized sedimentary basin in which several oil fields were subsequently discovered. Landsat images are also useful as base maps and for planning field operations. In forested areas of perennial cloud cover, radar images acquired from aircraft and satellites make it possible for geologists to map lithologic terrain and geologic structures that may be exploration clues for oil fields.

Hot springs and other surface indications of geothermal areas are recognizable in thermal IR images, but low-intensity thermal anomalies are generally obscured by variations in surface temperature.

QUESTIONS

1. You are employed by an international mineral exploration company that plans to explore for hydrothermal gold deposits in the southern third of the Andes Mountains of South America. Your assignment is to plan the remote sensing phase of the exploration campaign ranging from regional reconnaissance to definition of individual prospects. You can utilize the image-acquisition systems and image-processing systems described in this book. Prepare the remote sensing exploration campaign, with reasons and justification for each step.

2. As part of your Chile project you need to identify remotely the alteration minerals kaolinite, montmorillonite, illite, and alunite. Your company has an airborne scanner that can digitally record five bands of data in the 2.0-to-2.5-μm region. Each band has a spectral range of 0.05 μm, such as 2.10 to 2.15 μm. Use Figure 8.13 to select the five optimum bands for identifying the alteration minerals. List your reasons for selecting each band. Describe how you would digitally process the resulting airborne data to produce maps showing distribution of the different minerals.

3. The mineral industry is depressed, and you are now employed by an international oil exploration company. Your company plans to evaluate the petroleum potential of the western portion of the People's Republic of China in preparation for negotiating an exploration concession. Because of limited accessibility, deadlines, and competitor pressure, concession areas must be selected solely on the basis of remote sensing evaluations. Prepare such a plan for your management, giving reasons for each step.

4. Assume that western China is completely covered by Landsat TM images, SIR images, and LFC overlapping black-and-white photographs. List the advantages and disadvantages of each of these kinds of satellite images for your project.

REFERENCES

Abrams, M. J., and D. Brown, 1985, Silver Bell, Arizona, porphyry copper test site: The Joint NASA/Geosat test case study, section 4, American Association of Petroleum Geologists, Tulsa, Okla.

Abrams, M. J., J. E. Conel, and H. R. Lang, 1985, The joint NASA/Geosat test case study: American Association of Petroleum Geologists, Tulsa, Okla.

Albert, N. R. D., 1975, Interpretation of Earth Resource Technology Satellite imagery of the Nabesna quadrangle, Alaska: U.S. Geological Survey Miscellaneous Field Map MP 655J.

Ashley, R. P., 1974, Goldfield mining district: Nevada Bureau of Mines and Geology Report 19, p. 49–66.

Ashley, R. P., 1979, Relation between volcanism and ore deposition at Goldfield, Nevada: Nevada Bureau of Mines and Geology Report 33, p. 77–86.

Cannon, H. L., 1971, The use of plant indicators in groundwater surveys, geologic mapping, and mineral prospecting: Taxon, v. 20, p. 227–256.

Chenoweth, W. L., and D. N. Magleby, 1971, Mine location map, Cameron uranium area, Coconino County, Arizona: U.S. Atomic Energy Commission Preliminary Map 20.

Collins, W., S. H. Chang, G. Raines, F. Canney, and R. P. Ashley, 1983, Airborne biogeophysical mapping of hidden mineral deposits: Economic Geology, v. 78, p. 737–749.

Conel, J. E., and R. E. Alley, 1985, Lisbon Valley, Utah,

uranium test site report: The joint NASA/Geosat test case study, section 8, American Association of Petroleum Geologists, Tulsa, Okla.

Donovan, T. J., 1974, Petroleum microseepage at Cement field, Oklahoma—evidence and mechanism: Bulletin of the American Association of Petroleum Geologists, v. 58, p. 429–446.

Donovan, T. J., I. Friedman, and J. D. Gleason, 1974, Recognition of petroleum-bearing traps by unusual isotopic compositions of carbonate-cemented surface rocks: Geology, v. 2, p. 351–354.

Everett, J. R., and G. Petzel, 1973, An evaluation of the suitability of ERTS data for the purposes of petroleum exploration: Third Earth Resources Technology Satellite Symposium, NASA SP-356, p. 50–61.

Friedman, J. D., R. S. Williams, G. Pálmason, and C. D. Miller, 1969, Infrared surveys in Iceland: U.S. Geological Survey Professional Paper 650-C, p. C89–C105.

Goetz, A. F. H., 1984, High spectral resolution remote sensing of the land: Proceedings of the International Society for Optical Engineering, v. 475, p. 56–68.

Harvey, R. D., and C. J. Vitaliano, 1964, Wall-rock alteration in the Goldfield District, Nevada: Journal of Geology, v. 72, p. 564–579.

Horton, R. C., 1964, An outline of the mining history of Nevada, 1924–1964: Nevada Bureau of Mines Report 7, pt. 2.

Hunt, G. R., and R. P. Ashley, 1978, Spectra of altered rocks in the visible and near infrared: Economic Geology, v. 74, p. 1613–1629.

Hunt, G. R., J. W. Salisbury, and C. J. Lenhof, 1971, Visible and near-infrared spectra of minerals and rocks—III, oxides and hydroxides: Modern Geology, v. 2, p. 195–205.

Kappelmeyer, O., and R. Haenel, 1974, Geothermics with special reference to application: Geoexploration Monographs, series 1, no. 4, Gebrüder Borntraeger, Berlin.

Labovitz, M. L., E. J. Masuoka, R. Bell, A. W. Segrist, and R. F. Nelson, 1983, The application of remote sensing to geobotanical exploration for metal sulfides—results from the 1980 field season at Mineral, Virginia: Economic Geology, v. 78, p. 750–760.

Lang, H. R., W. H. Alderman, and F. F. Sabins, 1985, Patrick Draw, Wyoming, petroleum test case report: The NASA/Geosat test case project, section 11, American Association of Petroleum Geologists, Tulsa, Okla.

Lang, H. R., J. B. Curtis, and J. S. Kovacs, 1985, Lost River, West Virginia, petroleum test site: The NASA/Geosat test case project, section 12, American Association of Petroleum Geologists, Tulsa, Okla.

Lang, H. R., S. M. Nicolais, and H. R. Hopkins, 1985, Coyanosa, Texas, petroleum test site: The NASA/Geosat test case project, section 13, American Association of Petroleum Geologists, Tulsa, Okla.

Lowell, J. D., and J. M. Guilbert, 1970, Lateral and vertical alteration-mineralization zoning in porphyry ore deposits: Economic Geology, v. 65, p. 373–408.

Magnier, P., T. Oki, and L. W. Kaartidiputra, 1975, The Mahakam delta: World Petroleum Congress Proceedings, v. 2, p. 239–250.

Miller, J. B., 1975, Landsat images as applied to petroleum

exploration in Kenya: NASA Earth Resources Survey Symposium, NASA TM X-58168, v. 1-B, p. 605–624.

Miller, J. B., and J. Vandenakker, 1977, Sudan interior exploration project—planimetry and geology: American Association of Petroleum Geologists, Third Pecora Conference, Sioux Falls, S.D.

Moxham, R. M., 1969, Aerial infrared surveys at the Geysers geothermal steam field, California: U.S. Geological Survey Professional Paper 630-C, p. C106–C122.

Nicolais, S. M., 1974, Mineral exploration with ERTS imagery: Third ERTS-1 Symposium, NASA SP-351, v. 1, p. 785–796.

Offield, T. W., 1976, Remote sensing in uranium exploration: Exploration of uranium ore deposits, International Atomic Energy Proceedings, p. 731–744, Vienna, Austria.

Rowan, L. C., and P. H. Wetlaufer, 1975, Iron-absorption band analysis for the discrimination of iron-rich zones: U.S. Geological Survey, Type III Final Report, Contract S-70243-AG.

Rowan, L. C., A. F. H. Goetz, and R. P. Ashley, 1977, Discrimination of hydrothermally altered and unaltered rocks in the visible and near infrared: Geophysics, v. 42, p. 522–535.

Rowan, L. C., P. H. Wetlaufer, A. F. H. Goetz, F. C. Billingsley, and J. H. Stewart, 1974, Discrimination of rock types and detection of hydrothermally altered areas in southcentral Nevada by the use of computer-enhanced ERTS images: U.S. Geological Survey Professional Paper 883.

Schull, T. J., 1984, Oil exploration in nonmarine rift basins of interior Sudan (abstract): American Association of Petroleum Geologists, v. 68, p. 526.

Settle, M., 1985. The joint NASA/Geosat Test Case Project, executive summary: American Association of Petroleum Geologists, Tulsa, Okla.

Short, N. M., 1975, Exploration for fossil and nuclear fuels from orbital altitudes: Remote sensing energy related studies, p. 189–232, Hemisphere Publishing Corporation, Washington, D.C.

Spirakis, C. S., and C. D. Condit, 1975, Preliminary Report on the use of Landsat-1 (ERTS-1) reflectance data in locating alteration zones associated with uranium mineralization near Cameron, Arizona: U.S. Geological Survey Open File Report 75–416.

Stewart, J. H., G. W. Walker, and F. J. Kleinhampl, 1975, Oregon-Nevada lineament: Geology, v. 3, p. 251–268.

Valle, R. G., J. D. Friedman, S. J. Gawarecki, and C. J. Banwell, 1970, Photogeologic and thermal infrared reconnaissance surveys of the Los Negritos–Ixtlan de Los Hervores geothermal area, Michoacan, Mexico: Geothermics Special Issue 2, p. 381–398.

Vandenakker, J., and J. Ryan, 1983, Landsat applications for geophysical field operations (abstract): Geophysics, v. 48, p. 475.

Viljoen, R. P., M. J. Viljoen, J. Grootenboer, and T. G. Longshaw, 1975, ERTS-1 imagery—an appraisal of applications in geology and mineral exploration: Minerals Science and Engineering, v. 7, p. 132–168.

Watson, K., 1975, Geologic applications of thermal infrared images: Proceedings of the IEEE, n. 501, p. 128–137.

Wier, C. E., F. J. Wobber, O. R. Russell, R. V. Amato, and

T. V. Leshendok, 1973, Relationship of roof falls in underground coal mines to fractures mapped on ERTS-1 imagery: Third ERTS-1 Symposium, NASA SP-351, p. 825–843.

Wing, R. S., and J. C. Mueller, 1975, SLAR reconnaissance, Mimika-Eilanden Basin, southern trough of Irian Jaya: NASA Earth Resources Survey Symposium, NASA TM X-58168, v. 1-B, p. 599–604.

Yost, E., and S. Wenderoth, 1971, The reflectance spectra of mineralized trees: Proceedings of Seventh International Symposium on Remote Sensing of Environment, University of Michigan, v. 1, p. 269–284, Ann Arbor, Mich.

ADDITIONAL READING

Abrams, M. J., D. Brown, L. Lepley, and R. Sadowski, 1983, Remote sensing for porphyry copper deposits in southern Arizona: Economic Geology, v. 78, p. 591–604.

Goetz, A. F. H., B. N. Rock, and L. C. Rowan, 1983, Remote sensing for exploration—an overview: Economic Geology, v. 78, p. 573–590. (This issue includes a number of articles on remote sensing exploration.)

Offield, T. W., E. A. Abbott, A. R. Gillespie, and S. O. Laguercio, 1977, Structural mapping on enhanced Landsat images of southern Brazil—Tectonic control of mineralization and speculations on metallogeny: Geophysics, v. 42, p. 482–500.

Sabins, F. F., 1979, Oil occurrence and plate tectonics as viewed on Landsat images: Proceedings of the Tenth World Petroleum Congress, p. 105–109, Bucharest, Romania.

Schmidt, R., B. B. Clark, and R. Bernstein, 1975, A search for sulfide-bearing areas using Landsat-1 data and digital image-processing techniques: NASA Earth Resources Survey Symposium, NASA TM X-58168, v. 1-B, p. 1013–1027.

Reprinted by permission of Guilford Press from J. B.
Campbell, *Introduction to Remote Sensing*, 1987, p.
85-117.

CHAPTER FOUR

Image Interpretation

4.1. Introduction

Earlier chapters have defined our interest in remote sensing as focused primarily upon *images* of the earth's surface—maplike representations of the earth's surface based upon reflection of electromagnetic energy from the vegetation, soil, water, rocks, and structures that occupy the surface of our planet. From such images we can learn much that cannot be derived from other sources.

Yet such information is not presented to us directly—the information we seek is encoded in the varied tones and textures we see on each image. To "decode" the information, we must apply specialized knowledge that forms the field of *image interpretation*, which we can apply to derive useful information from the raw, uninterpreted images we receive from remote sensing systems. Proficiency in image interpretation comes from three separate kinds of knowledge, of which only one—the final one listed here—falls within the scope of this text.

Subject

Knowledge of the subject of our interpretation—the kind of information that motivates us to examine the image—must form the heart of the interpretation. Accurate interpretation requires familiarity with the subject of the interpretation. For example, interpretation of geologic information requires education and experience in the field of geology. Yet, narrow specializations are a handicap because each image records a complex mixture of many kinds of information, requiring application of broad knowledge that crosses traditional boundaries between disciplines. For example, accurate interpretation of geological information may require knowledge of botany and the plant sciences as a means of understanding how vegetation patterns reflect geologic patterns that may not be directly visible. As a result, image interpreters should be equipped with a broad range of knowledge pertaining to the subjects at hand and their interrelationships.

Geographic Region

Knowledge of the specific geographic region depicted on an image can be equally significant. Every locality has unique characteristics that influence the patterns recorded on an image. Often the interpreter's direct experience within the area depicted on the image can be applied to the interpretation. In unfamiliar regions the interpreter may find it necessary to make a field reconnaissance or to use maps and books that describe analogous regions with similar climate, topography, or land use.

Remote Sensing System

Finally, knowledge of the remote sensing system is obviously essential. The interpreter must understand how each image is formed and how each sensor portrays landscape features. Different instruments use separate portions of the electromagnetic spectrum, operate at different resolutions, and use different methods of recording images. The image interpreter must know how each of these variables influences the image to be interpreted and how to evaluate their effects on his ability to derive useful information from the imagery. The subject of this chapter, then, is an outline of how the image interpreter derives useful information from the complex patterns of tone and texture on each image.

4.2. The Context for Image Interpretation

Human beings are well prepared to examine images, as our visual system and experience equip us to discern subtle distinctions in brightness and darkness, to distinguish between various image textures, to perceive depth, and to recognize complex shapes and features. Even in early childhood, we apply such skills routinely in everyday experience, so that few of us encounter difficulties as we examine family snapshots or photographs in newspapers, for example. Yet the art of image interpretation requires conscious, explicit effort not only to learn about the subject matter, geographic setting, and imaging systems (as mentioned above) in unfamiliar contexts, but also to develop our innate abilities for image analysis.

Three issues distinguish interpretation of remotely sensed imagery from interpretation in everyday experience. First, remotely sensed images usually portray an overhead view—an unfamiliar perspective. Training, study, and experience are required to develop the ability to recognize objects and features. Second, many remote sensing images use radiation outside the visible portion of the spectrum; in fact, use of such radiation is an important advantage that we exploit as often as possible. Even the most familiar features may appear quite different in nonvisible portions of the spectrum than they do in the familiar world of visible radiation. Finally, remote sensing images often portray the earth's surface at unfamiliar scales and resolutions. Commonplace objects and features may assume strange shapes and appearance as scale and resolution change from those to which we are accustomed (Figure 4.1).

This chapter outlines the art/science of image interpretation. The student cannot

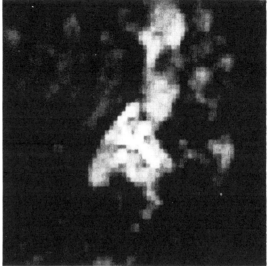

Figure 4.1. Familiar features may not be recognizable at coarse image resolution.

expect to become a proficient image analyst simply by reading about image interpretation. Experience forms the only sure preparation for skillful interpretation. Yet this chapter can highlight some of the issues that are important in the development of proficiency as an image interpreter.

In order to discuss this subject at an early point in the text, we must confine the discussion to interpretation of aerial photography—the only form of remote sensing imagery discussed thus far. But the principles, procedures, and equipment described here are equally applicable to other kinds of imagery acquired by the sensors described in later chapters.

Manual image interpretation is discussed in greater detail in the text by Avery and Berlin (1985); two older references that may also be useful are the text by Lueder (1959) and the *Manual of Photographic Interpretation*, published in 1960 by the American Society of Photogrammetry.

4.3. Image Interpretation Tasks

The image interpreter must routinely conduct several kinds of tasks, many of which may be completed together in an integrated process. Nonetheless, for purposes of clarification it is important to distinguish between these separate functions (Figure 4.2).

Classification

The first task is classification, assigning objects, features, or areas to classes based upon their appearance on the imagery. Often the distinction is made between three

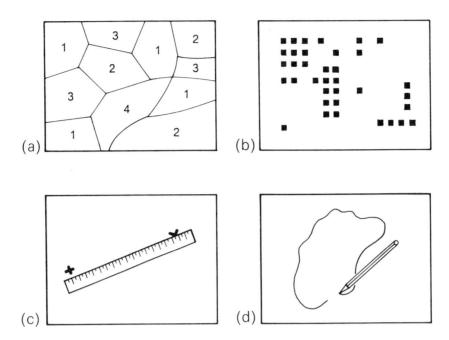

Figure 4.2. Image interpretation skills: (*a*) classification, (*b*) enumeration, (*c*) mensuration, (*d*) delineation.

levels of confidence and precision. *Detection* is the determination of presence or absence of a feature. *Recognition* implies a higher level of knowledge about a feature or object, such that the object can be assigned an identity in a general class or category. Finally *identification* means that the identity of an object or feature can be specified with enough confidence and detail to place it in a very specific class. Often an interpreter may qualify his or her confidence in an interpretation by specifying the identification as "possible" or "probable."

Enumeration

Enumeration refers to listing and counting discrete items visible on an image. For example, housing units can be classified as "detached single-family," "multifamily complex," "mobile home," "multistory residential," and so on, and then reported as numbers present within a defined area. Clearly the ability to conduct such an enumeration depends upon an ability to classify items accurately as discussed above.

Mensuration

Measurement, or mensuration, is an important function in many image interpetation problems. Two kinds of measurement are important. First is the measurement of distance and height, and by extension, of volumes and areas as well. The practice of making such measurements forms the subject of *photogrammetry*, which applies a knowledge of image geometry to derivation of accurate distances. Although strictly speaking, photogrammetry applies only to measurements from photographs, by extension it has analogues for derivation of measurements from other kinds of remotely sensed images.

A second form of measurement is quantitative assessment of image brightness. The science of *photometry* is devoted to measurement of the intensity of light, including estimation of scene brightness by examination of image tone, using a special instrument known as a *densitometer*, described below. If the measured radiation extends outside the visible spectrum, the term *radiometry* applies. Photometry and radiometry apply similar instruments and principles, so they are closely related to one another.

Delineation

Finally, the interpreter must often delineate, or outline, regions as they are observed on remotely sensed images. The interpreter must be able to separate distinct areal units that are characterized by specific tones and textures and to identify edges, or boundaries, between separate areas. Typical examples include delineation of separate classes of forest or of land use, both of which occur only as areal entities (rather than as discrete objects). Typical problems include selection of appropriate levels of generalization (when boundaries are intricate, or when many tiny but distinct parcels are present), and placement of boundaries when there is a gradation (rather than a sharp edge) between two units.

The image interpreter may simultaneously apply several of these skills in examining an image. Recognition, delineation, and mensuration may all be required as the interpreter examines an image. Yet specific interpretation problems may emphasize specific skills. Military photo interpretation often depends upon accurate recognition and enumeration of specific items of equipment, whereas land use inventory emphasizes delineation, although other skills are obviously important. Image interpreters therefore need to develop proficiency in all these skills.

4.4. Elements of Image Interpretation

By tradition, image interpreters are said to employ some combination of eight elements of image interpretation, which describe the characteristics of objects and features as they appear on remotely sensed images. Image interpreters obviously use these characteristics together in very complex, but poorly understood, processes as they examine images. Nonetheless, it is convenient to list them separately as a way of emphasizing their significance.

Image Tone

Image tone denotes the lightness or darkness of a region within an image (Figure 4.3). For black-and-white images, tone may be characterized as "light," "medium gray," "dark gray," "dark," and so on, as the image assumes varied shades of white, gray,

Figure 4.3. Image tone.

or black. For color or color infrared imagery, image tone refers simply to "color," described informally perhaps in such terms as "dark green," "light blue," or "pale pink." Image tone refers ultimately to the brightness of an area of ground as portrayed by the film in a given spectral region (or in three spectral regions for color or color infrared film).

Image tone can also be influenced by the intensity and angle of illumination and by processing of the film. Within a single aerial photograph, vignetting (Chapter 3) may create noticeable differences in image tone due solely to the position of an area within a frame of photography (the image becomes darker near the edges). Thus, the interpreter must employ caution in relying solely upon image tone for an interpretation, as it can be influenced by factors other than the absolute brightness of the earth's surface. Interpreters should also remember that very dark or very bright regions on an image may be exposed in the nonlinear portion of the characteristic curve (Chapter 3), so that they may not be represented in their correct relative brightnesses. Also, nonphotographic sensors may record such a wide range of brightness values that they cannot all be accurately represented on photographic film. In such instances digital analyses (Chapter 9) may be more accurate.

Experiments have shown that interpreters tend to be consistent in interpretation of tones on black-and-white imagery, but less so in interpretation of color imagery (Cihlar & Protz, 1972). As might be expected, interpreters' assessment of image tone is much less sensitive to subtle differences in tone than are measurements by instruments. For the range of tones used in the experiments, human interpreters' assessment of tone expressed a linear relationship with corresponding measurements made by instruments. The results imply that a human interpreter can provide reliable estimates of relative differences in tone, although they may not be capable of accurate description of absolute image brightness.

Image Texture

Image texture refers to the apparent roughness or smoothness of an image region. Usually texture is caused by the pattern of highlighted and shadowed areas as an irregular surface is illuminated from an oblique angle. Contrasting examples (Figure 4.4) include the rough textures of a mature forest and the smooth textures of a mature wheat field. The human interpreter is very good at distinguishing subtle differences in image texture, so that it is a valuable aid to interpretation—certainly equal in importance to image tone in many circumstances.

Image texture depends not only upon the surface itself, but also upon the angle

Figure 4.4. Image texture. Top: coarse texture. Bottom: smooth texture.

of illumination, so that it can vary as lighting varies. Good rendition of texture also depends upon favorable image contrast, so that images of poor or marginal quality may lack the distinct textural differences so valuable to the interpreter.

Shadow

Shadow is an especially important clue in the interpretation of objects. A building or vehicle, illuminated at an angle, casts a shadow that may reveal characteristics of its size or shape that would not be obvious from the overhead view alone. Because military photo interpreters are often primarily interested in identification of individual items of equipment, shadow has been of great significance in distinguishing subtle differences that might not otherwise be visible. By extension, we can emphasize this role of shadow in interpretation of any man-made landscape in which identification of separate kinds of structures or objects is significant.

Shadow is also of great significance in interpretation of natural phenomena, even though its role may not be as obvious. For example, Figure 4.5a depicts an open field occupied by scattered shrubs and bushes separated by areas of open land. Without

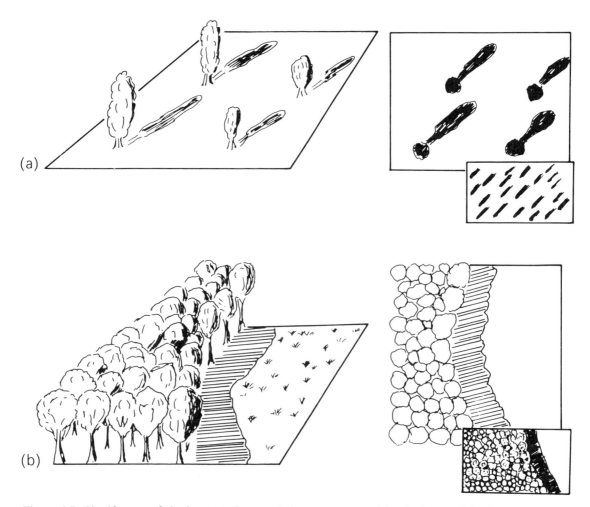

Figure 4.5. Significance of shadow: (*a*) characteristic pattern caused by shadows of shrubs cast on open field; (*b*) shadow at the edge of a forest enhances the boundary between two different land cover types.

shadows, the individual plants might be too small (as seen from above) and too similar in tone to their background to be visible. Yet their shadows are large and dark enough to create the streaked pattern on the imagery typical of this kind of land. In a second example (Figure 4.5b) at the edge between mature forest and open land, the forest often casts a shadow that, at small scale, appears as a dark strip that enhances the boundary between the two zones on the imagery.

Pattern

Pattern refers to the arrangement of individual objects into distinctive, recurring forms that permit recognition on aerial imagery (Figure 4.6). Pattern on an image usually follows a functional relationship between the individual features that compose the pattern. Thus the buildings in an industrial plant may have a distinctive pattern because they are organized to permit economical flow of materials through the plant from receiving raw material to shipping the finished product. The distinctive spacing of trees in an orchard arises from careful planting at intervals that prevent competition between individual trees and permit convenient movement of equipment through the orchard.

Association

Association specifies characteristic occurrence of certain objects or features, usually without the strict spatial arrangement implied by pattern. In the context of military photo interpretation, association of specific items has great significance, for example, when the identification of an object as a specific class of vehicle or radar implies that other, more important, items are likely to be found nearby.

Figure 4.6. Pattern.

Shapes

Shapes of features are obvious clues to their identities. For example, individual structures and vehicles have characteristic shapes, which, if visible in sufficient detail, provide the basis for identification. Features in nature often have such distinctive shapes that shape alone might be sufficient to provide clear identification. For example, ponds, lakes, and rivers occur in specific shapes that are unlike others found in nature. Often specific agricultural crops tend to be planted in fields that have characteristic shapes (perhaps related to constraints of equipment or the kind of irrigation that the farmer uses).

Size

Size is important in two ways. First is the relative size of an object or feature in relation to other objects on the image. This is probably the most direct and important function of size, as it provides the interpreter with an intuitive notion of the scale and resolution of an image, even though no measurements or calculations may have been made. This role is achieved by recognition of familiar objects (dwellings, highways, rivers, etc.), then extrapolation to relate sizes of these known features to estimate sizes and identities of those that might not be easily identified.

Secondly, absolute measurement can be equally valuable as an interpretation aid. Measurements of the size of an object can confirm its identification based upon other factors, especially if its dimensions are so distinctive that they form definitive criteria for specific items or classes of items. Furthermore, absolute measurements permit derivation of quantitative information, including lengths, volumes, or (sometimes) even rates of movement (of vehicles or ocean waves, for example, as they are shown in successive photographs).

Site

Site refers to topographic position. For example, sewage treatment facilities are positioned at low topographic sites, near streams or rivers, to collect waste flowing through the system from higher locations. Orchards may be positioned at characteristic topographic sites—often on hillsides (to avoid cold air drainage to low-lying areas) or near large water bodies (to exploit cooler spring temperatures near large lakes to prevent early blossoming).

4.5. Image Interpretation Strategies

An image interpretation strategy can be defined as a disciplined procedure that enables the interpreter to relate geographic patterns on the ground to their appearance on the image. Campbell (1978) defined five categories of image interpretation strategies.

Field Observation

Field observations, as an approach to image interpretation, are required when the image and its relationship to ground conditions are so imperfectly understood that the interpreter is forced to go to the field to make an identification. In effect the analyst is unable to interpret the image from knowledge and experience at hand and must gather field observations to ascertain the relationship between the landscape and its appearance on the image. Field observations are, of course, a routine dimension of any interpretation as a check on accuracy or a means of familiarization with a specific region. Here their use as an interpretation strategy refers to the fact that when they are required for the interpretation, their use reflects a rudimentary understanding of the manner in which a landscape is depicted on a specific image.

Direct Recognition

Direct recognition is the application of an interpreter's experience, skill, and judgment to associate the image patterns with informational classes. The process is essentially a qualitative, subjective analysis of the image using the elements of image interpretation as visual and logical clues. In everyday experience direct recognition is applied in an intuitive manner; for image analysis, it must be a disciplined process, with very careful systematic examination of the image.

Interpretation by Inference

Interpretation by inference is the use of a visible distribution to map one that is not itself visible on the image. The visible distribution acts as a surrogate, or proxy (a substitute) for the mapped distribution. An example: Soils are defined by vertical profiles that cannot be directly observed by remotely sensed imagery. But soil distributions are sometimes very closely related to patterns of landforms and vegetation that are recorded on the image. Thus, landforms and vegetation can form surrogates for the soil pattern; the interpreter infers the invisible soil distribution from patterns that are visible. Application of this strategy requires a complete knowledge of the link between the proxy and the mapped distribution; attempts to apply imperfectly defined proxies produce inaccurate interpretations.

Probabilistic Interpretations

Probabilistic interpretations are efforts to narrow the range of possible interpretations by formally integrating nonimage information into the classification process, often by means of quantitative classification algorithms. For example, knowledge of the crop calendar can restrict the likely choices for identifying crops of a specific region. If it is known that winter wheat is harvested in June, the choice of crops for interpretation of an August image can be restricted to eliminate wheat as a likely choice and thereby avoid a potential classification error. Often such knowledge can be expressed as a statement of probability. Possibly certain classes might favor specific topographic sites but occur over a range of sites, so a decision rule might express this knowledge as a 0.90 probability of finding the class on a well-drained site, but only a 0.05 probability of finding it on a poorly drained site. Several such statements systematically incorporated into the decision-making process can improve classification accuracy.

Deterministic Interpretation

A fifth strategy, deterministic interpretation, is the most rigorous and precise approach. Deterministic interpretations are based upon quantitatively expressed relationships that tie image characteristics to ground conditions. In contrast with the other methods, most information is derived from the image itself. Photogrammetric analysis of stereo pairs for terrain information is a good example. A scene is imaged from two separate positions along a flight path, and the photogrammetrist measures the apparent displacement. Based upon knowledge of the geometry of the photographic system, a topographic model of the landscape can be reconstructed. The result is therefore the derivation of precise information about the landscape using only the image itself and a knowledge of its geometric relationship with the landscape. Relative to the other methods, very little nonimage information is required.

Image interpreters, of course, may apply a mixture of several strategies in a given situation. Interpretation of soil patterns, for example, may require direct recognition to identify specific classes of vegetation, then application of interpretation by proxy to relate the vegetation pattern to the underlying soil pattern.

4.6. Collateral Information

Collateral, or ancillary, information refers to nonimage information used to assist in the interpretation of an image. Actually, all image interpretations use collateral information in the form of the implicit, often intuitive knowledge that every interpreter brings to the task in the form of everyday experience and formal training. In its narrower meaning, it refers to the explicit, conscious effort to employ maps, statistics, and similar material to aid in analysis of an image. In the context of image interpretation, use of collateral information is permissible, and certainly desirable, provided two conditions are satisfied. First, the use of such information is to be explicitly acknowledged in the written report. Second, the information must not be focused upon a single portion of the image or map to the extent that it produces uneven detail or accuracy in the final map. For example, it would be inappropriate for an interpreter to focus upon acquiring detailed knowledge of tobacco farming in an area of mixed agriculture if the result was highly detailed, accurate delineations of tobacco fields but less detail or accuracy in other fields.

Collateral information can come from books, maps, statistical tables, field observations, or other sources. Written material may pertain to the specific geographic area under examination, or if such material is unavailable, it may be appropriate to search for information pertaining to analogous areas—similar geographic regions (possibly quite distant from the area of interest) characterized by comparable ecology, soils, landforms, climate, or vegetation.

4.7. Image Interpretation Keys

Image interpretation keys are valuable aids for summarizing complex information portrayed as images and have been widely used for image interpretation (Coiner, 1972). Such keys serve either or both of two purposes: (1) a means of training inexperienced personnel in the interpretation of complex or unfamiliar topics, and (2) a reference aid for experienced interpreters to organize information and examples pertaining to specific topics.

An image interpretation key is simply reference material designed to permit rapid and accurate identification of objects or features represented on aerial images. A key usually consists of two parts, a collection of annotated or captioned images or stereograms and a graphic or word description, possibly including sketches or diagrams. These materials are organized in a systematic manner that permits retrieval of desired images by, for example, date, season, region, or subject.

Keys of various forms have been used for many years in the biological sciences, especially botany and zoology. These disciplines rely upon complex taxonomic systems that are so extensive that even experts cannot master the entire body of knowledge. The key, therefore, is a means of organizing the essential characteristics of a topic in an orderly manner. It must be noted that scientific keys of all forms require a basic familiarity with the subject matter. A key, then, is not a substitute for experience and knowledge, but a means of systematically ordering information so that an informed user can learn quickly.

Keys were first routinely applied to aerial images during World War II, when it was necessary to train large numbers of inexperienced photo interpreters in the identification of equipment of foreign manufacture and in the analysis of regions far removed from the experience of most interpreters. The interpretation key formed an effective way of organizing and presenting the expert knowledge of a few individuals. After the end of the war, interpretation keys were applied to many other subjects, including agriculture, forestry, soils, and landforms. Their use has been extended from aerial photography to other forms of remotely sensed imagery. Today interpretation keys may be used for instruction and training, but they may have somewhat wider use as reference aids. It is also true that construction of a key tends to sharpen one's interpretation skills and encourages the interpreter to think more clearly about the interpretation process.

Keys designed solely for use by experts are referred to as technical keys; nontechnical keys are those designed for use by those with a lower level of expertise. Often it is more useful to classify keys by their formats and organizations. *Essay keys* consist of extensive written descriptions, usually with annotated images as illustrations. A *file key* is essentially a personal image file with notes; its completeness reflects the interests and knowledge of the compiler. Its content and organization suit the

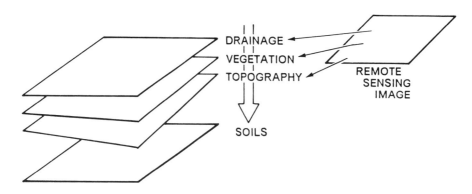

Figure 4.7. Interpretive overlays.

needs of the compiler, so that it may not be organized in a manner suitable for use by others.

4.8. Interpretive Overlays

In resource-oriented interpretations it is often necessary to search for complex associations of several related factors that together define the distribution or pattern of interest. For example, soil patterns may be revealed by distinctive relationships between separate patterns of vegetation, slope, and drainage. *Interpretive overlays* are a way of deriving information from complex interrelationships between separate distributions recorded on remotely sensed images. The correspondence between several separate patterns may reveal the other patterns not directly visible on the image (Figure 4.7).

The method is applied by means of a series of individual overlays for each image to be examined. The first overlay might show the major classes of vegetation, perhaps consisting of dense forest, open forest, grassland, and wetlands. A second overlay maps slope classes, including perhaps level, gently sloping, and steep slopes. Another shows the drainage pattern, and still others might show land use and geology. Thus, for each image, the interpreter may have as many as five or six overlays, each depicting a separate pattern. By superimposing these overlays, the interpreter can derive information presented by the coincidence of several patterns. From knowledge of the local terrain, the interpreter may know that certain soil conditions can be expected where steep slopes and dense forest are found together; and others are expected where the dense forest coincides with gentle slopes. From the information presented by several patterns, the interpreter can resolve information not conveyed by any single pattern.

4.9. Photomorphic Regions

Another approach to interpretation of complex patterns is the search for *photomorphic regions*—regions of uniform appearance on the image. The interpreter does not attempt to resolve the individual components within the landscape (as when using interpretive overlays), but instead looks for their combined influence on image pattern (Figure 4.8). For this reason photomorphic regions may be most applicable to small-scale imagery where the coarse resolution tends to average together separate components of the landscape. Photomorphic regions then are simply "image regions" of relatively uniform tone and texture.

In the first step the interpreter delineates regions of uniform image appearance, using tone, texture, shadow, and the other elements of image interpretation as a means of separating regions. In some instances, interpreters have used densitometers to measure image tone and variation in image tone quantitatively as an aid to more subjective interpretation techniques (Nunnally, 1969).

In the second step the interpreter must be able to match photomorphic regions to useful classes of interest. For example, the interpreter must determine if specific photomorphic regions match vegetation classes. This step obviously requires field observations or collateral information, because regions cannot be identified by image information alone. As the interpretation is refined, the interpreter may find that it is necessary to combine some photo regions or to subdivide others to produce an acceptable interpretation.

Delineation of photomorphic regions is a powerful interpretation tool, but it must be applied with caution. Photomorphic regions do not always correspond neatly to the categories of interest to the interpreter. The appearance of one region may be dominated by factors related to geology and topography, whereas that of

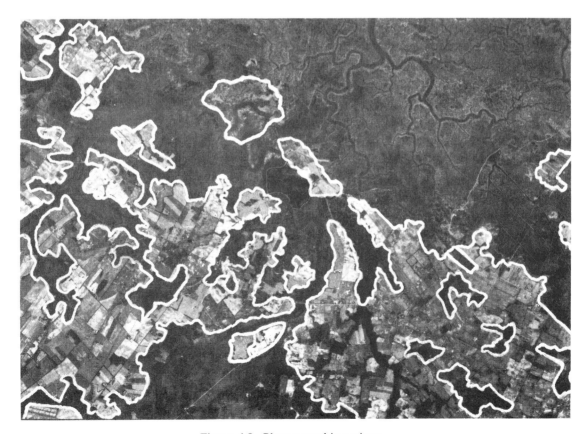

Figure 4.8. Photomorphic regions.

another region on the same image may be controlled by the vegetation pattern. And the image appearance of a third region may be the result of the interaction of several other factors.

4.10. Image Interpretation Equipment

Manual interpretation can usually be conducted with relatively simple, inexpensive equipment, although some of the optional items can be expensive. Typically, an image interpretation laboratory is equipped for storage and handling of images both as paper prints and as film transparencies. Paper prints are most frequently 9 in. × 9 in. contact prints, often stored in sequence in a standard file cabinet. Larger prints and indices must be stored flat in a map cabinet. Transparencies are available as individual 9 in. × 9 in. frames, but often they are stored as long rolls of film wound on spools and sealed in metal or plastic canisters as protection from dust and moisture.

Light Tables

A light table is simply a translucent surface illuminated from behind to permit convenient viewing of film transparencies. In its simplest form, the light table is simply a boxlike frame with a frosted glass surface. The viewing area can be desk-size in its largest form or as small as a briefcase. If roll film is to be used, light tables must be equipped with special brackets to hold the film spools and rollers at the edges to permit the film to move freely without damage. More elaborate models have dimmer switches to control intensity of the lighting, high-quality lamps (to control spectral properties of the illumination), and sometimes power drives to wind and unwind long spools of film.

Measurement of Length

Ordinary household rulers are not satisfactory for photo interpretation. Interpreters should use an engineer's scale or a ruler with accurate graduations. Both SI (metric) units (to at least 1 mm) and English units (to at least 1/20 in.) are desirable. Both measurement and calculation are more convenient if English units are subdivided into decimal divisions.

Stereoscopes

Stereoscopes are devices that facilitate stereoscopic viewing of aerial photographs. The simplest and most common is the *pocket stereoscope*. Its compact size and inexpensive cost make it one of the most widely used remote sensing instruments. Typically the pocket stereoscope (Figure 4.9) consists of a body holding two low-power lenses, attached to a set of collapsible legs that can be folded so that the entire instrument is only a bit larger than a deck of playing cards. The body is usually

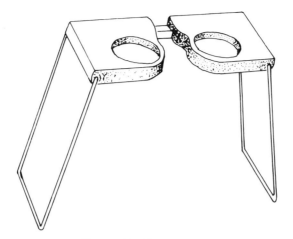

Figure 4.9. Stereoscope.

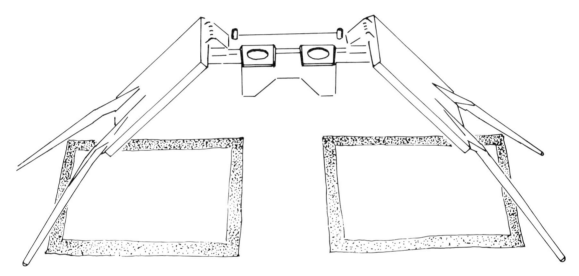

Figure 4.10. Mirror stereoscope.

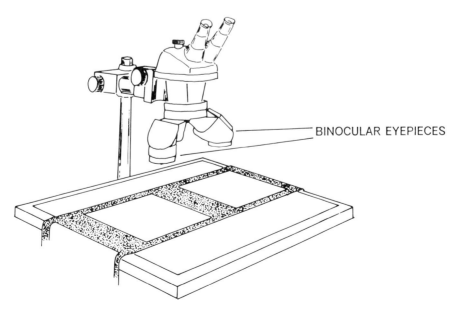

BINOCULAR EYEPIECES

Figure 4.11. Binocular stereoscope.

Figure 4.12. Tube magnifier.

formed from two separate pieces, each holding one of the two lenses, which can be adjusted to control the spacing between the two lenses to accommodate the individual user. Use of this instrument is described below.

Other kinds of stereoscopes include the *mirror stereoscope* (Figure 4.10), which permits stereoscopic viewing of large areas, usually at low magnification, and the *binocular stereoscope* (Figure 4.11), designed primarily for viewing transparencies on light tables. Often the binocular stereoscope has adjustable magnification that enables enlargement of portions of the image up to 20 or 40 times.

Magnification

Image interpreters almost always wish to examine images using magnification, although the exact form depends upon individual preference and the nature of the task at hand. A simple, hand-held reading glass is satisfactory in many circumstances. *Tube magnifiers* (Figure 4.12) are low-power lenses (2x to about 8x) mounted in a transparent tubelike stand. The base may include a reticule (Figure 4.13) calibrated

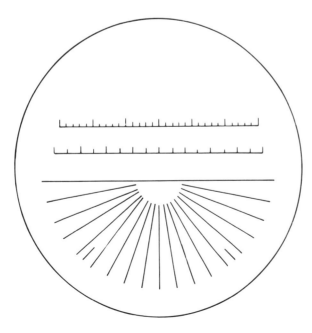

Figure 4.13. Reticule for tube magnifier.

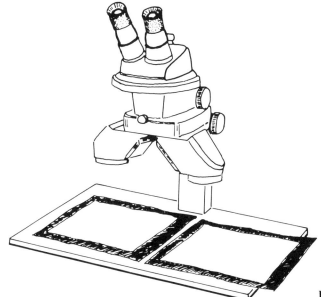

Figure 4.14. Binocular microscope.

in units as small as 0.001 ft. or 0.1 mm, to permit accurate measurement of small-scale images of objects. Sometimes it is necessary to use the much more expensive binocular microscopes (Figure 4.14) for examination of film transparencies; such instruments may have adjustable magnification to as much as 40×, which will approach or exceed the limits of resolution for most images.

Densitometer

Densitometry is the science of making accurate measurements of film density. In the context of remote sensing the objective is often to reconstruct estimates of brightness in the original scene or sometimes merely to estimate relative brightnesses on the film (Chapter 3). A *densitometer* (Figure 4.15) is an instrument that measures image density by directing a light of known brightness through a small portion of the image

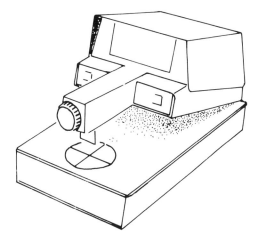

Figure 4.15. Densitometer.

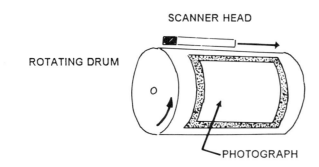

Figure 4.16. Scanning densitometer.

and then measuring its brightness as altered by the film. Typically the light beam might pass through an opening 1 mm in diameter; use of smaller openings (measured sometimes in micrometers) is known as *microdensitometry*. Such instruments find densities for selected regions within an image; an interpreter might use a densitometer to make quantitative measurements of image tone. For color or color infrared images, filters are used to make three measurements, one for each of the three additive primaries.

In principle, densitometric measurements can be used to estimate brightness in the original scene. However, several factors make such estimates difficult. The densitometer must be carefully calibrated; areas of known brightness must be represented on the film and subjected to the same processing as the image to be examined. Measurements of densities that fall in the nonlinear portion of the characteristic curve, of course, cannot be related to the brightness of the original scene. For these reasons and others, reliable estimates of scene brightness from densitometry is usually a very technical and difficult process.

An instrument known as a *scanning densitometer* (Figure 4.16) can be used to measure densities on an image by scanning systematically throughout the image, creating a set of quantitative measurements that represent the image pattern as an array of digital values. This process is the principal means of representing a pictorial image in numerical form, in which it is possible to use the methods of digital analysis discussed in Chapters 9–11.

Height Finder

The height finder (Figure 4.17) is an instrument designed for use with a stereoscope; it permits estimation of topographic elevation or heights of features from stereo aerial photographs. The height finder is a bar that attaches at the base of the stereoscope; the bar holds two plastic tabs, one under each lens of the stereoscope. Both tabs are marked with a small black dot, but one tab is fixed in position, whereas the other can be moved from side to side along a scale that measures its movement left to right parallel to the bar.

The operator aligns the photographs for stereoscopic viewing after marking their principal points and conjugate principal points, and the flight line. The stereoscope is positioned for stereoscopic viewing in the normal way (see below). Then the interpreter views the scene stereoscopically, positioning the movable dot over the

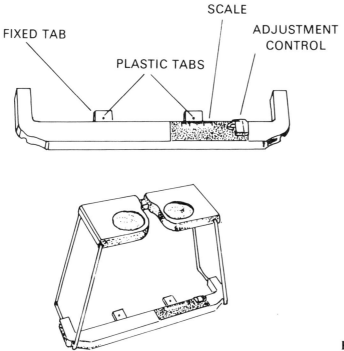

FIXED TAB

PLASTIC TABS

SCALE

ADJUSTMENT
CONTROL

Figure 4.17. Height finder.

feature of interest. Adjustment of the movable dot causes it to appear to float above the terrain surface; when it is positioned so that it appears to rest on the terrain surface, a reading of the scale gives the parallax measurement for that point. A parallax factor is found by using a set of tables. Readings for several points, then simple calculations, permit determination of elevation differences between the individual points (in feet or meters). If the photographs show a point of known elevation, then these points can be assigned elevation with reference to a datum (such as mean sea level); otherwise they provide only relative heights.

Zoom Transfer Scope

The Zoom Transfer Scope (ZTS)[1] (Figure 4.18), an instrument manufactured by Bausch and Lomb Corporation, is tailored for the visual matching of images. Separate maps or images at different scales can be manipulated optically so that they register to one another. The operator views both images through binocular eyepieces, and has control over magnification and orientation of one of the images. This control is used to bring the two images into registration. The operator can then trace detail from one image onto an overlay that registers to the second. The appendix describes use of the Zoom Transfer Scope.

1. "Zoom Transfer Scope" and "ZTS" are registered trademarks of Bausch and Lomb, Rochester, New York.

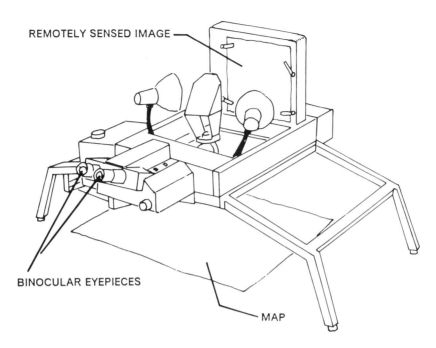

Figure 4.18. Zoom Transfer Scope.

4.11. Preparation for Interpretation

In general the interpreter should be able to work at a large, well-lighted desk or work table with convenient access to electrical power. Often it is useful to be able to control lighting with blackout shades or dimmer switches. Basic equipment and materials, in addition to the items described above, include a supply of translucent drafting film, an engineer's scale, together with protractors, triangles, dividers, and other drafting tools. Maps, reference books, and other supporting material should be available as required.

If the interpretation is made from paper prints, special care must be taken to prevent folding, tearing or rough use that will cause the prints to become worn. Usually it is best to mark an overlay registered to the print rather than the print itself. The drafting tape used to attach drafting film must be selected specifically for its weak adhesive quality, which will not tear the emulsion; the stronger adhesive used on many of the popular brands of paper tape will damage paper prints.

If transparencies are used, special care must be taken in handling and storage. The surface of the transparency must be protected by a transparent plastic sleeve or it must be handled only with clean cotton gloves. Moisture and oils naturally present on unprotected skin may damage the emulsion, and dust and dirt will scratch the surface. Even if such damage is invisible to the naked eye, it will later emerge as a major problem when high-power magnification is used.

For most interpretations, images should be oriented so that the shadows fall toward the analyst. Otherwise most individuals will see an apparent reversal in the relative elevations of terrain features— ridges will seem to be valleys, and valleys will appear as ridgelines.

4.12. Use of the Pocket Stereoscope

The pocket stereoscope is probably the most frequently used interpretation aid. As a result, every student should be proficient in its use. Although many students will require the assistance of the instructor as they learn to use the stereoscope, the following paragraphs may provide some assistance for the beginning student.

First, stereo photographs must be aligned so that the flight line passes left to right (as shown in Figure 4.19). Check the photo numbers to be sure that the photographs have been selected from adjacent positions on the flight line. Usually (but not always) the numbers and annotations on photos are placed on the leading edge of the image—the edge of the image nearest the front of the aircraft at the time the image was taken. Therefore, these numbers should usually be oriented in sequence from left to right, as shown in Figure 4.19. If the overlap between adjacent photos does not correspond to the natural positions of objects on the ground, then the photographs are incorrectly oriented.

Next, identify a distinctive feature on the image, within the zone of stereoscopic overlap. The photos should then be positioned so that the duplicate images of this feature (one on each image) are approximately 64 mm (2.5 in.) apart. This is the distance between the two pupils of the eyes of a person of average size (it is referred to as the "interpupilary distance"), but for many people it may be a bit too large or too small, so the spacing of photographs may require adjustment as the interpreter follows the procedure outlined here.

The pocket stereoscope should be opened so that the legs are locked in place to position the lens at their correct height above the photographs. The two segments of the body of the stereoscope should be adjusted so that the center of the eyepieces are

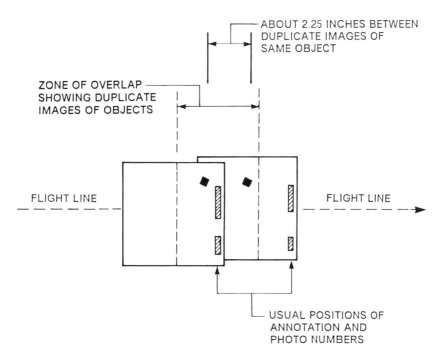

Figure 4.19. Positioning aerial photographs for stereoscopic viewing.

about 64 mm apart (or a slightly larger or smaller distance, as mentioned above). Then the stereoscope should be positioned so that the centers of the lenses are positioned above the duplicate images of the distinctive feature selected previously. Looking through the two lenses, the analyst sees two images of this feature; if the images are properly positioned, then the two images will apper to "float" or "drift." The analyst can with some effort control the apparent positions of the two images, so that they fuse into a single image; as this occurs, the two images should merge into a single image, which is then visible in three dimensions. Usually aerial photos show exaggerated heights, due to the large separation (relative to distance to the ground) between successive photographs as they were taken along the flight line. Although exaggerated height can prevent convenient stereo viewing in regions of high relief, it can be useful in interpretations of subtle terrain features that might not otherwise be noticeable.

The student who has successfully used the stereoscope to examine a section of the photo should then practice moving the stereoscope over the image to view the entire region within the zone of overlap. As long as the axis of the stereoscope is oriented parallel to the flight line, it is possible to retain stereo vision while moving the stereoscope. If the sterscope is twisted with resepct to the flight line, the interperter looses stereo vision. By lifting the edge of one of the photographs, it is possible to view the image regions near the edges of the photos.

4.13. Image Scale Calculations

Scale is a property of all images. As a result, all who work with aerial images and maps must be skilled in computation of image scale. Knowledge of image scale is essential for making measurements from images and for understanding the geometric errors present in all remotely sensed images. Work with image scale is not difficult, although it is easy to make simple mistakes that can result in serious errors.

Scale is simply an expression of the relationship between the *image distance* betweeen two points and the *actual distance* between the two corresponding points on the ground. This relationship can be expressed in several ways. The *word statement* sets a unit distance on the map or photograph equal to the correct corresponding distance on the ground. Examples: "1 in. equals 1 mi.," or just as correctly, "1 cm equals 5 km." The first unit in the statement in the expression specifies the map distance, the second, the corresponding ground distance. A second method of specifying scale is the *bar scale*, which simply labels a line with subdivisions that show ground distances.

The third method, the *representative fraction* (RF), is more widely used and is often the preferred method of reporting image scale. The representative fraction is the ratio between image distance and ground distance. It usually takes the form "1 : 50,000," or "1/50,000," with the numerator set equal to 1 and the denominator equal to the corresponding ground distance. The representative fraction has meaning in any unit of length, as long as both the numerator and the denominator are expressed in the same units. Thus, "1 : 500,000" can mean "1 in. on the image equals 50,000 in. on the ground," or "1 cm on the image equals 50,000 cm on the ground."

A frequent source of confusion is converting the denominator into the larger units that we find more convenient to use for measuring large ground distance. With metric units, the conversion is usually simple; in the example given above, it is easy to see that 50,000 cm equal 0.50 km, and that 1 cm on the map represents 0.5 km on the ground. With English units, the same process is not quite so easy. It is necessary to convert inches to miles to derive "1 in. equals 0.79 mi." from 1:50,000. For this reason, it is useful to know that 1 mi. equals 63,360 in. Thus, 50,000 in. is equal to 50,000/63,360 = 0.79 mi.

A typical scale problem requires estimation of the scale of an individual photograph. One method is to use the focal length and altitude method (Figure 4.20):

$$\text{Representative fraction} = \frac{\text{Focal length}}{\text{Altitude}} \qquad \text{(Eq. 4.1)}$$

Both values must be expressed in the same units. Thus, if a camera with a 6-in. focal length is flown at 10,000 ft., the scale is 0.5/10,000 = 1:20,000. (Altitude always specifies the flying height above the terrain, *not* above sea level.) Because a given flying altitude is seldom the *exact* altitude at the time the photography was taken, and because of the several sources that contribute to scale variations within a given photograph (Chapter 3), we must always regard the results of such calculations as an approximation of the scale of any specific portion of the image. Often such values are referred to as the "nominal" scale of an image, meaning that it is recognized that the stated scale is an approximation and that image scale will vary within any given photograph.

A second method is the use of a *known ground distance*. We identify two points on the aerial photograph that are also represented on a map. For example in Figure 4.21, the image distance between points A and B is measured to be approximately

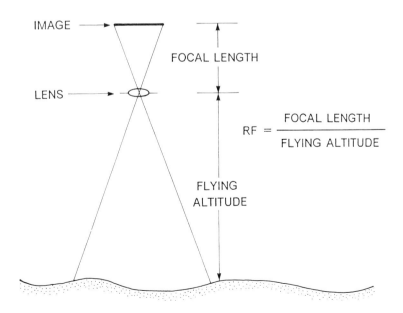

Figure 4.20. Estimating image scale by focal length and altitude.

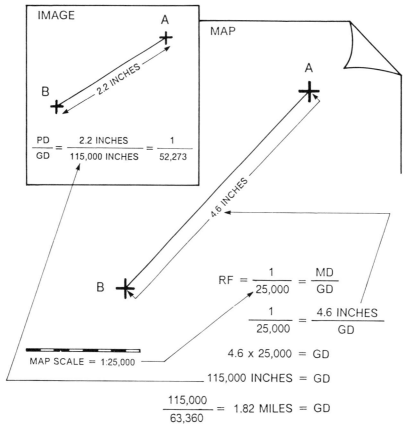

Figure 4.21. Measurement of image scale using map to derive ground distance.

2.2 in. (5.6 cm). From the map, the same distance is determined to correspond to a ground distance of 115,000 in. (about 1.82 mi.). Thus the scale is found to be:

$$RF = \frac{\text{Image distance}}{\text{Ground distance}} = \frac{2.2 \text{ in.}}{1.815 \text{ mi.}} = \frac{2.2 \text{ in.}}{115,000 \text{ in.}}$$

$$= \frac{1}{52,273} \qquad \text{(Eq. 4.2)}$$

In instances when accurate maps of the area represented on the photography may not be available, the interpreter may not know focal length and altitude. Then an approximation of image scale can be made if it is possible to identify an object or feature of know dimensions. Such features might include a baseball diamond or football field; measurement of a distance from these features, as they are shown on the image, provides the "image distance" value needed to use Equation 4.2. The "ground distance" is derived from our knowledge of the length of a football field or the distance between bases on a baseball diamond. Some photo interpretation manuals provide tables of standard dimensions of features commonly observed on aerial images including sizes of athletic fields (soccer, field hockey, etc.), lengths of railroad boxcars, distances between telephone poles, and so on, as a means of using the known-ground-distance method.

A second kind of scale problem is the use of a known scale to measure distance on the photograph. Such a distance might separate two objects on the photographs but not be represented on the map, or the size of a feature has changed since the map was compiled. For example, we know that image scale is 1 : 15,000. A pond not shown on the map is measured on the image as 0.12 in. in width. Therefore, we estimate the actual width of the lake to be:

$$\frac{1}{15,000} = \frac{\text{Image distance}}{\text{Ground distance}} \qquad \text{(Eq. 4.3)}$$

$$\frac{0.12 \text{ in.}}{\text{Unknown GD}} = \frac{\text{Image distance}}{\text{Ground distance}}$$

$$\frac{0.12}{\text{GD}} = \frac{1}{15,000}$$

$$\text{GD} = 0.12 \times 15,000 \text{ in.}$$

$$\text{GD} = 1,800 \text{ in., or } 150 \text{ ft.}$$

This example can illustrate two other points. First, because image scale varies throughout the image, we cannot be absolutely confident that our distance for the width of the pond is accurate; it is simply an estimate, unless we can have high confidence in our measurements and in the image scale at this portion of the photo. Second, measurement of short image distances is likely to have errors due simply to our inability to make accurate measurements of very short distances (such as the 0.12 in. distance measured above). As distances become shorter, our errors constitute a greater proportion of the estimated length. Thus an error of 0.005 in. is 0.08% of a distance of 6 in. but 4% of our distance of 0.12 in. Thus the interpreter should exercise a healthy skepticism regarding measurements made from images unless care has been taken to assure maximum accuracy and consistency.

4.14. Reporting the Results of Interpretation

An image interpreter possesses a specialized knowledge and experience that is not usually understood by those who use the information derived from remotely sensed images. These individuals may not appreciate the value and the reliability of the information, or they may not realize that the same information may have certain errors and limitations. Thus, the interpreter must be skilled not only in analysis of images, but also in the reporting of the results of the interpretation in a form that can be clearly understood by those who are not specialists in remote sensing.

Although the character of the report will vary greatly depending upon the topic, purpose, and audience, it is possible to suggest a few general features that might be important in most circumstances. The report itself might include five separate sections that outline (1) objectives, (2) equipment and materials, (3) regional setting, (4) procedure, and (5) results and conclusions.

Table 4.1. Objectives for Image Interpretation

A. For a land use interpretation:

   To classify the land use area of Montgomery County using five categories: Forest, Urban Land, Agricultural Land, Open Water, and Barren Land. Aerial Photographs at 1:20,000 dated April 1984 are to be used, with field surveys as necessary. Results are to be reported as a map at 1:50,000 and as a table reporting areas in each category.

B. For an inventory of housing units:

   To inventory housing units in Washington County, as of January 1985, then to report results by census subdivisions. To identify each structure on each image either as a dwelling unit or as a nondwelling unit. All structures must be assigned to categories that can be defined solely on the basis of their appearance on the image. Classification of dwelling units must be defined, then tested in early phases of the study, using field observations as necessary.

Objectives

The objectives of the interpretation should be specified clearly, so that there is no opportunity for misunderstanding of the scope or detail of the project. Examples are given in Table 4.1. Note that Example A in Table 4.1 requires that the results be plotted on an accurate map for measurement of areas, whereas Example B presents no such requirement, although it must be possible to mark census divisions boundaries on the photographs. The administrators who are to use the results may not realize the implications of these requirements, but the image interpreter must anticipate the problems that arise from these objectives (i.e., the requirements to register data to the map and to census units, respectively) and devise suitable procedures to assure that the objectives can be met.

Equipment, Materials, and Regional Setting

Equipment and materials can be reported in a simple list as shown in Table 4.2. The regional setting should briefly outline the geographic setting of the area considered in

Table 4.2. Equipment and Materials

Light tables

9 in. × 9 in. black-and-white aerial photographs (positive transparencies at 1:24,000 (dated Winter 1985)

Black-and-white paper enlargements at 1:8,000 (dated Winter 1985)

Descriptions of census division boundaries

Mylar drafting film

Magnifiers

Pens, pencils, drafting type, paper, etc.

Stereoscopes

IBM Personal Computer with dBase III

Table 4.3. Example of Regional Setting Section Report

Landscape and climate

Topography is characterized by a series of parallel valleys and ridges. Valleys, oriented northeast–southwest, are underlain by limestones and shales, and are separated by steep ridges underlain by sandstones and shales. Topography in valleys is rolling, with local relief of 3 to 30 m (10 to 100 ft.). . . .

In valleys, soils are generally deep and well drained, with moderate to gentle slopes. . . .

Climate here is characterized by an average annual rainfall of about 1,016 mm (40 in.), an average annual temperature of about 13°C (55°F), and about 200 frost-free days each year. Maximum precipitation is received during the summer months although late-summer moisture deficits are common. . . .

Economic and social characteristics

Population in this region is largely rural; about 75% of the land is farms, with only 2% to 3% urban land. . . .

In addition to agriculture, important industries include light manufacturing, recreation and tourism, manufacture of forest products. . . .

the interpretation, both for the benefit of the person who will use the information and as a means of preparing the interpreter for the conditions that will be encountered within the image. This section could include brief descriptions of local climate, topography, agriculture, soils, industry, population, and other factors that might pertain to interpretation of images of the region (Table 4.3).

Table 4.4. Example of Procedure for Image Interpretation

(This example corresponds to Objective B, Table 4.1.)

1. Plot census boundaries on photographs or on an overlay that registers to the photographs.

2. Each structure is interpreted separately and assigned a unique identifying number within its census division.

3. Interpretations usually will be made from the positive transparency, using the light table and the tube magnifier. Stereoscopes will be used to interpret paper prints in regions where detailed interpretation of multistory buildings is required. In urbanized areas, large-scale images are to be used to supplement the smaller scale photos.

4. As interpretations are made, the classifications are recorded on an overlay that registers to the enlarged photograph. On this overlay the interpreter outlines each structure and marks it with the identifying number and a symbol. That matches to one of the previously defined categories.

5. When interpretation of a photograph is complete, data are recorded on the computer. Each record will specify structure number, category, census identifier, photo identification, and the interpreter. The computer file permits recording of changes in interpretation, but all earlier interpretations are to be preserved.

6. Overlays to the large-scale photographs form the authoritative graphic record of the interpretation and are marked to record the latest change, if changes in classification are necessary.

7. For field use, the large-scale overlays are to be duplicated, then carried to the field and marked as necessary to record changes. Changes are to be recorded later on the computer (preserving all earlier interpretations). Field sheets are to be marked with a date and worker's initials and are to be retained for the duration of the project.

Interpreters to work in teams. One examines the image using magnification, the other records the results on the overlay. Periodically tasks are rotated. Each team participates in recording information on the computer and in field verification work.

The data base system permits retrieval of information by census unit, by dwelling unit type, by photo, or by interpreter. It also enables tabulation of data by census units at different levels and retrieval of data for each photo or each interpreter, if necessary to correct errors late in the process.

Table 4.5. Example of Results Section of Report

A. For inventory of land use and land cover:

Category	Area (ha)
Forested land	261
Coniferous	113
Deciduous	85
Mixed	63
Urban land	446
Residential	217
Commercial	93
Industrial	105
Transportation	31
Agricultural	1139
Pasture	318
Cropland, irrigated	97
Cropland, nonirrigated	453
Orchards	85
Market gardens	101

B. For count of housing units:

Category	Number
Single-family dwellings	251
Multi-unit dwellings	23
Hotels and motels	12
Mobile home units	86

Procedures

The procedures give a step-by-step description of the interpretation process in detail (Table 4.4). This section serves three purposes. First, it assures that the analyst has thought through the interpretation process in a way that is likely to anticipate problems before they arise. Second, a written discription of procedures is especially important if several people are working together on the same project, for to assure that results are consistent, all must follow the same procedure. Although written procedures are not in themselves sufficient to maintain high standards of uniformity, they are necessary. Finally, the written procedures form a record of how the interpretation was conducted, so that it is possible to answer questions that arise after the interpretation is complete, even if the individuals who actually did the work have moved to new jobs since the project was complete.

Results

Results are reported in a brief narrative section that is supported by tables, maps and diagrams as appropriate. Frequently the interpretation will result in an inventory or enumeration of specific items or categories that can be reported in a detailed tabulation (Table 4.5). These tabulations should, of course, match the requirements of those who will use the information. For example, in Part B of Table 4.3 the categories should correspond to those of interest to the users, and the values should be reported in appropriate units and with suitable detail to satisfy the intended use. The narrative can discuss points that may not be obvious from the table alone, to address such topics as relative accuracies or precision of values for the individual categories.

4.15. Summary

The fundamentals of manual image interpretation were developed for application to aerial photographs early in the history of aerial survey, although not until the 1940s and 1950s were they formalized into their present form. Since then, these techniques have been applied without substantial modification to other kinds of remote sensing imagery. As a result, we have a long record of experience in their application and knowledge of their advantages and limitations.

Manual image interpretation forms the most reliable, most practical method of deriving information from remotely sensed images. Although computer analysis of digital imagery can automate some interpretation tasks (Chapters 9–11), it is important to recognize that automated interpretations do not replace the need for the human interpreter. The human analyst is unequalled in ability to integrate the diverse information required to interpret aerial images and to make the detailed, subtle distinctions necessary for accurate interpretation. Although computer methods may be best for some tasks, others will always require the application of the skills and techniques described in this chapter.

Review Questions

1. A vertical aerial photograph was acquired using a camera with a 9-in. focal length at an altitude of 15,000 ft. Calculate the nominal scale of the photograph.

2. A vertical aerial photograph shows two objects to be separated by $6\frac{3}{4}$ in. The corresponding ground distance is $9\frac{1}{2}$ mi. Calculate the nominal scale of the photograph.

3. A vertical aerial photograph shows two features to be separated by 4.5 in. A map at 1 : 24,000 shows the same two features to be separated by 9.3 in. Calculate the scale of the photograph.

4. Calculate the area represented by a 9 in. × 9 in. vertical aerial photograph taken at an altitude of 10,000 ft. using a camera with a 6-in. focal length.

5. You plan to acquire coverage of a county using a camera with 6-in. focal length, and a 9 in. × 9 in. format. You require an image scale of 4 in. equal to 1 mi., 60% forward overlap, and sidelap of 10%. Your county is square in shape, measuring 15.5 mi. on a side. How many photographs are required? At what altitude must the aircraft fly to acquire these photos?

6. You have a flight line of 9 in. × 9 in. vertical aerial photographs taken by camera with a 9-in. focal length at an altitude of 12,000 ft. above the terrain. Forward overlap is 60%. Calculate the distance (in miles) between ground nadirs of successive photographs.

7. You require complete stereographic coverage of your study area, which is a rectangle measuring 1.5 mi. × 8 mi. How many 9 in. × 9 in. vertical aerial photographs at 1 : 10,000 are required?

8. You need to calculate the scale of a vertical aerial photograph, so you compile the following data:

Ground distance (as derived by map measurement)	Measured photo distance
3.03 mi.	8.0 in.
0.81 mi.	2.3 in.
0.21 mi.	0.43 in.
2.31 mi.	0.51 ft.
1.29 mi.	0.29 ft.

You are not convinced that all values are equally reliable but must use these data, as they are derived from the easily identifiable points visible on the photograph. After examining the data, you proceed to estimate the scale of the photo. Give your estimate, the procedure you used, and explain the rationale for your procedure.

9. You have very little information available to estimate the scale of a vertical aerial photograph, but are able to recognize a baseball diamond among features in an athletic complex. You use a tube magnifier to measure the distance between first and second base to be 0.006 ft. What is your estimate of the scale of the photo?

10. Assume you can easily make an error of 0.001 in your measurement for question 9. Recalculate the image scale to estimate the range of results produced by this level of error. Now return to question 3 and assume that the same measurement error applies (do not forget to consider the different measurement units in the two questions). Calculate the effect on your estimates of image scale. The results should illustrate why it is always better whenever possible to use long distances to estimate image scale.

References

Avery, T. E., & G. L. Berlin. 1985. *Interpretation of Aerial Photographs.* Minneapolis: Burgess, 470 pp.—Latest edition of standard text for photo interpretation.

Campbell, J. B. 1978. A Geographical Analysis of Image Interpretation Methods. *The Professional Geographer,* Vol. 30, pp. 264–269.—Outline of different strategies for image interpretation.

Cihlar, J., & R. Protz. 1972. Perception of Tone Differences from Film Transparencies. *Photogrammetria,* Vol. 8, pp. 131–140.—Study of interpreters' ability to distinguish image tone.

Coiner, J. C. 1972. *SLAR Image Interpretation Keys for Geographic Analysis.* Technical Report 177-19. Lawrence, KS: Center for Research, Inc., 110 pp.—Includes review of keys for photo interpretation.

Colwell, R. N., ed. 1960. *Manual of Photographic Interpretation*. Falls Church, VA: American Society of Photogrammetry, 868 pp.—Although outdated, this volume provides a complete survey of photo interpretation techniques and is useful as a reference.

Image Interpretation Handbook. 1967. Depts. of the Army, Navy, and Air Force. TM 30-245; NAVAIR 10-35-685; AFM 200-50. Washington, DC: U.S. Government Printing Office.—Comprehensive overview of skills for photo interpretation; includes useful tables and examples. Volume I is available to the public.

Lueder, D. R. 1959. *Aerial Photographic Interpretation: Principles and Applications*. New York: McGraw-Hill, 462 pp.—This volume was once the standard text for photo interpretation; although it has now been replaced, it is still useful.

Nunnally, N. R. 1969. Integrated Landscape Analysis with Radar Imagery. *Remote Sensing of Environment*. Vol. 1, pp. 1–6.—Gives a good example of use of photomorphic regions.

Photo Interpretation Handbook. 1954. Depts. of the Army, Navy, and Air Force. TM 30-245; NAVAIR 10-35-610; AFM 200-50. Washington. DC: U.S. Government Printing Office, 303 pp.—An earlier version of the 1967 *Image Interpretation Handbook* listed above; it may still be available in some libraries.

Reprinted by permission of Guilford Press from J. B.
Campbell, *Introduction to Remote Sensing*, 1987, p.
118-157.

CHAPTER FIVE

Land Observation Satellite Systems

5.1. Origins of the LANDSAT System

Today several nations operate or are planning to operate satellite remote sensing systems specifically designed for observation of earth resources, including crops, forests, water bodies, land use, and minerals. Satellite sensors offer several advantages over aerial photography. They provide a synoptic view (observation of large areas in a single image), as well as fine detail and systematic, repetitive coverage of most land areas. Such capabilities are well suited to monitoring many global environmental problems that the world faces today. Moreover, close relationships between these satellite systems and current developments in computer science, cartography, and image processing make this subject one of the most interesting topics within the field of remote sensing.

The design of land observation satellites has evolved from earlier systems that, although tailored for other purposes, provided design and operational experience necessary for successful operation of today's satellites. The first earth observation satellite, Television and Infrared Observation Satellite (TIROS), was launched in April 1960 as the first of a series of experimental weather satellites designed to monitor cloud patterns. TIROS formed a prototype for the operational programs that now provide meteorological data for daily weather forecasts throughout much of the world. Successors to the original TIROS vehicle have se n long service in several programs designed to acquire meteorological data.

When the concept for an earth resources observation satellite program was conceived in the 1960s, the operational feasibility of the satellite platform (i.e., telemetry, tracking, data transmission, etc.) had been established through experience with weather satellites. However, the early meteorological sensors had limited capabilities for land resources observation. Although these sensors were valuable for observing cloud patterns, most had rather coarse resolution (on the order of several kilometers), so that they could provide only the coarsest level of detail about land resources. Even the most rudimentary patterns of land use, vegetation, and drainage were not consistently visible (Figure 5.1).

LANDSAT ("Land Satellite") was designed in the 1960s and launched in 1972 as the first satellite tailored specifically for broad-scale observation of the earth's land areas—to accomplish for land resource studies what meteorological satellites had

118

Figure 5.1. TIROS 1 image. This image illustrates the nature of early meteorological satellite imagery. At best, such images permit visual separation of clouds, land, and water only. This image shows the northeastern portion of North America, 12 July 1961, although no landmarks are visible. Image from U.S. Department of Commerce.

Table 5.1. LANDSAT Missions

Satellite	Launched	Retired[a]	Sensors[b]
LANDSAT 1	23 June 1972	1 June 1978	MSS, RBV
LANDSAT 2	22 January 1975	30 September 1983	MSS, RBV
LANDSAT 3	5 March 1978	30 September 1983	MSS, RBV
LANDSAT 4	16 July 1982	[c]	TM, MSS
LANDSAT 5	1 March 1984	—	TM, MSS
LANDSAT 6	Funding requested	—	—

[a]Satellite systems typically operated on an intermittent or stand-by basis for considerable periods prior to formal retirement from service.

[b]Sensors are discussed in the text. Here "MSS" denotes the multispectral scanner, "RBV" the return beam vidicon, and "TM" the thematic mapper.

[c]Since March 1983, LANDSAT 4 has been operating at reduced power due to a failure in power cables from the solar array panels. After launch of LANDSAT 5, it has been used only for special acquisitions. Conceivably it could eventually be retrieved by a space shuttle mission, repaired, and then returned to service.

accomplished for meteorology and climatology. Today LANDSAT is important both in its own right as a remote sensing system that has contributed greatly to earth resources studies and also as an introduction to the study of more sophisticated satellites now planned and operated by several nations.

LANDSAT was proposed by scientists and administrators in the U.S. government who envisioned application of the principles of remote sensing to broad-scale, repetitive surveys of the earth's land areas. The first LANDSAT sensors recorded

Table 5.2. LANDSAT Sensors

Sensor	Band	Spectral sensitivity
		LANDSATs 1 and 2
RBV	1	0.475–0.575 μm (green)
RBV	2	0.58 –0.68 μm (red)
RBV	3	0.69 –0.83 μm (near infrared)
MSS	4	0.5 –0.6 μm (green)
MSS	5	0.6 –0.7 μm (red)
MSS	6	0.7 –0.8 μm (near infrared)
MSS	7	0.8 –1.1 μm (near infrared)
		LANDSAT 3
RBV		0.5– 0.75 μm (panchromatic response)
MSS	4	0.5– 0.6 μm (green)
MSS	5	0.6– 0.7 μm (red)
MSS	6	0.7– 0.8 μm (near infrared)
MSS	7	0.8– 1.1 μm (near infrared)
MSS	8	10.4–12.6 μm (far infrared)
		LANDSATs 4 and 5[a]
TM	1	0.45– 0.52 μm (blue-green)
TM	2	0.52– 0.60 μm (green)
TM	3	0.63– 0.69 μm (red)
TM	4	0.76– 0.90 μm (near infrared)
TM	5	1.55– 1.75 μm (mid infrared)
TM	6	10.4–12.5 μm (far infrared)
TM	7	2.08– 2.35 μm (mid infrared)
MSS	1	0.5 – 0.6 μm (green)
MSS	2	0.6 – 0.7 μm (red)
MSS	3	0.7 – 0.8 μm (near infrared)
MSS	4	0.8 – 1.1 μm (near infrared)

[a]On LANDSATs 4 and 5 MSS bands were renumbered although the spectral definitions remain the same.

energy in the visible and near infrared spectrum. Although these portions of the spectrum had long been used for aircraft photography, it was by no means certain that they would also prove practicable for observation of earth resources from satellite altitudes. Scientists could not be completely confident that the sensors would work as planned, that they would prove to be reliable, that detail would be satisfactory, or that a sufficient proportion of scenes would be free of cloud cover. Although many of these problems were experienced, the feasibility of the basic concept has been demonstrated, and LANDSAT now forms the model for similar systems operated by other nations.

During the first two years of operation the LANDSAT program was designated as the Earth Resources Technology Satellite (ERTS) system, a name that can still be encountered in older books and scientific papers. By convention, each satellite is designated by a letter before launch, then by a number once it is in orbit, so "LANDSAT D" became "LANDSAT 4" after launch. Thus far the LANDSAT program has employed five satellites (Table 5.1).

The LANDSAT system consists of spacecraft-borne sensors that observe the earth and transmit information by microwave signals to ground stations that receive and then process data for dissemination in both image and digital format to the user community. Early LANDSAT vehicles carried two sensor systems: the return beam vidicon (RBV) and the multispectral scanner (MSS) (Table 5.2). The RBV is a camera-like instrument designed to provide, relative to the MSS, high geometric accuracy but lower spectral and radiometric detail. That is, positions of features would be accurately represented, but without fine detail concerning their colors and brightnesses. In contrast, the MSS was designed to provide finer detail concerning spectral characteristics of the earth, but less positional accuracy. Because technical difficulties restricted RBV operation, the MSS soon became the primary LANDSAT sensor. The second generation of LANDSAT vehicles (LANDSATs 4 and 5) still carries the MSS, as well as the thematic mapper (TM), a more sophisticated version of the MSS.

5.2. Satellite Orbits

Satellite orbits are tailored to match capabilities of the sensors they carry and to the objectives of each satellite mission. For example, many meteorological and communications satellites are stationed at high altitudes (Table 5.3) in orbits designed to provide constant observation of the same portion of the earth. Satellites positioned in these *geosynchronous* orbits revolve at an angular rate that matches the earth's rotation, and therefore remain stationary with respect to the earth's surface. For weather satellites, such orbits permit continuous observation of changes in cloud patterns for an entire hemisphere. For communications satellites, geosychronous orbits provide simultaneous line-of-sight access to large areas of the earth, thereby enabling rapid relay of signals across and between continents. A satellite positioned in a geosynchronous orbit views a single region throughout the day, but also observes the full range of variation in angle and intensity of solar illumination. Scientists who practice remote sensing want images that record only small variations in illumination,

Table 5.3. Orbital Altitudes of Selected Earth Satellites

	mi.	km
Communications satellite (Westar)	22,300	35,680
Space shuttle	115–690	184–1,104
LANDSATs 1, 2, and 3	570	912
LANDSATs 4 and 5	423	705
SEASAT	480	800
TIROS 1 (weather satellite)	480	800
NIMBUS (weather satellite)	594	950
GOES (geostationary weather satellite)	21,480	35,800

so geosynchronous orbits are usually not advantageous for observation of earth resources by means of reflected solar radiation.

Therefore, many earth observation satellites are placed in *sun-synchronous* orbits that orient the plane of the satellite orbit in a position that maintains a constant angular relationship with the solar beam (Figure 5.2). The earth's mass and rotation carry the satellite track westward at about 15° of longitude each hour, matching the rate that the earth rotates within the solar beam. Therefore, to an observer on the

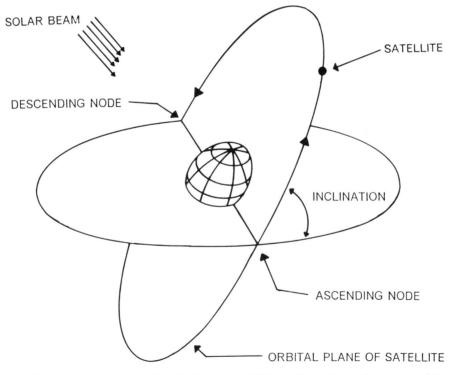

Figure 5.2. Inclination of satellite orbit (*i*). The nodes (*N*) are the points where the orbit crosses the equator. Here the ascending (south-to-north) track is shown; on the opposite side of the earth the satellite will pass from north to south.

earth's surface, such a satellite will always pass overhead at the same local sun time, minimizing variations in solar illumination. At a given latitude, the altitude of the sun above the horizon and the position of the sun along the horizon ("solar azimuth") will be approximately the same within similar seasons. However, for any two images acquired at different latitudes, or at different seasons, differences in the elevation of the sun *will* be observed because of the apparent motion of the sun from season to season. The passage of a sun-synchronous satellite over each line of latitude at constant local sun time produces similar seasonal illumination and shadowing for images of a given region, a quality that simplifies interpretations of satellite images. The *inclination* (*i*) of an orbit (Figure 5.2) is the angle that the orbital plane forms as it crosses the equatorial plane.

Satellites in sun-synchronous orbits pass through both the illuminated and the shadowed sides of the earth. Often the satellite passes from north to south on the sunlit side (the "descending" pass or node) and from south to north on the shadowed side (the "ascending" pass). Sensors that depend upon reflected solar radiation acquire data only during the descending pass, although radar and thermal sensors (Chapters 6 and 7) can acquire data independently of solar illumination, thereby observing the earth's surface during both passes.

Exact timing of the LANDSAT overpass has varied with the different satellites, but most orbits carried the satellite over the equator about 9:30 a.m. local sun time, a time that was judged to be optimal with respect to sun angle and cloud coverage for acquiring high-quality imagery. Locations north of the equator will observe a slightly later time of overpass, and those south of the equator will experience a slightly earlier time. Of course, local clock time will not be constant for LANDSAT coverage.

5.3. LANDSATs 1, 2, and 3

Orbit and Ground Coverage: LANDSATs 1, 2, and 3

Although the first generation of LANDSAT sensors is no longer in service, they acquired a large library of images that will be available in the future as a baseline reference for environmental conditions in land areas throughout the world. Therefore, knowledge of these early LANDSAT images is important both as an introduction to later satellite systems, and as a basis for work with the historical archives of images from LANDSATs 1, 2, and 3.

From 1972 to 1983, various combinations of LANDSATs 1, 2, and 3 orbited the earth in sun-synchronous orbits every 103 minutes—14 times each day. After 252 orbits—completed every 18 days—LANDSAT passed over the same place on the earth to produce repetitive coverage (Figure 5.3). When two satellites were both in service, their orbits were tailored to provide repetitive coverage every 9 days. Of course, sensors were activated to acquire images only at scheduled times, so this capability was not always used. In addition, equipment malfunctions and cloud cover sometimes prevented acquisition of planned coverage.

Because of the earth's rotation on its axis from west to east, each successive north-to-south pass of the LANDSAT platform was offset to the west by 2,875 km

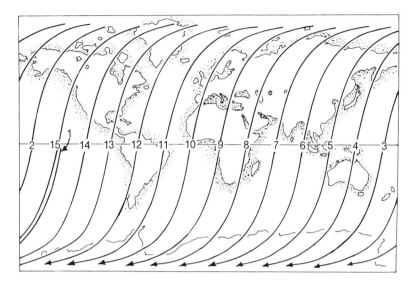

Figure 5.3. Coverage cycle, LANDSATs 1, 2, and 3. Each numbered line designates a northeast-to-southwest pass of the satellite. In a single 24-hour interval the satellite completes 14 orbits; the first pass on the next day (orbit 15) is immediately adjacent to pass 1 on the preceding day. Based upon NASA diagram.

(1,786 mi.) at the equator (Figure 5.4). Because the longitudinal shift westward of adjacent orbital tracks at the equator was approximately 159 km (99 mi.), gaps between tracks are incrementally filled during the 18-day cycle. Thus, on day 2, orbit number 1 was displaced 159 km to the west of the path of orbit number 1 on day 1. On the 18th day, orbit number 1 was identical to that of orbit number 1, day 1. The first orbit on the 19th day coincided with that of orbit number 1 on day 2, and so on.

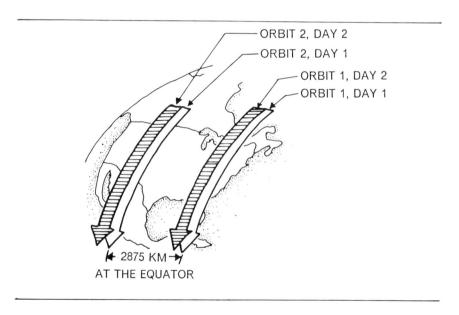

Figure 5.4. Incremental increases in LANDSAT coverage. On successive days, orbital tracks begin to fill in the gaps left by the displacement of orbits during the preceding day. After 18 days, progressive accumulation of coverage fills in all gaps left by coverage acquired on day 1. Based upon NASA diagram.

Therefore, the entire surface of the earth between 81° north and 81° south latitude was subject to coverage by LANDSAT sensors once every 18 days (every 9 days if two satellites were in service). Use of an orbit with 90° inclination would have permitted coverage of the polar regions (which are of interest because of varying extent of sea ice and other factors), but would have required sacrificing the sun-synchronuous orbit, because a direct north-to-south orbital track would not permit the westward progression of the orbital track that matches the westward movement of the solar beam.

Support Subsystems

The design of the platform for LANDSATs 1, 2, and 3 (Figure 5.5) is based upon that of the NIMBUS weather satellites first launched in 1967. The NIMBUS design was chosen because it had already demonstrated its ability to maintain stable orientation, a capability essential for acquiring high-quality images. The design for LANDSATs 4 and 5 differs considerably from the initial LANDSAT vehicle, because it incorporates numerous improvements in both the sensors and the orbital platform.

Figure 5.5. Platform for LANDSATs 1, 2, and 3. Diagram based upon General Electric Company, *LANDSAT 3 Reference Manual*, p. 2-2.

Although our primary interest here is the sensors that these satellites carried, it is important to mention briefly the support subsystems, units that are necessary to maintain proper operation of the sensors. The *attitude control subsystem* (ACS) maintained orientation of the satellite with respect to the earth's surface and with respect to the orbital path. The *orbit adjust subsystem* (OAS) maintained the orbital path within specified parameters after the initial orbit was attained. The OAS also made adjustments throughout the life of the satellite to maintain the planned repeatable coverage of imagery. The *power subsystem* supplied electrical power required to operate all satellite systems by means of two solar array panels and eight batteries. The batteries were charged by energy provided by the solar panels while the satellite was on the sunlit side of the earth and then provided power when the satellite was in the earth's shadow. The *thermal control subsystem* controlled the temperatures of satellite components by means of heaters, passive radiators (to dissipate excess heat), and insulation. The *communications and data handling subsystem* provided microwave communications with ground stations for transmitting data from the sensors, commands to satellite subsystems, and information regarding satellite status and location.

Data from sensors were transmitted, in digital form, by microwave signal to ground stations equipped to receive and process data. Direct transmission from the satellite to the ground station as the sensor acquired data was possible only when the

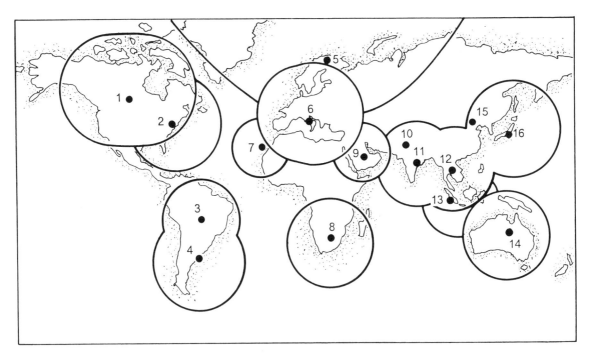

Figure 5.6. LANDSAT ground stations. Each circle shows the approximate range for direct communication with the satellite. Because the map projection distorts sizes of areas at high latitudes, ranges for northern stations appear here as large, unevenly shaped regions. On a globe, all would have the same size and shape. Key to LANDSAT ground stations: (1) Prince Albert, Saskatchewan; (2) Greenbelt, Maryland; (3) Cuiaba, Brazil; (4) Mar Chiquita, Argentina; (5) Kiruna, Sweden; (6) Fucino, Italy; (7) Maspalomas, Spain; (8) Johannesburg, South Africa; (9) Riyadh, Saudi Arabia*; (10) Islamabad, Pakistan*; (11) Hyderabad, India; (12) Bangkok, Thailand; (13) Djakarta, Indonesia; (14) Alice Springs, Australia; (15) Beijing, China; (16) Tokyo, Japan. (*Planned but not yet in operation.)

satellite had direct line-of-sight view of the ground antenna (a radius of about 1,800 km from the ground stations). In North America, stations at Greenbelt, Maryland; Fairbanks, Alaska; Goldstone, California; and Prince Albert, Saskatchewan, provided this capability for most of the United States and Canada. (Some of these stations are no longer in service.) Elsewhere a network of ground stations has been established over a period of years through agreements with other nations (Figure 5.6). Areas outside receiving range of a ground station could be imaged only by use of the two tape recorders on board each of the early LANDSATs. Each tape recorder could record about 30 minutes of data; then as the satellite moved within range of a ground station, they could transmit the stored data to a receiving station. Thus these satellites had, within the limits of their orbits, a capability for worldwide coverage. Unfortunately the tape recorders proved to be one of the most unreliable elements of the LANDSAT system, so that when they failed, the system was unable to image areas beyond range of the ground stations. These unobservable areas became smaller as more ground stations were established, but still there remained areas that LAND-SAT could not observe. Later sections in this chapter explain how LANDSATs 4 and 5 are able to avoid this problem by use of a relay satellite.

Return Beam Vidicon

The return beam vidicon camera (RBV) system generated high-resolution television-like images of the earth's surface. On LANDSATs 1 and 2 the RBV system consisted of three independent cameras which operated simultaneously, each sensing a different segment of the spectrum (Table 5.2). All three instruments were aimed at the same region beneath the satellite, so that the images they acquired registered to one another to form a three-band multispectral representation of a 185 km × 170 km ground area, known as a LANDSAT *scene* (Figure 5.7). This area matches the area represented by the corresponding MSS scene, as explained below.

The RBV shutter was designed to open briefly, like a camera shutter, to view the entire scene simultaneously (Figure 5.8a). Solar radiation reflected from the ground passed through a lens and was focused on the photo-sensitive surface of the vidicon camera. After the shutter closed, the photo-sensitive surface was then scanned by an electron beam to convert the scene's reflectance values into a video signal for transmission to a ground station. RBV design is therefore analogous to that of a camera, but with a photo-sensitive plate substituted in the focal plane where the film would normally be. The RBV system could acquire one image every 25 seconds, to produce a series of images with a small forward overlap. Because the RBV was designed with an optical geometry analogous to that of a camera, scientists hoped to be able to use RBV imagery as the source of accurate cartographic measurements, in much the same way that aerial photographs had long been used to compile accurate maps. Thus the function of the RBV was primarily to provide the basis for accurate measurements of position and distance.

Because of electrical failures on LANDSAT 1, the RBV was turned off less than 1 month after launch. A similar malfunction caused the LANDSAT 2 RBV to be shut down shortly after launch, so the first two satellites acquired relatively little RBV imagery.

Figure 5.7. Schematic diagram of LANDSAT scene.

On LANDSAT 3 the RBV was modified to include two panchromatic cameras instead of the three-camera system of LANDSATs 1 and 2 (Figure 5.8b). The two panchromatic cameras sensed reflected solar radiation from 0.505 to 0.750 μm, a spectral region analogous to that of color infrared films used in aerial photography (Chapter 3). The LANDSAT 3 RBV cameras were aimed to cover adjacent areas on the ground that, with some overlap, match about one-half the area covered by the previous RBV (Figure 5.8b). A pair of RBV images show a ground area 99 km × 185 km in size. Thus, successive pairs (i.e., a total of four) of these RBV *subscenes* combine to represent a ground area that matches the same area covered by earlier RBVs. An increase in image detail from approximately 80 m to 40 m is achieved by increasing the focal length to twice that of earlier RBVs. Because of limits on the amount of data that could be transmitted, this increase required loss of spectral detail. The LANDSAT 3 RBV images show only a single broad spectral region compared to the three bands of the RBVs on LANDSATs 1 and 2.

Some scientists found the increased detail of the LANDSAT 3 RBV imagery to be of interest, but the sensor experienced technical problems that seemed to preclude routine acquisition of high-quality imagery. As a result, LANDSAT RBV imagery was not used in the role that was envisioned at the start of the LANDSAT program.

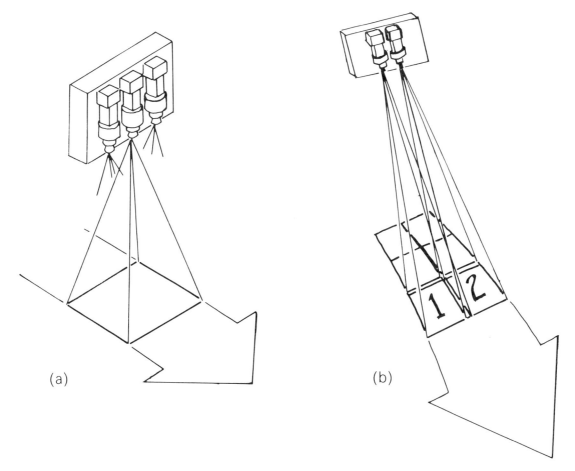

(a)   (b)

Figure 5.8. RBV configuration, (*a*) LANDSATs 1 and 2, and (*b*), LANDSAT 3.

Multispectral Scanner

As a result of malfunctions in the RBV sensors early in the missions of LANDSATs 1 and 2, the multispectral scanner (MSS) became the primary LANDSAT sensor. Whereas the RBV was designed to capture images with known geometric properties, the MSS was tailored to provide multispectral data without as much concern for positional accuracy. In general, MSS imagery and data have been of good quality—much better than many expected—and have clearly demonstrated the merits of satellite observation of earth resources. The economical, routine availability of MSS digital data has formed the foundation for a sizable increase in the number and sophistication of digital image processing facilities available to the remote sensing community. A version of the MSS is still in operation on LANDSATs 4 and 5.

The MSS (Figure 5.9) is a scanning instrument that uses a flat, oscillating mirror to scan from west to east to produce a ground swath of 185 km (100 nautical mi.) perpendicular to the orbital track. The satellite motion along the orbital path provides the along-track dimension to the image. Solar radiation reflected from the earth's surface is directed by the mirror to a telescope-like instrument that focuses the energy onto fiber optic bundles located in the focal plane of the telescope (Figure 5.9).

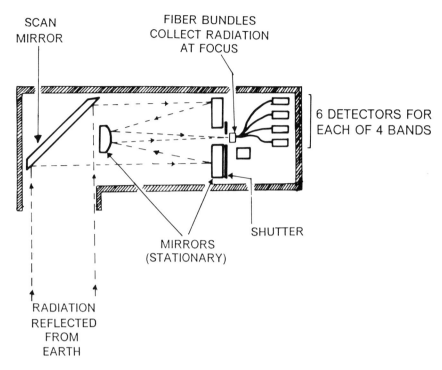

Figure 5.9. Schematic diagram of LANDSAT multispectral scanner. Based upon NASA diagram.

The fiber optic bundles then transmit energy to detectors sensitive to four spectral regions (Table 5.2).

Each west-to-east scan of the mirror represents a strip of ground approximately 185 km long in the east–west dimension, and 474 m wide in the north–south dimension (Figure 5.10). The 474-m distance corresponds to the forward motion of the satellite during the interval required for the west-to-east movement of the mirror, and its inactive east-to-west retrace to return to its starting position. The mirror returns to start another active scan just as the satellite is in position to record another line of data at a ground position immediately adjacent to the preceding scan line. Each motion of the mirror corresponds to *six* lines of data on the image because the fiber optics split the energy from the mirror into six contiguous segments (Figure 5.10).

The instantaneous field of view (IFOV) of a scanning instrument can be informally defined as the ground area (ground cell) viewed by the sensor at a given instant in time. The nominal IFOV for LANDSAT scanning is 79 m × 79 m. Slater (1980) provides a detailed examination of MSS geometry that shows the IFOV to be approximately 76 m × 76 m (about 0.58 ha, or 1.4 acres), although small differences exist between LANDSATs 1, 2, and 3, and within orbital paths as satellite altitude varies. The brightness from each IFOV is displayed on the MSS as a "pixel" (picture element) formatted to correspond to a ground area said to be approximately 79 m × 56 m in size (about 0.45 ha, or 1.1 acres). The term "pixel" is used variously to refer to (1) the actual pixel on an image, (2) the digital value that represents the pixel, and (3) the ground area represented by a pixel. Pixel dimensions in the along-track dimension are determined by the optical geometry of the MSS, whereas the across-track dimension is determined by the rate at which the continuous electrical

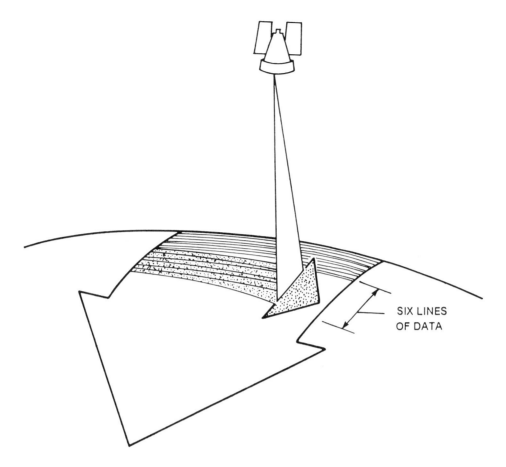

Figure 5.10. Scan pattern for LANDSAT MSS.

signal from the sensor is sampled to provide discrete values for each pixel. Note that display of a single pixel on the image in this form does not alter the fact that the information it conveys was collected within an IFOV of about 76 m × 76 m, as mentioned above. In everyday terms the ground area corresponding to an MSS pixel can be said to be somewhat less than that of an American football field (Figure 5.11). Those who work routinely with MSS data know that, as a practical matter, images matching the stated levels of detail are observed only under ideal circumstances.

For the MSS instruments on board LANDSATs 1 and 2, the four spectral channels are located in the green, red, and infrared portions of the spectrum:

- *Band 4*: 0.5–0.6 μm (green)
- *Band 5*: 0.6–0.7 μm (red)
- *Band 6*: 0.7–0.8 μm (near infrared)
- *Band 7*: 0.8–1.1 μm (near infrared)

The LANDSAT 3 MSS includes an additional band in the far infrared from 10.4 to 12.6 μm. Because this band includes only two detectors, energy from each mirror movement is subdivided into two segments, each 234 m wide. The IFOV for the thermal band is therefore 234 m × 234 m, producing much coarser resolution than images for the other bands.

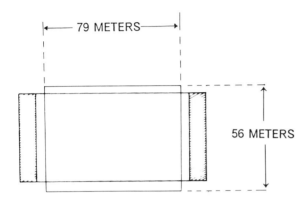

Figure 5.11. Nominal MSS pixel shown in relation to a U.S. football field. (Here the *orientation* of pixel is rotated 90° to match to the usual rendition of a football field; on an image, the narrow ends of MSS pixels are oriented approximately north–south.)

5.4. MSS Images

Definition of an MSS Scene

Over the years there have been many modifications in the manner in which MSS images have been processed. This section describes current procedures for the LANDSAT 4 MSS, which are similar, although not identical to, methods used for LANDSATs 1, 2, and 3. The MSS scene is defined as an image representing a ground area approximately 185 km in the east–west (across-track) direction, and 170 km in the north–south (along-track) direction (Figure 5.12). (The along-track dimension for LANDSAT 3 was somewhat smaller.) The across-track dimension is defined by the side-to-side motion of the MSS scanner; the along-track dimension is provided by the forward motion of the satellite along its orbital path. If the MSS were operated continuously for an entire descending pass, it would provide a continuous strip of imagery representing an area 185 km wide. The 170-km north–south dimension simply divides this strip into segments of convenient size.

The MSS scene, then, is an array of pixel values (in each of four bands) consisting of about 2,400 scan lines, each composed of 3,240 pixels (Figure 5.12). The location of the center point and the dimensions of each scene correspond approximately with RBV scenes that may have been acquired at the same time. Although the center points of scenes acquired at the same location at different times are intended to register with each other, there is often in fact a noticeable shift (the problem of "temporal registration") in ground locations of center points, due to uncorrected drift in the orbit.

There is a small overlap (about 5%, or 9 km) between scenes to the north and south of a given scene. This overlap is generated by repeating the last few lines from the preceding image, *not* by stereoscopic viewing of the earth. Overlap with scenes to

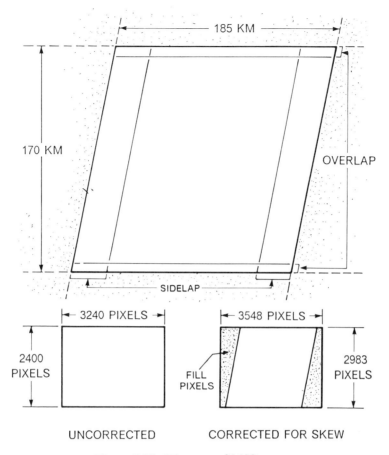

185 KM

170 KM

OVERLAP

SIDELAP

3240 PIXELS

2400 PIXELS

UNCORRECTED

3548 PIXELS

FILL PIXELS

2983 PIXELS

CORRECTED FOR SKEW

Figure 5.12. Diagram of MSS scene.

the east and west depends upon latitude; sidelap will be a minimum of 14% (26 km) at the equator, and increases with latitude to 57% at 60°, then to 85% at 80° north and south latitude. Because this overlap is created by viewing the same area of the earth from different perspectives, the area within the overlap can be viewed in stereo. At high latitudes, this area can constitute an appreciable portion of a scene.

Image Format

MSS data are available in several formats that have been subjected to different forms of processing to adjust for geometric and radiometric errors. This section describes some of the basic forms for LANDSAT 5 MSS data, which are similar, but not identical, to those for other MSS data. Individual users can of course conduct additional processing as desired once they have the data in hand (Chapter 10).

In its initial form an MSS image consists of a rectangular array of pixels in each of four bands (Figure 5.12). In this format, however, the image does not provide for the combined effects of spacecraft movement and rotation of the earth as the MSS acquires the image line by line. When these effects are considered, the image assumes the shape of a parallelogram (Figure 5.12). For convenience in processing of data, extra ("fill") pixels are added to preserve the rectangular shape of the image. The fill pixels of course convey no information, as they are simply assigned values of zero as necessary to attain the desired shape.

LANDSAT MSS images are generated from video data that have been processed to remove some geometric and radiometric errors, and to frame and annotate each image as described below. Each image is generated by an instrument that displays the digital values from each pixel as brightnesses on a 70-mm film master. This film image represents each MSS scene with dimensions of about 56 mm at a scale of 1:3,369,000. Data from each band are represented as separate black-and-white film negatives, which form the masters from which all other images are generated.

The most common image products are paper prints at scales of 1:1,000,000, 1:500,000, and 1:250,000, which represent enlargements of the master image by factors of 3.4, 6.7, and 13.5, respectively. The 1:1,000,000 image, printed on 9-in. (241-mm) paper or film, is one of the most economical and convenient image products for many users. Each of the four bands of the MSS generates a separate image, each emphasizing landscape features that reflect specific portions of the spectrum (Figures 5.13 and 5.14). These separate images, then, record in black and white the spectral reflectance in the green, red, and infrared portions of the spectrum.

Figure 5.13. LANDSAT MSS image, band 5, New Orleans, Louisiana, 16 September 1982. Scene ID: 40062-15591. Image reproduced by permission of EOSAT.

Bands 4, 5, and 7 can be combined into a single color image (Plate 8), known as an MSS "false-color composite." Band 7 is projected onto color film through a red filter, band 5 through a green filter, and band 4 through a blue filter. The result is a false-color rendition that uses the same assignment of colors used in conventional color infrared aerial photography (Chapter 3). Strong reflectance in the green portion of the spectrum is represented as blue on the color composite, red as green, and infrared as red. Thus, living vegetation appears as a bright red, turbid water as a blue color, and urban areas as a gray or sometimes pinkish-gray.

The annotation block at the lower edge gives essential information concerning the identification, date, location, and characteristics of each image. Since the launch of LANDSAT 1, the content and form of the annotation block have been changed several times, but some of the basic information can be shown in a simplified form to illustrate key items (Figure 5.15). The *date* has obvious meaning. The *format center* and *ground nadir* give, in degrees and minutes of latitude and longitude, the ground location of the center point of the image. The *spectral band* is given in the form "MSS 6," or "RBV 1" (meaning multispectral scanner, band 6, and return beam vidicon, band 1, respectively). *Sun angle* and *sun elevation* designate, in degrees, the solar

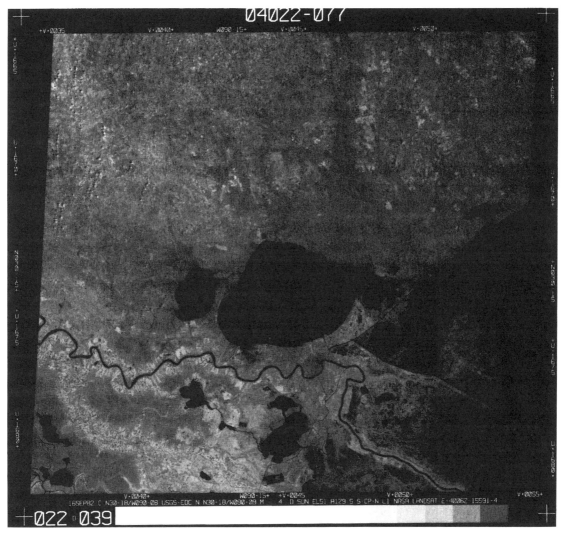

Figure 5.14. LANDSAT MSS image, band 7, New Orleans, Louisiana, 16 September 1982. Scene ID: 40062-15591. Image reproduced by permission of EOSAT.

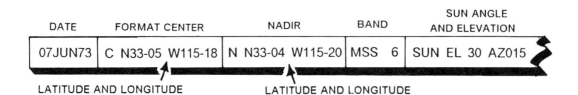

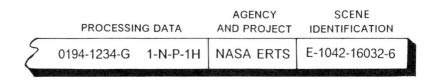

Figure 5.15. LANDSAT annotation block. The annotation block has been changed several times, but all show essentially the same information.

elevation (above the horizon) and the azimuth of the solar beam from true north at the center of the image. Of the remaining items on the annotation block, the most important for most users is the *scene ID*, a unique number that specifies the scene and band. The scene ID uniquely specifies any MSS scene, so that it is especially useful as a means of cataloging and indexing MSS images as explained below. Note also that each image is annotated on the margins to show latitude and longitude tick marks. If corresponding marks are connected by straight lines, a portion of the graticule can be reconstructed to serve as a locational reference.

MSS Digital Data

The nominal MSS image is composed of approximately 2,400 scan lines, each with about 3,240 pixels per line. Data for each band, therefore, is composed of over 7,000,000 pixel values. These values, in four spectral bands for each scene, are available to the user in the form of computer compatible tapes (CCTs). The CCT is a magnetic tape that presents the MSS digital data in a form suitable for use on image processing computers that most applications scientists will use to examine and analyze the data. The CCT is a user-oriented form of the data contained on high-density archival tapes that form the permanent reference record of the data. The high-density archival tapes are generated from video data received from the satellite; they include a number of basic corrections for radiometric errors. CCTs are available to the public in a variety of forms that can include additional radiometric and geometric corrections. Data provided by CCTs form the basis for digital analysis and classification of LANDSAT data (Chapters 9–12).

Worldwide Reference System

The worldwide reference system (WRS) is a concise designation of nominal center points of LANDSAT scenes used to index LANDSAT scenes by location. The

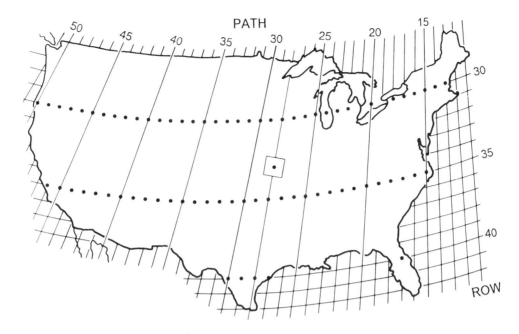

Figure 5.16. WRS path–row coordinates for the United States. The square represents approximate coverage of a single LANDSAT scene (path 28, row 34).

reference system is based upon a coordinate system in which there are 233 north–south paths corresponding to orbital tracks of the satellite, and 119 rows representing latitudinal center lines of LANDSAT scenes. The combination of a path number and a row number uniquely identifies a nominal scene center (Figure 5.16). Because of the drift of satellite orbits over time, actual scene centers may not match path–row locations exactly, but the method does provide a convenient and effective means of indexing locations of LANDSAT scenes.

5.5. LANDSATs 4 and 5

Thematic Mapper

Even before LANDSAT 1 was launched, it was recognized that existing technology could improve the design of the MSS, and efforts were made to incorporate improvements into a new instrument modeled on the basic design of the MSS. LANDSATs 4 and 5 carry a replacement for MSS known as the *thematic mapper* (TM), which can be considered an upgraded MSS. In these satellites, both the TM and an MSS are carried on an improved platform (Figure 5.17) that can maintain a high degree of stability in orientation as a means of improving geometric qualities of the imagery. In addition, the satellite is designed to permit access and service by the space shuttle. The TM is essentially an improved MSS; its design and operation are based upon the same principles as the MSS, but its design is considerably more complex. It provides finer spatial resolution, improved geometric fidelity, greater radiometric detail, and more detailed spectral information in more precisely defined spectral regions.

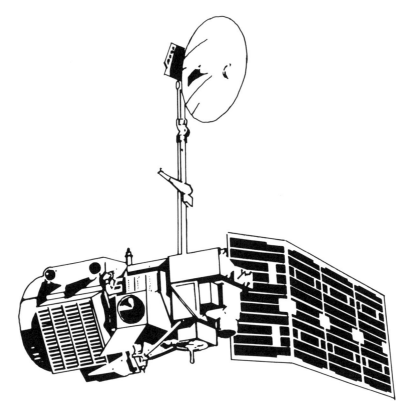

Figure 5.17. Platform for LANDSATs 4 and 5. Diagram based upon *LANDSAT Data Users Notes*, No. 23, July 1982, p. 3.

Improved satellite stability and orbit adjust subsystem are designed to improve positional and geometric accuracy. The objectives of the second generation of LANDSAT instruments are to assess the performance of the TM, to provide continued availability of MSS data, and to continue foreign data reception. Like the earlier LANDSATs, LANDSATs 4 and 5 are experimental programs intended to lead to an operational system, which is as yet uncertain with respect to both design and funding.

Despite the historical relationship between the TM and the MSS, the two sensors are distinct. Whereas the MSS has four broadly defined spectral regions, the TM records seven spectral bands;

- *Band 1*: 0.45–0.52 μm (blue-green; separation of soil and vegetation)
- *Band 2*: 0.52–0.60 μm (green; reflection from vegetation)
- *Band 3*: 0.63–0.69 μm (red; chlorophyll absorption)
- *Band 4*: 0.76–0.90 μm (near infrared; delineation of water bodies)
- *Band 5*: 1.55–1.75 μm (mid infrared; vegetative moisture)
- *Band 6*: 10.4–12.5 μm (far infrared; hydrothermal mapping)
- *Band 7*: 2.08–2.35 μm (mid infrared; plant heat stress)

These spectral bands have been carefully tailored to record radiation of interest to specific scientific investigations, as suggested above. Spatial resolution is said to be about 30 m × 30 m (about 0.09 ha, or 0.22 acre), compared to the 76 m × 76 m IFOV of the MSS. (TM band 7 has coarser spatial resolution of about 120 m × 120 m.) The

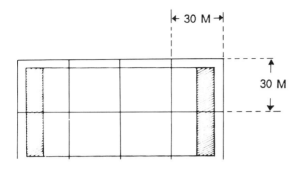

Figure 5.18. Nominal size of thematic mapper pixel.

finer spatial resolution provides a noticeable increase (relative to the MSS) in spatial detail recorded by each TM image (Figure 5.18). Digital values are represented using a wider range of brightness values (256 different levels of brightness) than are available using the MSS (which can record only 128 different levels). These kinds of improvements produce images with much finer detail than those of the MSS (Figures 5.19 and 5.20).

Each scan of the TM mirror acquires 16 lines of data (4 lines for band 6). Unlike the MSS, the TM scan acquires data as it moves in both east–west and west–east directions. This feature permits engineers to design a slower speed of mirror movement, thereby improving the length of time that the detectors can respond to brightness in the scene. However, this design requires additional processing to reconfigure image positions of pixels to form a geometrically accurate image (Figure 5.21). TM detectors are positioned in an array in the focal plane (it does not use the fiber optics employed in the MSS to assure registration of the MSS bands); as a result, there may be slight misregistration of TM bands.

TM imagery is analogous to MSS imagery with respect to areal coverage and organization of data into several sets of multispectral digital values that overlay to form an image. In comparison with MSS images, TM imagery has much finer spatial and radiometric resolution, so that TM images show relatively fine detail of patterns of the earth's surface. Geometric properties of the imagery are said to be improved relative to MSS images.

Use of seven rather than four spectral bands and use of smaller pixel sizes within the same image area means that TM images consist of many more data values than do MSS images. (Four bands of an MSS scene require about 31,000,000 pixels; seven bands of a TM scene include over 230,000,000 pixels!) For the analyst to use all TM bands is clearly impractical routinely; even for small areas, time and expense would greatly exceed practical limits. As a result, each analyst must determine those TM bands that are likely to provide the required information. Because the "best" combinations of TM bands will vary according to the purpose of the study, season, geographic region, and other factors, it is unlikely that a single selection of bands will be equally effective in all circumstances. As a result, each analyst will probably find it necessary to spend more time in preparation for each study. The analyst who used MSS data did not really face this problem, as there are fewer choices and fewer data, and therefore the decisions were not as difficult.

A few combinations of TM bands appear to be effective for general purpose use.

EOSAT 307-5

16SEP82 C N30-17/W090-08 USGS-EDC N N30-18/W090-08 T 3   SUN EL51 A129 S S CP N   NASA LANDSAT E-40062-15591-3

+022 ᴿ039

Figure 5.19. Thematic mapper image (band 3). This image shows New Orleans, Louisiana, 16 September 1982. Scene ID: 40062-15591. Image reproduced by permission of EOSAT.

Use of TM bands 1, 3, and 4 can be used to make an image that is analogous to the MSS false-color composite (Plate 9). TM bands 1 (blue-green), 2 (green), and 3 (red) can be combined to form a natural color composite, approximately equivalent to a color aerial photograph in rendition of colors. Experiments with other combinations have shown that bands 2, 3, and 5; 2, 7, and 4; and 5, 2, and 4 are also effective for visual interpretation. Of course, there are a large number of combinations of the seven TM bands that may be useful for special interpretations, and examination of individual bands as single black-and-white images is also effective in some situations.

LANDSATs 4 and 5 both have a reconfigured version of the MSS to assure continuity of the MSS coverage that dates back to 1972. Because of the lower orbit for LANDSATs 4 and 5, it was necessary to modify the optics to approximate the resolution of the earlier instruments. Due to changes in the satellite orbit, the re-configured MSS is said to have a resolution of about 82 m in the along-track dimension and 58 m in the cross-track dimension. The coverage of a scene is about 197 km along-track, and 188 km cross-track. It should be emphasized that despite the improvements offered by the TM, the MSS is still a valuable instrument. Many scientists will wish to use the MSS data to assure comparability with earlier data, especially for studying changes over an interval of many years. Others may prefer the lower cost of the MSS data, especially if they do not require the finer detail of the TM.

Shortly after launch in 1982, failures of several components in the electrical system for LANDSAT 4 caused system engineers to deactivate the satellite. LANDSAT 5 was launched in 1984 to assure continued coverage. LANDSAT 4 remains in

Figure 5.20. Thematic mapper image (band 4). This image shows New Orleans, Louisiana, 16 September 1982. Scene ID: 40062-15591. Image reproduced by permission of EOSAT.

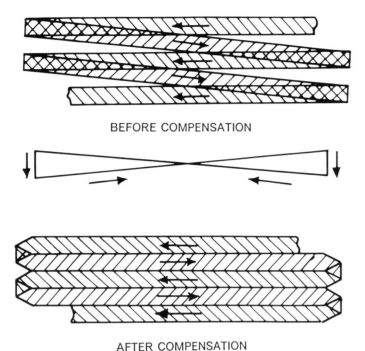

BEFORE COMPENSATION

AFTER COMPENSATION

Figure 5.21. Thematic mapper scan pattern. Based upon NASA diagram.

orbit, but with only restricted capability for operational use. At present it remains inactive, but available for limited use in an emergency if LANDSAT 5 fails. Even though LANDSAT 4 was designed to be retrieved and repaired by the space shuttle, no plans have been made to do so because of the expense.

Orbit and Ground Coverage: LANDSATs 4 and 5

LANDSATs 4 and 5 have been placed into orbits that resemble those of earlier LANDSATs. Sun-synchronous orbits bring the satellites over the equator at about 9:45 a.m., thereby maintaining approximate continuity of solar illumination with imagery from LANDSATs 1, 2, and 3. Data are collected as the satellite passes northeast to southwest on the sunlit side of the earth. The image swath remains at 185 km. In these respects coverage is compatible with that of the first generation of LANDSAT systems.

However, there are important differences. The finer spatial resolution of the TM is achieved in part by a lower orbital altitude, which requires several changes in the coverage cycle. Earlier LANDSATs produced adjacent image swaths on successive days. However, LANDSATs 4 and 5 acquire coverage of adjacent swaths at intervals of 7 days. LANDSAT 4 completes from 9 to 16 orbits per day, with an average of about 14; 233 orbits complete a coverage cycle of 16 days. Successive passes of the satellite are separated at the equator by 2,752 km (Figure 5.22a); gaps between successive passes are filled in over an interval of 16 days (Figure 5.22b). Adjacent passes are spaced at 172 km. At the equator, adjacent passes overlap by about 7.6%; overlap increases as latitude increases.

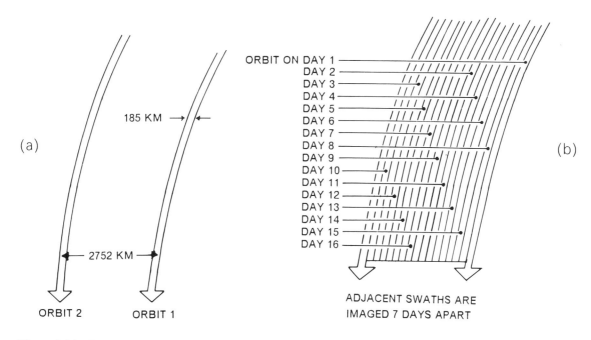

Figure 5.22. Coverage diagram, LANDSATs 4 and 5. (*a*) Successive passes are spaced at 2,752 km at the equator. (*b*) Gaps between successive passes are filled in over a 16-day interval. Based upon NASA diagram.

A complete coverage cycle is achieved in 16 days—233 orbits. Because this pattern differs from earlier LANDSATs, LANDSATs 4 and 5 require a new WRS indexing system for labeling paths and rows (Figure 5.23). Row designations remain the same as before, but a new system of numbering paths is required. In all there are 233 paths and 248 rows, with row 60 positioned at the equator.

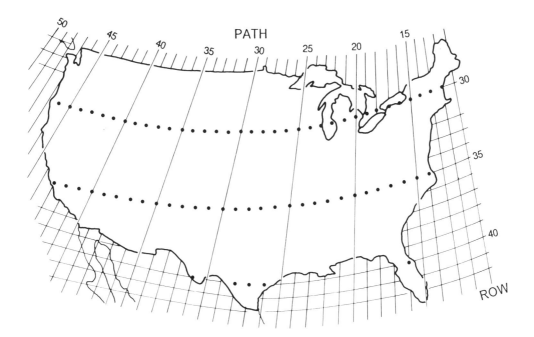

Figure 5.23. WRS for LANDSATs 4 and 5. The rows are the same as those for the WRS for LANDSATs 1–3 (Figure 5.16) but there are fewer paths.

105

Data Flow for LANDSATs 4 and 5

Communications for LANDSATs 4 and 5 are also quite different from those for LANDSATs 1, 2, and 3. The new LANDSATs are teamed with two tracking and data relay satellites (TDRSs). TDRSs are in geosynchronous orbits that permit direct transmission of sensor data to a central ground receiving station near White Sands, New Mexico. When both TDRSs are on station, LANDSAT 4 and 5 will have capabilities for coverage of large areas of the earth without the requirement for the tape recorders that were so troublesome for earlier LANDSATs. TDRSs also can relay commands from mission control to the LANDSATs and telemetry from the LANDSAT to the ground. In addition, LANDSATs 4 and 5 will be able to use several navigational satellites to acquire accurate information concerning location and velocity. At present, the full complement of TDRSs and navigational satellites is not yet in place, so that the system is operating without all its planned capabilities using existing receiving stations and temporary facilities.

After initial processing at White Sands, data are transmitted via a communications satellite for subsequent processing at the Goddard Space Flight Center in Greenbelt, Maryland (Figure 5.24). There the image data undergo initial processing. MSS and thematic mapper data receive separate treatment. MSS data are transmitted via a communications satellite to the EROS Data Center, Sioux Falls, South Dakota, where they receive further processing to create master film images (241 mm, or about 9 in., in size) and archival computer tapes used to generate CCTs. Thematic mapper data receive processing at Greenbelt and then the master images and tapes are shipped by air freight to Sioux Falls for storage and distribution.

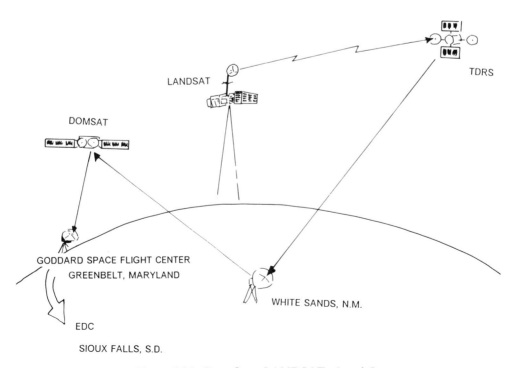

Figure 5.24. Data flow, LANDSATs 4 and 5.

5.6. Computer Searches

Because of the unprecedented amount of data generated by the LANDSAT system, it is necessary to use a computerized data base to record essential data concerning each scene. The data base, located at the EROS Data Center (EDC) in Sioux Falls, South Dakota, includes basic information concerning date, location, and quality of all LANDSAT imagery. The public has access to this data base through requests mailed or telephoned to EDC as described in the information distributed by EDC, the U.S. Geological Survey, and other organizations. In general, a user can obtain, free of charge, information regarding availability of aircraft imagery, so that this service is especially valuable to all who use remote sensing imagery, whether acquired by LANDSAT or aircraft.

The computer search is based upon user-supplied information regarding the area of interest, desirable dates of coverage, and the minimum quality of coverage (Figure 5.25). The response is a computer tabulation of all coverage meeting the constraints specified by the user (Figure 5.26). LANDSAT products are identified by scene identification numbers, listed in the annotation block of each image (Figure 5.15).

The computer listing will vary in length according to the nature of the user's request. A conservative, restrictive request closely tailored to the user's needs will result in a much shorter listing than will a loosely defined, very broad request. Unless the user suspects that there is very little coverage of an area, it is usually best to define the request to match his or her needs closely, as it can be very tedious to search through a long list to identify the best images.

An example (Figure 5.26) illustrates information on the computer listing. The path–row designation indicates that the scenes that follow show essentially the same geographic area at different dates. Each scene is then listed, giving the LANDSAT vehicle and sensor (e.g., "LANDSAT 1 MSS"), the scene identification, and image characteristics. The "BSW-02.1" indicates that the master image is a black-and-white transparency in a 2.1-in. (70-mm) format (the actual image that is ordered could be, if desired, in color at a variety of sizes). The image quality gives a subjective quality rating from 0 (poor) to 9 (excellent) for each of 5 MSS bands, listed in order. "M" designates missing data for a given band; an asterisk signifies the absence of the thermal infrared band for LANDSATs 1 and 2. Estimated cloud cover, exposure date, and location of the center of the scene and its corner points (in degrees and minutes of latitude and longitude) are also listed. A detailed explanation that accompanies each listing provides details of all the symbols and codes. Note that to prepare the request for a geographic computer search and to interpret the resulting computer listing, it is necessary to have access to medium or large-scale maps with accurate representation of latitude and longitude.

5.7. Administration of the LANDSAT Program

LANDSAT was originally operated by the U.S. National Aeronautics and Space Administration (NASA) as a part of its mission to develop and demonstrate applications of new technology related to aerospace engineering. Although NASA had primary responsibility for LANDSAT, other federal agencies, including the U.S.

Inquiry Form

Earth Observation
Satellite Company

Name _____ Company _____

Address _____

City/State _____ Zip _____

Phone (Home) _____ (Business) _____

This data will be used to initiate a computer geosearch which will be returned to you as a computer listing with a listing key from which Landsat products can be selected and ordered. The preferred manner for inquiry is to identify the appropriate Landsat Worldwide Reference System (WRS) path/row scene centers. A WRS map for your area of interest will be sent to you on request. If WRS information is not available, complete the required information for either the point search or area rectangle inquiry.

PREFERRED TYPE OF COVERAGE

	Black & White	False Color
□ Landsat 1-3	□	□
□ Landsat 4-5	□	□

□ MSS   □ RBV   □ TM

PREFERRED TIME OF YEAR

JAN-MAR	□	□	ALL COVERAGE
APR-JUNE	□	□	LATEST COVERAGE
JULY-SEPT	□	□	SPECIFIC DATES _____
OCT-DEC	□		

MINIMUM QUALITY RATING ACCEPTABLE

0 □	2 □	5 □	8 □
(VERY POOR)	(POOR)	(FAIR)	(GOOD)

NOTE

Classification of percent of cloud cover is subjective and is relative to the amount of clouds appearing on the imagery and not on their location.

MAXIMUM CLOUD COVER ACCEPTABLE

□ 10%   □ 30%   □ 50%   □ 70%   □ 90%

Important

WRS maps: Landsat 1,2, and 3 have different paths/rows than Landsat 4 and 5. USE CORRECT SIDE OF MAP.

Landsat 1,2,3

Single Scene				Contiguous Area	
PATH ___ ROW ___	PATH ___ ROW ___	PATH ___ ROW ___	PATH ___ TO PATH ___	PATH ___ TO PATH ___
PATH ___ ROW ___	PATH ___ ROW ___	PATH ___ ROW ___	ROW ___ TO ROW ___	ROW ___ TO ROW ___
PATH ___ ROW ___	PATH ___ ROW ___	PATH ___ ROW ___	PATH ___ TO PATH ___	PATH ___ TO PATH ___
PATH ___ ROW ___	PATH ___ ROW ___	PATH ___ ROW ___	ROW ___ TO ROW ___	ROW ___ TO ROW ___

Landsat 4,5

Single Scene				Contiguous Area	
PATH ___ ROW ___	PATH ___ ROW ___	PATH ___ ROW ___	PATH ___ TO PATH ___	ROW ___ TO PATH ___
PATH ___ ROW ___	PATH ___ ROW ___	PATH ___ ROW ___	ROW ___ TO ROW ___	PATH ___ TO ROW ___
PATH ___ ROW ___	PATH ___ ROW ___	PATH ___ ROW ___	PATH ___ TO PATH ___	ROW ___ TO PATH ___
PATH ___ ROW ___	PATH ___ ROW ___	PATH ___ ROW ___	ROW ___ TO ROW ___	PATH ___ TO ROW ___

POINT SEARCH

Imagery with any coverage over the selected point will be included.

POINT NO. 1	POINT NO. 2	POINT NO. 3
LAT. _____ N or S	LAT. _____ N or S	LAT. _____ N or S
LONG. _____ E or W	LONG. _____ E or W	LONG. _____ E or W

AREA RECTANGLE

LAT.
LAT.
LONG.   LONG.

Imagery with any coverage within the selected area will be included.

AREA NO. 1	AREA NO. 2	AREA NO. 3
LAT. _____ N or S to	LAT. _____ N or S to	LAT. _____ N or S to
LAT. _____ N or S	LAT. _____ N or S	LAT. _____ N or S
LONG. _____ E or W to	LONG. _____ E or W to	LONG. _____ E or W to
LONG. _____ E or W	LONG. _____ E or W	LONG. _____ E or W

Comments: _____

Return completed form to .... *c/o Earth Observation Satellite Company [EOSAT]*
EROS Data Center, Sioux Falls, South Dakota 57198
For additional information or to receive the appropriate WRS map for your area of interest ..... *Please call* ... *1 800 367-2801*

EOSAT 205-6

Figure 5.25. Request for computer search.

```
                                   E R O S   D A T A   C E N T E R                                   REPORT NO. DI029-1
                                   SIOUX FALLS, SOUTH DAKOTA  57198                                  DATE       08/24/81
                         CONTACT NUMBER 0645135003                                                   TIME       23:39
                              RJB CAMPBELL                                                           PAGE           4

                                          DATA TYPE  LANDSAT
                                                                                                              GA NO PP2
  IMAGERY-TYPE              SCENE  ID   FILM-SOURCE QUALITY CLOUD EXPO-DATE  SCENE-CENTER-POINT   SCENE-SCALE  MICROFORM CCL  IN DE CCT

PATH= 17  ROW= 34  LANDSAT  - - - - - - - - - - - - - - - - - - - - - - - - - - - - - - - - - - - - - - - - - - - - - - - - - - -
LANDSAT-2 (MSS)        8281514563500 B&W-02.1"   8288*   10%  04/16/77 N37D28M00S W078D54M00S 1:3,369,000 82100291077 P  L  C   Y
  CORNER POINT COORDINATES=#1:N38D03M54S W077D37M04S #2:N38D22M36S W079D58M44S #3:N36D51M19S W080D09M46S #4:N36D33M07S W078D10M26S

LANDSAT-2 (MSS)        8276114591500 B&W-02.1"   8888*   20%  02/21/77 N37D22M00S W078D57M00S 1:3,369,000 82100280184 P  L  C   P
  CORNER POINT COORDINATES=#1:N37D58M05S W077D39M59S #2:N38D16M43S W079D41M54S #3:N36D45M08S W080D12M50S #4:N36D26M59S W078D13M16S

LANDSAT-1 (MSS)        8566214195500 B&W-02.1"   8888*   10%  02/09/77 N37D25M00S W078D59M00S 1:3,369,000 81100550219 P  L  C   P
  CORNER POINT COORDINATES=#1:N38D01M39S W077D42M32S #2:N38D19M14S W079D44M35S #3:N36D47M35S W080D14M17S #4:N36D30M28S W078D14M36S

LANDSAT-2 (MSS)        8274314595500 B&W-02.1"   5888*   10%  02/03/77 N37D27M00S W078D53M00S 1:3,369,000 82100270227 P  L  C   P
  CORNER POINT COORDINATES=#1:N38D02M35S W077D35M25S #2:N38D21M22S W079D37M21S #3:N36D50M38S W080D08M25S #4:N36D32M20S W078D09M48S

LANDSAT-2 (MSS)        8265315032500 B&W-02.1"   5888*   10%  11/05/76 N37D21M00S W078D52M00S 1:3,369,000 82100240412 P  L  C   P
  CORNER POINT COORDINATES=#1:N37D56M54S W077D34M13S #2:N38D16M18S W079D36M43S #3:N36D44M18S W080D08M36S #4:N36D25M24S W078D08M28S

LANDSAT-2 (MSS)        8263515040500 B&W-02.1"   5888*   10%  10/18/76 N37D15M00S W078D57M00S 1:3,369,000 82100231019 P  L  C   P
  CORNER POINT COORDINATES=#1:N37D51M06S W077D39M52S #2:N38D09M53S W079D41M51S #3:N36D38M07S W080D12M58S #4:N36D19M49S W078D13M19S

LANDSAT-2 (MSS)        8259915051500 B&W-02.1"   8588*   00%  09/12/76 N37D27M00S W078D51M00S 1:3,369,000 82100220790 P  L  C   Y
  CORNER POINT COORDINATES=#1:N38D02M22S W077D34M43S #2:N38D21M09S W079D35M03S #3:N36D50M52S W080D06M08S #4:N36D32M34S W078D08M05S

LANDSAT-2 (MSS)        8258115054500 B&W-02.1"   8888*   30%  08/25/76 N37D35M00S W078D47M00S 1:3,369,000 82100211413 P  L  C   P
  CORNER POINT COORDINATES=#1:N38D10M17S W077D30M38S #2:N38D29M08S W079D31M03S #3:N36D58M57S W080D02M13S #4:N36D40M35S W078D04M06S

LANDSAT-2 (MSS)        8254515064500 B&W-02.1"   8888*   20%  07/20/76 N37D24M00S W078D50M00S 1:3,369,000 82100201155 P  L  C   P
  CORNER POINT COORDINATES=#1:N37D59M53S W077D32M47S #2:N38D18M52S W079D34M43S #3:N36D47M20S W080D06M03S #4:N36D28M50S W078D06M27S

LANDSAT-2 (MSS)        8252715071500 B&W-02.1"   8888*   10%  07/02/76 N37D23M00S W078D50M00S 1:3,369,000 82100200376 P  L  C   O
  CORNER POINT COORDINATES=#1:N37D59M07S W077D32M23S #2:N38D18M08S W079D34M59S #3:N36D46M06S W080D06M25S #4:N36D27M34S W078D06M12S

LANDSAT-1 (MSS)        8541314410500 B&W-02.1"   8888*   10%  06/05/76 N37D26M00S W079D05M00S 1:3,369,000 81100500344 P  L  C   Y
  CORNER POINT COORDINATES=#1:N38D02M51S W077D48M16S #2:N38D20M24S W079D50M50S #3:N36D48M22S W080D20M32S #4:N36D31M17S W078D20M21S

LANDSAT-2 (MSS)        8247315084500 B&W-02.1"   5888*   00%  05/09/76 N37D30M00S W078D44M00S 1:3,369,000 82100180835 P  L  C   P
  CORNER POINT COORDINATES=#1:N38D05M15S W077D27M26S #2:N38D24M21S W079D27M59S #3:N36D53M58S W079D59M25S #4:N36D35M22S W078D01M10S

LANDSAT-1 (MSS)        8537714433500 B&W-02.1"   8888*   20%  04/30/76 N37D31M00S W079D02M00S 1:3,369,000 81100490282 F  L  C   P
  CORNER POINT COORDINATES=#1:N38D07M34S W077D46M11S #2:N38D24M41S W079D47M30S #3:N36D53M41S W080D16M38S #4:N36D37M01S W078D17M40S

LANDSAT-2 (MSS)        8245515090500 B&W-02.1"   8888*   10%  04/21/76 N37D33M00S W078D45M00S 1:3,369,000 82100180054 P  L  C   P
  CORNER POINT COORDINATES=#1:N38D08M17S W077D28M44S #2:N38D27M05S W079D29M01S #3:N36D56M57S W080D00M07S #4:N36D38M38S W078D02M07S

LANDSAT-1 (MSS)        8535914444500 B&W-02.1"   8888*   00%  04/12/76 N37D33M00S W078D58M00S 1:3,369,000 81100480593 P  L  C   Y
  CORNER POINT COORDINATES=#1:N38D09M16S W077D42M13S #2:N38D26M39S W079D43M10S #3:N36D55M58S W080D12M36S #4:N36D39M03S W078D13M59S

LANDSAT-2 (MSS)        8243715094500 B&W-02.1"   8888*   00%  04/03/76 N37D30M00S W078D46M00S 1:3,369,000 82100170808 P  L  C   P
  CORNER POINT COORDINATES=#1:N38D05M22S W077D29M24S #2:N38D24M21S W079D30M06S #3:N36D53M52S W080D01M27S #4:N36D35M21S W078D03M03S

LANDSAT-1 (MSS)        8532314471500 B&W-02.1"   8888*   00%  03/07/76 N37D31M00S W078D55M00S 1:3,369,000 81100470352 F  L  C   Y
  CORNER POINT COORDINATES=#1:N38D07M20S W077D33M21S #2:N38D25M19S W079D40M16S #3:N36D53M54S W080D10M27S #4:N36D36M23S W078D10M54S
```

Figure 5.26. Results of computer search for LANDSAT imagery

Geological Survey and the National Oceanographic and Atmospheric Administration (NOAA) contributed to several aspects of the program. Although these federal agencies operated the LANDSAT system for many years, LANDSAT was officially considered to be "experimental" because of NASA's mission to develop new technology. In 1983 NOAA assumed administrative responsibility for the LANDSAT system, although NASA continued to provide some of the facilities, personnel, and technical services required for daily operation.

Over a period of many years both the Carter and the Reagan administrations and the U.S. Congress participated in long discussion and debate concerning the future of many federal services, including many of the weather satellites and LANDSAT. In essence, the debate centered on the advantages and disadvantages of public operation (by agencies of the federal government) compared to operation by a private corporation. Those favoring private operation emphasized the prospects for more efficient operation of the system and more aggressive pursuit of new applications and new technologies. Those who favored continued government operation stressed the need for supervision to maintain continuity of data flow and data format, the importance of providing public access to data, and the significance of a public archive of data as a historical record.

In June 1984 the U.S. Congress passed the Land Remote Sensing Act of 1984, which established the process by which the LANDSAT system passed from public to private operation. Title I of the act summarizes the rationale for the transfer to private operation and established certain constraints including assurances of public access to data. It also defines the fundamental distinction between "unenhanced" data—raw data that have been subjected to only minimal processing—and "enhanced" data that have been processed and analyzed to derive usable information. Title II establishes the Department of Commerce (the parent organization for NOAA) as the monitor for supervising the performance of a private contractor and provides for selection of a private contractor to market unenhanced data.

Title III assures continuity of data as private contractors assume responsibility for operation of the existing system and development of new components. Title IV establishes government regulation and licensing of private remote sensing systems operated in space. Title V provides for continued research and development by agencies of the federal government, and Title VI outlines general provisions of the act, including maintenance of an image archive by the federal government. Finally, Title VII prohibits commercialization of weather satellites—an issue that assumed political significance during the congressional debate.

In 1985 the first phases of this policy were implemented by the transfer of LANDSAT from NOAA to Earth Observation Satellite Company (EOSAT), a private firm formed as a partnership between Hughes Aircraft Company and the Radio Corporation of America (RCA). EOSAT operates existing LANDSAT components, receives and disseminates data and images, conducts research to define new applications for LANDSAT data, and designs new spacecraft and instruments to continue development of LANDSAT.

At the time of this writing plans are being made for LANDSAT 6, now scheduled for launch in late 1988. LANDSAT 6 will carry a thematic mapper redesigned to provide a panchromatic band with spatial resolution of about 15 m × 15 m. The

MSS will not be carried on LANDSAT 6, but the system will provide a means of simulating the MSS data from the TM data. LANDSAT 7, planned for launch in 1991, is expected to be similar to LANDSAT 6, except that additional infrared bands are under consideration. Both satellites will carry tape recorders necessary to store the high volumes of data acquired by the new sensors; they will not use TDRSs to acquire images over foreign areas.

5.8. SPOT

The success of the U.S. LANDSAT system in the early 1970s stimulated the interests of scientists in other nations. The French in particular have designed an earth observation satellite modeled after some of the fundamental features of the LANDSAT program. French scientists report that French users were impressed with the LANDSAT system, and that the French space program has been strongly influenced by the U.S. program as they designed their own system.

The French observation program, SPOT—"Le Système Pour l'Observation de la Terre" ("Earth Observation System")—was initiated in 1977; the first satellite was launched 21 February 1986. SPOT was conceived and designed by the Centre National d'Etudes Spatiales (CNES) in Paris, with the cooperation of other European groups. Current plans call for a series of satellites, starting with SPOTs 1 and 2, and continuing with SPOTs 3 and 4 tentatively planned for subsequent use in the late 1980s and early 1990s, respectively.

The SPOT system is planned to have capabilities for land use studies, assessment of renewable resources, exploration of geologic resources, and cartographic work at scales of 1 : 50,000 to 1 : 100,000. It is the first commercial remote sensing satellite designed to provide high-quality service and data for an operational user community worldwide. SPOT data, like those from LANDSAT, promise to meet the needs of a wide variety of customers with diverse technical, scientific, and commercial needs. Design requirements included provision for complete world coverage, rapid dissemination of data, stereo capability, high spatial resolution, and sensitivity in spectral regions responsive to reflectance from vegetation. It is still too soon to assess the success of the SPOT program, but the overall design and results of simulation studies indicate that the program will have capabilities of great interest to scientists and commercial users throughout the world. If all goes as planned, the SPOT program will have a lasting impact upon earth resources remote sensing.

The SPOT "bus" is the basic satellite vehicle, designed to be compatible with a variety of sensors (Figure 5.27). The bus provides basic functions related to orbit control and stabilization, reception of commands, telemetry, monitoring of sensor status, and other functions. The bus, with its sensors, is placed in a sun-synchronous orbit at about 832 km, with a 10:30 a.m. equatorial crossing time. For vertical observation, successive passes occur at 26-day intervals, but because of the ability of SPOT sensors to view areas at the oblique, successive imagery can be acquired, on the average, at $2\frac{1}{2}$-day intervals. (The exact repeat coverage interval varies with latitude.)

The SPOT payload consists of two identical sensing instruments, a telemetry transmitter, and magnetic tape recorders. The two sensors are known as HRV ("high

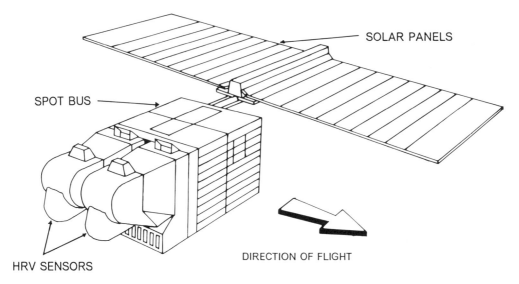

Figure 5.27. SPOT bus. Based upon SPOT Image Corporation illustration.

resolution visible") instruments. HRV sensors use "pushbroom" scanning technology, based upon charge coupled devices (CCDs) that can simultaneously image an entire line of data (in the cross-track axis) without mechanical movement (Figure 5.28). SPOT linear arrays consist of some 6,000 detectors for each scan line in the focal plane; the array is scanned electronically to record brightness values (8 bits; 256 brightness values) in each line. An important advantage over conventional technology is the absence of moving parts, which in principle should provide greater reliability, more uniform scanning speed across the image swath, and greater stability of the satellite, thereby assuring high geometric accuracy. (The moving mirror in mechanical scanning instruments is subject to wear, displays variable speed of movement

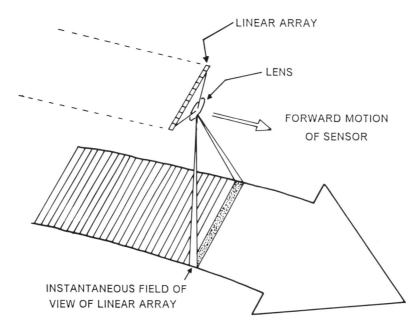

Figure 5.28. Pushbroom scanning.

across the image, and contributes to instability of satellite orientation.) Radiation from the ground is reflected to the two arrays by means of a movable plane mirror. An innovative feature of the SPOT satellite is the ability to control the orientation of the mirror by commands from the ground—a capability that enables the satellite to acquire oblique images, as described below.

The HRV can be operated in either of two modes. In the *panchromatic mode* the sensor is sensitive across a broad spectral band from 0.51 to 0.73 μm. It images a 60-km swath with 6,000 pixels per line, for a spatial resolution of 10 m × 10 m (Figure 5.29a). In this mode the HRV instrument provides fine spatial detail, but of course records a rather broad spectral region. In the panchromatic mode the HRV instrument provides coarse spectral resolution, but fine spatial resolution.

In the other mode, the *multispectral configuration*, the HRV instrument senses three spectral regions:

- *Band 1*: 0.50–0.59 μm (green)
- *Band 2*: 0.61–0.68 μm (red; chlorophyll absorption)
- *Band 3*: 0.79–0.89 μm (near infrared; atmospheric penetration)

In this mode the sensor images a strip 60 km wide using 3,000 samples for each line, at a spatial resolution of about 20 m × 20 m (Figure 5.29b). Thus, in the multispectral mode the sensor records fine spectral resolution but coarse spatial resolution. The

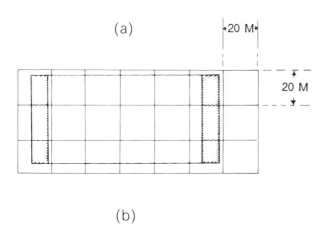

(a)

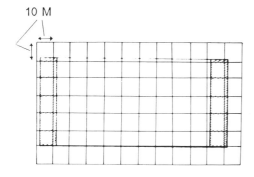

(b)

Figure 5.29. Nominal sizes of SPOT HRV pixels. The top diagram (*a*) depicts the pixel size of the HRV employed in the multispectal mode. The panchromatic mode (*b*) achieves finer resolution. An outline of a U.S. football field is shown for approximate scale.

three images from the multispectral mode can be used to form false-color composites, in the manner of CIR, MSS, and TM images (Plate 10). In some instances, it is possible to "sharpen" the lower spatial detail of multispectral images by superimposing them on the fine spatial detail of high resolution panchromatic imagery of the same area.

Each of the HRV instruments can be positioned in either of two configurations (Figure 5.30), which produce different geometric characteristics. For *nadir* viewing (Figure 5.30a) both sensors are oriented in a manner that provides coverage of adjacent ground segments. Because the two 60-km swaths overlap by 3 km, the total image swath is 117 km. At the equator, centers of adjacent satellite tracks are separated by a maximum of only 108 km, so in this mode the satellite can acquire complete coverage of the earth's surface.

An *off-nadir* viewing capability is possible by pointing the HRV field of view as much as 27° relative to the vertical in 45 steps of 0.6° each (Figure 5.30b) in a plane perpendicular to the orbital path. Off-nadir viewing is possible because the sensor observes the earth through a pointable mirror that can be controlled by command from the ground (Figure 5.30). (Note that although mirror orientation can be changed upon command, it is *not* a scanning mirror as used by the MSS and the TM.) With this capability, the sensors can observe any area within a 950-km swath centered on the satellite track. Options for sensor pointing provide scene centers that are 10 km apart. When SPOT uses off-nadir viewing, swath width of individual images varies from 60 to 80 km, depending upon viewing angle. Alternatively the same region can be viewed from separate positions (different satellite passes) to acquire stereo coverage. (Such stereo coverage depends, of course, upon cloud-free weather

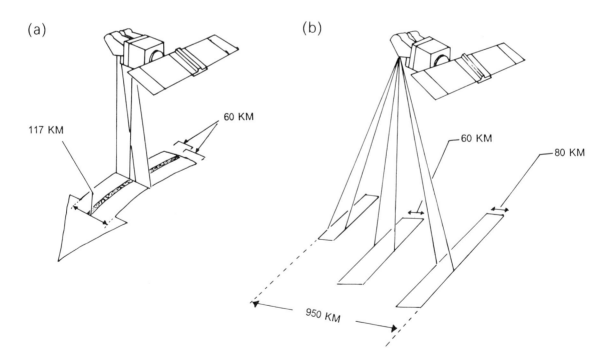

Figure 5.30. Geometry of SPOT imagery: (*a*) nadir viewing, and (*b*) off-nadir viewing. Based upon SPOT Image Corporation illustrations.

114

during *both* passes.) The twin sensors are not required to operate in the identical configuration; that is, one HRV can operate in the vertical mode while the other images obliquely. Using the off-nadir viewing capability, SPOT can acquire repeat coverage at intervals of 1 to 5 days, depending upon latitude.

Ground control is provided by a station in Toulouse, France, with a number of ground stations around the world providing relay and backup capability as required. Direct reception of image data is possible within a 2,600-km radius of participating ground image receiving stations. In addition, tape recorders provide the capability for recording data acquired beyond this range. An on-board computer provides the capability for programming a sequence of viewing modes and angles for later execution. Reception of SPOT data is possible at existing LANDSAT stations if they are appropriately modified. Current plans call for establishment of reception capabilities at eight sites positioned on all continents except Antarctica.

Data for North America are received by two stations in Canada. In the United States, SPOT data are distributed by SPOT Image, a company established to service customers in the United States. SPOT Image maintains an image data base and indexing system analogous to that previously described for the LANDSAT system. Customers can request a listing of available imagery, and procedures are available to request coverage of specific areas. Like LANDSAT images, SPOT data are available to all customers without restrictions based upon nationality. Customers in the United States can contact any of several SPOT offices; the primary center for U.S. operations is located in Reston, Virginia.

Data processing will be conducted by CNES and the Institut Géographique National (IGN); image archives will be maintained by the Centre de Rectification des Images Spatiales (CRIS), also operated by CNES and IGN. Four levels of processing are planned; they are listed below in ascending order of precision:

- *Level 1*: Basic geometric and radiometric adjustments:
  - *Level 1a*: Sensor normalization.
  - *Level 1b*: Level 1a processing, with the addition of simple geometric corrections.
- *Level 2*: Use of ground control points to correct image geometry; no correction for relief displacement.
- *Level 3*: Further corrections using digital elevation models.

Products supplied to the user community include CCTs and film or print images at 1:400,000 and larger. An important objective is to speed image processing to provide timely service to ultimate users of the image data. Estimated costs per scene are said to be higher than those for LANDSAT MSS data, but due to the finer spatial resolution, the cost "per pixel" is estimated to be comparable to that of LANDSAT data.

Planning for additional SPOT missions is now in progress. SPOT 2 is planned to include additional bands in the near infrared. Present plans for SPOTs 3 and 4 include consideration of the additional near infrared bands, an improved channel in the green region, and elimination of the existing 10-m panchromatic channel.

5.9. Other Land Observation Satellites

Several other nations have operated, or have initiated planning for, earth observation satellites. Many are still in early stages of planning, and for others only limited information is available, so that only brief sketches of their capabilities can be given. All follow the general model given here for the LANDSAT and SPOT programs in respect to general features of orbits, sensors, and other features. Details of these systems may change substantially as planning progresses during the next few years, so that the information given here may not correspond to the actual systems placed in operation. However, this section presents a perspective on the variety and scope of interest in satellite observation of earth resources.

The Japanese government plans to operate two earth observations satellites in the late 1980s. The first system is tailored for observations of marine resources. It will be placed in a sun-synchronous orbit, with a 14-day coverage cycle. Three sensors have been designed to detect radiation in several portions of the spectrum. For observation of sea-surface color, a sensor will be sensitive to four bands in the visible and near infrared. This sensor will image a 200-km swath with an IFOV of 50 m × 50 m. A second sensor will measure ocean-surface temperature by detecting energy in several bands, including the far infrared. It will have coarser resolution of 90 m for some bands and 2,700 m for those in the far infrared. Finally, a passive microwave sensor will detect energy in two bands as the basis for measurement of atmospheric moisture.

A second satellite program is designed to provide information for geologic exploration, land use analysis, and vegetation mapping. It will have a stereoscopic capability and use linear array technology similar to that employed for the SPOT sensors. The Japanese plan to use four spectral regions in the visible and near infrared as well as a radar system similar to those to be described in Chapter 6. Images in the visible and near infrared are planned to have a resolution of about 30 m; the radar images will have resolution near 25 m.

The European Space Agency, a cooperative agency of several European nations, plans to launch an earth observation satellite in the late 1980s. This satellite will have a variety of sensors, including radars and sensors sensitive to visible and near infra-red radiation. Most of the sensors are designed for observation of oceanographic phenomena.

The Soviet Union has operated earth observation satellites since 1977. It has used several systems based upon large format cameras (analogous, no doubt, to aerial mapping cameras described in Chapter 3) with a variety of lenses and filters. Images from these cameras can provide multispectral data, possibly with detail of about 30 m for ground areas about 335 km × 335 km. Film from these cameras must be recovered for processing and use on earth, and thus they differ greatly in concept from the LANDSAT and SPOT systems, which send all data to earth by microwave signal.

In 1980 the Soviet Union began operation of a series of METEOR weather satellites. These satellites include three sensors designed for experimental observations of earth resources. The MSU-E is a solid-state scanning instrument (as described for the SPOT sensors) sensitive to three spectral regions in the visible and near infrared. It is said to have an IFOV of 30 m × 30 m.

The MSU-SK is an optical-mechanical scanner (analogous to the LANDSAT MSS in principle) that is sensitive to four spectral regions. The IFOV is about 170 m × 170 m. Finally, a third sensor, known as "Fragment," has an IFOV of 80 m × 80 m and is sensitive to eight spectral regions in the visible and infrared. So far, little of the imagery from these sensors has been released to the international scientific community.

The People's Republic of China has operated several earth observation satellites since 1975 but has not released imagery to the public. For the future, a number of satellite missions are planned to gather data in the visible, infrared, and micro-wave regions, to be used for applications in meteorology and earth resources. Other satellites have been planned by India, the Netherlands (in cooperation with Indonesia), Canada, and Brazil.

5.10. Summary

Satellite observation of the earth has greatly altered the field of remote sensing. Since the launch of LANDSAT 1, a larger and more diverse collection of scientists than ever before have participated in remote sensing research and applications. Public knowledge of, and interest in, remote sensing has increased. Digital data for satellite images has contributed to the growth of image processing, pattern recognition, and image analysis (Chapters 9–12). Satellite observation systems have increased international cooperation through activities such as building of receiving stations and training of scientists from nations that previously were without substantive capabilities for remote sensing.

This chapter has an important role in building a framework for development of topics to be presented in subsequent chapters. Much of the information presented here in relation to LANDSAT is equally important as a basis for understanding other satellite systems that operate in the microwave (Chapter 6) and far infrared (Chapter 7) regions of the spectrum.

Finally, it can be noted that the discussion thus far has emphasized *acquisition* of data. Little has been said about *analysis* of these data and their *applications* to specific fields of study. Both topics will be covered in subsequent chapters (Chapters 9–12 and 13–17, respectively).

Review Questions

1. Outline the procedure for identifying and ordering LANDSAT images for a study area near your home. Identify for each step the *information* and the *materials* (maps, etc.) necessary to complete that step and proceed to the next. Can you anticipate some of the difficulties you might encounter?

2. In some instances it may be necessary to form a mosaic of several LANDSAT scenes, by matching several images together at the edges. List some of the problems you expect to encounter as you prepare such a mosaic.

3. What are some of the advantages (relative to use of aerial photography) of using satellite imagery? Can you identify disadvantages?

4. Manufacture, launch, and operation of earth observation satellites is a very expensive undertaking—so large that it requires the resources of a national government to support the many activities necessary to continue operation. Many people question whether it is necessary to spend government funds for earth resource observation satellites and have other ideas for use of these funds. What arguments can you give to justify the costs of such programs?

5. Why are orbits of land observation satellites so low relative to those of communications satellites?

6. Would it be feasible to design an earth observation satellite with a sun-synchronous orbit to provide coverage of the poles? Explain.

7. Discuss problems that would arise as engineers attempt to design LANDSAT sensors with smaller and smaller pixels. How might some of these problems be avoided?

8. Can you suggest some of the factors that might be considered as scientists select the observation time (local sun time) for a sun-synchronous earth observation satellite?

9. LANDSAT does not continuously acquire imagery, but only those individual scenes as instructed by mission control. List factors that might be considered in planning scenes to be acquried during a given week. Design a strategy for acquiring LANDSAT images worldwide, specifying rules for deciding which scenes are to be given priority.

10. Using information given in the text, calculate the number of pixels for a single band of an MSS scene, for a TM scene, and for a SPOT HRV image. (For SPOT assume the image is acquired at nadir.) Recompute the numbers to include all bands available for each sensor.

11. Estimate the number of aerial photographs at a scale of 1:15,470 that would be required to show the land area represented on a single LANDSAT MSS scene. Assume end lap of 20% and side lap of 10%.

12. Explain why a LANDSAT MSS scene and an aerial mosaic of the same ground area are not equally useful, even though image scale might be the same.

13. How many pixels are required to represent a complete LANDSAT MSS scene (one band only)? A single band of a TM scene?

14. Prepare a template showing (at the correct scale) the dimensions of MSS, TM, and SPOT pixels for an aerial photograph of a nearby area, or other images provided by your instructor. Position the template at various sites throughout the aerial photograph, and assess the effectiveness of the sensors in recording various components of the landscape, including forested land, agricultural land, urban land, etc. (If pixels are composed of only a single category or feature, they tend to be recorded more effectively than if pixels are composed of two or more classes.)

15. On a small-scale map (such as a road map, or similar map provided by your instructor) plot at the correct scale the outlines of an MSS scene centered on a nearby city. How many different counties are covered by this area?

References

Begni, G. 1982. Selection of the Optimum Spectral Bands for the SPOT Satellite. *Photogrammetric Engineering and Remote Sensing*, Vol. 48, pp. 1613–1620.—Describes rationale for the definition of SPOT bands.

Chen, H. S. 1985. *Space Remote Sensing Systems*. New York: Academic Press, 257 pp.—Overview and brief descriptions of a wide variety of satellite systems for earth observation.

Chevrel, M., M. Courtois, & G. Weill. 1981. The SPOT Satellite Remote Sensing Mission. *Photogrammetric Engineering and Remote Sensing*, Vol. 47, pp. 1163–1171.—Description of the SPOT satellite.

EOSAT LANDSAT Application Notes. Lanham, MD: Earth Observation Satellite Company.—Newsletter describing uses of LANDSAT data; the first issue is dated June 1986.

General Electric Company. no date. *Data Users Handbook*. Philadelphia: Space Division, General Electric Company.—Basic reference for LANDSATs 1 and 2; possibly available in some libraries. Much of its contents presented in other publications listed here.

General Electric Company. no date. *LANDSAT 3 Reference Manual*. Philadelphia: Space Division, General Electric Company.—Basic reference for LANDSAT 3; much of its contents included in other publications listed here.

LANDSAT Data Users Notes. Sioux Falls, SD: NOAA LANDSAT Customer Services.—A quarterly newsletter published until 1986; provided current information on the LANDSAT system and numerous related matters. Back issues are available on microfiche; issue 35 (March 1986) has an index.

McClain, E. P. 1980. Environmental Satellites. In *McGraw-Hill Encyclopedia of Environmental Science*. New York: McGraw-Hill, pp. 15–30.—A concise summary of satellite programs for observing the earth; emphasis upon meteorological satellites.

Satellite Technology Serving Earth. 1977. *Aviation Week and Space Technology*, 11 October.—A brief overview of the LANDSAT system with illustrations and examples.

Sheffield, C. 1981. *Earthwatch: A Survey of the World from Space*. New York: Macmillan, 160 pp.—LANDSAT images with accompanying text.

Sheffield, C. 1983. *Man on Earth: How Civilization and Technology Changed the Face of the World—A Survey From Space*. New York: Macmillan, 166 pp.—LANDSAT images with accompanying text.

Short, N. M. 1976. *Mission to Earth: Landsat Views the World*. Washington, DC: NASA, 459 pp.—A collection of LANDSAT images accompanied by brief explanations.

Slater, P. N. 1979. A Re-examination of the LANDSAT MSS. *Photogrammetric Engineering and Remote Sensing*, Vol. 45, pp. 1479–1485.—Detailed discussion of the LANDSAT MSS, with emphasis upon its optical characteristics, spatial resolution, and spectral sensitivity.

Slater, P. N. 1980. *Remote Sensing: Optics and Optical Systems*. Reading, MA: Addison-Wesley, 575 pp.—Chapter 14 includes detailed, comprehensive material on LANDSAT system.

SPOT Image Corporation. *SPOTLIGHT*. Reston, VA: SPOT Image Corporation.—Quarterly newsletter providing current information concerning SPOT.

Taranick, J. V. 1978. *Characteristics of the LANDSAT Multispectral Data System*. U.S. Geological Survey Open File Report 78-187, 76 pp.—A brief overview of the MSS from the perspective of a user of MSS digital data.

Plate 1. Panoramic aerial photograph. The complete image is almost 1.3 m ($4\frac{1}{2}$ ft.) long and 11.4 cm ($4\frac{1}{2}$ in.) wide, much too large to show as a single illustration. Here sections from the right hand edge (top image), nadir (center image), and left hand edge (bottom image) illustrate qualities of a panoramic photograph. Although at the edges panoramic photography resembles an oblique photograph and at the nadir it resembles a vertical photograph, additional positional errors are introduced by the combined effects of the forward motion of the aircraft and the side-to-side scan of the camera lens. The flight direction is from right to left. Color infrared imagery of a region in central Pennsylvania, June 1981. NASA photo courtesy of U.S. Department of Agriculture Forest Service.

Plate 2. Color aerial photograph: Wayson's Corner, Maryland, 15 July 1972. Photograph courtesy of NASA.

Plate 3. Color infrared photograph: Wayson's Corner, Maryland, 15 July 1972. Photograph courtesy of NASA.

Plate 4. High oblique aerial photograph: Manhattan Island, New York City. U.S. Geological Survey photograph.

Plate 5. Low oblique aerial photograph. Photograph courtesy of North Pacific Aerial Surveys, Anchorage, Alaska.

Plate 6. Image from the large-format camera: Great Barrier Reef, Australia, 11 October 1984. LFC Frame 4107-1696. This image shows the center section from the original, which is rectangular in shape, as explained in the text. Image from U.S. Geological Survey EROS Data Center, Sioux Falls, South Dakota.

Plate 7. High-altitude aerial photograph: Trenton, Nebraska, 18 September 1980. Image from U.S. Geological Survey EROS Data Center, Sioux Falls, South Dakota.

Plate 8. LANDSAT MSS color composite of the same region depicted by Figures 5.13 and 5.14, formed from MSS bands 4, 5, and 7, New Orleans, Louisiana, 16 September 1982. Scene ID: 40062-15591. Image reproduced by permission of EOSAT.

Plate 9. Thematic mapper color composite (bands 1, 3, and 4). This image forms a color composite that is approximately equivalent to the MSS color composite. This image shows the same region depicted by Figures 5.19 and 5.20: New Orleans, Louisiana, 16 September 1982. Scene ID: 40062-15591. Image reproduced by permission of EOSAT.

Plate 10a. SPOT color composite, New Orleans, Louisiana. This image is a false-color composite image formed from the superimposition of the 20-m resolution multispectral data and the 10-m panchromatic data from the two HRV sensors. The image shows a region approximately 60 km × 60 km. Copyright © 1986 by CNES; provided by SPOT Image Corporation, Reston, Virginia.

Plate 10b. An enlargement of a portion of Plate 10a showing the central business district of New Orleans, Louisiana, in greater detail. Copyright © 1986 by CNES; provided by SPOT Image Corporation, Reston, Virginia.

Plate 10c. A further enlargement of the area shown in Plates 10a and 10b. Individual structures, including the Huey P. Long Bridge and the Superdome, are now visible. Copyright © 1986 by CNES; provided by SPOT Image Corporation, Reston, Virginia.

Plate 11. Thermal image of houses showing thermal properties of separate elements of a wooden structure (March 1979, 11 p.m.; air temperature = 2 C). The top image shows the original thermal image (white = hot; dark = cool). The bottom image is a color-coded version of the same data, with each change in color representing a change in temperature of about 1.5 C, over a range from 2 C to 11 C. Colors represent increasing temperatures in the sequence black, magenta, blue, cyan, green, yellow, red, and white. Thermal images provided by Daedalus Enterprises, Inc., Ann Arbor, Michigan.

Plate 12. Applications Explorer Mission 1 (HCMM) image of northeastern United States and southeastern Canada, 11 May 1978. Image courtesy of NASA.

Plate 13. Example of a classified image. This is a portion of a LANDSAT MSS scene of the Roanoke, Virginia, region. Each pixel has been assigned to a land cover class and coded with a color.

Plate 14. CIR image illustrating distinctness of vegetation classes: southeastern Maryland, July 1972. Image courtesy of NASA.

Plate 15. Color composite AVHRR image. This image has been formed from two dates, as explained in the text. Courtesy of the Environmental Research Institute of Michigan, Ann Arbor, Michigan.

Plate 16. Turbidity patterns as recorded by color aerial photography. Photograph courtesy of North Pacific Aerial Surveys, Anchorage, Alaska.

Plate 17. Color aerial photograph of Patuxent River near Potts Point, Maryland, 15 July 1972. Image courtesy of NASA.

Plate 18. Color infrared photograph of the same area shown in Plate 17 and Figures 14.15 and 14.16. Image courtesy of NASA.

Plate 19. Color ratio image of the Cuprite, Nevada, region. Variations in color represent variations in ratios between spectral bands, as described in the text. Image courtesy of GeoSpectra Corporation, Ann Arbor, Michigan. From Vincent *et al.* (1984, p. 226). Reproduced by permission of the Environmental Research Institute of Michigan, Ann Arbor, Michigan.

Plate 20. Land cover map compiled from AVHRR data. Data from numerous AVHRR composite images gathered during 1982 and 1983 are summarized here, such that seasonal variation in vegetation cover is considered in the classification. Tan: desert and semidesert; light green: semiarid wooded grassland and bushland; purple: woodland and grassland; dark blue: mixed grassland and tropical forest; red: tropical rainforest and montane forest; green: woodland; light blue: bushland and thicket; yellow: wooded grassland and deciduous bushland. See Tucker *et al.* (1985) for details. Image courtesy of NASA–Goddard Space Flight Center, Greenbelt, Maryland.

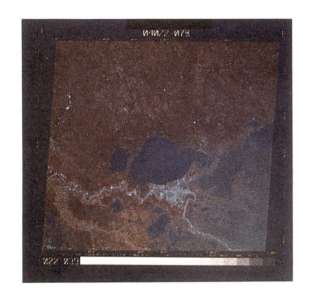

Plate 8   Plate 9

Plate 10a

Plate 10b

Plate 10c

121

Reprinted by permission of the American Society for Photogrammetry and Remote Sensing from *Manual of Remote Sensing*, ed. Robert N. Colwell, second edition, 1983, v. 2, p. 1797-1802.

Petroleum Exploration

Introduction to the Use of Remote Sensing Technology in the Search for Oil and Gas

Since 1945, petroleum geologists have frequently used aerial photographs as an important tool in exploring for surface manifestations of structural features that might lead to the discovery of new oil and gas reservoirs (see also the "Photogeology and Geological Remote Sensing" section of this chapter). Stereoscopic pairs of aerial photographs have been especially helpful in estimating the dip of strata to an accuracy of 1° to 2°. Trollinger (1968) was one of the first to describe the advantage of the synoptic view provided from space altitudes, when he studied color satellite photographs of the Delaware Basin of west Texas acquired by the Gemini astronauts (see Table 31-6). He found that for the first time he could see surface manifestations of deep crustal structures that, until then, had been recognized only in geophysical and borehole data of the region.

Saunders et al. (1973) conducted systematic visual studies of ERTS 1 (Landsat 1) image mosaics covering several oil and gas provinces in west Texas, New Mexico, Colorado, and Montana-Wyoming. They demonstrated that, in many cases, there was a correlation between surface lineaments expressed in the images and the locations of subsurface oil and gas fields. Similar results were obtained by Coilins et al. (1974) in western Oklahoma.

Floyd Sabins, in his textbook, *Remote Sensing, Principles and Interpretation,* devotes several pages to oil exploration (Sabins, 1978b). Bentz and Gutman (1977) discussed the importance of Landsat to petroleum exploration in foreign areas. Peterson (1979) addressed the significance of lineaments in oil and gas exploration. Venkataramanan (1979) discussed the application of Landsat imagery to petroleum exploration in India. Gathright (1982) and Blodget (1981) addressed the use of Landsat images in the search for oil and gas fields in the Appalachian Mountains.

Miller (1977) demonstrated the value of visual analysis of Landsat images to petroleum exploration in foreign, less well-mapped regions, such as in the Sudan and Kenya. Lineaments in Sudan and Kenya that he interpreted from Landsat images, he defined as part of a regional fracture system at the north end of the Lamu Embayment. He

associated these features with the East African rift system and interpreted them to be a failed, immature triple junction with one arm opening into the Rudolf Trough and a second arm extending into the Ogaden Basin. Although much of the area is covered by Quaternary deposits of merging alluvial fans, a swampy area along the Ewaso N'Giro drainage system suggested subsidence along what may have been a structural trough or downwarp. The presence of a local trough along the Ewaso N'Giro was confirmed by magnetometer and gravity surveys as well as reflection seismic surveys. Unconfirmed information made available by J. Vandenakker (oral communication, 1982) indicates that the Chevron Oil Company has drilled and successfully developed oil resources in the region; reportedly it is also planning a pipeline to Port Sudan.

Maurin and Riguidel (1978), of the French Petroleum Company, TOTAL, of Paris, France, prepared a comprehensive treatise (in French) on the use of various digital enhancement and analysis methods to assist geomorphic and structural mapping of petroleum exploration targets in Tunisia. A Landsat 1 image (1199-09305; 7 February 1973) of the Kef Si AEK oil field area, Tunisia, was contrast-stretched and filtered by a 5 by 5 pixel array "boxcar" filter. This technique enhanced the Turonian fold-and-fault system which trends northeast across the more northerly trending Djibel Mrhila anticline. Oil and gas deposits were found to occur at the intersection of these two structural features. The treatise by Maurin and Riguidel (1978) provides much of the mathematical background that geologists need in order to work with Landsat CCT's. Similarly, Taranik (1978b) prepared a paper entitled "Principles of Computer Processing of Landsat Data for Geologic Applications." A detailed publication on digital image analysis methods was published by Johannes Moik (1982) of the NASA Goddard Space Flight Center (See also the "Analysis Techniques" section of this chapter).

Halbouty (1976), in a comprehensive article in the *Bulletin of the American Association of Petroleum Geologists*, provided many examples of the visual analysis of Landsat data in petroleum exploration. A few examples of the applications of experimental digital image analysis were also cited. Four years later, Halbouty (1980) published a second study of 15 giant oil and gas fields of the world, and pointed out that if Landsat data had been available earlier, the data could have been of significant help in discovering and developing at least 13 of them. He indicated that the systematic study of Landsat images, especially those of remote and poorly mapped areas of the world, could help cut the costs of exploration significantly. These articles contributed significantly to the exploration process by encouraging major oil companies, and some independents, to invest considerable time and effort in developing their own expertise in the use of Landsat data and in establishing special in-house research groups and digital image processing laboratories. Furthermore, it created interest that led to the development of the Geosat Committee, Inc., a consortium of more than 100 mineral and petroleum exploration firms and consultants who participate in joint government (NASA and USGS) and industry development of remote sensing techniques. The objective of the Geosat Committee, Inc., is to foster the development of remote sensing technology and to prepare its constituency for the effective use of new data that have higher resolution and more or different spectral bands, such as the Landsat 4 Thematic Mapper data, first available in July 1982. The Geosat Committee, Inc., in association with NASA, contracted with the Jet Propulsion Laboratory, for a major remote sensing study of three petroleum fields, three copper deposits, and two uranium deposits, as part of the "Joint NASA/Geosat Test Case Project" (Abrams et al., 1983). This report of more than 2000 pages, illustrated with over 200 color plates and maps, will become available by mid-1983. Synopses of two of the joint NASA/Geosat test case projects are presented in the "Economic Geology" section of this chapter. These are (1) an analysis of airborne remote sensing of a porphyry copper test site in Arizona (see the preceding "Mineral Exploration" section of this chapter), and (2) the following description from Harold Lang's work (Lang, 1982) on analysis of airborne remote sensing of the Lost River gas field, West Virginia:

"The Lost River gas field is typical of many fields in the Ridge and Valley Province of the Appalachian Mountains. The Devonian sandstone and shale reservoir has fracture porosity and forms an anticlinal trap. Analysis of a 1:500,000-scale Landsat image (2815-14560; 16 April 1977) by lineament interpretations prepared by two independent interpreters demonstrates that subjectivity in recognizing individual lineaments may be obviated by lineament density (number of lineaments per unit area) mapping. Such maps show, for example, that the Lost River gas field is located beneath an area of high lineament density. Lineament density mapping at 1:48,000 scale, using aircraft-acquired Landsat 4 Thematic Mapper Simulator (TMS) data, yields lineament density isopleths that mimic subsurface structural contours of the Lost River gas field. These results suggest an approach for using lineament density mapping as an exploration tool and demonstrate that lineament density mapping could have been used to help find the Lost River gas field, if such information had been available prior to field development" (Lang, 1982).

The following two sections describe the use of radar and Landsat images in the quest for petroleum reservoirs. The first section presents exam-

ples of the use of Seasat SAR and SIR-A images in the Appalachian Mountains and in Wyoming for petroleum exploration. The second section discusses the use of a specially processed Landsat image, in conjunction with seismic surveys, to discover a concealed fault in the Bay County area of Michigan.

The Use of Radar in the Search for Hydrocarbons

The primary advantage of using radar images in exploring for hydrocarbons lies in the ability of the imaging systems to: (1) enhance subtle topographic features in heavily vegetated areas; and, (2) to penetrate cloud cover. Radar images, therefore, have been used for geological exploration mainly in areas of tropical rainforest. For example, the regional distribution of faults and linear patterns in western Irian Jaya, Indonesia, was first seen on airborne radar images acquired in 1974 with a radar operating in the X-band (2.8-cm wavelength). Analysis of the regional structural pattern suggested that plate tectonism was responsible for the fragmentation that has separated the oil-producing Salawati Island area from the mainland (Froidevaux, 1978 and 1980). For a general reference on the geological interpretation of radar images, the work by Mekel (1972), constituting one of the chapters in the ITC Textbook of Photo-Interpretation, is recommended.

The synthetic-aperture radar (SAR) experiment on the Seasat satellite provided in 1978 the first spaceborne radar images of North America available to the scientific community. The imaging radar was operated at the 23.5-cm wavelength (L-band) with a look angle (the supplementary angle to the depression angle) of about 20° and a maximum resolution of 25 m (when images have been digitally processed). Images of the heavily forested terrain of the southern Appalachians (Figures 31-126 and 31-127) reveal a pronounced enhancement of linear topographic features in the area. The potential for hydrocarbon traps is favorable in zones of fracture porosity, particularly at the intersection of lineaments. The enhancement on the images of extensive linear topography, and of short linear features less than 10 km long, results from the high sensitivity of the Seasat SAR to change of surface slope. This high sensitivity is also responsible for the strong geometric distortion on the images. Foreslopes having a magnitude equal to or greater than the radar look angle are obliterated by layover. Linear features that have substantial topographic relief from end to end are geometrically rotated on the images. Corresponding images from the Landsat multispectral scanner (MSS) and return beam vidicon (RBV) camera show no geometric distortion of this type. Linear topography on images acquired with the Landsat sensor systems is enhanced by

Fig. 31-126. Seasat radar image of the folded Appalachians and Pine Mountain overthrust sheet, Tennessee-Kentucky-Virginia. The image has been optically correlated from digitally recorded data and is a digitally mosaicked composite from two Seasat radar images: Seasat Rev 407, acquired 25 July 1978, and Seasat Rev 163, acquired 12 June 1978. The spatial resolution is approximately 30 m; the scene center is at 36°15′ N. latitude, 83°47′ W. longitude. The illumination direction is N. 67°30′ E., and the orbital altitude is 795 km. The Seasat radar operated at a 23.5-cm wavelength; polarization was parallel or horizontal-horizontal (HH). The two separate images are archived by NOAA's National Environmental Satellite Data and Information Service (Table 31-14). Image mosaic courtesy of the Jet Propulsion Laboratory.

Fig. 31-127. Seasat radar image of the folded Appalachians and Pine Mountain overthrust sheet, Tennessee-Kentucky-Virginia (Seasat Rev 874, acquired on 27 August 1978). Optically correlated from digitally recorded data, the digitally mosaicked image corresponds to Seasat coverage in part of Figure 31-126. The spatial resolution is 30 m; the illumination direction is N. 67.5° W. Image courtesy of the Jet Propulsion Laboratory.

solar shadowing, when the Sun elevation is below about 30° (see Figures 31-117B, 31-204, and 31-225). This condition occurs only for images acquired between November and February in temperate latitudes. Mapping from a suitable Landsat MSS image (Figure 31-128) and from the area common to the Seasat SAR images (Figures 31-126 and 31-127) has shown that the small-scale linear topographic features are preferentially enhanced on the radar images (Ford, 1980). The extent of the enhancement is shown on the histograms of lineament frequency versus strike in Figure 31-129. In this instance, the preferred orientations of the lineaments are more readily interpreted from the SAR data. The histograms show in each case that lineament perception is comparatively reduced in the direction of scene illumination. In the case of Seasat this deficiency is offset by the dual directions of scene illumination that were obtained with the imaging radar system, as a result of passes having been made over an area during the descending and ascending orbits of the spacecraft. The regional distribution and

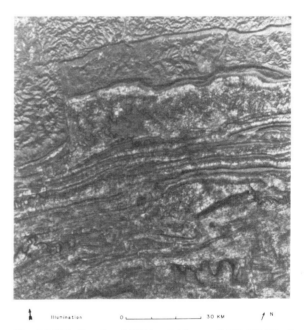

Fig. 31-128. Landsat MSS band 6 image (1858-15300) of the folded Appalachians and Pine Mountain overthrust sheet, acquired on 28 November 1974, corresponding to Seasat radar coverage in Figure 31-126. Image is digitally processed and contrast-stretched to enhance linear features. Pixel resolution is 79 m, Sun elevation is 25°, and the azimuth is 151° (N. 29° W.). Specially processed Landsat image courtesy of the Jet Propulsion Laboratory.

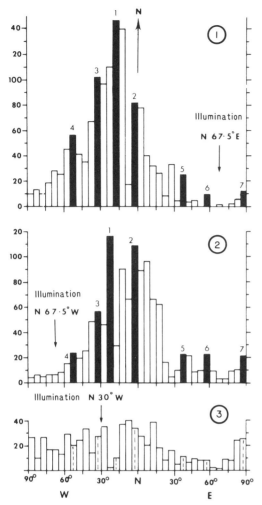

Fig. 31-130. Seasat radar image subscene of Bitter Creek area, Patrick Draw oilfield, southwest Wyoming. Image is digitally correlated from digitally recorded data (Seasat Rev 789; acquired on 21 August 1978). Resolution is 25 m; subscene center is at 41°45' N., 108°30' W.; the illumination direction is N.66° W. Image courtesy of the Jet Propulsion Laboratory.

Fig. 31-129. Histograms of lineament frequency versus strike for short lineaments mapped from corresponding areas on Seasat SAR and Landsat MSS images: (1) from the Seasat SAR image in Figure 31-126, (2) from the Seasat SAR image in figure 31-127, and (3) from Landsat MSS image in Figure 31-128. Frequency on vertical axis, strike on 5° increments east and west of north. Solid bars are frequency maxima interpreted from the two Seasat SAR images. These maxima are not apparent from the Landsat MSS image (Figure 31-128).

alignment of the short lineaments mapped from the SAR images, and the relationship of the lineaments to known structures and geophysical trends, provide a basis for further locating faults and fracture traces.

In the Patrick Draw Oilfield in the Bitter Creek area of southwest Wyoming the Seasat SAR backscatter is dominated by slope effects from the numerous small drainages of low relief near the Continental Divide and by surface scattering from the vegetation and rocks in the interchannel areas. The slopes produce bright returns that have a high spatial frequency on the radar image (Figure 31-130). Variations in the density of the predominantly sagebrush vegetation cover tend to pro-

duce medium to dark returns that have a low spatial frequency on the image. Systematic changes in the low-frequency distribution of the medium- to dark-gray levels are obscured on the image by the high-frequency distribution of the bright returns.

SAR images consist of three basic spectral components (Figure 31-131). Daily (1983) has shown a method of enhancing both the high- and the low-frequency components that contain useful information by filtering and color encoding. Low-frequency spatial variations on the image that represent changes in vegetation density are displayed by hue. High-frequency spatial variations associated with the small drainages are retained as intensity (Color Figure 31-132). The changes in vegetation represented by the hues on the image are influenced by soil moisture, soil composition, slope, and subtle surface characteristics of the Tertiary bedrock section. Thus the hues outline the plunging structure of the Wamsutter Arch; this feature is not evident on the unenhanced Seasat SAR image (Figure 31-130). The structure is not readily apparent at the surface in the field, though it is known from subsurface records. Retention of the high-frequency spatial

126

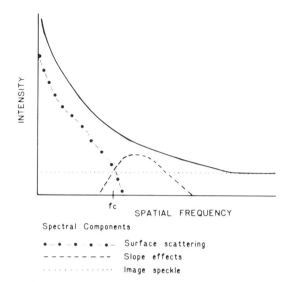

Spectral Components

•—•—•—•—•— Surface scattering

– – – – – – –   Slope effects

· · · · · · · · · · · ·   Image speckle

Fig. 31-131. Typical power spectrum of a SAR image, showing spatial distribution of spectral components. Cutoff frequency, fc, that provides the basis for filtering surface scattering components from slope components, is scene dependent.

variations, as intensity, enhances linear topographic features on the image.

The Shuttle Imaging Radar (SIR-A) experiment provided a second generation of spaceborne scientific radar images of the Earth in 1981 (Elachi and others, 1982; McCauley and others, 1982). The images were acquired at the same wavelength that was used to operate the Seasat SAR, but with a look angle of about 47° and resolution of 40 m. SIR-A coverage of the oil and gas producing Appalachian Plateau area in southeast Kentucky, and the adjacent Ridge and Valley area of southwest Virginia (Figure 31-133) enhances extensive linear topography and short linear features on the image. The strike and distribution of the latter relative to known faults and magnetic trends suggest that many of them are structural in origin (Ford, 1982). On corresponding Seasat SAR images the extensive features are perceptible but the short linear features are strongly distorted or impossible to locate. This contrast results primarily from the difference in the SIR-A and the Seasat SAR look angle. The look angle and the slope of the terrain govern the local incidence angle, which determines the factors that dominate the radar backscatter (Figure 31-134). At the steeply sloping surfaces in southeast Kentucky the SIR-A backscatter is dominated by slope effect. This effect enhances the small-scale topography of the area. In contrast, the Seasat SAR backscatter is dominated by layover. This layover obliterates the small-scale topography. Lineament mapping from a corresponding Landsat 3 RBV image (Figure 31-135) shows equivalent perception of the major and minor features seen on the SIR-A image

(Figure 31-133), but the small-scale topography that strikes near-parallel to the scene illumination of the RBV image is suppressed.

Analysis of SIR-A images of foreign areas has already yielded some important scientific results. Charles Elachi of Jet Propulsion Laboratory and his colleagues have discovered that Seasat SIR-A images can penetrate from 1 to 5 m of dry sand in Egypt and the Sudan, thereby revealing geomorphic and structural characteristics of the concealed bedrock (Elachi et al., 1982). This discovery has important implications for the geological exploration of the sand seas of the Earth and those of Mars. In another paper, Elachi (1982) summarized the use of radar images of the Earth from space and provided several excellent examples of such images, including color-enhanced radar images similar to Color Figure 31-132.

Application of Landsat Imagery to Petroleum Exploration in Bay County, Michigan

Introduction

The following case history of the use of Landsat in petroleum exploration is excerpted from a previously published paper by Vincent and Coupland (1980). It is a good example of the use of computer enhancement of a Landsat image to emphasize a subtle lineament in an area of the Michigan Basin where 200 m of glacial drift blanket the underlying bedrock. Seismic data confirmed the existence of a fault and a favorable structure for oil and gas accumulations along this fault on the southeastern margin of a graben structure.

Geologic Setting in Relation to Oil and Gas Potential of the Michigan Basin

The Michigan Basin, an intracratonic basin within the North American lithospheric plate, underlies most of the Southern Peninsula of the State of Michigan. Most of the consolidated sedimentary rocks in the basin are of Paleozoic age (600-270 million years old), though some are of the late Jurassic Period (180-160 million years old), and lie unconformably on Pennsylvania sediments (Permian and Triassic rocks are missing). Overlying these consolidated sedimentary rocks are unconsolidated glacial deposits of Pleistocene age (≤ 1 million years old). The depth of sedimentary rocks overlying crystalline basement rocks of Precambrian age (≥ 600 million years) in Gratiot County, near the central part of the basin, was found to be approximately 5300 m by a McClure Oil Company deep test hole in 1976. Although oil or gas has been found in every formation in the Michigan Basin (including small pockets of oil and gas in the glacial drift), the most prolific formations have been the Devonian-aged Traverse, Dundee, and Detroit River Formations, the Silurian-aged Niagran Formation (reefs), and

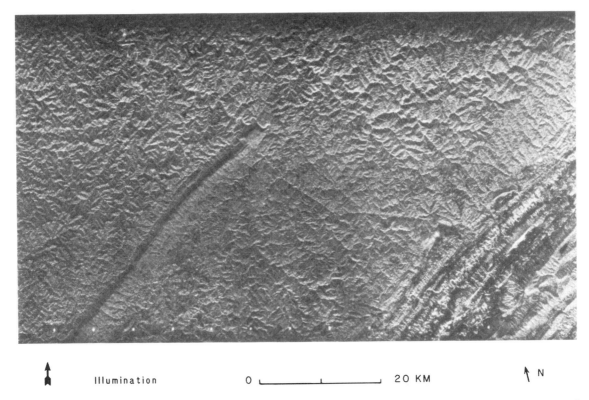

↑ Illumination   0 └────────────┘ 20 KM   ↑ N

Fig. 31-133. Shuttle radar image of the Appalachian Plateau in southeast Kentucky and southwest Virginia. Image is optically correlated from optically recorded data, acquired on SIR-A Data Take 24A on 13 November 1981. The spatial resolution is 40 m; the scene center is at 37°10′ N., 82°16′ W.; the illumination direction is N.18°E. Image courtesy of the Jet Propulsion Laboratory.

the Ordovician-aged Trenton and Black River Formations. All these are almost exclusively carbonate rocks. The Trenton-Black River Group are the producing horizons of the Albion-Scipio Trend in southern Michigan, which has produced over 116 million barrels of oil and over 180 billion cubic feet of gas since its discovery in 1957.

It has long been assumed that Pleistocene glacial deposits, which blanket virtually all the bedrock in the Southern Peninsula of Michigan to varying thicknesses (up to approximately 300 m), would render faults and other structural features in the bedrock invisible. This assumption has previously discouraged the use of satellite imagery or aerial photography in reconnaissance exploration for underlying structures favorable for the accumulation of hydrocarbons.

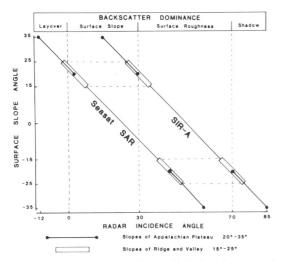

Fig. 31-134. SIR-A and Seasat SAR incidence angles plotted against range of surface slope angle common in the Appalachian Plateau and Ridge and Valley area, Kentucky-Virginia, showing range of factors that dominate backscatter.

Reprinted by permission of Colorado School of Mines
Press from L. W. LeRoy, D. O. LeRoy, and J. W.
Raese, eds., *Subsurface Geology: Petroleum, Mining,
Construction*, 4th edition, 1977, p. 767-787.

1

Integration of Geological Remote-Sensing Techniques in Subsurface Analysis

J. V. Taranik and C. M. Trautwein

Geological remote-sensing techniques are employed to minimize costs and maximize results of ground-based geologic investigations. Prediction of subsurface geological relationships from analysis of remote-sensor data depends on the expertise of the data analyst who evaluates variations in electromagnetic energy emanating from the earth's surface and extrapolates to the subsurface using these surficial attributes and predictive, conceptual geologic models.

Geological remote sensing can be defined as the study of the earth utilizing electromagnetic radiation (EMR) which is either reflected or emitted from the land surface in wavelengths ranging from ultraviolet (0.3 micrometer) to microwave (3 meter). The geological remote-sensing spectrum thus includes only a portion of the entire electromagnetic spectrum (fig. 1). In contrast, geophysical remote sensing can be defined as the study of the earth using electromagnetic radiation of wavelengths shorter than ultraviolet (X-rays, gamma rays) and longer than microwave (radio). Some workers include magnetic, gravity, sonic, and seismic methods as geophysical remote-sensing techniques, but these methods do not detect electromagnetic radiation.

Electromagnetic Energy and the Earth

Electromagnetic radiation is radiant energy in wave and particulate form that is propagated through space at the speed of light. It is classified according to the number of wavecrests which pass an arbitrary point each second (frequency) or according to the distance between wavecrests (wavelength).

Electromagnetic radiation originates in the sun, in the earth, or can be generated by artificial sources. The sun's spectral energy distribution has a maximum at a wavelength of 0.5 micrometer (a wavelength close to the peak sensitivity of our eyes). Solar radiation is largely unaffected as it travels through space, but is selectively scattered and absorbed by the earth's atmosphere. The atmosphere scatters blue wavelengths of visible light four times more than red wavelengths. The earth's surface is illuminated by EMR from the sun (fig. 2) that is not scattered or absorbed by the atmosphere (sunlight), and by atmospherically scattered solar EMR (skylight). The dominant spectral component of sunlight is yellow, whereas, that of skylight is blue. Sunlight and skylight (daylight) combine to produce white light that has spectral characteristics that change throughout the day. Because the sun's radiation must travel through a longer atmospheric path at sunrise and sunset, the atmospheric scattering causes the sun to appear more red and the sky overhead to appear more blue at those times of the day. Scattered blue skylight illuminates topography that is not directly illuminated by the sun (topography in shadow).

Electromagnetic radiation from the sun that is incident on the earth's surface is absorbed, reflected, and in some cases, transmitted through materials. Natural EMR emitted by the earth's surface comes mostly from solar energy absorbed in the visible and near-visible portions of the electromagnetic spectrum (EMS), and secondarily from local geothermal sources (volcanoes and hot springs). Electromagnetic energy is also radiated through the earth's skin as heat is conducted from the earth's hot interior, however, this amount of thermal EMR is small compared to that furnished by the sun. Absorbed solar energy is mostly reradiated from the earth's surface in the thermal (3 to 5 micrometer) portion of the EMS. Except for geothermal areas, variations in the earth's surface temperature are mostly related to the angle at which the sun's radiation strikes the earth's surface and the manner in which this energy is absorbed and reradiated by earth materials.

The process of "reflection" occurs within one-half wavelength of a material's surface, in the molecular structure of the material, and results in the instantaneous reradiation of EMR. The amount of EMR reflected from the earth's surficial materials depends on the spectral distribution and intensity of incident energy, the angle of incidence of incoming radiation, the orientation of topography, the roughness of the topographic surface with respect to incident radiation, and the absorptive characteristics of the materials.

767

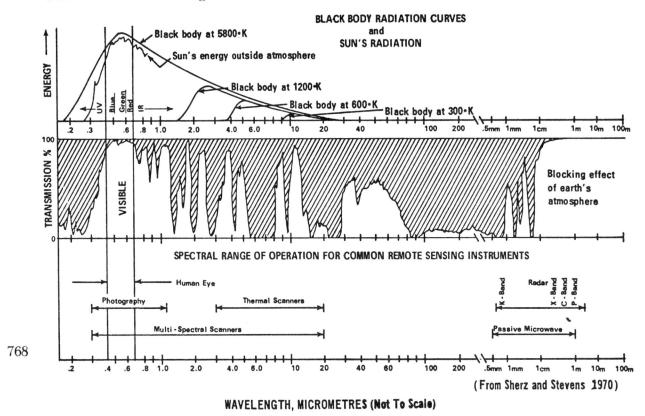

BLACK BODY RADIATION CURVES
and
SUN'S RADIATION

(From Sherz and Stevens 1970)

WAVELENGTH, MICROMETRES (Not To Scale)

768

Figure 1.—Electromagnetic energy spectrum.

Specular (mirror) reflection occurs when incident EMR strikes a surface with irregularities (roughness) of dimensions many times smaller than the wavelengths of incident EMR. If a surface has irregularities of dimensions close to or larger than those of the wavelengths of incident EMR, diffuse (hemispherical) reflection may occur. At visible and near-visible wavelengths most natural surfaces behave as mixed reflectors and have the properties of both specular and diffuse reflectors. On horizonal surfaces some incident EMR is scattered forward, away from the direction of illumination, and some incident EMR is scattered back in the direction of illumination; the forward scatter is dominant for most natural materials. The amount of EMR reflected to a remote-sensing system is related to the orientation of small (with respect to incident wavelengths) specular reflection surfaces, the angle of incident radiation, and the angle of observation. Consider a relatively flat, granite outcrop; most of the individual reflecting surfaces may lie almost horizontally, some may dip gently away from or toward the source of illumination, and a few may face directly away from or toward the source of illumination. The distribution of reflected radia-

tion from small surfaces oriented in this manner is shown in figure 3. The longest vector represents energy that is specularly reflected from the almost horizonatal surface.

Detection of Electromagnetic Radiation

The most commonly used geological remote-sensing tools only detect EMR emanating from the upper millimeter of the earth's surface (fig. 4), and, except for snow, ice, clear water, grass blades, brush, leaves on trees, and very dry materials in deserts, there is almost no transmission of EMR through earth materials. In the visible and near visible portions of the electromagnetic spectrum, remote sensors detect reflected radiation, whereas, the earth is the source of passive EMR detected in the thermal infrared through the microwave (3.0 micrometer to 3 meter) portion of the electromagnetic spectrum.

Active microwave (radar) systems produce their own EMR which is transmitted from an antenna toward the earth's surface at an angle less than 90 degrees. Radar EMR is scattered, absorbed, and reflected back to a re-

130

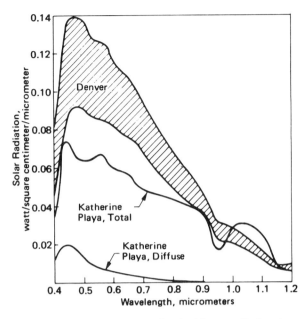

Figure 2.—Variations in solar incident radiation between 0940 and 1430 hours in the spring near Denver, Colorado (shaded); upper curve is for 1430 hours; curve for Katherine Playa, New Mexico, shows total solar radiation (daylight) for a typical clear winter day in a dry atmosphere; skylight curve for Katherine Playa is typical for the scattered (diffuse) part of solar radiation. From NASA Skylab Earth Resources Data Catalogue, Johnson Space Center, Houston, Texas, 1974.

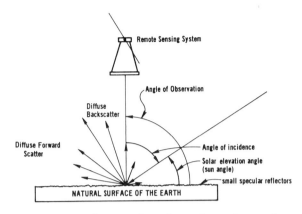

Figure 3.—Reflection of incident solar radiation from natural surfaces.

ceiving antenna, the returns depending on the angle of incidence, orientation of topography relative to the transmitter, surface roughness, and the electrical properties of the earth-materials present.

The radiation type, wavelength, frequency, usual source, and the usual methods of detection of the EMS are summarized in table 1.

769

Table 1.—Electromagnetic spectrum for geologic remote sensing

Kinds of Waves	Wavelength (cm)	Frequency (Hz)	Usual Source	Usual Method of Detection
Ultraviolet	$3-4 \times 10^{-5}$		Disturbance of intermediate electrons	Fluorescence chemical effect
		$f = \dfrac{c}{\lambda}$		
Visible		$3 = 3 \times 10^{10}$ cm/sec	Disturbance of valence electrons	Eye, photochemical effect; photo detectors
Blue	$4-5 \times 10^{-5}$			
Green	5-6 "			
Red	6-7 "			
Infrared			Disturbance of atoms and molecules Vibration and rotation molecular motion	Photochemical effect for near visible semiconductor detector
Near (photo)	$0.7-1.5 \times 10^{-4}$			
Thermal				
(1)	3.5-5.5 "			
(2)	8.0-14 "			
Microwave			Molecular motion Electrical resonance of tuned circuits antenna	Diodes Solid state crystals, antenna
Passive (Thermal)	0.05-100	600-0.3 GHz		
Active (Radar)	0.10-300	300-0.1 GHz		

Units commonly used in geological remote sensing are summarized in table 2.

Material Responses

Remote detection of a spectrum of different EMR wavelengths emanating from the earth's surface allows different properties of earth materials to be evaluated (fig. 5).

Topographic Responses

The topographic arrangement of the earth's surface can have a significant affect on the amount of electromagnetic radiation reflected and absorbed by it, especially when incoming illuminating radiation does not strike the surface at right angles. Consider a ridge illuminated by solar radiation. If the ridge is along the ground track of the sun, both sides of the ridge will be equally illuminated throughout the day. If the ridge is oriented at right angles to the ground track of the sun, in the morning the side facing the sun will be brightly illuminated while the side facing away will be in the shade. These conditions will change until both sides will be equally illuminated at noon, and conditions will be reversed at sunset. In the early morning and late afternoon the sun's rays have a small angle with the earth's surface. Photography taken under these conditions is referred to as low-sun-angle photography. Topographic trends not oriented along the sun's ground track will be subjected to varying degrees of differential solar illumination. Areas at latitudes progressively north or south of the location where the sun's rays strike the earth perpendicularly at noon are illuminated at progressively smaller (lower) sun angles for a particular time of day. At higher latitudes, during winter months, the sun's rays strike the earth's surface at lower angles than in summer months. Solar-elevation angle relationships for the Landsat satellite system are shown in figures 11 and 12. Low-sun-angle illumination of sloping topography is shown in figure 13. Topography in shadow is illuminated only by scattered blue skylight and by backscattered radiation from adjacent sloping topography.

770

Table 2.—Units commonly used in geological remote sensing

Unit	Use (Abb.)	Relationship To Other Units
Wavelength		Meters
Nanometer	Standard Unit (nm)	10^{-9} meter
Micrometer	Standard Unit (μm)	10^{-6} meter
Centimeter	Standard Unit (cm)	10^{-2} meter
Meter	Standard Unit (m)	meter
(Angstrom)	By physicists in gamma-ray, X-ray, and visible (A)	10^{-10}
(Millimicron)	By physicists, and electrical engineers (mμ) in visible. Discont.	10^{-9} meter
(Micron)	By physicists, and electrical engineers (μ) in thermal infrared use discontinued	10^{-6} meter
Frequency		Cycles per second 1 cps = 1 Hertz
(Picohertz)	Standard Unit (PHz)	10^{10} cps
Gigahertz	Standard Unit (GHz)	10^{9} cps
Megahertz	Standard Unit (MHz)	10^{6} cps
Kilohertz	Standard Unit (KHz)	10^{3} cps
Hertz	Standard Unit (Hz)	1 cps
Photometric and Radiometric Quantities		
Radiance	Used in reference to the amount of electromagnetic energy received, reflected, or emitted.	Watts per cm^2
Radiant spectral emittance	Used in reference to the amount of energy emitted over a definite wavelength interval	Watts per cm^2 per micrometer

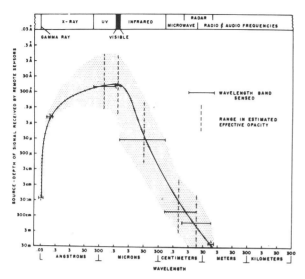

Figure 4.—Generalized absorption-coefficient curve for dry natural materials derived from terrestrial data; indicates estimated source-depth of information in received signal expected from lunar surface materials. In part after Badgley and Lyon 1964, N. Y. Acad. Sci., and Second Symposium on Remote Sensing of Environment, October 15-17, 1962, U. of Michigan.

Representative visible and near-visible spectral characteristics of natural surficial materials are summarized in figures 6, 7, 8, 9, and 10.

Analysis of Remotely Sensed Data

Most remote-sensor data used in geologic analysis consist of imagery, whether it is aerial photography, a multiband

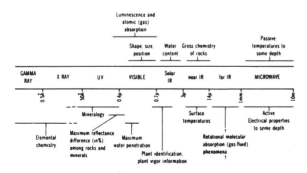

Figure 5.—Summary of types of information and properties of materials that may be interpreted form observations of various parts of the electromagnetic spectrum.

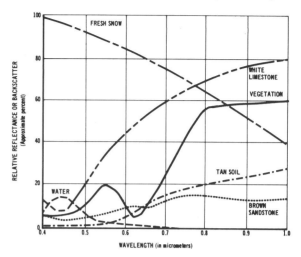

Figure 6.—General spectral characteristics of main cover types; some data from G. Moore, personal communication. Idealized curves.

color-additive display, or a digitally-enhanced cathode-ray-tube color display. Imagery displays the colors or tones representing detected spectroradiometric responses of surficial materials.

771

Image Formation

The Landsat system provides an example of how radiometric responses of the earth's surface are sampled, recorded, and displayed on imagery. The multispectral scanner (MSS) in Landsat measures portions of the flux of EMR from a 79 meter by 79 meter area at any instant in time; this measurement is made over 3,240 times along a 185 kilometer line. The resulting areas (taking overlap

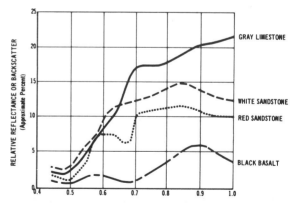

Figure 7.—Spectral characteristics of rock types; data in unpublished form furnished by G. D. Orr and from visual analysis. Idealized curves.

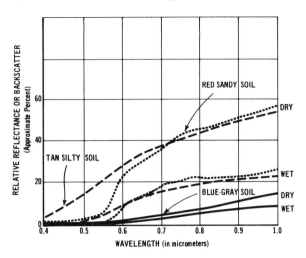

Figure 8.—Spectral characteristics of soils; data from H. R. Condit, 1970. Idealized curves.

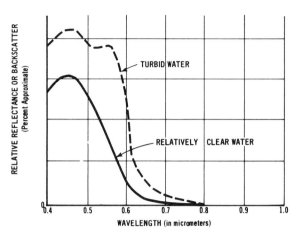

FIGURE 10.—Spectral characteristics of backscattered EMR from water. Data from Polcyn and Lyzenga, 1973 and G. Moore personal communication.

Figure 10.—Spectral characteristics of backscattered EMR from water; data from Polcyn and Lyzenga 1973, and G. Moore, personal communication. Idealized curves.

772

into account) are approximately 79 meters by 57 meters and each area is referred to as a *picture element (pixel)*. Approximately 2,256 lines are required to produce a single frame of Landsat imagery 185 km by 178 km; some 7.3 million pixels are present as the smallest resolution element on a single Landsat image covering 32,930 square km. In each pixel, the MSS has recorded spectral reflectance in four wave length bands:

Band 4: 0.5 to 0.6 μm (visible green)
Band 5: 0.6 to 0.7 μm (visible red)
Band 6: 0.7 to 0.8 μm (reflected solar infrared)
Band 7: 0.8 to 1.1 μm (reflected solar infrared)

The amount of energy recorded for each pixel (pixel brightness) in any one band is a function of the percent of each surface material within the area of the pixel and the amount of EMR reflected by each surface material.

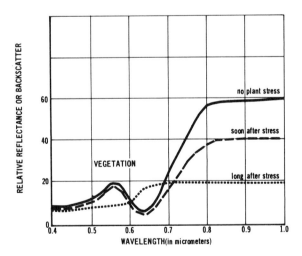

Figure 9.—Spectral characteristics of vegetation showing decrease in infrared response because of stress; data from *Remote Sensing in Ecology*, Philip L. Johnson, ed., 1969. Idealized curves.

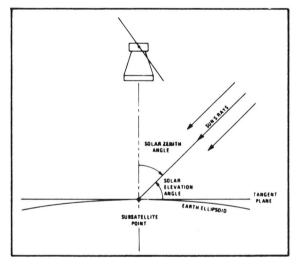

Figure 11.—Solar elevation angle diagram; from NASA ERTS Data Users Handbook.

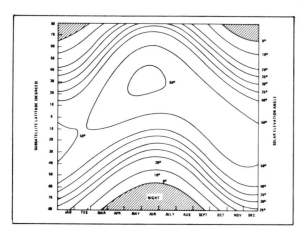

Figure 12.—Solar elevation angle history for the Landsat satellite system as a function of latitude. Satellite crosses at 9:42 a.m. (descending node); data from NASA Data Users Handbook.

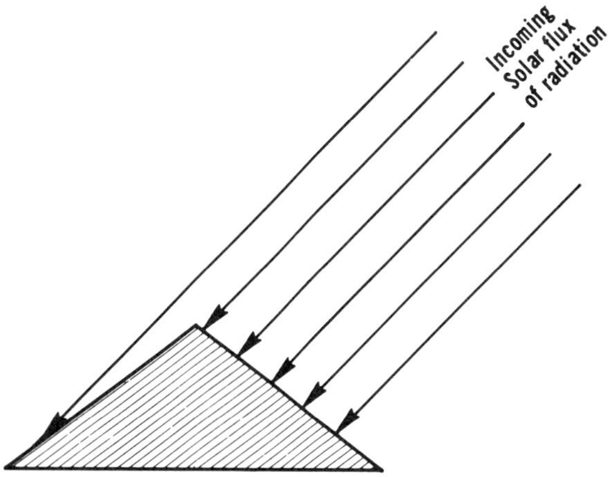

Figure 13.—Incoming solar flux of radiation striking topography in slope; slope facing the source of illumination receives four times the energy per unit area as the slope facing away.

Image Scale and Image Resolution

Image scale is defined as the ratio of the measured distance between two points on imagery to the measured distance between the same two points on the ground; the smaller the value of the ratio, the smaller the scale of the imagery. A 1:1,000,000 scale Landsat image is referred to as small-scale in contrast to large-scale aerial mapping photographs at 1:20,000 scale. Areal coverage of an 18.5 by 17.8 cm Landsat image is 32,930 square kilometers, compared to over 18 square kilometers covered by a standard 9x9 inch (nominal size) aerial-mapping print. The combination of aircraft and satellite imagery now available provides a variety of different scales of imagery for analysis of the earth's surface.

Image resolution is defined as a measure of the smallest ground radiometric element that can be recorded on an image. Resolution of a radiometric element on imagery is a function of:

1. Dimensions of the ground element.
2. Difference in reflected or emitted radiometric energy between the ground element and its background.
3. Shape and orientation of the ground element with respect to illuminating radiation.
4. Resolving characteristics of the imaging system. The system's spectral, radiometric, and spatial (areal) resolving power.
5. Location of ground element with respect to nadir.

Detection of a resolved radiometric element by an analyst depends on the following:

1. Contrast in reflected or emitted radiometric energy between the radiometric element and surrounding radiometric background.
2. Radiometric uniformity of surrounding background against which the radiometric element is imaged.
3. Areal extent of the radiometric background.
4. Regularity of the shape of the radiometric element.
5. Ratio of the radiometric element's length to width.
6. Regularity of groups of similar radiometric elements.

Table 3 shows the size of areas resolved on the ground increases as image scale decreases for photographic systems. Radiometric elements resolved on imagery are not necessarily detected by analysts. Even when an interpreter recognizes a radiometric element, it does not directly follow that they are easily identified objects. Detected radiometric elements that are of a size and radiometric nature close to the limits of system resolution are usually difficult to identify.

Integration of Spectral Response

The size of the minimum area resolved by an imaging system can be important in the detection of spectral characteristics of surficial materials. If brown sandstone, tan soil, and green vegetation were present in equal amounts in an area covered by a Landsat pixel, figure 6 shows that reflectance from green vegetation would mostly account for the digital value (brightness value) of EMR recorded in MSS bands 4 and 7. However, the digital value of the pixel would be more if there were 100 percent vegetative cover. Similarly, if the pixel contained

773

Table 3.—Ground resolution on imagery available for most natural cover types

Scale of Unenlarged Data	Type of Imagery	Usual Minimum Area Detected by Interpreters
1:20,000	Black and white mapping photography[1]	1 square meter
1:70,000	Black and white mapping photography[1]	8 square meters
1:120,000	Color reconnaissance photography[1]	20 square meters
1:950,000	Color Skylab photography (S-190B)[2]	400 square meters
1:3,370,000	Landsat Imagery (MSS)[3]	10,000 sq. meters

(1) Ground resolution is primarily limited by camera and film resolution.
(2) Skylab S-190B camera system has better resolution than most aerial mapping camera systems.
(3) Landsat system resolution is limited by the instantaneous field of view of the multispectral scanner.

equal amounts of only soil and sandstone then sandstone would mostly account for the value of EMR recorded in band 4, but soil would mostly account for the value recorded in band 7. In each case, the digital value in each band would be less than if the entire pixel area consisted of the material with the highest reflectance in that band.

Consider the same two conditions described above for bands 5 and 6. All three materials (vegetation, soil, and sandstone) have about the same average reflectance (about the same area under the curve). Thus, any mixture of these materials in the pixel area would have the same digital value, in bands 5 and 6, as a pixel area covered entirely by one material.

All imagery can be expected to possess the characteristics summarized above, but larger scale, higher-resolution imagery may more accurately define the spectral characteristics of surficial materials when the same wavelength bands are used; this is because the detected EMR is reflected from fewer different surface material types as the resolved ground area decreases.

Surface Geologic Analysis

Basic techniques for landscape analysis were developed by photogeologists over 50 years ago and were patterned for stereoscopic analysis of black and white, low-altitude aerial photographs. An automatic data processing system that will successfully analyze the enormous variety of surficial features has yet to be constructed. Therefore, geologists must rely on their experience and understanding of the fundamental principles of landform develop-

ment to successfully evaluate the significance of the earth's surface features controlled by stratigraphy, structure, weathering, erosion, and hydrology.

Cover Types

It has previously been stated that geological remote-sensing tools generally detect electromagnetic radiation that emanates from the upper millimeter of the earth's surface; interpreters of remote-sensor data cannot "see through" soils, grass, and unconsolidated earth materials in mapping consolidated rocks with geologic remote-sensing techniques.

The natural surface of the earth is composed of a diversified combination of cover types. Rarely are unweathered rock materials exposed at the surface. More frequently, consolidated surface rocks are altered by chemical and biological agents, are covered by unconsolidated materials, contain or are covered by water, or have soil mantles. Lichens often veneer bare rocks, or grasses, shrubs, and trees obscure the soils on which they have developed. Man often obliterates natural surface cover and, in its place, erects structures or plants crops. The ground-based geologist maps geological units throughout an area of interest by: (1) interpolating between rock exposures, (2) using rock debris in soils, (3) using residual soil associations, (4) using plant associations, and (5) projecting geometric attitudes of exposed rock strata through areas dominated by other cover types. *Because geological remote-sensing techniques can only measure EMR reflected or emitted from the earth's surface, geologists should understand the following about landscape surface cover:*

1. The physical and physiological characteristics of landscape cover types.
2. The effects of different climatic and physiographic environments on landscape cover types.
3. How electromagnetic radiation interacts with, and emits from, different landscape cover types.
4. The association of landscape cover types with geological relationships.

The major types of surface cover of geologic interest are:
1. Natural surface cover
   a. Consolidated rock cover
   b. Unconsolidated rock cover (alluvium, colluvium, etc.)
   c. Soil cover
   d. Vegetation cover
   e. Water cover (including snow)
2. Cultural surface cover
   a. Man-made structures (cities, roads, houses, etc.)
   b. Agricultural surface cover (crops, pasture, etc.)
   c. Altered surface cover (logging, stream channelization)

774

Topography

Differential illumination of the earth's surface topography by EMR has several major effects that influence the interpretation of remotely sensed data:

1. Topography facing the source of EMR will have a higher reflectance than topography facing away; this is because the amount of sunlight incident per unit area (radiance) is less on slopes facing away from the sun (see fig. 13). Note that the slope facing away from the sun will receive the same amount of skylight as the slope facing the sun. Both slopes will also be illuminated by wavelengths of EMR scattered from adjacent sloping topography. Thus, not only is the total amount of incident EMR (flux) different on opposed slopes, but the spectral composition of EMR incident on opposed slopes is usually different as well. Topography covered by a single landscape cover type that is differentially illuminated by EMR may appear to be covered by two spectrally different cover types. The effect of differential illumination can mask real reflectance differences for materials on opposed slopes. The amount of radiation absorbed on opposed slopes will also be different under the above conditions, and thus the same relationships can exist for emitted heat radiation.

2. Differential illumination by the sun at higher latitudes causes different ground-cover types to develop on slopes facing away from the sun, as opposed to slopes facing the sun.

3. The shadowing effect caused by differential illumination of topography by EMR may greatly assist interpretation of remotely sensed data, particularly if these data are not in a format that may be viewed stereoscopically; this is particularly true for topographic elements not oriented parallel to the direction of illumination. Detection and identification of landforms oriented parallel to the azimuth of incoming, illuminating EMR may be more difficult because both sides of the features are equally illuminated; thus, an interpretive bias may be introduced when differentially illuminated terrain is to be evaluated.

Landscape Patterns

Landscape patterns are composed of elements that indicate physical, biological, and cultural components of the landscape. Similar conditions in similar environments produce similar landscape patterns, and unlike conditions are expressed by different patterns. Image interpretation involves a step procedure summarized as follows:

1. Detect, delineate, and classify radiometric data displayed on an image.

2. Recognize symmetrical distribution of radiometric elements (patterns) on imagery.

3. Identify landscape surface characteristics through systematic pattern analysis of relief, landforms, drainage, and cover types.

Some image characteristics which allow geologists to evaluate the radiometric attributes of remote-sensor data are summarized below:

1. *Tone: The degree of brightness, ranging from dark to light.* A degree of brightness is often referred to as a "gray level" and is considered to be a relative measure of the amount of EMR reflected or radiated from materials on the earth's surface. The average interpreter can distinguish about eight "gray" levels on imagery, and up to 16 levels if the tone changes are abrupt.

2. *Color: Visible color results from the interaction of specific wavelengths of visible EMR with the eye.* When blue, green, and red wavelengths of visible EMR are presented to the eye in equal amounts, the result is white. Hue (magenta, purple, etc.), brightness (light blue vs. dark blue, etc.), and saturation or color purity (green vs. blue-green, etc.) are three variables used to describe colors. The human eye can distinguish about 8,000 hues of color.

3. *Texture: Texture on imagery is created by the frequency of tonal or color change, and the abruptness of changes.* Texture is produced on imagery because some resolved radiometric elements, that are individually too small to be identified, produce subtle variations in the amount of EMR emanating from surface cover.

These radiometric attributes are individually uninformative with respect to geological analysis of imagery; however, their arrangements (their patterns) on imagery allow geologists to deduce the following landscape characteristics:

1. *Relief*: degree of dissection of the landscape.

2. *Landforms*: size, shape, position, and association of topographic elements in the landscape.

3. *Drainage*: drainage patterns, drainage density (drainage texture), cross-sectional geometry of valleys and width of stream channels.

4. *Cover types*: spatial arrangement, symmetry, variability, and association of landscape cover types.

To those experienced in photo analysis, these listings are not new. The importance of understanding the significance of landscape pattern elements cannot be overemphasized when useful geological information is to be extracted from remote-sensor data. A partial list of key

775

references on the subject of systematic landscape pattern analysis are included at the end of this paper.

Bases for Geologic Interpretation from Image Data

Geologic interpretation of remotely sensed image data ultimately depends on four factors:

1. The geologist's understanding of the fundamental aspects of image formation.
2. The geologist's ability to detect, delineate, and classify radiometric image data; recognize patterns; and identify landscape surface characteristics as expressed on imagery.
3. The geologist's ability to interpret geomorphic processes from their static, surface expression as landscape characteristics on imagery.
4. The geologist's ability to conceptualize dynamic processes both above and below the surface responsible for the evolution of features expressed on imagery.

776 In the interpretive process, the first two factors in the foregoing list are directly related to static elements displayed on the imagery. Correct identification of landscape characteristics depends on an understanding of the physical principles governing the interaction of EMR with the earth and its detection by a remote-sensor system. Interpretation of the static elements displayed on imagery, is referred to as an *image interpretation* and defines the surface characteristics of the coverage area. In contrast to the first two factors, the last two are related to both static and dynamic elements that are not detected by the sensor and are not displayed on imagery. If all four factors are applied in interpreting imagery, a *geologic interpretation* is derived. A geologic interpretation integrates *surface* characteristics of the landscape with *subsurface* geologic relationships. Differences between the two types of interpretation are primarily related to the differences between identification of static surface relationships on imagery and interpretation of surface processes and subsurface relationships. The third factor in geologic interpretation from imagery relates relief, landforms, drainage, and cover types to the temporal aspects of fluvial, glacial, eolian, extrusive igneous, or surficial gravity processes. The last factor relates all of the factors previously considered in image interpretation and stratigraphic, structural, intrusive, and metamorphic attributes to the temporal aspects of tectonic, igneous, diagenetic, and metamorphic processes.

Interpretation of remotely sensed data on imagery is, in many respects, analogous to interpretation of ground-based data. Geologic interpreters detect, delineate, clas-

sify, recognize, identify, project, and conceptualize in an orderly fashion. The geologist can generally "see" no deeper into surficial cover types than a remote-sensing device; the only difference is the *relative scale* of observation. Remote-sensing imagery instantly provides data from areas many times larger than the ground-based geologist can observe. Regardless of what or who has collected the data, a geologic interpretation of what cannot be seen can be derived only by an individual trained in the fundamentals of geology.

Integration of Different Scales of Analysis

Remotely sensed data are analyzed through recognition of symmetrical distributions of image elements (tone, color, that is, data) displayed on imagery; symmetrical distributions of the elements are referred to as *patterns*. Patterns are differentiated on the basis of their size, shape, and internal organization or homogeneity. If "similar patterns reflect similar static conditions," then the similarity of static conditions reflected by two similar patterns must include pattern sizes.

During the course of analysis, it is both convenient and efficient for the analyst to proceed by first looking for the largest patterns displayed on the imagery, and then within these patterns, attempt to recognize the next successive set of smaller patterns. Through this analytical approach, it is possible to associate each larger pattern with the set of smaller patterns that comprise it. The association of a larger pattern with its constituent smaller patterns is important for identification of the larger pattern. An example of this association applied to the identification and interpretation of patterns is as follows. A flat-bottomed valley, identified on the basis of a large pattern's organization, shape, and size may not be interpreted as a floodplain until the analyst identifies a river system with natural levees from smaller patterns within the valley. The necessity of identifying patterns within patterns, etc., and associating patterns, to derive a valid interpretation of the larger pattern, is evident.

The term "scale," when applied to the analysis of patterns on imagery, is different from the scale of the imagery itself. Several "scales of analysis" are possible on imagery of a given scale. The closer an analyst is to an image, the more detail he observes; the farther he is from the image, the less detail he observes. The amount of data on the image is constant and is limited by the resolving capabilities of the remote-sensor-image system. In contrast, the amount of data on the image detected by the analyst is determined by the resolving power of the analyst's eyes, his distance from the image, and by his position and angle of viewing.

It is important to note that image scale has no bearing on the "scale of analysis"; however, dimensions of the image and the smallest image data element impose upper and lower limits, respectively, on the "scale of analysis." If a pattern on an image cannot be identified and cannot be subdivided into smaller constituent patterns, either an enlarged image or higher-resolution larger-scale imagery may be required. See figure 14 for several types (and scales) of imagery in obtaining recognizable patterns that may be continuously analyzed on successively smaller scales of analysis.

Geologic Importance of Temporal Analysis

In some areas, landscape characteristics change throughout the year primarily because surface conditions change with season. Prior to the launch of Landsat the importance of the time of year in data acquisition was not fully appreciated by geological analysts. The Landsat system currently consists of two data acquisition satellites that follow each other in orbit. If weather conditions permit, the two satellites may be used to collect ground data over approximately the same area every 9 days. Using this type of repetitive coverage, phenological changes in vegeta-

tion, changes in the amount of moisture contained in soils, rocks, and plants, and in some areas, changes in snow cover can be observed. Identification of lineaments is greatly assisted by low-sun-angle illumination in winter months and can be aided by thin, but continuous, snow cover. In arid regions, annual vegetation often thrives in locations where groundwater is near the surface. In the midwestern United States, gross Quaternary material types, formed during periods of continental glaciation, are best detected on imagery acquired in May, whereas, landforms are best detected in early January; this is because soil moisture variations are greatest in May and soil is tilled and bare at that time of year. In some areas, plant species closely associated with certain rock types and plant reflectance differences may be greatest in spring or fall. Remote-sensor data acquired at one particular time of year may best define certain surface characteristics while data acquired at another time best define others; both types of data may be required for a correct geologic interpretation.

Development of Geologic Models

In interpretation of geologic relationships developed from a particular scale of analysis, geologists are confronted with the problem of inferring relationships they cannot directly observe from any one point in the field; this problem is usually approached by constructing a three-dimensional, conceptual geologic model. The field geologist plots strikes and dips of rock strata on a map, and usually constructs structural cross sections in the field so as to predict where to establish important relationships. The development of a geologic model in the field usually requires a number of time-consuming, widely spaced observations. The model is developed from observations of outcrops and consists of a geologic map which displays an interpretation of geologic data in plan view, and structural cross sections which display interpretations of surface geologic data in cross-sectional view (below the surface). Remote-sensing imagery allows the geologist to view the landscape in a synoptic format, and if surface characteristics of the landscape are well defined on imagery and patterns of geological significance can be identified, geologic models can often be rapidly developed.

The development of geological models enables analysis of field data (and imagery) to identify key areas that require analysis in greater detail (at higher resolution and larger scale). An efficient approach to the analysis of an area that is ultimately to be evaluated by ground-based methods is to employ several scales of image analysis, and eventually to integrate information derived from these analyses with data derived from field investigations. Usually geological models are developed and employed in a variety of different ways in geological analysis. Often the analyst has a

777

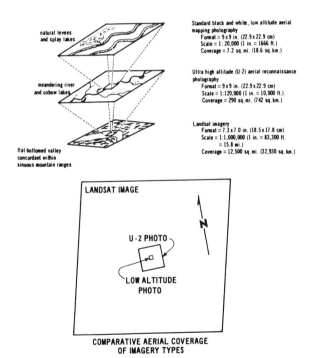

Figure 14.—Illustration of different scales of imagery that may be used to recognize patterns on successively smaller scales of analysis.

preconceived geological model in mind prior to analyzing the imagery. If the preconceived geological model was developed from suitable ground data and from experience in similar geologic settings, landscape patterns identified on imagery can be interpreted in terms of their correlation with the model.

The analysis of small-scale imagery that covers large areas sometimes presents problems to analysts with preconceived geological models in mind; this is because their geological experience and available ground data often are limited to small areas. An interpreter with a preconceived model in mind may introduce bias in his interpretation and incorrectly determine geological relationships, unless care is taken to carefully consider all factors of the interpretation process. Most geologists recall the statement, "you can usually go out into the field and find data to substantiate your preconceived point of view"; this statement applies also to the interpretation of imagery. A reasonable approach of eliminating an interpretive bias from an analysis of imagery is to begin with the smallest scale imagery available for the area to be analyzed. This imagery should be analyzed objectively by using a systematic pattern recognition procedure, and without the influence of ancillary geologic information (published maps, etc.); the analysis should first define regional trends and relationships (a model) and then should proceed to larger scales, using larger-scale imagery in key areas identified in the previous analysis, if this imagery is available.

Sometimes the interpretation of small-scale imagery can be revised (the model can be refined) through analysis of large-scale imagery. Eventually a point is reached where ground data or additional remote-sensing data are required in key areas to resolve conflicts in, or confirm the validity of, the interpretation. At this point the analyst should bring as much ancillary geologic information as possible (maps, etc.) to bear on the analysis and should determine if additional information can be extracted from the imagery. The careful image analyst will plot this additional information on a separate overlay because it is easy to interject information from an existing, and not necessarily accurate, map into that derived from interpretation of imagery; this separation also allows information not appearing on the maps, etc. (new information) to be evaluated.

At this point in the analysis, a conceptual geologic model should be defined, and key ground analysis areas should be identified. Remote-sensor data should be acquired over these key areas if such data will support field work and will provide additional information. Field visits to key areas may require that imagery interpretations and, consequently, the geological interpretation (the model) be revised. Often the entire analysis proceeds in an iterative fashion and culminates in large-scale, ground-based

778

geologic mapping of site-specific target areas. The technique economically conserves resources because the entire area under analysis does not have to be completely "walked out" on the ground.

Groundwater Exploration in Arid Environments

This section is included to illustrate how remote-sensing data might be used to develop a geohydrologic model for groundwater exploration in arid environments. Tucson, Arizona, was selected because ancillary data on groundwater occurrence is well documented. Landscape characteristics of the area that are identifiable on imagery and have geohydrologic significance are the following:

1. *Relief*: The area is characterized by mountainous bedrock areas and intermontane alluvial valley areas.
2. *Landforms*: Linear and curvilinear valleys with steep gradient occur in bedrock areas. Alluvial fans are present at the bases of mountains and a few playa lakes occur in the valleys.
3. *Drainage*: Bedrock areas have coarse, rectangular-dendritic drainage patterns. In some areas, adjacent to well developed bedrock valleys, alluvial fans show medium- to coarse-textured, collinear and parallel drainage patterns. The central portions of basins have very coarse, dendritic drainage patterns and floodplains are well developed.
4. *Surface cover*: Riparian vegetation is localized in some basinal areas, but patterns of relatively homogeneous vegetation occur on alluvial fans.

Figure 15 is a Landsat scene covering the Tucson area. The imagery was acquired under conditions of low-sun-angle illumination that occur during winter months. Note that relief, landforms, and drainage are well displayed on this imagery. Image interpretations of these landscape characteristics appear on figures 16, 17 and 18. Figure 19 is a geological interpretation of the probable extension of lineaments into basin areas. Figure 20 is a Landsat band 5 image which displays riparian vegetation, and anomalous vegetation patterns on alluvial fans. Figure 21 is an image interpretation of these vegetative surface cover types.

At this point in the analysis we can go no further without bringing in ancillary information, if only Landsat imagery is employed. This information can be summarized as follows:

A. Ancillary information ("facts"):

1. Bedrock areas were uplifted and basins were downdropped by normal faulting during the Cenozoic.

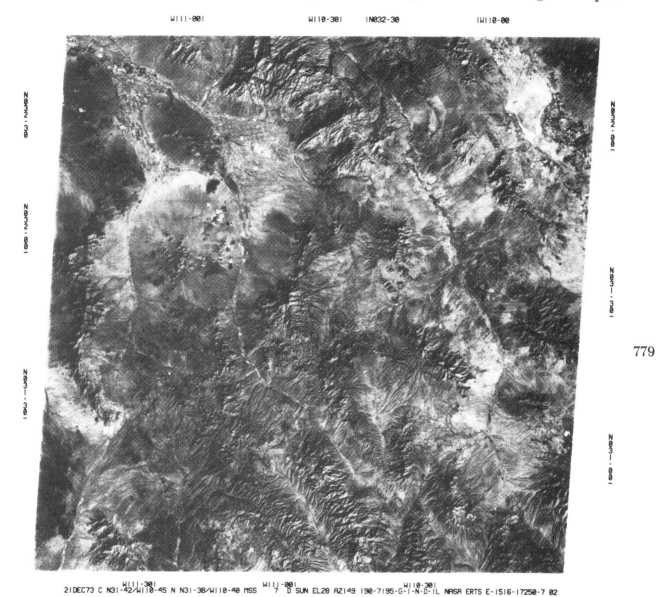

21DEC73 C N31-42/W110-45 N N31-38/W110-40 MSS 7 D SUN EL28 AZ149 190-7195-G-I-N-D-1L NASA ERTS E-1516-17250-7 02

Figure 15.—Landsat 1, band 7 (reflected solar infrared). Tucson, Arizona, appears near the upper left corner and Wilcox Playa in the upper right; image was acquired on December 21, 1973, under conditions of low sun angle illumination (28°). Note particularly the lineaments, drainage patterns, and drainage texture.

2. Valleys have been partially filled with alluvial gravel, sand, silt, and clay, mostly during the Tertiary.
3. Climate of the area was more humid in the Pleistocene than today.
4. Both stream flow and groundwater recharge come almost entirely from the melting of snowpack on the upper slopes of mountain areas.

5. Largest amounts of groundwater occur in coarse-grained alluvial valley fill.
6. Areas having abundant soil moisture are covered by dense healthy vegetation.

B. Assumptions:
1. Coarse-grained aquifers are now mostly covered by

Other Subsurface Investigations

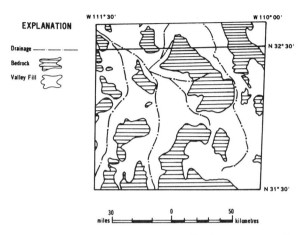

Figure 16.—Image interpretation of bedrock areas, valley fill, and major stream drainage. From Landsat band 7 image, 1:1,000,000 scale. Image acquired December 21, 1973. Landsat image identification: E-1516-17250. Refer to figure 15.

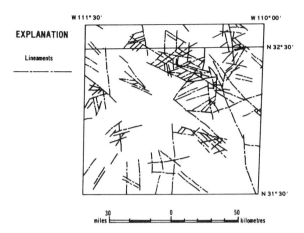

Figure 18.—Image interpretation of lineaments in bedrock and adjacent valley fill. From Landsat band 7, 1:1,000,000 scale. Image acquired December 21, 1973. Landsat image identification: E-1516-17250. Refer to figure 15.

780

finer-grained materials which were deposited in the present day arid environment.

2. Joints and faults influenced the distribution of coarse-grained materials in basin areas.

3. Lineaments mark the locations of joints and faults.

4. Coarser-grained sediments occur on the land surface in areas where drainage has a coarse- or medium-texture on Landsat imagery.

5. Most recharge occurs where relatively coarse-grained, porous alluvium contacts bedrock.

6. Groundwater moves down the lower mountain slopes, and down through basin areas in the same directions as surface streams.

7. Dense vegetation indicates areas where the water-table may be near the surface.

In development of a model for groundwater occurrence and movement in arid environments, it is important to separate facts from assumptions. By making this separation it is relatively easy to revise assumptions, as additional data become available.

Figure 22 is a geohydrologic interpretation of landscape patterns recognized on Landsat imagery. The figure is a two-dimensional display, based on a three-dimensional, conceptual geologic model, that can be used to predict groundwater recharge, movement, and discharge in the

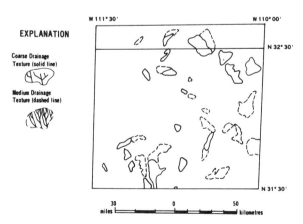

Figure 17.—Image interpretation of drainage density (texture). From Landsat band 7, 1:1,000,000 scale. Image acquired December 21, 1973. Landsat image identification: E-1516-17250. Refer to figure 15.

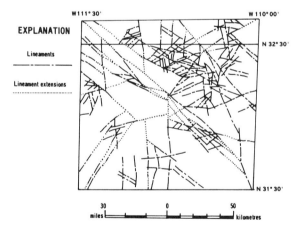

Figure 19.—Interpretation of lineaments by extending lineaments into valley fill along trends. This is a geological interpretation of image information; from figure 18.

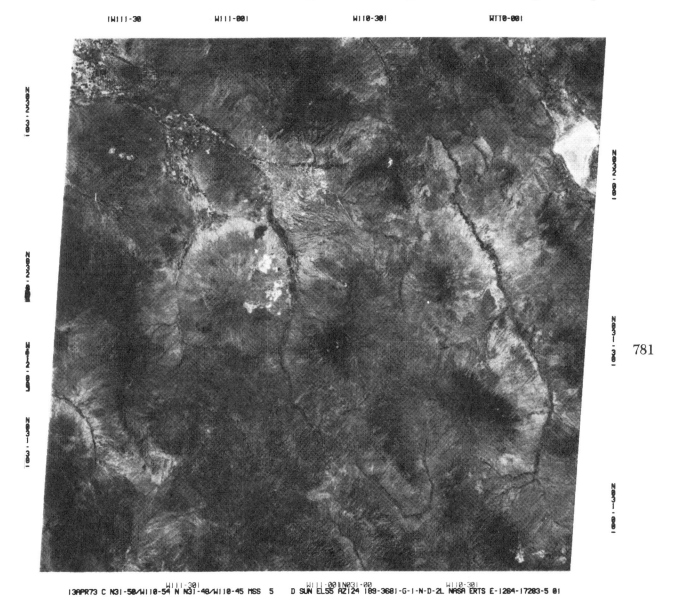

781

Figure 20.—Landsat band 5 image of Tucson, Arizona. Date of acquisition, April 13, 1973. Image identification: E-1264-17283. Riparian vegetation shows as dark, sinuous lines in basin areas. Vegetation anomalies on alluvial fans show as darker toned patterns. Actual photo interpretation of vegetation (fig. 21) was done using a Landsat color composite from the same date.

Tucson area of Arizona. Based on the model, two areas have been identified for exploration. The orientation of the proposed lines of exploration wells should define important relationships for further analysis. It should be noted that this model probably only approximates very gross groundwater relationships, and that it was developed primarily for purposes of illustration. The actual groundwater conditions in the Tucson area are much more complex. The proposed exploration plan, however, would have encountered major supplies of groundwater (Davidson 1973).

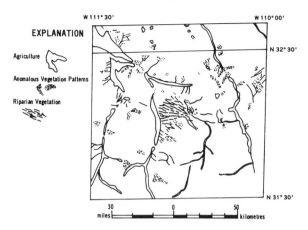

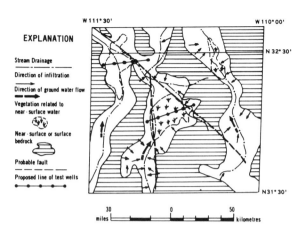

Figure 21.—Image interpretation of vegetation patterns. From Landsat color composite, 1:1,000,000 scale. Date of image acquisition, April 13, 1973. Landsat image identification: E-1264-17283. Note: figure 20, Landsat 5 band substituted for this image in this paper.

Figure 22.—Geohydrologic interpretation of landscape patterns recognized on Landsat imagery; interpreted from figures 16, 17, 19, and 21.

Targeting Mineral Exploration Efforts in Southwestern Idaho

782

The utility of remotely sensed imagery in providing data covering a broad area is demonstrated by this example of its application to base- and precious-metal exploration in the Pacific Northwest. A specific problem was defined in which a 36,974 square kilometer (approximately 14,443 square mile) area was to be evaluated with respect to volcano-tectonic environments potentially favorable for localization of mineralization.

In order to determine the scale, or scales, of analysis to be undertaken in evaluating the area of interest, the specific types of information required for an interpretation were considered. The most evident required information was:

1. Location of volcanic centers within the area.
2. Location of the major structural elements within the area.

With respect to the discrimination of environments potentially favorable for base- and precious-metal mineralization, the required information was further refined to include:

1. Location of *felsic* volcanic centers that indicate some degree of magmatic differentiation (the location of mafic and felsic centers that are either superposed or in close proximity).
2. Location and orientation of *ruptural* structures (faults, fractures, joints, etc.) associated with, or in close proximity to, felsic volcanic centers.

The effectiveness of methods employed in data acquisition, analysis, and interpretation depends on four parameters:

1. Type of data required.
2. Amount of data and areal extent over which data was to be acquired.
3. Time required to collect, analyze, and interpret the data.
4. Cost of manpower and materials required for the collection, analysis and interpretation of the data.

The type of data and area over which data were to be acquired was specified by the problem. The amount of data that could be directly acquired from surficial exposures was fixed in the area being evaluated. Regardless of the method by which the data were collected, they had to be relevant to the specific problem and had to be displayed in formats (map and image, respectively) that could be analyzed and interpreted.

Time required to collect, analyze, and interpret data varies considerably between ground-based techniques and remote-sensing methods; the difference is almost entirely a function of the time it takes to acquire the data in a format that may be analyzed. Remotely sensed imagery is currently available at several scales, in several formats, for any season of the year and may be acquired within a month (dependent on order processing and shipping times). Field acquisition of data and compilation of previously published data are considerably more involved methods of collection and usually do not result in an unbiased synoptic presentation of the data. In this respect, remotely sensed imagery was considered to be a time-saving method of acquiring relevant data in a format which could be efficiently analyzed for the defined problem.

The cost of two, black and white, 1:1,000,000 scale prints of Skylab photography covering the area of interest is $6.00. The cost of transparent overlay materials, pens, stereoscope, etc., for analyzing the data should not exceed $20.00. Seven man-days expended on analysis and interpretation of the data plus seven, subsequent, man-days spent in field checking critical analytical assumptions should not exceed $1,000 plus transportation costs. It was considered doubtful that the ground-based acquisition and formating of data alone (not including analysis, interpretation, and field checking) would cost less than the entire acquisition and evaluation expense of remotely sensed imagery for the specified problem.

The area under investigation in southwestern Idaho and southeastern Oregon is displayed on Landsat imagery, Skylab photography, NASA (U-2) reconnaissance photography, and U.S. Geological Survey standard aerial mapping photography. Both Landsat imagery and Skylab photography display the entire area on two adjoining prints, thereby affording the most synoptic (or condensed) coverage of the area. Regional structural trends and associated volcanic centers are resolved by both systems and may be recognized and identified on 1:1,000,000 scale prints. Skylab photography is more advantageous in structural analysis than Landsat imagery because it can be stereoscopically analyzed in its entirety due to the 50 percent overlap of adjacent frames. Skylab photography from August 1973 also provides greater contrast (because of low-sun-angle shadowing) of structural features. In southwestern Idaho, late summer and autumn photography is optimal for detection and consideration of structurally related vegetation and soil moisture patterns.

With regard to the foregoing considerations, two overlapping frames of Skylab photography were chosen for analysis. Black and white, 1:1,000,000 scale prints of adjoining August 8, 1973, scenes emphasized the surficial structural features required in the evaluation.

The analysis of the Skylab photographs required two man-days and involved the recognition and identification of patterns which were indicative of surficial volcanic materials and associated structural features. Skylab photos of the study area in southwestern Idaho and adjacent southeastern Oregon are shown in figure 23.

Volcanic features identified from the association of recognized relief, landform, drainage, and cover-type patterns included calderas, cones, shields, and both fissure- and vent-type flows. Only one caldera area was characterized by light tonal elements that may be indicative of felsic volcanic materials. This area is shown on figure 24 as a cluster of circular anomalies above the center of the illustration.

Structural elements, collectively termed lineaments, are also illustrated on figure 24. The sequence of ruptural discontinuities expressed by these "lineaments" is determined by their cross-cutting and offsetting relationships with identified landscape features. Only apparent strike-slip and approximated apparent dip-slip displacements could be determined from the photography. An analysis of structural trends within the area is shown on figure 25. Three major alignments were noted: N40W, N5W, and N85W. A weakly developed N40E trend was also observed.

Interpretation of volcanic and associated structural features within the area involved the correlation of the caldera with the structural symmetry, sequence, and apparent displacements. The resultant interpretation is illustrated in synoptic form on figure 26. Symbols in the figure represent the probable major directions of movement. The symbolized thrust elements include high-angle reverse faulting.

Ground confirmation of the interpretation derived from the analysis of Skylab photography is given in figure 27. Previous workers in the field have identified the major fault trends, but displacements are not well defined. 783

The integration of several additional scales of analysis using remotely sensed imagery resulted in the association of regional and local volcano-tectonic features as illustrated in figure 28. Diagram 3, within figure 28, represents the conceptualized, geologic model of features determined from Skylab analysis.

In this example of the utility of remotely sensed imagery, a specific problem was defined; the types of information required for the solution of the problem were deduced; the most efficient means of obtaining relevant data were decided upon; special conditions imposed by the data source, the area of interest, and types of information required of the data were considered; specific data were acquired; the data were analyzed in a systematic manner with special reference to the types of information and consequent scales of analysis required; and the resultant information was interpreted in terms of the static subsurface relationships and dynamic processes.

Other types of data and scales of analysis could have been acquired, evaluated, interpreted, and, subsequently, correlated with or contrasted against this initial interpretation. Additional confirmation of the geologic model could have been obtained if need, time, and money permitted. It should be realized, however, that any dynamic interpretation of static data must be substantiated by "ground truth" or field checking of critical surficial and subsurface relationships that have not been anticipated or

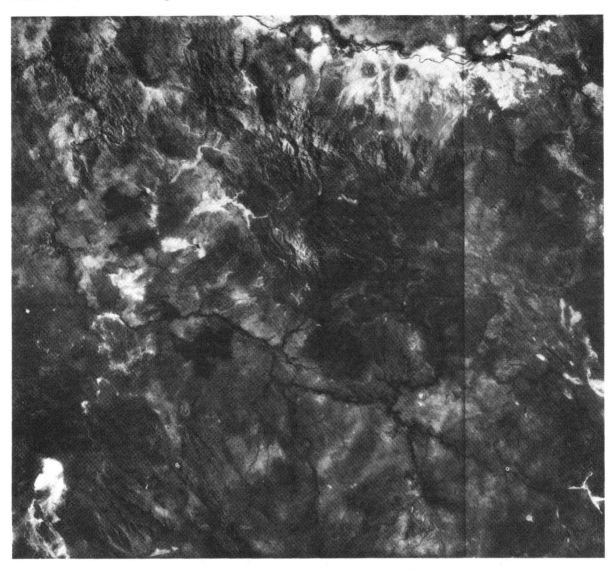

784

Figure 23.—Overlapped Skylab photos, identification numbers G30A020315 and G30A020316. Major rivers displayed on the photos are the Snake (top center) and Owyhee (top left to lower right); photos taken August 1973, by Skylab astronauts.

detected prior to interpretation. This field checking is required by both ground-based and remotely sensed data interpretation. The final test of any interpretation of a geologic model is not defined in terms of its simplicity, amount of data collected and evaluated, logical argument, or scientific principle, but rather by how accurately it describes the present three-dimensional geologic environment in its application.

Acknowledgments

Several colleagues made useful suggestions during the development of this manuscript. We are especially grateful to Gerald K. Moore and Donald G. Orr who provided some of the illustrations. The writers are thankful for the critical manuscript review of Robert G. Reeves, Gerald K. Moore, and Wayne G. Rohde.

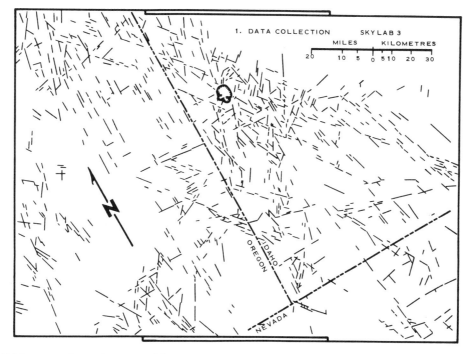

Figure 24.—Caldera complex and associated linear structural elements.

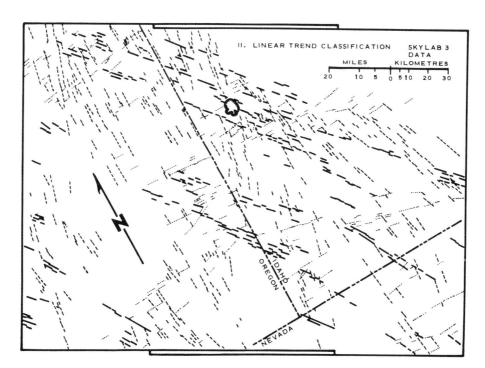

Figure 25.—Accentuated major structural trends; N40W = heavy, NSW = medium, N85W = light.

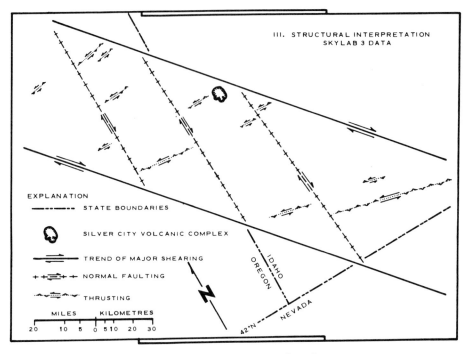

786

Figure 26.—Simplified volcano-tectonic interpretation showing dominant movement directions along major zones of weakness.

IV. GROUND CONFIRMATION

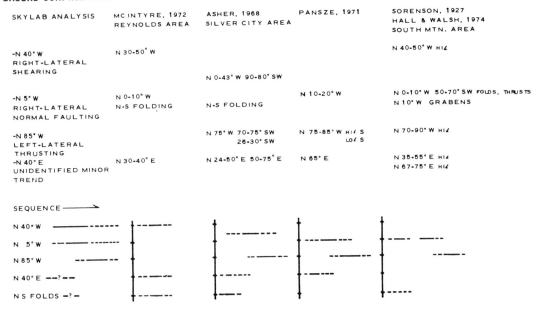

Figure 27.—Correlation of interpreted fault geometries and documented field evidence.

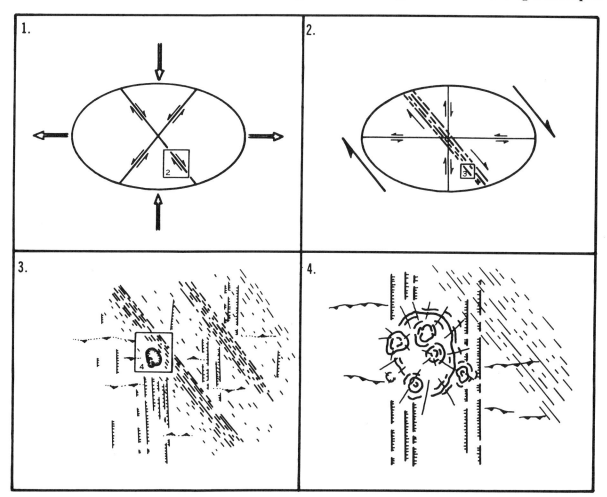

Figure 28.—Synoptic geologic model of interpretations based on several scales of remotely sensed data analysis: (1) small-scale regional pure shear system based on analysis of 1:5,000,000 scale Landsat band 7 mosaic of conterminous U.S.; (2) large-scale regional simple shear system based on analysis of four adjacent 1:1,000,000 scale Landsat band 7 images; (3) Skylab analysis of this example; (4) local analysis based on NASA reconnaissance color-infrared photography (four 1:120,000 scale prints).

Uncited References

The following references include articles on image formation, image interpretation, pattern recognition, and the use of remote-sensing techniques in geologic investigations. These references contain extensive bibliographies which can be effectively used to identify additional background materials.

Colwell, R. N., ed., 1960, Manual of photographic interpretation: Falls Church, Virginia, Am. Soc. Photogrammetry, 868 p.

Lueder, D. R., 1959, Aerial photographic interpretation: New York, McGraw-Hill, 462 p.

Miller, V. C., and Miller, C. F., 1961, Photogeology: New York, McGraw-Hill, 248 p.

Ray, R. G., 1960, Aerial photographs in geologic interpretation and mapping: U.S. Geol. Survey Prof. Paper 373, 230 p.

Reeves, R. G., ed., 1975, Manual of remote sensing: Falls Church, Virginia, Am. Soc. of Photogrammetry, 2144 p.

Von Bandat, W. F., 1962, Aerogeology: Houston, Texas, Gulf Publishing Co., 350 p.

Way, D. S., 1973, Terrain analysis: Stroudsburg, Pennsylvania, Dowden, Hutchinson, and Ross, Inc., 392 p.

Reprinted from Nicholas M. Short and Robert W. Blair, Jr, eds., *Geomorphology from Space: A Global Overview of Regional Landforms*: National Aeronautics and Space Administration NASA SP-486, 1986, p. 679-688.

Appendix A
Remote Sensing Principles Applied to Space Imagery

Nicholas M. Short *

To provide a better understanding of the interpretation and limitations of space imagery, this appendix presents a brief survey of relevant principles of remote sensing and of those spaceborne sensors that provide images in this book (Table A-1). For a more indepth discussion of remote sensing, the reader is referred to Lillesand and Kieffer (1979), Lintz and Simonett (1976), Sabins (1978), Short (1982), Siegal and Gillespie (1980), and the second edition of the Manual of Remote Sensing (1983).

Table A-1
Satellite Systems Providing Useful Geomorphic Images

Vehicle/Sensor	Spectral Bands (μm)	Nominal Spatial Resolution (m)	Areal Coverage (km)	Frequency of Coverage	Data Center
Landsat 1, 2, 3, 4, 5/MSS	0.5–0.6 0.6–0.7 0.7–0.8 0.8–1.1	79	34 000 km^2	Once every 18 days	EROS Data Center (EDC)
Landsat 3 RBV	0.50–0.75	30	98 × 98 km		EDC
Landsat 4, 5/TM	0.45–0.52 0.52–0.60 0.63–0.69 0.76–0.90 1.55–1.75 2.08–2.30 10.40–12.55	30	185 × 185 km	16 days	EDC
Heat Capacity Mapping Mission (HCMM)	0.5–1.1 10.5–12.5	500 600	700-km Swath	3 days	NASA/GSFC
Seasat SAR	1.35 GHz (L-Band)	25	100-km Swath	As scheduled	JPL/NOAA
STS (Shuttle) SIR-A	L-Band	25–100	100–200-km Swath	As scheduled	JPL/NOAA
Tiros N/AVHRR	4, 5 Visible Bands, IR, Thermal IR	1100–4000	Subcontinental	12–24 hr	NOAA/NESS
Large-Format Camera	Panchromatic, Stereo	10	Continental (480 × 180 km)	As scheduled	EDC/NSSDC

*Geophysics Branch (Code 622), Goddard Space Flight Center, Greenbelt, Maryland, 20771.

REMOTE SENSING AND ELECTROMAGNETIC RADIATION

Remote sensing as a technology refers to the acquisition of data and derivative information about objects, classes, or materials located at some distance from the sensors by sampling radiation from selected regions (wavebands) of the electromagnetic (EM) spectrum. For sensors mounted on moving platforms (e.g., aircraft and satellites) operating in or above the Earth's atmosphere, the principal sensing regions are in the visible, reflected near-infrared, thermal infrared, and microwave/radar regions of the EM spectrum (Figure A-1). The particular wavelengths (or frequencies) detectable by visible/infrared sensors depend in large measure on the extent to which the waveband radiation is absorbed, scattered, or otherwise modified by the atmosphere ("windows of transparency" concept). The radiation measured from space platforms is usually secondary in that it is reflected or emitted energy generated from molecular interactions between incoming radiation (irradiance) and the Earth material being sensed. Common primary energy sources include the Sun or active radiation-generating devices such as radar; sensed thermal radiation from the Earth's surface results from both internal heat sources and the heating effect of solar radiation. Because most materials absorb radiation over the sensed parts of the EM spectrum, only fractions of the incoming radiation (typically, 1/20th (for water) to 4/5ths (sand) in the reflected region) are returned to the sensor.

The spectral character of the source radiation depends on how it is generated. A spectral distribution plot shows the variation with wavelength of irradiance levels, usually measured as intensity or power functions (illustrated for solar irradiance in Figure A-2). This distribution is initially modified as incoming radiation interacts with the atmosphere. It is then further changed through interaction with the surficial materials (to depths ranging from micrometers to a few meters, depending on wavelength),

and the returned fraction is altered once more as it passes back through the atmosphere. Finally, the sensor itself modifies the returned radiation according to the response characteristics of the radiation-sensitive detectors. The end result is a spectral signature for each sampled section of the sensed surface, which is made by plotting the intensity or power variations of the final signal as a function of wavelength (Figure A-3). For the wavebands commonly used, the greatest modification is imparted by the interactions involving the ground materials, so that the signature is generally diagnostic of the particular substances or of objects composed of them. Targets of interest at the Earth's surface are usually an intimate mix of several materials (e.g., soil clays and rock particles, as well as water, air, and organic substances) or even several classes of materials such as soil plus trees plus grass

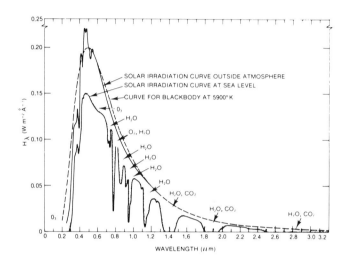

Figure A-2. Solar irradiation curves, showing location of atmospheric absorption bands; from Handbook of Geophysics and Space Environments, S. Valley (Ed.), copyright © 1965 McGraw-Hill (published with permission of McGraw-Hill Book Company).

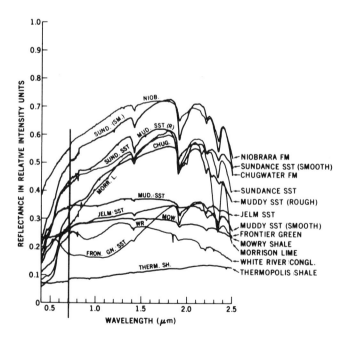

Figure A-3. Reflectance spectra of Wyoming rock stratigraphic units.

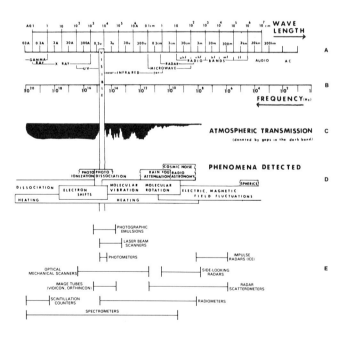

Figure A-1. The electromagnetic spectrum, atmospheric windows, and spectral operating range of sensors; modified from R. Colwell (upper diagram) and from Remote Sensing of Environment, J. Lintz, Jr., and D. S. Simonett (Eds.), 1976, with permission of the publisher, Addison-Wesley, Reading, Massachusetts (lower diagram, E).

plus manmade objects that are grouped together in an area of the ground. The size of the sampled area (target) is specified by the spatial resolution limit of the sensor (its instantaneous field of view as determined by the sensor's optics, electronics, etc.). On an image, this "area" is represented by the picture element (called a pixel); the pixel size for a Landsat Multispectral Scanner (MSS) scene is 79 m, and each full scene (for a band) is made up of 7.5 million pixels.

Spectral signatures obtained with a spectrometer, which uses a grating or prism to disperse a radiation continuum into discrete wavelengths, appear as continuous plots. More commonly, the detector/counter system operating in a moving sensor is capable of measuring only the radiation distributed over a finite wavelength interval or band. Band limits are determined by the transmission characteristics of a filter that passes only radiation of certain wavelengths. The detector integrates the distribution of spectroradiances into a single intensity/power value. If the radiation distribution is sampled at several intervals, the plots of these single values for their respective wavebands resemble a histogram (Figure A-4) that crudely approximates the spectrometer-produced signature. The more spectral bands sampled over a given spectral region, the closer the resultant signature will be to the characteristic signature of the object or material sensed.

REMOTE SENSING INSTRUMENTATION

The most common types of remote sensors are radiometers, multispectral scanners, spectrometers, and film cameras. The first two convert radiation (photons) emanating from each surface target into electrical signals whose magnitudes are proportional to the spectroradiances in the intervals (bands) sensed. The surface is usually sampled sequentially, as by a mirror scanning from side to side while the sensor moves forward on its platform. Each pixel contributes the collective radiation from the materials within it to the record as a single discrete quantity so that the converted signal is a measure of the combined ground target variation from one successive pixel area to the next. After being recorded, the signals can be played back into a device that generates a sweeping light beam whose intensity varies in proportion to the photon variations from pixel to pixel. The beam output is, in a sense, a series of light pulses, each equivalent to an average value for an individual ground target. Because the signal pulses were collected as an array of XY space in relation to their successively sensed target positions, the individual values can be displayed as a sequential series of points (pixels) of varying intensity. The image display may be on a television monitor, a film (pixels are represented by diffuse clusters of silver grains), or a sheet of paper

on which pixel intensities are indicated by alphanumeric characters or by spots of variable densities or sizes (exemplified by a newspaper photograph). Any of these displays produces an image of the sensed scene comprising the variations in tonal densities of the different objects or classes within it in their correct relative positions. Normally, scanning spectrometers must dwell on a single target long enough for the full spectral interval to be traversed and hence cannot be operated from a moving platform unless the target is tracked (as was done by an astronaut on Skylab). This type of spectrometer presents the spectral signature as a continuous curve on a strip chart or plate. Fixed prism or grating spectrometers usually show discrete spectral wavelengths as a dispersed sequence of lines (images of a slit aperture). The Jet Propulsion Laboratory has developed an airborne imaging spectrometer that uses a slit to pass radiation from the ground onto a multilinear array charge-coupled detector (CCD) to sense successive areas along a moving track. The film camera differs from the sequential sensor types in that it allows the radiation from all surface targets sensed at the same instant to strike the film (recorder) simultaneously in their correct positions as determined by the optics of the system.

A set of multispectral images is produced by breaking the image-forming radiation into discrete spectral intervals through the use of waveband filters (or other light dispersion or selection devices). If a surface material has high reflectance or emittance in some given interval, it will be recorded as a light (bright) tone on a positive film-based image. Conversely, a dark tone represents a low reflectance or emittance. Because the same material normally has varying values of reflectance or emittance in different spectral regions, it will produce some characteristic gray level (on film) in the image for each particular waveband. Different materials give rise to different gray levels in any set of waveband images, thus creating the varying tonal patterns that spatially define classes, objects, or features. Multispectral images of the same scene are characterized by different tonal levels for the various classes from one band to the next.

REMOTE SENSING DATA DISPLAYS

Before Landsat and similar multispectral systems were developed, the principal remote sensing data displays were nearly always aerial photographs. Aerial cameras typically employ panchromatic films that use the visible region of the spectrum from about 0.40 to 0.70 micrometers (μm). Use of a yellow haze (minus blue) filter, which prevents energy transmission below 0.51 μm, narrows the actual waveband interval to 0.51 to 0.70 μm in black and white aerial photography. Color aerial photographs are recorded on natural color or false-color infrared film. These operate on the color subtraction principle, in which the three color substrates or layers of a negative on development are yellow, magenta, and cyan, being sensitive to blue, green, and red light, respectively. However, when producing a "color composite" from individual waveband images, the color additive principle applies. Three such images are needed to make a multispectral color composite. Multiband photography utilizes several cameras, each consisting of a bore-sighted lens and a color filter that transmits a specific spectral interval or band through the optical train onto black and white film.

For each band, the film records the scene objects as various gray tones related to the visible colors (or other radiation) variably transmitted and absorbed by the particular filter. Suppose two objects, one red and the other green, are photographed by three bore-sighted cameras. Each camera's lens would be focused on its own film, with one fronted by a blue filter, the second by a green filter, and the third by a red filter. When all three lens shutters are triggered together, light from the red object (mostly in the 0.6- to 0.7-μm interval) will only pass through the red filter (being absorbed to varying extents by the blue and green filters).

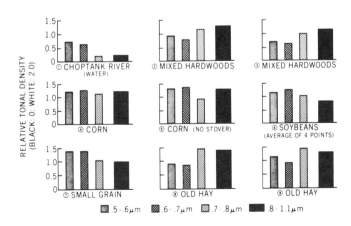

Figure A-4. Relative densities of ground class MSS signatures of nine cover types in the Choptank River, Maryland, area.

In the red filter camera, the shape of the object is reproduced as a light-toned pattern set off against dark (equivalent to non-red) surroundings. For cameras that record this object through green and blue filters, the red light object is absorbed, reproducing its presence on film as a dark tone. The green object is likewise recorded as light-toned only on the green filter/film combination. Obviously, a blue object, if in the scene, would have appeared in light tones only on the blue filter/film product. Non-primary colors would likewise be rendered as various gray tones in the three images, at levels depending on their relative transmission through each filter (e.g., yellow light normally will be only partially absorbed by red and green filters).

Color-composite photographs can be produced by passing white light successively through a primary color filter and each respective black and white transparency after all three images are superimposed and registered to one another on a color-sensitive film. The red-band image activates red color on the color film if projected through a red filter: light tones (clear in a transparency) representing red objects pass red-filtered light onto the film while screening out blue and green objects (dark or opaque in a transparency). Analogous results for blue and green objects (or a color mix) are obtained with blue and green filters. The resulting composite is a natural color photograph. When an infrared band transparency is projected through a red filter, and red and green bands through green and blue filters, respectively, vegetation, in particular, which is highly reflective (very bright) in the infrared, moderately reflective in the green, and low in the red because of absorption of red light by chlorophyll, will appear red in the color composite (little blue and almost no green contribution). Thus, in a false-color composite, red is almost always a reliable indicator of vegetation. Light-colored rocks or sand, which are generally bright in the infrared, red, and green bands, will be rendered whitish (with color tints) on false-color film because about equal amounts of blue, green, and red light (additively producing white) are transmitted through the light film tones associated with their spatial patterns. Specific gray tones in each of several multispectral band images or diagnostic colors in natural or false-color images can be used along with shape or textural patterns to identify particular classes of surface features or materials that compose them, as summarized in Table A-2.

MULTISPECTRAL SCANNER IMAGES

Images produced by the MSS on Landsats 1 through 5 and the Thematic Mapper (TM) on Landsats 4 and 5 use the image production system previously described. The MSS senses four contiguous spectral bands that cover sequentially the wavelength intervals from 0.5 to 1.1 μm; the TM includes three bands that cover nearly the same intervals as the MSS, together with a blue band and two additional bands in the near infrared (wavelengths not overlapping) and one in the thermal infrared. In those new near-infrared band intervals, many rock/soil materials are more reflective than vegetation, but certain materials (e.g., clays) show absorption in one or both bands. Examples of several TM band images reproduced in Figure A-5 are typical of multispectral images. Various combinations of three bands and color filters can produce a variety of color composites, some rather exotic and unfamiliar to most geoscientists (e.g., Panel 6 in Figure A-5).

Reflectances, emittances, and other radiation parameters measured by spaceborne sensors, after conversion to electrical signals, are commonly digitized on board before transmission to receiving stations. The digital numbers representing the radiance values can then be reconverted to analog signals introduced into image-writing devices that generate the individual band (black and white) images; the numbers can also be retained in digital format on computer-compatible tapes (CCTs). Minicomputer processing of the digitized data, using a variety of software-based special functions, yields new insights into the nature of the Earth's

surface materials. If numerical reflectance values for any two MSS or TM bands are ratioed, new sets of numbers result that often indicate the identities of the materials. Thus, red-colored mineral alteration zones should produce a high value when their red band digitial numbers (DNs) are ratioed to (divided by) their green band values. Other ratio values can be used to vary a beam intensity to generate a film product whose gray levels are proportional to the values. Combinations of band-ratio images and color filters give rise to distinctive color composites in which certain materials tend to stand apart in distinctive colors (see Figure T-8.1, for an example). A similar approach can be followed with images produced from Principal Components data. (See the *Landsat Tutorial Workbook* (Short, 1982) for details of the above techniques.)

Experience has shown that the larger geomorphic landforms are about equally well displayed in any of the four Landsat MSS band images and probably any of the six reflectance TM band images. However, expression of landforms in the TM thermal band image may be notably different, with lower overall contrast and lower resolution. On those images, shadows and Sun-facing slopes in mountainous terrain generally correspond to cool and warm (dark and light-toned) patterns. The tonal patterns in the reflectance band images show up best in semiarid to arid country, where vegetation is sparse to absent. Small-scale landforms and associated surface materials, such as fan outwash, can often be better discriminated from other landforms and materials by using select spectral bands, different color-composite combinations, or special process (e.g., ratio) images. In general, because the two infrared bands (6 and 7) on the MSS commonly show the best tonal contrast, they are used for the bulk of the black and white Landsat images comprising most of the gallery in this book. Bands 5 and 7 on TM frequently are even better, particularly in accentuating contrast, and are exemplified in several Plates. Specific information on the acquisition dates and conditions, band(s) used in the black and white and color images, and other characteristics of the Landsat MSS and TM, Heat Capacity Mapping Mission (HCMM), Seasat SAR, Shuttle Imaging Radar (SIR-A), and other space images shown in this book are documented in Appendix B. Guidelines for characterizing and interpreting thermal and radar space images that appear in this book are briefly surveyed in the following paragraphs; the reader is directed to Table A-1 for information on the systems pertinent to those images and to the *Landsat Tutorial Workbook* (Short, 1982) for a fuller discussion of the nature of these images.

THERMAL INFRARED

Thermal images are derived from sensors that detect emitted radiation within the 3- to 5- and 8- to 14-μm regions of the EM spectrum. The sensors measure radiant rather than kinetic temperatures; the values are less than direct contact (thermometer) temperatures by amounts determined by the emissivity (μ) of the surface materials. (Most rocks have emissivities ranging from 0.80 to 0.95.) The perceived radiant temperatures represent the effects of diurnal (daily) heating by the Sun's rays and subsequent cooling at night; internal sources of heat add only a small thermal contribution. The temperature variations during a heating/cooling cycle are largely controlled by the thermal inertia* of each ground constituent in the top meter or so (of soil, rock, or water), plus the influence of vegetation. Low thermal inertias result in large temperature differences over the cycle; high inertias involve small changes. Thermal inertia decreases with decreasing conductivity, density, and heat capacity. By convention, low radiant temperatures are shown as dark tones in a thermal image, with higher

---

*Thermal inertia is defined as the resistance of a material to temperature change during a full heating/cooling cycle.

154

Category	Best MSS Bands	Salient Characteristics
a. Clear Water	7	Black tone in black and white and color.
b. Silty Water	4, 7	Dark in 7; bluish in color.
c. Nonforested Coastal Wetlands	7	Dark gray tone between black water and light gray land; blocky pinks, reds, blues, and blacks.
d. Deciduous Forests	5, 7	Very dark tone in 5, light in 7; dark red.
e. Coniferous Forest	5, 7	Mottled medium to dark gray in 7, very dark in 5, and brownish-red and subdued tone in color.
f. Defoliated Forest	5, 7	Lighter tone in 5, darker in 7, and grayish to brownish-red in color, relative to normal vegetation.
g. Mixed Forest	4, 7	Combination of blotchy gray tones; mottled pinks, reds, and brownish-red.
h. Grasslands (in growth)	5, 7	Light tone in black and white; pinkish-red.
i. Croplands and Pasture	5, 7	Medium gray in 5, light in 7, and pinkish to moderate red in color depending on growth stage.
j. Moist Ground	7	Irregular darker gray tones (broad); darker colors.
k. Soils—Bare Rock—Fallow Fields	4, 5, 7	Depends on surface composition and extent of vegetative cover. If barren or exposed, may be brighter in 4 and 5 than in 7. Red soils and red rock in shades of yellow; gray soil and rock dark bluish; rock outcrops associated with large landforms and structure.
l. Faults and Fractures	5, 7	Linear (straight to curved), often discontinuous; interrupts topography; sometimes vegetated.
m. Sand and Beaches	4, 5	Bright in all bands; white, bluish, to light buff.
n. Stripped Land—Pits and Quarries	4, 5	Similar to beaches—usually not near large water bodies; often mottled, depending on reclamation.
o. Urban Areas: Commercial Industrial	5, 7	Usually light-toned in 5, dark in 7; mottled bluish-gray with whitish and reddish specks.
p. Urban Areas: Residential	5, 7	Mottled gray, with street patterns visible; pinkish to reddish.
q. Transportation	5, 7	Linear patterns; dirt and concrete roads light in 5; asphalt dark in 7.

temperatures being lighter. The HCMM thermal sensor produces day and night temperature distribution images of the Earth's land and sea surfaces, as well as thermal inertia images derived from these and the visible band image.

RADAR

Radar (radio detection and ranging) operates as an active system that provides its own illumination (thus it is all-weather and nighttime capable) as discrete pulses of energy in frequencies that lie within the microwave region of the EM spectrum. Wavebands in common use are the K-band (wavelength: 1.1 to 1.7 cm), X-band (2.4 to 3.8 cm), and L-band (15.0 to 30.0 cm). The effective resolution of a radar sensor depends on the mode of operation, physical dimensions of the antenna that transmits and receives signals, and subsequent data processing. Airborne systems usually use a linear real aperture antenna (5 to 6 meters long) and direct the radar beam off to the side of the aircraft (normal to flight path), hence the term "Side Looking Airborne Radar" (SLAR). Spaceborne systems, and some that are mounted on aircraft, use a smaller antenna that functions on the synthetic aperture principle, hence Synthetic Aperture Radar (SAR). (The SAR applies the Doppler effect to analyze variable frequencies that arise from relative motions between the sensor platform and ground targets.)

The typical mode of operation and character of signal return for a radar system are depicted in Figure A-6. The outward-sweeping beam scans a strip of surface elongated normal to the azimuthal direction of flight. Its length is set by the depression angle (measured from the horizontal) downrange along the look direction. Its complement, the incidence angle, is measured from the vertical. Photons in the energy burst interact with the ground

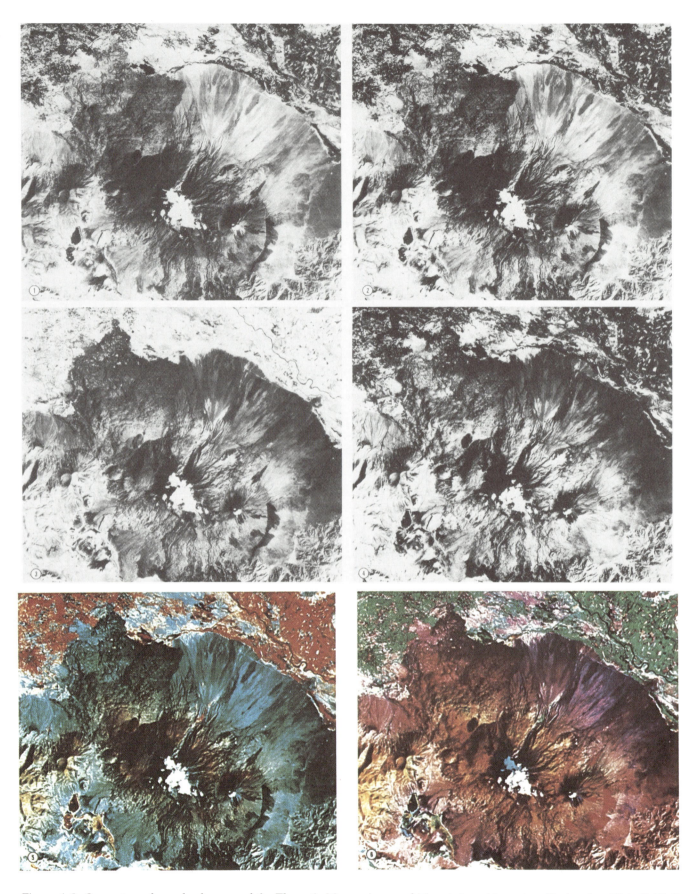

Figure A-5. Computer-enhanced subscenes of the Thematic Mapper image of Mount Ararat in eastern Turkey (see Plate V-17) illustrating different bands and color-composite images: (1) Band 2; (2) Band 3; (3) Band 4; (4) Band 5; (5) Bands 2 (blue), 3 (green), and 4 (red); (6) Bands 1 (blue), 4 (green), and 5 (red).

156

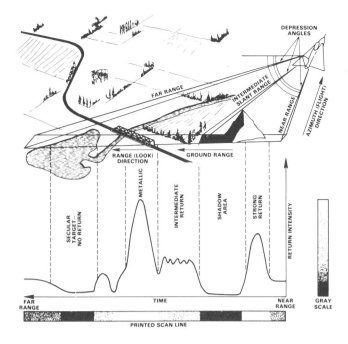

Figure A-6. Schematic diagram showing radar beam terminology and characteristics of returned signals from different ground features (modified from Sabins, 1978).

targets, creating new radiation, some of which is returned to the antenna where it is converted to an amplified electrical signal. The strength of that signal depends on a number of variables, mainly the geometry (shape) of the surface, the physical roughness of the surface relative to the wavelength of the pulse transmitted, and the dielectric constant of each material present in the target area. In the schematic diagram, the sensor-facing slope of the hill sends back considerable energy to the radar receiver, but the opposite slope is not illuminated by the beam, which causes a dark shadow. Plants and other vegetation scatter the radiation from their leafy surfaces, but a moderate amount is returned to the radar. The metal bridge consists of planar surfaces and corners, some oriented to efficiently return a high fraction of the beam. However, water, if not churned up by waves, acts as a specular reflector that redirects the radiation away from the receiver.

If surface roughness (as from pebbles or pits) has average dimensions much different (usually larger) than the radar wavelength, it acts to produce considerable backscatter that results in a strong (bright) signal return; a smooth surface relative to the radar wavelength generates a weak (dark) signal because of significant specular reflection away from the sensor. Likewise, because leaves may or may not interfere with the radar waves, depending on leaf size and on the radar wavelength, some tree canopies can be penetrated. Clouds are "transparent" because cloud-vapor droplets are too small to interact with most radar wavebands. However, ice crystals and raindrops may backscatter a signal. Reflection of radar radiation back to the antenna increases as the relative dielectric constant becomes smaller (3 to 16 for rocks and soils and 80 for pure water); a dry soil or sand (very low dielectric) permits penetration to depths of meters in some cases.

Radar images can be generated on recording devices from the electrical signals whose magnitudes are proportional to the returned radiation intensities. By convention, strong signal returns are printed as light tones and weak ones as dark tones (Figure A-7). Because of the proximity of airborne radars to the ground

Figure A-7. SAR image of central Pennsylvania, aquired during an ascending orbit (1260) on September 28, 1978, as processed on the digital correlator system at the Jet Propulsion Laboratory.

(as contrasted to the far greater distance of the Sun to a local surface on Earth), the geometry of radar-sensed features (such as hills) on the ground is more prone to distortions than that in images obtained with natural illumination. Distortions also vary as the depression angle changes from near to far range. Close in, foreshortening is expressed by an imposed asymmetry on forms such as ridges, so that the radar-facing slopes appear to steepen (the bright pattern becomes narrower) and the back slope broadens (dark pattern wider); this effect diminishes with decreasing depression angle. In the extreme, the facing slope appears to "layover" if its foreslope is greater than the look angle.

IMAGE ENHANCEMENT OF GEOMORPHIC SCENES

Many of the space images appearing throughout this book have been subjected to special computer-processing to more sharply define the geomorphic features and other geologic information that led to their selection. Although various techniques and operations can be applied to the data from which these images are constructed (see Condit and Chavez (1979) for a succinct summary of digital image processing or Short (1982) for a more in-depth review), several known collectively as *image enhancements* are generally the most useful to geomorphologists simply because they tend to improve the spatial display and characteristics of landforms in a scene. These are described briefly in this section, with emphasis on integrating the computer into the enhancement option.

Any experienced photographer is fully aware of the value of imparting an optimal *contrast*, or levels and spread of tone distribution, to a photograph. This usually involves expanding the number of discernible gray levels; the process is called contrast-stretching. The result is a picture both pleasing to the eye and, in scientific photography, effective in increasing the information content. At one or more stages in the entire photographic process, contrast can be influenced by various factors: at the time of picture-taking, such conditions as film type used, lens filters, illumination, and other exposure variables; in processing and printing, film development conditions, filters, properties of the printing paper, and exposure times.

The printing of a Landsat image is also affected by similar factors, but the importance of the film negative generated from the digital data is often paramount. Superior contrast can be introduced at some stage of negative production by manipulating the range of radiances (usually as reflectances) emanating from the scene and modified by the sensor. These radiances, expressed as digital numbers (DNs), fall within a range defined by gain settings and other sensor characteristics. For a Landsat MSS data set, the levels of brightness that can be detected by the sensor are digitized over a DN range of $(2^n - 1; n = 1$ to $8)$ or 0 to 225. Now, suppose the normally gaussian distribution of DNs representing the radiances from a given scene, after those are quantified and digitized, is 40 to 110. A photographic negative made to image this spread of values might have a limited range of gray levels within a particular film used, depending on the gamma (density transfer function) of the characteristic curve (X–Y plot of density D versus log exposure E) chosen; in other words, the negative may be "flat" and would produce a low-contrast positive print. Using an appropriate program, a computer can systematically expand (or contract, for a scene marked by a wide spread of radiances) the DN range so that a new negative will contain more of the gray levels that potentially are available within the density response capability of the film or print paper. (Of course, the benefits of such a stretch can be reduced by ineffective photoprocessing afterward.) "Before" and "after" images following stretch-processing by computer are exemplified in Figures A-8a–b. Various kinds of numerical stretching are possible in computer-based processing, including linear, stepwise linear, logarithmic, and probability distribution function expansions or contractions. Similar principles of stretching underlie the moderation of contrast on electronic image displays (such as a television monitor).

Another enhancement approach involves combining sequential DNs within a specified range into a single gray level or density. Adjacent DN ranges over the total distribution are treated in like manner, thus reducing the variation of many individual brightness values to a new set of much fewer values. This method, known as density slicing, produces a simplified pattern of varying tones in a black and white image; each composite gray level can likewise be assigned a discrete color to visually enhance the image display by color coding. Sometimes a surface feature, such as a landform type, has a unique or characteristic tonal signature and a narrow range of gray tones, allowing it to be separated as a particular pattern by slicing its mean level and spread from other levels.

A powerful enhancement technique that sharpens an image and can selectively bring out or delineate boundaries is addressed under the general term of *spatial filtering*. In any X–Y array of brightness values, like that of pixels in an image, the changing value can be considered to vary in a spatial as well as a radiance sense. This spatial variability can be expressed as frequencies (number of cycles of change over a given distance). A low spatial frequency represents gradual changes in the quantity (e.g., DNs) over a large areal extent of contiguous pixels; high spatial frequency results from rapid changes as only a few pixels in an area of the array are traversed. Variable line spacings in resolution test patterns are an artificial example of differing frequencies. (This can also be related to the concept of Modulation Transfer Function (MTF) that is a fundamental property of films or images.) Any image (or a complex harmonic wave) can be separated into discrete spatial frequency components by the mathematical technique known as Fourier Analysis.

Spatial filtering of an image can be done by scanning vidicons equipped with special functions; alternatively, the input data (DNs) from which images are derived after scanning (as by an MSS) are run through algorithm-based "filters" in a computer program designed to screen out or diminish certain frequencies and pass or emphasize other frequencies. To do this, a traveling "window" or "box" consisting of an array of n × m (line and sample) pixels is set up. This, in effect, creates a new value for each pixel as it passes during the computations into the center point of the moving array; the new DN for each such pixel depends on the brightness value frequency distribution of its neighboring pixels in the array size chosen. A low-pass filter ("band-pass" when the image data come from a spectral interval or band) tends to respond to features (including separated natural landforms such as dunes, divides, streams, fracture-controlled lakes, and series of folds) whose sizes and recurrence intervals (spacing) are larger than the averaging array (i.e., high-frequency spacings are not picked up). A high-pass filter reacts to features having dimensions and spacings smaller than the array. The new set of pixel values resulting from this will enhance (sharpen) those features whose periodicity or scale allows them to be enhanced. This filtered data set can be converted directly into a new image or can be combined (restored) with another image (such as the original one with its particular tonal patterns). The new image is usually contrast-stretched in the process to bring out density differences among the pixels in the array. Linear features such as rock or stratigraphic contacts and lineaments thus emphasized are said to be edge-enhanced. An example from the spatial filtering process is presented in Figure A-8c.

Although not an enhancement process in the strict sense, the ratioing of brightness values in one spectral band to those of equivalent pixels in another band yields a new set of DNs from which another image can be generated (Figure A-8d). For MSS data, this allows comparisons of relative reflectances between spectral intervals. Consider a ratio of bands A to B: a high value for A reflectances and a low for B produces a high ratio whose DNs would give rise to light-gray tones; conversely, low A and high B values cause low ratios and dark tones; similar A and B values lead to intermediate tones. Three sets of ratio images (e.g., A/B, C/A, D/B) can be combined into color composites with various colors often diagnostic of particular materials.

Identification of objects or features and materials by one of several methods of computer-directed *classification* can be treated as another means of the information extraction that is the ultimate goal of enhancement. The essence of the concept underlying classification is this: each identifiable class of object/material is considered to have one or more distinguishing properties with certain statistical parameters (usually means and

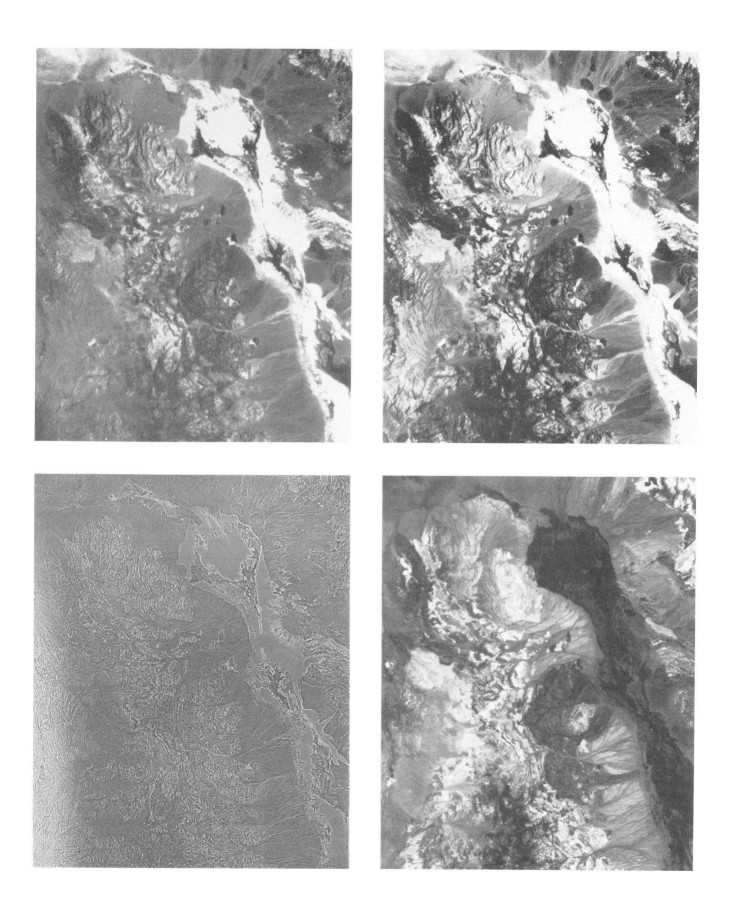

Figure A-8. Four products of image enhancement of a Landsat-5 Thematic Mapper subscene (50114-17550, June 23, 1984) of the Death Valley, California, area (see Plate T-5): (a) band 3 "raw" product (minimal enhancement), (b) band 3 gaussian stretch, (c) band 3 high-pass filter plus stretch, and (d) band 3/4 ratio.

variances) that can be demonstrated to be different (statistically) for other classes set up or recognizable in the data set being analyzed (such as natural terrains or land cover on a planetary surface). The classification program clusters property data into separable numerical sets (unsupervised classification) or obtains data characteristic of each known class in the scene by sampling those data at specific training sites (supervised classification). Once the specific classes are established from a fraction of the total sample points (e.g., pixels in an image), then all other (still unknown) sample points are assigned to some given class, identified by comparing their parametric characteristics (within the bounds of the statistical limits chosen) to those of the classes defined initially. Each unknown point is thus matched up with the class whose selected property or properties (e.g., radiance in an MSS image) is stochastically closest to it.

Classification of a Landsat image is usually based on spectral properties. An example of a geologic scene classified into rock/stratigraphic units by sampling the reflectances of each recognizable formation appears in Figure A-9. Other classifications can be devised to include spatial information (pattern recognition); although this has seldom been done yet for the geomorphic content of space imagery, in principle, it could be readily accomplished.

SITE 9

ALLUVIUM
MASUK SH
EMERY SST
BLUEGATE Sh
FERRON Sst
TUNUNK Sh
DAKOTA Sst
MORRISON Fm
SUMMERVILLE/
SALT WASH
ENTRADA Sst
CARMEL Fm
NAVAJO Sst
KAYENTA Fm
WINGATE Sst
UPPER CHINLE Fm
LOWER CHINLE Fm
SHINARUMP Cong L.
UPPER MOENKOPI Fm
LOWER MOENKOPI Fm
VEGETATION
WATER

SUPERVISED CLASSIFICATION IMAGE FOR TM
SUMMER WATERPOCKET SUBSCENE: 7 BANDS USED:
IN THIS VERSION NO ALLUVIUM WAS MAPPED ON
THE BLUEGATE MEMBER.

Figure A-9. Classification of a Landsat-5 Thematic Mapper subscene of the Waterpocket Fold (monocline) between Circle Cliffs and the Henry Mountains (see Plate F-6); classes are stratigraphic formations; training sites selected from several U.S. Geological Survey maps (Short and Marcell, 1985).

CLOSING REMARK

This necessarily brief exposition of the principles of remote sensing was designed to introduce those unfamiliar with interpreting sensor-created images to those main ideas needed to appreciate the varieties of space and aircraft multispectral, thermal, and radar images appearing throughout this book. For a fuller understanding, consult any of the references cited in the introductory paragraph of this Appendix.

REFERENCES

Condit, C. D., and P. S. Chavez, Jr., Basic concepts of computerized digital image processing for geologists, *U.S. Geol. Surv. Bull.*, **1462**, 16 pp., 1979.

Lillesand, T., and R. W. Kieffer, *Remote Sensing and Image Interpretation*, 612 pp., John Wiley, New York, 1979.

Lintz, J., Jr., and D. S. Simonett, *Remote Sensing of Environment*, 694 pp., Addison-Wesley Publ. Co., Reading, Massachusetts, 1976.

Manual of Remote Sensing, 2nd Ed. (R. N. Colwell, Editor-in-Chief), Amer. Soc. of Photogrammetry, Falls Church, Virginia, 1983.

Sabins, F., *Remote Sensing: A Better View*, 426 pp., W. H. Freeman & Co., San Francisco, California, 1978.

Short, N. M., *The Landsat Tutorial Workbook*, 553 pp., NASA RP-1078, U.S. Govt. Printing Office, Washington, D.C., 1982.

Short, N. M., and R. Marcell, New results for geologic units mapping of Utah test sites using Landsat TM data, *ERIM Fourth Thematic Mapper Conference*, San Francisco, Calif., in press, 1985.

Short, N. M., and L. M. Stuart, Jr., *The Heat Capacity Mapping Mission (HCMM) Anthology*, 264 pp., NASA SP-465, U.S. Govt. Printing Office, Washington, D.C., 1983.

Siegal, B. S., and A. R. Gillespie (Eds.), *Remote Sensing in Geology*, 702 pp., John Wiley, New York, 1980.

Geological Remote Sensing:
Landsat, Shuttle, and Beyond

JAMES V. TARANIK

*Dean, Mackay School of Mines
University of Nevada—Reno
Reno, Nevada*

The land satellite remote sensing program is now a decade old. Four land observation satellites (Landsats) have been successfully launched by the United States Government since the summer of 1972. The data produced by the highly sensitive earth-scanning instruments on these satellites are now routinely applied to a wide spectrum of natural resource problems by a diverse user community. Even though Landsat sensors were not developed specifically for geological applications, the energy and mineral industry is the most skilled and active group of industrial users of Landsat data.

The first three Landsat satellites were launched into polar orbits about 900 km above the earth's surface. From that altitude, their Multispectral Scanner (MSS) instrument detected reflected solar radiation in four wavelength bands from football field (80 m) sized ground areas. Multispectral Scanner data are formatted so entire images cover 185 by 185 km areas. A video camera flown on the third satellite with the MSS produced images with 40-m resolution (Taranik, 1978a). Data from the Landsats are distributed by the Department of the Interior's Earth Resources Observation Systems Data Center in Sioux Falls, South Dakota, under management of the National Oceanic and Atmospheric Administration, Department of Commerce.

In July 1982, a fourth Landsat was launched into a 705-km orbit with an advanced seven wavelength band sensor called

185

the Thematic Mapper. This multispectral scanning sensor covers the same ground areas as the MSS, but at higher spatial resolution—30-m ground resolution. The additional wavelength bands provide a valuable measurement capability for geological applications (Taranik, 1981).

The Carter Administration proposed that the operational portions of the Landsat program be transferred to private industry over a ten-year period; however, when the Reagan Administration took over, the Office of Management and Budget (OMB) decided that the Government's role in the land satellite program should end in 1985, if not sooner.

Experimentation, using the Space Shuttle as a platform for observations, has recently documented significant breakthroughs in geological mapping from space. Recently the funding for these civil remote sensing research programs has declined to a point where even small payloads cannot be developed for test using the Space Shuttle. Foreign governments have been following the results and fortunes of the United States space program closely, and when they recognized that the land satellite program was not going to be continued, they developed aggressive programs leading to the launch of major new land satellite systems in this decade (Taranik, 1982).

Significant Results of the Landsat Program for Geologic Applications

Landsat-1, -2, and -3 data have now been used for ten years in energy and mineral exploration. Many larger petroleum companies have a worldwide data base of Landsat images, and interpretation of these data is now a routine procedure used in identifying exploration targets. Both mineral and energy exploration companies have found that the large area coverage, the multispectral characteristics, and the seasonal coverage of Landsat data can be used to great advantage in delineating new structural features to identify abundances of rock materials and anomalous vegetation patterns which are indicators of mineralization and potential areas for hydrocarbon accumulations.

When Landsat data were first distributed to the geological

community, geoscientists discovered linear features displayed by Landsat imagery which extended for tens, if not hundreds, of km. The extent and pervasive nature of these linear features had not been previously recognized by geologists, and no geological models existed which could be used to satisfactorily explain them. Lineament mapping became the vogue. Intersections of lineaments were postulated to control mineralization or fracturing in the subsurface. Little geological expertise was needed to map and interpret the interconnectability of these features on imagery. Skilled photogeologists found that they could not duplicate the linear features delineated by others from imagery. Skepticism set in, and conservative geologists rejected lineament analysis as a useful analytic technique. Although the geological significance of these landscape features is still not completely understood, now many of them have been recognized as the surface expressions of zones of fracturing in the earth's crust that have been active for long periods of geological time. Often the displacements on these features are slight, and they are difficult to recognize on the basis of subsurface geologic indicators alone. Repeated movements throughout geologic time cause these fundamental fractures to be maintained, and as a geological basin evolves, they often control the alignment of streams and the distribution of reservoir rocks and create fracture porosity in sedimentary sequences. Explorationists have now found that many of these surface features display trends in the subsurface which are also substantiated by gravity and magnetic data, but are not well delineated by seismic data. Systematic linear features mapping from Landsat data is now accepted practice when used with geophysical and/or subsurface data to define exploration targets (Taranik and Trautwein, 1977).

For the past fifteen years, the U.S. Geological Survey (USGS) and the Jet Propulsion Laboratory (JPL) of the California Institute of Technology have been jointly investigating the applicability of Landsat data to geologic mapping for mineral resource investigations. This research has shown that iron oxide coatings can be detected on rocks and soils using computer processing of Landsat Multispectral Scanner data. The USGS and industry geologists now routinely use iron oxide abundance maps, derived from Landsat data, to guide their field investigations (Taranik, 1978b).

In 1976, a group of industry geologists formed a committee to advise the U.S. Government on their needs for global remote sensing data. This organization, called the Geosat Committee, now includes over 140 energy and minerals firms that contribute more than 25 percent of the gross national product of the United States. That same year the Geosat Committee and USGS geologists organized a workship in Flagstaff, Arizona, to document their remote sensing data requirements for the next decade. A similar Flagstaff workshop is planned for the summer of 1983. At the 1976 workshop, geologists realized that NASA (National Aeronautics and Space Administration) planned to launch Landsat-4 primarily for agricultural inventory, and they were concerned that the needs of geoscientists would be overlooked. Research by the USGS and JPL in 1975 had shown that if another infrared band was added to the new Landsat-4 sensor, the Thematic Mapper, it would be possible to discriminate not only iron oxides but clays as well.

In 1977, NASA decided to include the additional channel on the Landsat-4 sensor. To further understand the requirements of the geological community, NASA offered to acquire data using aircraft and satellites over eight industry test sites in the United States selected by the Geosat Committee. In return for the data, industry was asked to evaluate the usefulness of the data for geological investigations. This project, called the Joint NASA/Geosat Test Case Project, has now been completed, and its results will be jointly published, possibly by the American Association of Petroleum Geologists. Three sites were evaluated in Arizona to determine the applicability of remote sensing data for targeting porphyry copper mineralization, two sites in Utah and Wyoming evaluated deposits of uranium, and three sites in Texas, West Virginia, and Wyoming were oil/gas sites.

One of the early and most striking results of the joint industry/NASA investigations was in Silver Bell, Arizona. There, aircraft scanner data, simulating those to be acquired by the new sensor on Landsat-4, clearly delineated the alteration zone characterized by iron oxide stain and clay alteration.

Simulated Landsat-2 Multispectral Scanner data were acquired at 8-m resolution, ten times better than those acquired by the satellite scanner. These aircraft data were processed to determine if additional information could be extracted with higher resolution data, but no additional rock materials were delineated. A most important conclusion of this study was that increased spectral resolution (narrower wavelength band) resolution was much more important than increased spatial resolution. ASARCO geologists who have spent many years mapping and analyzing the Silver Bell deposit were surprised at the good correlation between the alteration mapped by ground techniques and laboratory analyses and the clay and iron oxide abundances detected and mapped through processing and analysis of aircraft scanner data. Another very important conclusion of this research was the development of methods for mapping mineral abundances that geologists do not normally observe in the field. In fact, clay abundances are usually determined using a petrographic microscope (Taranik, 1982).

In many areas of the world, vegetative cover obscures rock and soil materials, and remote sensing techniques must use variations in vegetation patterns as indicators of the underlying geologic materials. In some areas, such as Australia, rock types are not particularly well exposed, and even outcrops do not produce characteristic weathering patterns. However, in Australia, plant species are very selective in associating with particular rock materials, and geological mapping can be almost completed without actually observing the rock materials themselves. With repetitive Landsat data, the optimal times of year for "geobotanical" mapping can be selected and the data can be computer enhanced to bring out subtle variations in plant cover.

Often anomalous concentrations of elements can cause vegetation to be stressed and stunted, and these areas can be readily identified from Landsat data. The U.S. Geological Survey and U.S. Steel studied an area in Indonesia to determine if areas having nickel laterite duricrusts could be identified. Not only could the duricrusts be detected and mapped in known areas on the basis of the vegetation patterns, but the analysis could be extended to unknown areas to delineate potential new sites for exploration. In this study, the areas analyzed were covered with dense, triple-canopy vegetation. Although much geobotanical research remains to be done, most geologists are encouraged that these techniques will greatly aid exploration in tropical areas (Taranik, et al., 1978).

Geological Remote Sensing on the Second Space Shuttle Flight

In 1976, NASA asked for proposals for instruments to be flown as a part of the Shuttle engineering test program. Eventually six earth-viewing experiments were selected, including two geological remote sensing experiments—Shuttle Multispectral Infrared Radiometer and Shuttle Imaging Radar (Taranik and Settle, 1981). The flight of both instruments in November of 1982 resulted in spectacular breakthroughs in geological remote sensing (Taranik, 1982).

Analysis of the radiometer data demonstrated that direct mineralogic identification of carbonates and specific clay minerals was possible by narrow wavelength band measurements from earth orbit. On an orbital pass over Baja California, analysis of the data showed that clay and iron oxides were abundant. Subsequent field work documented that a zone of hydrothermal alteration had been detected. When the Mexican Government analyzed the samples, they confirmed that a previously unknown deposit of silver and molybdenum had been discovered (Goetz, et al., 1982). The possibilities for global assessment of and exploration for mineral deposits from space seem promising, provided the technology continues to be tested.

On the basis of the results from the analysis of the Shuttle Multispectral Infrared Radiometer data, the Jet Propulsion Laboratory has proposed that NASA fund the development of a Shuttle Imaging Spectrometer. This instrument would image a 10-km ground swath with 30-m ground resolution in 128 different spectral wavelength bands. This instrument could be flown on the Space Shuttle in 1987 if funds were available, but currently NASA has funds only for engineering feasibility studies.

The Shuttle Imaging Radar provided remote sensing image

data not previously available to geoscientists. The radar actually penetrated dry sand cover of the Sahara Desert and detected bedrock more than 2 m below the surface. The radar images from this mission revealed previously unknown buried valleys, geologic structures, and possible Stone Age archaeological sites. Because the radar is sensitive to variations in roughness and soil moisture and its illumination geometry is independent of the sun, analysis of its data provides information not available from Landsat data. Scientists from the U.S. Geological Survey have concluded that the radar data collected on the second flight of the Shuttle represent a major breakthrough in geological remote sensing (McCauley, et al., 1982).

The development of the Space Shuttle provides a technological breakthrough which should allow the United States to maintain its technological lead in geological remote sensing. Entire geoscience payloads can be developed quickly and flown at a fraction of the costs associated with free-flying satellites. Standards for instrument design can be relaxed because the instruments can be returned and reflown. Data can be acquired over specific sites in diverse global environments, eliminating the requirements for large, elaborate ground data facilities to process and archive large quantities of unwanted data.

Launch of the Shuttle into polar orbits will permit complete global coverage. Approximately twenty-five launches per year are planned by the middle of the decade. Thus, multitemporal data could be acquired to analyze seasonal, annual, or long-term changes in the landscape which are clues to the geology.

Simultaneous flights of the shuttle could permit simultaneous data collection and electromagnetic sounding of the earth using very low frequencies. Also, radar contouring of terrain may be possible using inferometry. NASA has proposed that the Shuttle be used to fly a low-altitude magnetometer on a tether, and a joint project with the Italian Government was approved until recently. Finally, manned space platforms, serviced by the Shuttle, will allow development of large antenna structures and large optical systems that will ultimately lead in the next decade to a new generation of geological exploration tools. Unfortunately, NASA funds for these projects have been curtailed, and there is considerable doubt that the tremendous technological opportunity the Shuttle provides will be exploited (Taranik, 1982).

Other Recent Developments in Geological and Geophysical Remote Sensing

Research in the thermal infrared also shows great promise for geological applications. When thermal mappers were first made available to geologists, research indicated that the manner in which geological materials heated up in response to the sun (and cooled down) was related to body properties of materials—heat capacity, density, and thermal conductivity. Thermal imagery was used to locate faults in unconsolidated alluvium, to map structures in areas having thin sand cover, and to separate dolomites from limestones.

Recently NASA developed a multiband thermal mapper which the agency has been flying over sites in Nevada. The data collected were recently analyzed, and the feasibility of mapping free silica in rocks and identifying carbonates seems likely. This new capability is significant because quartz and calcite are major constituents of rocks and soils. The combination of being able to map abundances of iron, carbonate, and silicate minerals and being able to identify specific clays using remote sensing data represents a very large technological breakthrough for global geological investigations. NASA has proposed development of a Shuttle multiband thermal mapping instrument; however, funds are not approved for even the engineering studies for such a sensor (Taranik, 1982).

For almost twenty years, NASA has been studying the earth's magnetic and gravity fields using satellite techniques. Since 1971, scientists have discovered that variations in these fields at satellite altitudes can be measured and inferences can be made regarding the earth's crustal composition and structure on a global basis. NASA has proposed repeatedly, but unsuccessfully, a mission called the geopotential field mapping mission, which would fly vector and scalar magnetometers for mapping the earth's magnetic field to an accuracy of 1/50,000 of the total field (1 gamma) and would map the earth's gravity field to 1 milligal in ½ by ½ degree blocks of latitude and longitude. Because the measurements will be made from an altitude of 160 km, detection of compositional (density and magnetic susceptibility differences in rocks) changes will be possible at a spatial resolution of 80 km. Such data will be extremely valuable for recognition of regional trends in airborne and ground-based potential field data. These regional trends can then be systematically removed to identify local anomalies which may be related to structures and compositional differences on a local scale (Taranik and Thome, 1980).

Subtle variations in the ocean surface may also be detected and mapped with sensitive altimeters to determine the ocean geoid—the equipotential gravitational surface that the earth's oceans would adopt if they were solely under the influence of gravitational and rotational forces. NASA has unsuccessfully proposed a topography experiment (TOPEX) designed to measure repeatedly the sea surface to +2 cm over 30 to 40 km distances and 10 cm over 3,000 to 5,000 km distances. At this level of accuracy, it may be possible to detect subtle variations in the marine geoid that are related to subsurface continental shelf structures, such as salt domes (Settle and Taranik, 1982).

Several new space-related techniques have been developed for determining distances between points and elevations on the surface of the earth and permitting dynamic movements of crustal plates to be evaluated on a global basis. Very long baseline inferometry (VLBI) using extragalactic radio sources and laser ranging to satellites like lageos (5,000-km altitude orbit) and the moon have now made it possible to measure distances over thousands of km to the +2 cm accuracy level. This capability will allow systematic establishment and maintenance of global geodetic grids. The new global positioning satellite (GPS) system will allow location of points on the ground through inferometry to an accuracy of 1 m with a few simple in-the-field calculations. These capabilities will revolutionize global geodesy and cartography (Taranik and Thome, 1980).

In 1978, the Geosat Committee recommended that a global stereoscopic data collection capability (Stereosat) be developed to support global mineral and petroleum exploration.

Congress directed reprogramming of NASA funds to study Stereosat feasibility when the Office of Management and Budget refused to let NASA initiate the project. Although the studies by the Jet Propulsion Laboratory were highly successful, the eventual response of the executive branch was to ask the energy and mineral industry to pay for development, launch, and operation of the system, and the concept died a bureaucratic death in 1980. When the French Government was certain that the United States was not going to develop a global digital stereoscopic capability, it approved development of a satellite called SPOT, a solid state imaging system with 10-m ground resolution, which will be launched in 1984. The U.S. Geological Survey tried to obtain Office of Management and Budget approval for a system called Mapsat (very similar to Stereosat) in 1981 and 1982, but they were also frustrated by OMB. These two systems would have allowed the United States to have a worldwide data digital base for production of topographic maps at 1:50,000 scale with 20-m contour accuracy.

Commercialization of the Civil Land and Weather Satellite Program

Soon after the present Administration was established in 1980, the Office of Management and Budget recommended termination of the Government's role in the Landsat program beyond Landsat-5, if not sooner. This decision deleted two additional Landsats (Landsat-6 and Landsat-7) and initiated a plethora of Government studies on how to transfer the operational aspects of the program to private industry. The Government has recommended that both the land and the weather satellite programs be considered for transfer and is entertaining proposals from industry to maximize cost savings to the Government. Draft legislation is being prepared for Congress, and congressional approval is expected to take months if not years.

In the meantime, most proposals by civilian government agencies for new experimental systems, for fundamental research using either satellites or the Space Shuttle, have been postponed. Such research will probably not be approved until the Government determines what kind of research will be done in private industry and where within the Government the longer term, high-risk, but potentially high payoff research will be done. Development and testing of new technology has been postponed out of a concern that the new technology, like that of Landsat, may become less experimental and more operational. Unfortunately, lead times for the development of satellite technology are from six to eight years, so it is doubtful that new free-flying systems can be introduced by the United States in this decade.

Currently Landsat-3 has an inoperable MSS, and the video camera system is being used to finish collection of a worldwide data set. Landsat-4 experienced a control failure soon after launch, and a backup system is being used to command the spacecraft. Although Landsat-5 has been built, it is not approved for launch. It will probably take one to two years for a decision to be reached and legislation to be enacted on a transfer of all or a part of the land remote sensing program to industry. Therefore a break in data continuity will most likely occur beginning in 1987.

Balanced against these trends in the United States is the fact that aerospace remote sensing has emerged as a major technological thrust on an international scale. France, the Soviet Union, Japan, Germany, Italy, and India are already moving aggressively in government-supported satellite remote sensing programs. The French and Japanese governments do not expect "a return on their space segment investments." Rather, they view their programs as the first step toward the creation of an entire new information technology which will stimulate the growth and development of new industries for their economies.

At the present time, rapidly maturing geological remote sensing techniques and the availability of the Space Shuttle as a platform for research combine to form a significant technological breakthrough for the United States of America. There is considerable uncertainty whether the United States will organize civil remote sensing to take advantage of this large opportunity and thus stimulate growth and development of our industrial economy.

REFERENCES

Goetz, A.F.H., Rowan, L.C., and Kingston, M.J., 1982, Mineral Identification From Orbit: Initial Results From the Shuttle Multispectral Infrared Radiometer: Science, Vol. 218, pp. 1020–1024, Dec. 3, 1982, American Association for the Advancement of Science, Washington, D.C.

McCauley, J.F., et al., 1982, Subsurface Valleys and Geoarchaeology of the Eastern Sahara Revealed by Shuttle Radar: Science, Vol. 218, pp. 1004–1020, Dec. 3, 1982, American Association for the Advancement of Science, Washington, D.C.

Settle, M., and Taranik, J.V., 1982, Mapping the Earth's Magnetic and Gravity Fields From Space: Current Status and Future Prospects: in Proceedings Symposium on Remote Sensing and Mineral Exploration, GOSPAR XXIV, Ottawa, Canada, May 1982, 10p.

Taranik, J.V., and Trautwein, C.M., 1977, Integration of Geological Remote-Sensing Techniques in Subsurface Analysis: in Subsurface Geology, Petroleum and Mining Constructions, LeRoy and LeRoy, Editors, Colorado School of Mines, pp. 564–586.

Taranik, J.V., 1978a, Characteristics of the Landsat Multispectral Data System: U.S. Geological Survey Open-File Report 78–187, 76p.

Taranik, J.V., 1978b, Principles of Computer Processing of Landsat Data for Geological Applications: U.S. Geological Survey Open-File Report 78–117, 42p.

Taranik, J.V., Reynolds, C.D., Sheehan, C.A., and Carter, W.D., 1978, Targeting Exploration for Nickel Laterites in Indonesia With Landsat Data: in Proceedings 12th International Symposium of Environment, Vol. II, Manila, Philippines, Environmental Research Institute of Michigan, Ann Arbor, Michigan, pp. 1037–1051.

Taranik, J.V., and Thome, P.G., 1980, Development of Space Technology for Resource Applications in the 1980's: 14th Congress, International Society of Photogrammetry, Hamburg, Germany, 22p.

Taranik, J.V., 1981, Advanced Technology for Global Resource Applications: in 15th International Symposium Remote

Sensing of the Environment, Vol. I, Environmental Research Institute of Michigan, Ann Arbor, Michigan, pp. 1–15.

Taranik, J.V., and Settle, M., 1981, Space Shuttle: A New Era in Terrestrial Remote Sensing: Science, Vol. 214, pp. 619–626, Nov. 6, 1981, American Association for the Advancement of Science, Washington, D.C.

Taranik, J.V., 1982, Geological Remote Sensing and Space Shuttle: A Major Breakthrough in Mineral Exploration Technology: Mining Congress Journal, Vol. 68, No. 7, pp. 18–25, American Mining Congress, July 1982.

Reprinted by permission of the Environmental
Research Institute of Michigan from *Proceedings of
the Third Thematic Conference on Remote Sensing for
Exploration Geology*, 1984, v. 1, p. 251-271.

STRUCTURAL ANALYSIS OF LOW RELIEF BASINS USING LANDSAT DATA*

by

Zeev Berger

Exxon Production Research Company
Houston, Texas 77001

ABSTRACT

This paper documents a new
technique for structural analysis of
low relief basins. The technique is
based on integration of Landsat data
with other geological data sets such
as gravity, magnetic, subsurface, and
production data. Five analytical
steps are recommended and are
illustrated by a series of examples
supported by both surface and subsurface controls. These analytical
steps are: (1) Analysis of exposed
structures that form the margin of
the basin; (2) Recognition of
structural trends within the basin;
(3) Recognition of buried and obscured
structures within the basin; (4)
Construction of an exploration model;
and (5) Generation of new leads for
the entire region. Examples sited
are from various low relief basins
such as the Paris, Powder River, and
the Central Basin Platform of West
Texas.

The mechanisms that cause surface
expressions of buried and obscured
structures are being attributed to
differential compaction, loading,
structural reactivation, and to other
processes related to abnormal flows
of ground and surface waters in the
vicinity of buried and obscured
structures. These are well recognized
processes that occur under various
climatic and surface conditions.

*Presented at the International Symposium on Remote Sensing of Environment,
Third Thematic Conference, Remote Sensing for Exploration Geology, Colorado
Springs, Colorado, April 16-19, 1984.

251

The study concludes that Landsat
data can be used in low-relief frontier
areas as a reconnaissance tool to
identify regional trends and structural
styles as well as potentially prospective structures. These data can also
be used in low-relief mature areas to
locate subtle structure not identified by other exploration techniques.

INTRODUCTION

This paper documents a new technique for structural analysis of low-
relief basins that has been developed, tested, and successfully applied for
exploration in several basins throughout the world. The technique is based
on integration of Landsat data with other geological data sets such as
gravity, magnetic, subsurface, and production data. The term low-relief
basin is used to describe those sedimentary basins whose structures and
related hydrocarbon traps are too subtle to be detected by conventional
seismic techniques. These basins are also characterized by very gentle
topography in which either the surface expressions of structures are obscured
by a thick cover of sediments, soil, and vegetation or the structures are
completely buried under undeformed sedimentary units.

First, examples from various basins are used to illustrate the procedures
involved in the analysis of low-relief basins. Then, the mechanisms that
cause the surface expressions of buried structures and their unique depiction
on remote sensing data are discussed. Emphasis in this part of the paper is
placed on recognition of ground water and surface water as important geomorphic agents that enhance the surface expressions of buried structures.
Finally, types of sedimentary basins that are most favorable for such analysis
are mentioned, and the need to constrain Landsat interpretation with all
available geologic data is emphasized.

PROCEDURES

The basic concept in the analysis of low relief basins is illustrated in
Figure 1. A sedimentary basin can be divided into two main parts: (1) the
basin margin, consisting of areas of exposed structures that can be mapped
and analyzed in detail from surface data; and (2) the basin interior, where
structures are commonly obscured by a thick cover of vegetation and soil or
are completely buried under sedimentary cover. Such structures manifest only
subtle and indirect topographic expressions, and their recognition requires
careful analysis that involves the integration of various data sets.

Structural analysis of low-relief basins should begin with the exposed
part and then be carried into the central part of the basin. Available geological and geophysical data for the basin area and its surrounding margins
should be collected and entered into an image processing system. This data
set should include Landsat, gravity, magnetic, subsurface, and surface geologic data, as well as production data. Five major steps in the analytical
procedure are recommended and can be summarized as follows:

(1) Analysis of exposed structures that form the margins of the basin.
    The first step in the analysis is to map and examine structures in the
    exposed areas surrounding the basin. During this step it is possible to

252

176

establish the style and trend of the structure, and the timing of events of all tectonic units surrounding the basin that may have affected its tectonic evolution. Figure 2 illustrates this procedure. Here, a Landsat image and surface geology maps were used to map and identify the structural setting of the Ouachita structural belt in an area where the Appalachian orogen makes a reentrance into the North American platform (see Wielchowsky and Davidson, 1984, in this volume).

(2) Recognition of structural trends within the basin.
Once the structural trends around the basin are defined, it is then possible to identify structural elements that can be traced from the exposed margins into the basin area. The buried extensions of structural trends into the basin area is typically expressed at the surface as composite alignments of streams, topography, vegetation, textural changes, and color changes that together make up Landsat lineaments (Gary et al., 1974).

Probably the most common though problematic step in the analytical procedures of Landsat data is the recognition and analysis of Landsat lineaments (Wise, 1982). Many linear features of various directions, lengths, and densities can be identified on the imagery. These linear features may or may not reflect the surface expression of subsurface structures. Thus, prior to further investigation of these features, it is necessary to constrain the analysis with other available geological data.

I suggest that Landsat lineaments may be related to the surface manifestations of buried structures only if they meet one or more of the following criteria: (1) The lineaments are aligned with known structural trends in the exposed margins of the basin; (2) they are spatially coincident with other data sets that reflect subsurface structures such as gravity, magnetic, subsurface, or production data; or (3) surface features that are commonly associated with major subsurface structures, such as truncated structures, sets of parallel faults, or structurally controlled drainage patterns, can be documented along these lineaments (Fig. 3). These surface features can be either directly identified on the imagery or traced from other data sets such as high-resolution areal photography, geologic maps, or topographic maps.

An excellent example of Landsat lineament interpretation that meets all the criteria mentioned above has recently been documented in the Powder River Basin, U.S.A. (Slack, 1981). On the imagery (Fig. 4), northeast-trending lineaments can be traced from the exposed basement outcrops in the Big Horn Mountains across the Powder River Basin and into the exposed basement in the Black Hills. In the exposed basement areas the lineaments reflect the surface expression of major fault systems. In the Powder River Basin the lineaments are spatially coincident with major steep slopes of gravity and magnetic data, and they appear to correlate well with production data. Typical surface features that commonly form along major lineaments can also be identified on the imagery in the basin area (See Fig. 4).

A case study from the Meggen area, West Germany, demonstrates how geologic and topographic maps can be used to scrutinize the interpretation of Landsat lineaments in areas of poor bedrock exposure. Here, northeast-trending lineaments cut the Hercynian fold belt obliquely. The lineaments are believed to reflect the reactivation of Devonian fault systems that may have localized the deposition of sediment-hosted

253

177

massive sulfides in the area (Krebs 1971; Gwosdz and Kruger, 1972). The lineaments can be identified on Landsat imagery only as subtle topographic features; however, surface features that support their relationship to subsurface fault systems can be clearly documented from large-scale geologic and topographic maps (Figs. 5,6).

(3) Recognition of buried and obscure structures within the basin. The next step in the analytical procedure is to investigate the surface expression of buried and obscured structures within the basin. These may be related to the major structural trends identified in Step (2), or they may be surface expressions of solitary features.

One method of recognizing buried and obscured structures is through the application of a series of morphotectonic models such as those shown in Figures 7 and 8. Morphotectonic models are three-dimensional block diagrams that describe the relationships among surface features observed on Landsat imagery, subsurface structures observed from seismic data, and basement configuration deduced from gravity and magnetic data. Morphotectonic models are developed on the basis of experience gained in the analysis of surface expressions of various structures in low-relief basins.

Figure 7 is a morphotectonic model of the typical outcrop pattern of a structurally undisturbed basin. That is, the basin forms roughly circular to elliptical outcrop patterns, with beds dipping toward the center of the basin and younger rock units cropping out at the center. Figure 8 illustrates a morphotectonic model for a basin that was disrupted by an uplifted basement block. The most common erosional features that are related to such structural settings are (1) topographic highs that form above drape folds, (2) topographic scarps that develop along faulted block boundaries, and (3) breached areas that reflect folds that have draped over the faulted boundaries of uplifted blocks.

The Paris Basin in France provides a good example of the outcrop patterns of a structurally undisturbed basin. As illustrated in Figures 9 and 10, the outcrop patterns of this basin can be observed on geologic maps as well as on remote sensing data. The light brown areas at the center of the basin are Eocene and Oligocene rock units. The rimming white areas are the famous Cretaceous chalk of the Paris Basin, and, the outer most dark brown areas are outcropping Jurassic units.

In contrast with the Paris Basin area, the Crockett Arch area in the Central Basin Platform of west Texas (Figs. 11,12), is a large Permian basin that was structurally inverted by tectonic reactivation. The arch is made up of several large uplifted basement blocks that are bounded in part by major fault systems (Galley, 1958; Hills, 1968).

Analysis of Landsat, magnetic, and other data sets of the western edge of the Crockett Arch shows that although the structures in the region are completely buried under undeformed Cretaceous rock units, the region exhibits a morphotectonic setting reminiscent of uplifted basement block areas. That is, large and broad topographic highs with well-developed radial drainage are depicted on Landsat data. These surface features suggest the presence of subtle folds located over the crest of a northeast-trending basement high that is reflected on magnetic data as a northeast-trending magnetic high (A in Figs. 13 and 14). Smaller and more tightly breached folds with well-developed central valleys appear

254

178

to be located along the faulted margin of uplifted blocks that are expressed as steep and linear slopes on magnetic data (B in Figs. 13 and 14).

(4) Construction of an exploration model.
An exploration model can now be constructed. One way of presenting such a model is through a schematic block diagram that illustrates the relationships among surface features observed on Landsat data, subsurface structures observed with seismic and well data, basement configurations predicted from gravity and magnetic data, and production data. An example of an exploration model that was developed for exploration of massive sulfides in Meggen, West Germany, is shown in Figure 15. Sediment-hosted massive sulfides are believed to be deposited in several Devonian age subbasins. These are located at intersections of major lineaments that reflect buried fault systems.

Another way of presenting an exploration model is through a series of maps that describe the relationship between the structures identified in the area and known production. This type of presentation enables the interpreter to identify specific potential plays that can be delineated in the study area through an integrated Landsat analysis (e.g., Berger, 1983, p. 588).

(5) Generation of new leads for the entire region.
Landsat data can now be used in conjunction with other available data sets to produce a map of new leads associated with different plays for the entire basin or for areas of high interest. The leads can be ranked from high to low on the basis of available data and in conjunction with the exploration model developed for the region.

CAUSE OF SURFACE EXPRESSIONS OF OBSCURED AND BURIED STRUCTURES

Because obscured structures are exposed at the surface it is easy to understand how they might influence surface conditions such as topography, drainage, vegetation health and type, and soil moisture content. Buried structures, however, influence surface conditions in a more indirect manner. The exact mechanisms by which this influence is manifested are not fully understood. The most common explanations can be summarized as follows:

- Differential Loading - Differences in sediment thickness and type in the vicinity of the buried structure can result in differential vertical stresses. These stresses may cause partial reactivation of structures and lead to local increases in topographic relief or increased fracture density at the surface.

- Differential Compaction - Lateral variations in the sedimentary column covering buried structures may result in differential compaction and subsidence and lead to local changes in topographic relief or surface fracturing. This process may be repeated several times as newly deposited sediments dewater and differentially compact over and adjacent to a buried structure.

- Reactivation of Geologic Structure - New, renewed, or continued stresses may cause reactivation of buried structures. New stresses refer to

255

changes in orientation of the principal stresses that generated the buried structure. Renewed stresses refer to an increase in stress magnitude to values that cause permanent deformation. Continued stresses refer to a stress field with similar orientation to that which generated the buried structure, but of a reduced magnitude such that the overlying cover is not significantly deformed.

- Combination - Structures may produce surface expressions by combinations of the previously mentioned mechanisms. For example, differential loading around the buried structure may cause differential compaction and result in increased fracturing over a structure.

- Disruption of Near-Surface Ground-Water Flow - Disruption in the uniform flow of ground water may occur over and in the vicinity of buried structures (Fig. 16). This disruption in flow occurs when significant permeability contrasts exist between the rock involved in the structure and the cover sediments, and when increased fracturing causes an upward movement of ground water (Kudryakov, 1974; Toth, 1980). The differences in depth of the water table over a buried structure can alter near-surface soil moisture content. This, in turn, produces an observable contrast in vegetation, drainage, and soil type (Berger and Aghassy, 1982, 1984). Recognition of ground-water flow as a mechanism that produces surface expressions of buried structures is particularly important in humid and subhumid low-relief basins where the water table is high and frequently fluctuates during wet and dry seasons or after heavy rains. In such areas, changes in ground moisture conditions over buried structures are typically depicted on remote sensing data as brightness variations. In general, local bright areas are caused by reduced ground moisture and vegetation, whereas darker areas are caused by excessive soil moisture and increased vegetation density (Fig. 17).

CONCLUSION

The presence of surface features that reflect buried and obscured structures in low-relief basins is not unique to the case studies described in this paper. Integrated Landsat analysis has been successfully applied by the author for exploration in various basins throughout the world. The geological settings most favorable for such integrated Landsat analysis are plate interiors, passive margins, and foreland basins where reservoir targets are less than 15,000 feet deep and where potential traps are developed in association with reactivated basement structures.

The approach to Landsat structural analysis presented here is one that recognizes the resolution limitations of Landsat data and emphasizes careful scrutiny of Landsat interpretation in conjunction with all available geological data. The mechanisms that cause these surface features are attributed to differential compaction, loading, structural reactivation, and other processes related to abnormal flow of ground and surface waters in the vicinity of buried and obscured structures. These are well recognized processes that occur under various climatic and surface conditions. The results of this and many other studies clearly demonstrate that the surface manifestations of buried and obscured structures can be detected and analyzed with Landsat data even in low-relief and poorly exposed bedrock areas. Landsat structural analysis can be used as an effective exploration tool that allows rapid assessment of major structural style and trend, as well as prospective structures in areas of poor bedrock exposure. Consequently, such studies are more effective when done in conjunction with basin analysis.

256

REFERENCES

Berger, Z., and Aghassy, J., 1982, Geomorphic manifestations of salt dome stability, in Craig, R. G., and Cradt, J. L., eds., Applied geomorphology: 11th Ann. Binghampton Geomorphology Symposium, p. 72-84.

Berger, Z., and Aghassy, J., 1984, Near-surface moisture and evolution of structurally controlled drainage in soft sediments, in LeFleur, R. G., ed., Groundwater as a geomorphic agent: 13th Ann. Binghampton Geomorphology Symposium.

Berger, Z., 1983, The use of Landsat data for detection of buried and obscured geologic structures in the East Texas Basin, U.S.A: International Symposium on Remote Sensing of the Environment, Second Thematic Conference, Fort Worth, Dallas, p. 577-589.

Gary, M., McAffee, R., Jr., and Wolf, C. L. (eds.), 1974, Glossary of geology: Am. Geol. Inst., Washington, D. C., 805 p.

Galley, J. E., 1958, Oil and geology in the Permian basin of Texas and New Mexico, in Habitat of oil: Am. Assoc. Petroleum Geologists, p. 395-446.

Gwosdz, W., and Kruger, H., 1972, Meggener Schichten (Devon Sauerland Rheinisches Schiefergebirge): Neues Jahrb Geologie u. Palaontologie Monatsh., v. 2, p. 85-94.

Hills, J. M., 1968, Gas in Delaware and Val Verde basins, West Texas and southeastern New Mexico, in Natural gases of North America: Am. Assoc. Petroleum Geologists Mem. 9, p. 1394-1432.

Krebs, W., 1971, Devonian reef limestones in the eastern Rhenish Schiefergebirge, in Miller, G., ed., Sedimentology of parts of Central Europe. Guidebook: VIII. Internat. Sed. Cong., Heidelberg, p. 45-81.

Kudryakov, V. A., 1974, Piezometric minima and their role in the formation and distribution of hydrocarbon accumulations: Doklady, Acad. Sci. USSR, v. 207, p. 240-242.

Slack, P. B., 1981, Paleotectonics and hydrocarbon accumulation, Powder River Basin, Wyoming: Am. Assoc. Petroleum Geologists Bull., v. 65, p. 730-743.

Toth, J., 1980, Cross-formational gravity-flow of groundwater: a mechanism of the transport and accumulation of petroleum, in Roberts, W. H., and Cordell, R. J., eds., Problems of petroleum migration: Am. Assoc. Petroleum Geologists, Studies in Geology, No. 10, p. 121-167.

Wise, D. U., 1982, Linesmanship and the practice of Linear Geo-art: Geol. Soc. America Bull., v. 93, p. 886-888.

257

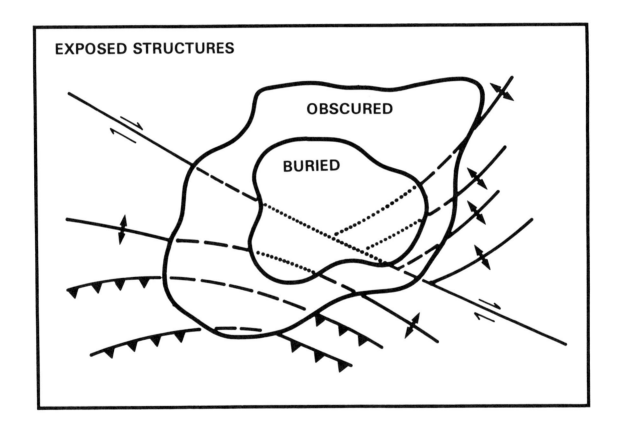

Fig. 1 - A conceptual model for structural analysis of a low-relief basins. Geological structures are first interpreted in the exposed margins of a basin and later carried into the center of the basin, where the surface expression of structures is obscured by soils and vegetation or where the structures are completely buried under sediments.

258

40 KM

Fig. 2 - Landsat imagery of the Ouachita Mountains, U.S.A., showing the
structural setting of the Ouachita structural belt in areas where
the Appalachian orogen makes a reentrance into the North American
platform.

259

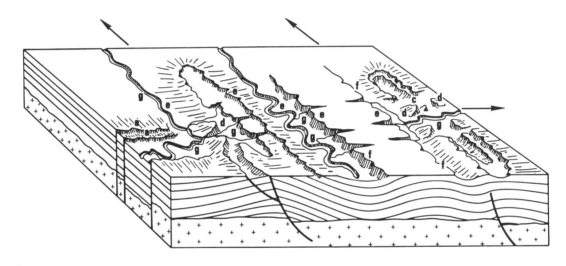

a - FAULTS

b - TRUNCATED STRUCTURES WITH OBSERVABLE OFFSET (NOT SHOWN)

c - TRUNCATED STRUCTURES WITH NO OBSERVABLE OFFSET

d - BREACHED STRUCTURE (EROSIONAL)

e - LINEAR LITHOSTRATIGRAPHIC UNITS (PARALLEL TO LINEAMENTS)

f - LINEAR TOPOGRAPHIC RIDGES (PARALLEL TO LINEAMENTS)

g - STRUCTURALLY CONTROLLED STREAMS

——→ MAJOR LINEAMENTS

Fig. 3 - Block diagram showing the types of surface features that are often
developed along buried fault systems. Alignments of such surface
elements produce large-scale Landsat lineaments.

260

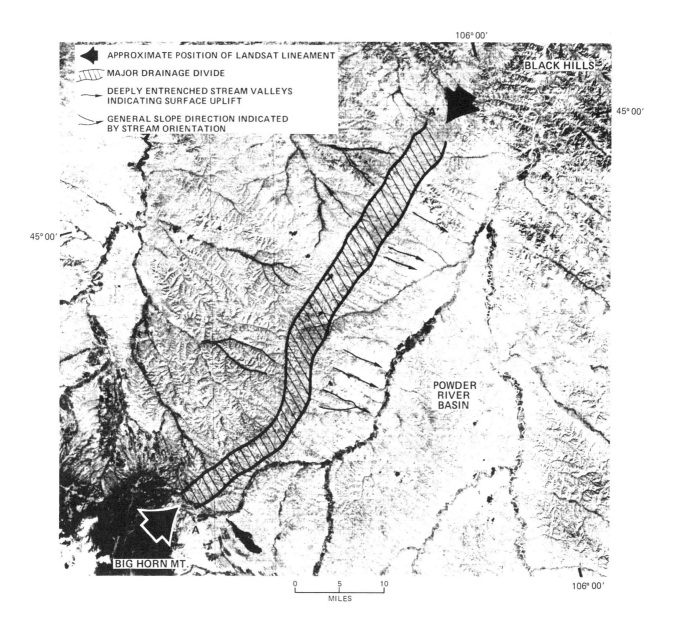

Fig. 4 - Landsat image subscene of the northwestern margins of the Powder
River Basin, U.S.A., showing surface expression of northeast-
trending lineaments that reflect reactivated basement fault systems
(A-A').

261

185

Fig. 5 - Standard false color Landsat image subscene of the Meggen area,
West Germany, showing examples of major north-northwest - and
east-trending lineaments that obliquely cut the Hercynian fold
belt. Geological evidence of two sets of lineaments is documented
in Fig. 6. The image subscene in approximately 75 km square.

(Original figure in color)

262

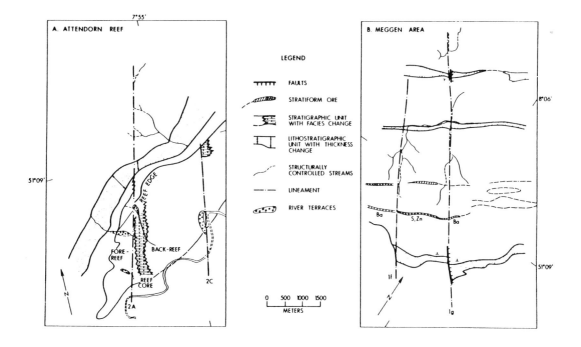

Fig. 6 - Geological and surface evidence complied from geologic and topographic maps along two major sets of Landsat lineaments identified
in the Meggen area (shown in Fig. 5).

263

187

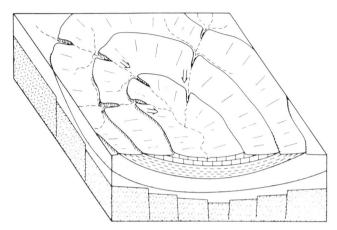

⬅ **REGIONAL SLOPE DIRECTION**

⋙ **TOPOGRAPHIC SCARPS (CUESTAS) FORMED ALONG LITHOSTRATIGRAPHIC CONTACTS**

Fig. 7 - Morphotectonic model showing typical erosional features of a structurally undisturbed sedimentary basin.

A TOPOGRAPHIC HIGH FORMED ABOVE DRAPE FOLDS
B TOPOGRAPHIC SCARPS DEVELOPED ALONG FAULTED BLOCK BOUNDARIES
C BREACHED FOLDS FORMED ALONG FAULTED BLOCK BOUNDARIES

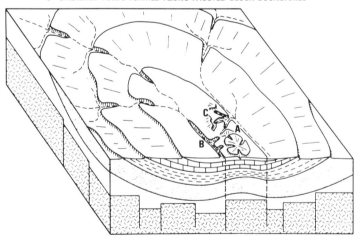

Fig. 8 - Morphotectonic model showing erosional features caused by uplifted basement blocks in a sedimentary basins.

264

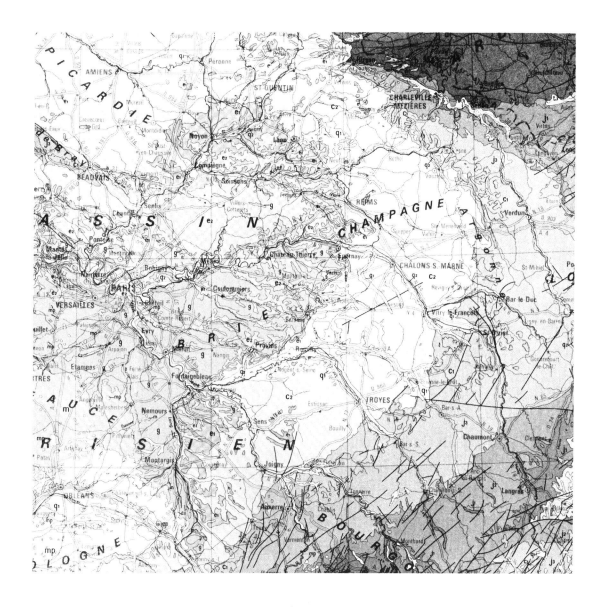

Fig. 9 - Geologic map of the Paris Basin, France, illustrating the outcrop patterns of a structurally undisturbed basin.

(Original figure in color)

265

Fig. 10 - Landsat mosaic of the Paris Basin, France, showing the outcrop
patterns of a structurally undisturbed basin.
(Original figure in color)

266

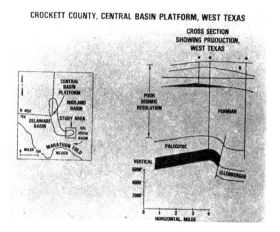

Fig. 11 - Location map and generalized cross section of the Crockett Arch
area major structural traps.

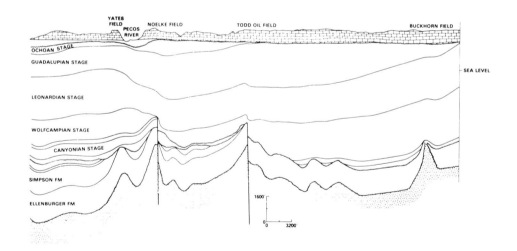

Fig. 12 - Schematic cross section of the study area showing (A) the close
correspondence between basement faults and drape structures, and
(B) the complete burial of Permian and older structures beneath
undeformed Cretaceous sediments.

267

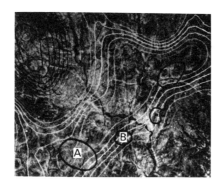

Fig. 13 - A Landsat image and total intensity magnetic data of the western
edge of the Crockett Arch area, showing the location of two types
of folds that are recognized in the study area. The surface
expressions of these folds is illustrated in Fig. 14.

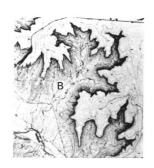
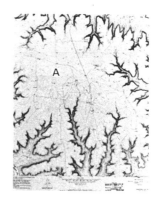

Fig. 14 - Topographic maps showing the surface expression of two types of
folds that are recognized in the study area.

268

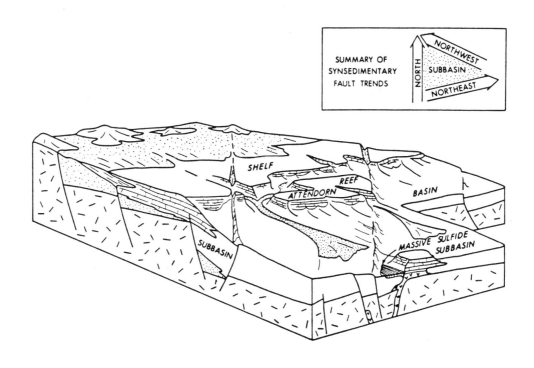

SUMMARY OF
SYNSEDIMENTARY
FAULT TRENDS

NORTH

NORTHWEST

SUBBASIN

NORTHEAST

SHELF

ATTENDORN

REEF

BASIN

SUBBASIN

MASSIVE SULFIDE
SUBBASIN

Fig. 15 - Conceptual model of Middle Devonian basin development and associated
sediment-hosted massive sulfide mineralization in the Meggen area,
West Germany. North, northwest, and northeast lineaments interact
to produce triangular-shaped subbasins.

269

193

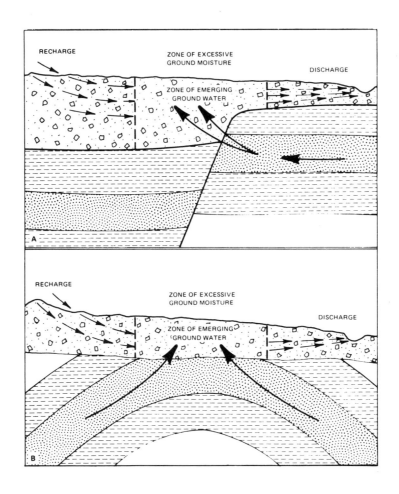

Fig. 16 - Schematic illustration showing common geological settin wherein
cross-formational ground-water flows cause zones of eme   _ng movement of ground water and areas of excessive ground moisture
(modified from Kudrykov, 1974). (A) Excessive ground moisture
occurs along a buried fault scarp. (B) Excessive ground moisture
outlines the position of a buried breached anticline.

270

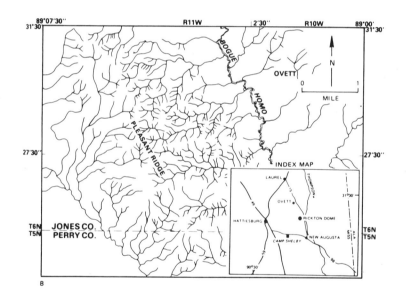

Fig. 17 - An aerial photographic mosaic and drainage network map showing the
relationships between ground moisture and drainage networks as they
appear along Pleasant Ridge in Jones County, Mississippi. This
long, straight ridge of the Williana deposits is flanked along its
length by aligned tributaries and small ponds, suggests the existence
of a structure (probably a fault) that is buried under this topographic ridge. (A) Excessive ground moisture is shown by the
dark-toned area. (B) Drainage networks in the area of excessive
moisture exhibit angular patterns that are typically formed by
piping and sapping (after Berger and Aghassy, 1984).

271

The American Association of Petroleum Geologists Bulletin

V. 56, No. 5 (May 1972), P. 903-915, 13 Figs.

Afar Tectonics Analyzed from Space Photographs[1]

DIETER BANNERT[2]

Hannover, German Federal Republic

Abstract The Afar region is the subject of recent activity of the International Upper Mantle Project because of its position at the triple junction of the East African rift, the Gulf of Aden, and the Red Sea. Extended aerial and ground surveys in the past 3 years revealed tectonic deformation strongly influenced by plate tectonics. Single blocks at the eastern edge of Africa seem to undergo northeastward-directed drifting motions, following the east- or northeastward-drifting Arabian Peninsula. Analysis of Apollo 9 space photographs supplemented the existing ground and aerial surveys and extended the investigations to the region north of the Gulf of Tadjura. Evidence suggests that this region is a block with drifting motion similar to that of the Arabian Peninsula. There are both advantages and disadvantages to using space photographs for geologic exploration.

Introduction

Geophysical investigations of the ocean floors during the past 20 years have brought new and better data on continental movement and sea-floor spreading (Hurley, 1968; Internat. Upper Mantle Project, 1966). Rotation of the Arabian Peninsula along the Red Sea rift in a counterclockwise direction, away from Africa, has been suggested by Girdler (1966) and Laughton (1966), and substantiated recently by Phillips and Ross (1970). According to these authors the opening of the Red Sea took place after the Miocene Epoch. Spreading now has reached an amount of 7° around an imaginary center located in the Mediterranean Sea south of Cyprus. It has been suggested that right-lateral slip faults along the Jordan Rift Valley (Fig. 1) compensate for the expected strong distortion near this imaginary center, the motions along these faults causing counterclockwise rotation of different blocks in the northern Negev and Judean Mountains west of the Dead Sea (Freund, 1965, p. 195; Bannert and Kedar, 1971).

At the southern end of the Red Sea apparent local motions of different blocks could be the result of the rotation of the Arabian Peninsula. The region affected is on the African continent east of the Ethiopian Highland and north of the Somalia Plateau. This region is called the Afar.

The Afar region is a large triangular lowland, partly below sea level, with numerous volcanoes, some of them still active (Varet, 1971). The morphology of the Afar region is influenced mainly

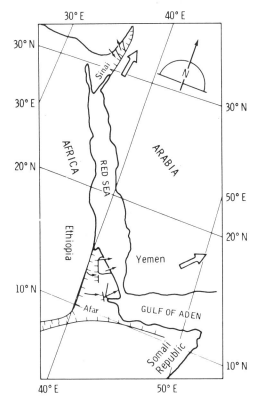

FIG. 1—Red Sea, Gulf of Aden, and adjacent parts of Africa and Arabia. Small black arrows indicate local tectonic movements. Large white arrows indicate rotation of Arabian Peninsula. Lines with short tick marks indicate major faults, with tick marks pointing to downfaulted side (after Bannert and Kedar, 1971).

by the recent tectonic development. The influence of tectonics on the morphology is demonstrated dramatically on aerial and space photographs (Figs. 2-4). After an analysis of more than 2,000 aerial photographs and numerous field observations (Bannert et al., 1970; Bannert, 1972),

[1]Manuscript received, January 8, 1971; revised and accepted, September 17, 1971.

[2]Federal Geological Survey.

This report was started while the writer was assigned to the NASA Manned Spacecraft Center in Houston, Texas, as a senior postdoctoral research associate of the National Research Council, National Academy of Sciences, on leave from the Federal Geological Survey in Hannover, Germany. The writer gratefully thanks John L. Kaltenbach, NSAS MSC; John D. Haun, Golden, Colo.; H. Tazieff, Paris; and J. Varet, Addis Ababa, for review and comment on the manuscript, and for contributing field observations.

903

FIG. 2—Apollo 9 photograph AS 9-23-3539. Gulf of Tadjura and Tadjura block (right), overlain by black Afar Basalt. Intensity of tectonic movement increases southwestward (left). See Figure 5 for orientation of Figures 2-4. Photograph by NASA.

space photo graphs have been used to extend the investigations from the northern Afar toward the east (Fig. 6, insert).

Earth-scanning space photographs, taken during NASA's manned orbital missions, seem to be very helpful in obtaining new data from the earth's surface. One of the main target areas during the photographic experiments of the Gemini and Apollo flights was the Red Sea region, because of the various geologic problems involved (Lowman, 1964, 1967; Lowman and Tiedemann, 1971). Geologic interpretation of selected Gemini photographs of the Red Sea region was undertaken by Abdel-Gawad (1969a, b; 1970), who compared major tectonic and morphologic features on both sides of the Red Sea. The results support the hypothesis of a drifting motion of the Arabian Peninsula.

During the Apollo 9 mission in March 1969, NASA astronauts McDivitt, Scott, and Schweickart took a sequence of 70-mm color photographs (numbered AS 9-23-3535 through AS 9-23-3540) over the northern and eastern Afar region (Fig. 1). These views allow an analysis of the fracture pattern of this part of Africa. The original 70-mm frames are on a scale of 1:2,000,000. From these views color paperprints 16 × 16 in. were prepared and then were covered with a

semitransparent overlayer for drawing purposes. The investigations have been done on high precision 8 × 8 in. color transparencies, providing a scale of approximately 1:675,000, which have been enlarged from the original 70-mm space photographs. As a result of overlapping, the investigations could be done stereoscopically for most of the region. Results were traced on the overlayers of the 16 × 16 in. paperprints.

GENERAL GEOLOGY OF NORTHERN AFAR

The geology of the Afar is strongly affected by plate tectonics, as could be demonstrated in recent papers by Mohr (1970), Tazieff (1970, p. 37), Bannert et al. (1970, p. 436-438), and Barberi et al. (1970, p. 310). On the Ethiopian side of the Red Sea small blocks of sialic crust seem to have drifted eastward following the Arabian plate (Laughton, 1966, p. 78; Tazieff and Varet, 1969; Bannert and Kedar, 1971). These motions caused the narrowing in the Strait of Bab el Mandeb to a width of 30 km of open sea. Between the Ethiopian Highland in the west and the Danakil Mountains in the east the Afar region appears to be split into single blocks of crust, with smaller crevasses filled with upwelling oceanic magma building up major volcanic ranges in the northern part of the Afar triangle (Barberi et al., 1970;

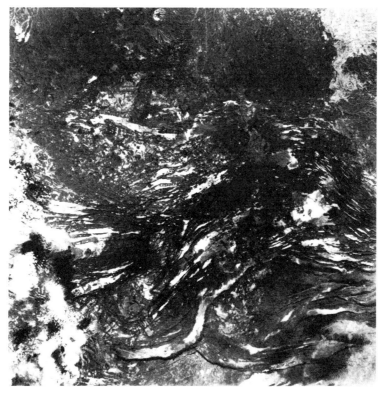

FIG. 3—Apollo 9 photograph AS 9-23-3538, overlapping region shown in AS 9-23-3539. Investigation could be done in stereoscopic view for entire region. Photograph shows complex of annular grabens. In center photograph are youngest basalts of Aden Volcanic Series (black). Acid extrusive rocks of Aden Volcanic Series have rougher surface and lighter color than basalts of Aden Volcanic Series. See Figure 5 for orientation. Photograph by NASA.

Tazieff and Varet, 1969).

Between the Gulf of Zula in the north and Lake Abbé in the south Precambrian, Paleozoic (slightly metamorphosed), and transgressive Jurassic sandstone and limestone compose the eastern and western flanks of a late Tertiary graben, the Danakil graben (Figs. 6, 7). Clastic and carbonatic-sulfatic marginal sediments of Pliocene and perhaps Miocene age, called the Danakil Series (Bannert *et al.*, 1970, p. 415-419), are found in the graben (Fig. 7). In the center of the graben a "rift-in-rift" structure contains more than 3,000 m of Quaternary saline deposits, which have been drilled (Holwerda and Hutchinson, 1968). The deepest point in the graben, 130 m below sea level, indicates recent downfaulting. During the late Tertiary huge masses of plateau-forming Afar Basalt (Mohr, 1967a, p. 12) flowed over the southern extension of the Danakil graben. These basalt masses could be ascribed to the marginal sediments in the graben region. Magma of the Pleistocene Aden Volcanic Series (Blanford, 1870) penetrated the entire region, creating an outpouring of olivine basalts (Mohr, 1962; Tazieff *et al.*, 1970) and acid ignimbrites and lavas. In the Danakil graben, recent volcanic activity is restricted almost completely to the cen-

tral rift-in-rift and the Alaito Range. In the northern part of the rift-in-rift, basement rocks border the western flank of the Ethiopian Highlands (Fig. 6, cross section *A-A'*). On the east, many steps approach the basement of the Danakil Mountains. These facts indicate the presence of an asymmetric graben with movement of its eastern flank.

The Danakil graben opens on the southeast in a funnel-like shape and can be traced southward to 13°N lat. (Tazieff *et al.*, 1970; Bannert *et al.*, 1970). The adjacent region on the south is characterized by the common occurrence of acid extrusive rocks of the Aden Volcanic Series and by the interaction of faults with different strikes. The acid extrusive rocks cannot be ascribed to the northern extension of the Wonji fault belt of Mohr (1967), because this fault belt is not present in the northern part of the Afar region. Southward, at approximately 12°N lat., a graben is present, with Lake Abbé at its lowest point. This graben (Figs. 6, 7) does not seem to be as deep as the Danakil graben. West of this shallow graben is a region only slightly affected by tectonic movements and extending to the escarpment of the Ethiopian Highlands. Eastward, to the Gulf of Tadjura, several smaller grabens are present,

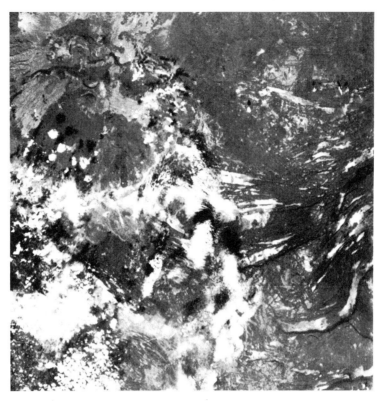

FIG. 4—Apollo 9 photograph AS 9-23-3537, overlapping AS 9-23-3538. In lower right corner is part of annular Immino graben. Bidu Mountains, formed by 5-mi diameter caldera, are shown in upper left. On mountain slopes, light-colored acid tuffs, ignimbrites, and lavas can be seen. At extreme upper left are basement and Mesozoic rocks of southeastern Danakil Mountains. Basaltic central volcanoes of Danakil graben are shown in lower left corner. See Figure 5 for orientation. Photograph by NASA.

and some have an annular shape (Mohr, 1967b, 1968; Bannert, 1969; and Tazieff, 1971). The region of annular grabens is composed of faulted Afar Basalt that has been partly covered by lavas of the Pleistocene Aden Volcanic Series (Figs. 8, 9).

ANALYSIS OF SPACE PHOTOGRAPHS
Rock Type Classification

As shown in Figures 7 and 8, the region northwest of the Gulf of Tadjura is covered by Afar Basalt of Tertiary (probably Miocene) age (Mohr, 1967a), which is partly overlain by basaltic and acid lavas of the Pleistocene Aden Volcanic Series. North of the Gulf of Tadjura lighter colored rocks crop out from beneath the Afar Basalt. These rocks appear to have lithologic characteristics similar to those observed in the Danakil Mountains on the Apollo 9 frame AS 9-23-3535. The Danakil Mountains are composed of crystalline rocks with overlying Mesozoic sediments (Mohr, 1962, p. 53; Bannert et al., 1970, p. 430).

The international geologic map of Africa (ASGA-UNESCO, 1963, sheet 6) shows only a small outcrop of basement rocks near the town of Tadjura. More information is provided by the geologic map of French Somalia (Besairie, 1946). The whole region east of the Afar Basalt is shown as rhyolitic rocks. Recent observations (J. Varet, written commun., 1971) seem to verify this statement.

However, on space photographs (AS 9-23-3539; AS 9-23-3538) two rock units can be differentiated. A northern light-colored unit (Figs. 2,

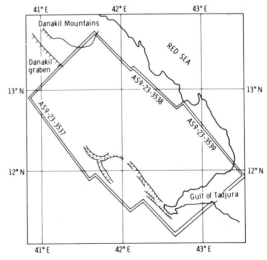

FIG. 5—Sketch map showing part of northern Afar region and coverage by Apollo 9 photographs.

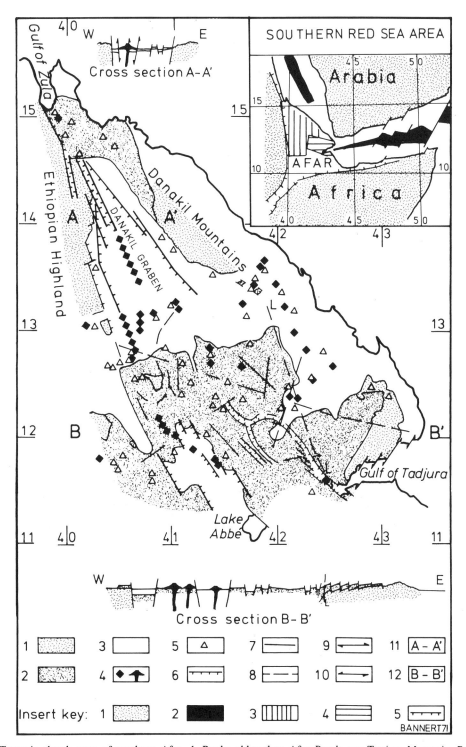

FIG. 6—Tectonic sketch map of northern Afar. *1*, Rocks older than Afar Basalt, pre-Tertiary Mesozoic, Paleozoic, pre-Paleozoic, and crystalline rocks; *2*, main area of Tertiary plateau-forming Afar Basalt; *3*, upper Tertiary to Holocene sediments and Pleistocene lavas of Aden Volcanic Series (no data available east of Gulf of Zula and along Red Sea coastline); *4*, main volcanoes and fissures of Pleistocene to Holocene basalts of Aden Volcanic Series; *5*, acid volcanoes of the Aden Volcanic Series; *6*, main graben faults, with tick marks pointing to downfaulted side; *7*, major faults and structural trends; *8*, long lineations observed on space photographs; *9*, shear fault zone, indicating right-lateral displacement; *10*, shear fault zone, indicating left-lateral displacement; *11*, cross section *A* along 14°N lat.; *12*, cross section *B* along 12°N lat.; *L-L*, lineation, probably western limit of Tadjura block.

Key for insert: *1*, regions bordering rift structures; *2*, rough zone of probably basaltic rocks located in center of Red Sea and Gulf of Aden; *3*, geology compiled from aerial photography and ground surveys; *4*, region investigated on space photographs; *5*, major graben faults, with tick marks pointing to down-faulted side.

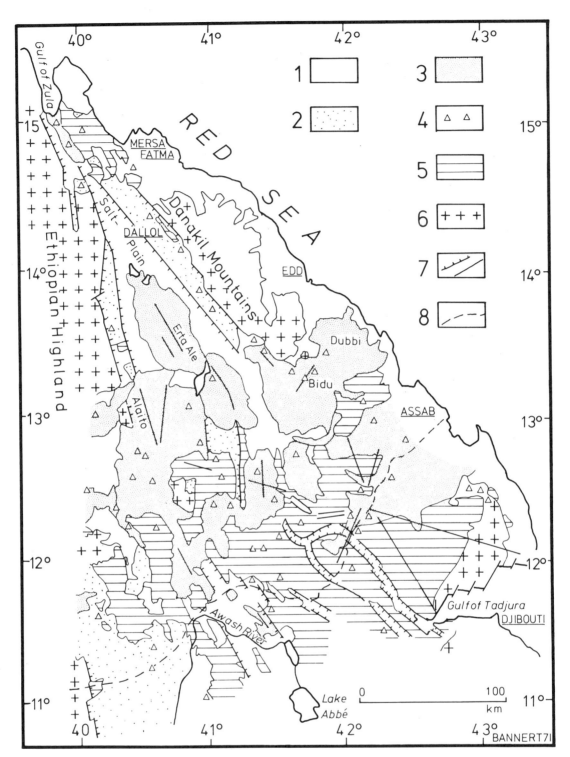

FIG. 7—Geologic sketch map of northern Afar. *1*, Recent fluviatile and lacustrine deposits; *2*, upper Tertiary and Quaternary sediments; *3*, basaltic effusive rocks, Aden Volcanic Series, Pleistocene; *4*, acid alkali-rich effusive rocks, Aden Volcanic Series, Pleistocene (west of Danakil Mountains, partly older intrusives); *5*, plateau-forming Afar Basalt. Tertiary; *6*, Mesozoic and older rocks; north of Gulf of Tadjura probably rhyolites; *7*, faults; *8*, road (*see* Centre Recherche Sci. and Consiglio Naz. Recherche, 1971).

FIG. 8—Geologic map of region north of Gulf of Tadjura, prepared on basis of information from Apollo 9 space photographs AS 9-23-3537 through AS 9-23-3540.

QUATERNARY: *1*, fans; *2*, clastic debris; *3*, fine-grained sediments with high reflection in periodically inundated lakes; *4*, white salina (?) deposits; *5*, coastal sediments on shorelines; *6*, undetermined—light-brown sediments overlying Afar Basalt.

LATE TERTIARY-PLEISTOCENE:—Aden Volcanic Series: *7*, basalts of Aden Volcanic Series with flow pattern; *8*, acid extrusives; *9*, cinder cones; *10*, caldera; *11*, volcano.

MIOCENE: *12*, Plateau forming Afar Basalt; *13*, rocks underlying Afar Basalt, probably rhyolite; *14*, shoreline; *15*, wadi; *16*, stratigraphic contact; *17*, boundaries within a geologic unit; *18*, major and minor faults with tick marks pointing to downfaulted side; *19*, fractures and joints; *20*, lineations; *21*, shear fault, right-lateral displacement; *22*, shear fault, left-lateral displacement. →

202

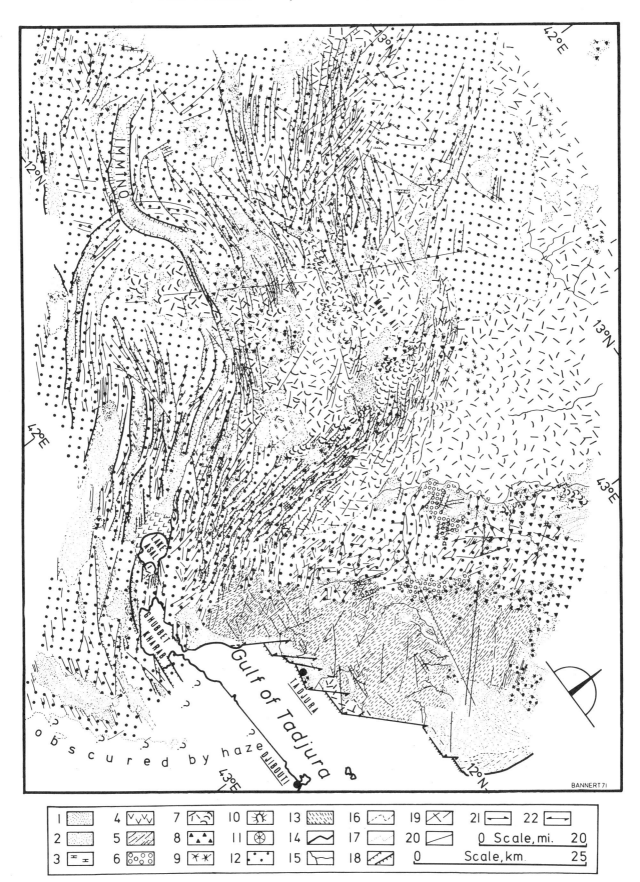

BANNERT 71

1		4		7		10		13		16		19		21		22	
2		5		8		11		14		17		20			Scale, mi.		
3		6		9		12		15		18			0 Scale, km 25			

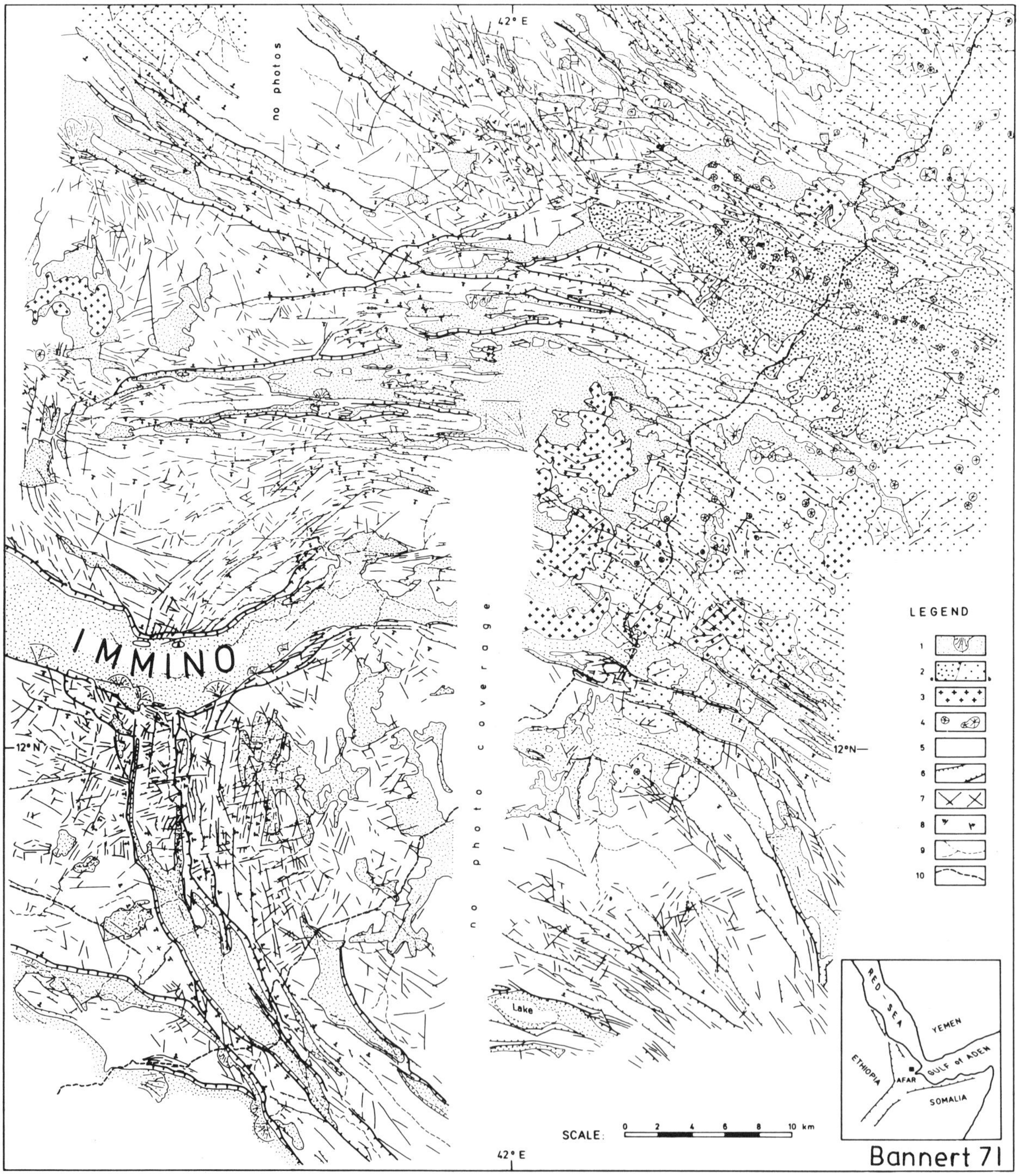

LEGEND:
1 Recent deposits
2 Basaltic volcanoes a) younger, b) older } Aden Volcanic Series
3 Acid volcanoes
4 Cinder cones
5 Plateau forming basalts Afar Basalts
6 Faults
7 Photolineations
8 Dip
9 Wadi
10 Road

FIG. 9—Annular grabens in northern Afar. Photogeologic map prepared on basis of information from aerial photographs; scale, 1:60,000.

10), shows smaller domes and indications of out-flow patterns similar to those observed in the rhyolitic extrusives of the rest of northern Afar (Tazieff *et al.*, 1970, Fig. 10; Bannert, 1971, Fig. 9). This complex also lacks the many north-northwest-striking joints. These joints dominate the southern darker-colored rhyolitic unit, which underlies the Afar Basalt (Fig. 11). They also seem to underlie parts of the lighter rhyolite. The darker rhyolite looks very similar to the basement rocks of the Danakil Mountains.

An explanation is necessary for the different morphology of the two rock types. This problem demonstrates the difficulty of interpreting space photographs. Jointing and fracturing can mask completely the normal morphology of a specific rock unit. The small scale (in this case approximately 1:700,000) reduces the observable data so that most of the details usable for identification vanish. Rock-type identification from space photographs *per se* is possible only in very few cases (*e.g.*, volcanoes, alluvial deposits). In this example of the rhyolite, which has two modes of expression on the space photograph, two explanations can be given. One of the rhyolites may have a different fracture pattern formed by regional tectonism that is unexplained, with transitions between the two types. In Figure 11 there are

Fig. 11—Enlargement from Figure 2 (NASA no. AS 9-23-3539) showing strongly jointed rock unit in lower part of picture near *D*. Rock unit appears similar to basement rocks in Danakil Mountains. Unit probably is composed of rhyolite. It underlies lighter colored rhyolite shown in Figure 10 and Afar Basalt near *E*. *F* indicates Afar Basalt. *G* is lighter colored unit of undetermined origin, overlying Afar Basalt. This series may be composed of acid tuff of nearby rhyolitic volcanoes. Near *H*, circular structure of unknown origin, rocks are slightly lighter than near *D*. Dark tone at bottom center probably indicates stronger vegetation at higher elevations.

indications that rocks in the arcuate area belong to the lighter colored type of rhyolite. Second, there may be two different rhyolites, one of them older and underlying the Afar Basalt (therefore shown as "basement" on the maps) and the other younger and overlying the older. This explanation seems to be more likely at the present stage of geologic exploration.

The overlying Afar Basalt forms extensive plateaus in the Afar triangle. The basalt is very dark on aerial and space photographs, which show all major tectonic features, such as joints and faults (Figs. 8, 9). Afar Basalt and the volcanoes of the Pleistocene Aden Volcanic Series presented no difficulties in identification. In the Aden Volcanic Series rhyolitic and basaltic extrusives are discernible. The rhyolitic extrusives compare with the large rhyolitic complex (upper right of Fig. 2) which therefore was mapped on Figure 8 as belonging to the Aden Volcanic Series. The nature and origin of the light-colored rocks outlined by small circles in Figure 8 are undetermined. They cover the Afar Basalt about 40 km north of the town of Tadjura. They have a comparatively smooth surface and no major drainage pattern

Fig. 10—Enlargement from Figure 2 (NASA no. AS 9-23-3539) showing light-colored rock unit which has been interpreted as rhyolite of Aden Volcanic Series. Near *A*, centers of volcanic eruptions. Near *B*, indications of flow pattern. Near *C*, older underlying unit which is shown in Figure 11.

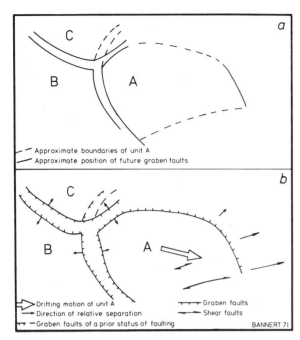

FIG. 12—Suggestion of motions forming annular grabens. (a) Outlines of different blocks before separation; (b) present situation with black arrows indicating relative separation. White arrow indicates drifting motion of unit A.

can be observed which would help in their identification. It is assumed that these rocks are acidic tuffs produced by the extensive rhyolitic volcanism in the adjacent areas on the northeast.

Afar Basalt and Aden Volcanic Series are both very significant on aerial photographs, as well as on space photographs. Their identification has been verified in extended field surveys (Tazieff *et al.*, 1970; Bannert *et al.*, 1970) mostly north of 13°N lat., but also on field trips along the road from Assab to Dessie, crossing the regions shown in the frames AS 9-23-3538 and -3539 (Figs. 3, 4) and in several other places (Besairie, 1946; Tazieff, 1971).

Tectonics

Two different tectonic systems dominate the region northwest of the Gulf of Tadjura, the Tadjura block in the east and several annular grabens in the west. As seen in Figures 2 and 8, the strike of the joints in older rhyolitic rocks ("basement" underlying the Afar Basalt) is about north-northwest. The northern shoreline of the Gulf of Tadjura is very conspicuous and is formed by east-west-striking faults, along which the gulf is downfaulted. These faults are cut by northeast-striking left lateral transcurrent faults, which indicate a northeast motion vector. The transcurrent faults seem to extend offshore for some distance, as recorded in recent bathymetric and geophysical surveys (Peter and DeWald, 1969, Fig. 1).

The southern shoreline of the gulf looks completely different. This shoreline strikes strictly east-west, and only two transcurrent faults with a minor amount of offset are observed in the western part. The difference between the northern and southern shoreline is obvious, and apparently, in addition to its downfaulted characteristics, near the southern shoreline an east-west-striking right-lateral wrench fault is present (Mohr, 1968, Fig. 3), which separates the Tadjura block from a southern uplift composed of crystalline rocks and called Aisha horst by Mohr (1967b, p. 664). The western part of the Gulf of Tadjura is called Ghubbet Kharab and is shaped by a northwest-striking graben. This graben extends to the annular Immino graben (Fig. 8). Numerous right-lateral transcurrent faults strike east and southeast. These features, together with the left-lateral transcurrent faults north of the Gulf of Tadjura, indicate a northeastward movement of the whole Tadjura block, including the western part that is covered by Afar Basalt. The western border of the drifting Tadjura block may be indicated by a long lineation detected on Apollo 9 photographs (Figs. 2, 3). In Figure 6, this lineation is shown as the line *L-L*. Where this lineation crosses two other (east-west and north-south striking) major lineations, a volcanic field of the youngest basalts of the Aden Volcanic Series is present (Figs. 7, 8). Two facts indicate that, in the south, this lineation could be a flexure fault. East of it, toward the Tadjura uplift, the overlying Afar Basalt is faulted into tilted blocks that dip slightly southwestward. The decrease in the thickness of the Afar Basalt and the decrease in the intensity of faulting indicate the approach of the rhyolitic "basement" rocks of the Tadjura block to the surface. West of the lineation, the Afar Basalt is dissected into many horst and graben structures (cross section B-B', Fig. 6), which generally strike northwestward, as mentioned previously. These features may indicate a greater depth of the basement rock and a zone of inhomogeneity and differential movement caused by the drifting of the Tadjura block.

One unique feature in the Afar region is a system of conspicuous annular grabens west of the Tadjura block (Figs. 3, 4, 8, 9). These grabens were first mentioned by Mohr (1967b, 1968) and were mapped on aerial photographs at the scale of 1:60,000 (Fig. 9; Bannert, 1969, p. 523). If the tectonic features are considered to be caused by crustal movements, different and individual movements of the adjoining blocks bordering the annular grabens can be assumed (Fig. 12). This hypothesis could best explain the lack of graben

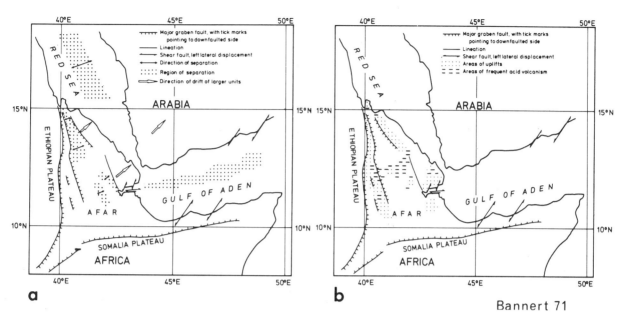

FIG. 13—Tectonic units in southern Red Sea and Gulf of Aden area;
(a) area of strong drifting; (b) horst areas.

structures just north of region A in Figure 12. Block A, approximately 25 km across and trending northwest-southeast, lacks major fractures and faults, as shown on both aerial and space photographs (Figs. 3, 4, 9). It is, therefore, assumed that this region is an individual block that is rotating and following the northeastward-moving Tadjura block. A zone of right-lateral east-northeast-striking transcurrent faults indicates this motion. These faults are distinguished by the offset of minor grabens and horsts, and by the offset of tilted blocks, and can be observed only on space imagery. Examination of aerial photographs at the west end of the zone of the right-lateral transcurrent faults only revealed single photolineations in the same direction.

There are two points of particular interest regarding the tectonic movements of the northern part of the Afar region. Between the Ethiopian Highlands in the west and the Danakil Mountains and Tadjura block in the east, several large horst regions, partly with outcrops of Mesozoic rocks and crystalline rocks, are found (Figs. 6, 7). It is suggested that these horsts are isolated blocks, each drifting separately in the space opened as crustal separation progressed. This open space is filled continuously by upward-moving basalt from the deep oceanic crust (Tazieff and Varet, 1969). In addition, the distribution of acid alkali-rich extrusive rocks of the Aden Volcanic Series is very conspicuous. Tazieff (1970) and Barberi et al. (1970) noted that, according to chemical composition, the acid and basaltic material that forms ranges in the center of the Da-

nakil graben apparently has an origin different from the acid and basaltic materials in the western part of Danakil graben. By studying aerial and space photographs, it may be assumed that acid extrusive rocks are not present where strong drifting is suggested (Fig. 13a), except for some acid volcanoes, which have a different morphology probably caused by different magmatic composition (Bannert, 1971, Fig. 10).

In the horst areas adjacent to regions of strong drifting (Fig. 13b), acid volcanoes are numerous, but are not aligned in a structure like the hypothetical Wonji fault belt (Mohr, 1967b).

DIFFERENCES IN INTERPRETATION OF AERIAL AND SPACE PHOTOGRAPHS

Figure 9 shows a photogeologic map of the region of annular grabens. Ground surveys have been made in the southwestern part of this region (Bannert et al., 1970; Brinckmann and Kürsten, 1970). As basaltic and acidic volcanoes are relatively easy to distinguish on aerial photographs, a greater problem is the differentiation between the Afar Basalt and the superjacent, older basalt of the Aden Volcanic Series. It is commonly impossible to distinguish these two types of basalt if there are not typical lava flows on the slopes of volcanoes of the Aden Volcanic Series. Even in ground surveys, distinguishing between the two is nearly impossible. Space photographs, however, show a clear differentiation between the two basalt lavas (Figs. 2-4), apparently because space photographs have less resolution and greater tonal color differences than black and white aerial photographs. The surface of both the Afar Basalt and the older basalts of the Aden Volcanic Series are partly covered by outwash and have many small depressions filled with recent deposits of material that probably has been transported by wind. Generally, it is impossible to differentiate between the two types of basalt on aerial photographs with a scale of 1:60,000. On space photographs, however, the small depressions cannot be seen because of the lower resolution and smaller scale, and only the large depressions are discernible. The number of depressions per square unit are quite different in the two types of basalt; therefore, with the additional aid provided by the tonal color contrasts in space photographs, it is possible to outline the boundaries between the two basalt units.

Another advantage of space photography is that a synoptic view of large areas is provided by a single photograph. Thus, it is possible to determine that the intensity of tectonic movements decreases toward the north. This decrease is not clearly visible on the map prepared from aerial photographs (Fig. 9). From photograph to photograph, the number and density of the fractures decrease in small steps. The interpreter of aerial photographs may change his scale of values regarding tectonic features (*e.g.*, large, medium, or small faults) in the same steps. The result is that important grabens with major downfaulting are found north of the annular grabens (Fig. 9) and are comparable with the Immino graben in the south. However, on the geologic map (Fig. 8) prepared from space photographs, this interpretation does not show the true proportions of the tectonic features.

Conclusions

In remote regions, space photography can provide a rather detailed knowledge of rock distribution, if the circumstances are adequate to allow geologic interpretation. The knowledge of rock distribution and rock identification, together with clearly visible fracture and joint patterns, have been the bases for tectonic investigations leading to more detailed analysis of drift motions along the southern end of the Red Sea rift. The analysis suggests that parts of the African craton (or African plate) are drifting between Africa and the northeastward-moving Arabian Peninsula. In the future, space photography will help to increase knowledge about general geologic features in many poorly mapped regions of the world and will provide more detailed knowledge about plate motions on the land surfaces.

References Cited

Abdel-Gawad, M., 1969a, New evidence of transcurrent movements in Red Sea area and petroleum implications: Am. Assoc. Petroleum Geologists Bull., v. 53, p. 1466-1479.

——— 1969b, Geological structures of the Red Sea area inferred from satellite pictures, *in* Hot brines and recent heavy metal deposits in the Red Sea: New York, Springer-Verlag, p. 25-37.

——— 1970, Interpretation of satellite photographs of the Red Sea and Gulf of Aden: Royal Soc. London Philos. Trans., v. 267, p. 23-40.

ASGA-UNESCO, 1963, Geological map of Africa: Paris, Assoc. Serv. Géol. Africains, UNESCO, scale 1:5,000,000, sheet 6.

Bannert, D., 1969, Geologie auf Satellitenbildern: Naturw. Rundschau, v. 22, p. 517-524.

——— 1971, Photogeologische Untersuchungen an den Vulkanen des nördlichen Afar (Äthiopien): Bildmess. und Luftbildwesen, v. 39, p. 77-84.

——— in press, Fotogeologische Untersuchungen im nördlichen Afar (Äthiopien): Beihefte Geol. Jahrb. 116, 52 p.

——— J. Brinckmann, K. Ch. Käding, G. Knetsch, M. Kürsten, and H. Mayrhofer, 1970, Zur Geologie der Danakil-Senke (nördliches Afar-Gebiet, NE-Athiopien): Geol. Rundschau, v. 59, p. 409-443.

——— and E. Y. Kedar, 1971, Plate tectonics in the Red Sea region as inferred from space photography: NASA Tech. Note D-6261, 16 p.

Barberi, F., S. Borsi, G. Ferrara, G. Marinelli, and J. Veret, 1970, Relations between tectonics and magmatology in the northern Danakil depression (Ethiopia): Royal Soc. London Philos. Trans., v. 267, p. 293-311.

Besairie, H., 1946, Carte géologique de la Côte française des somalis: Paris, scale 1:400,000.

Blanford, W. T., 1870, Observations of the geology and zoology of Abyssinia, made during the progress of the British expedition to that country in 1867-68: London, MacMillan, 487 p.

Brinckmann, J., and M. Kürsten, 1970, Geological sketchmap of the Danakil depression: Hannover, Germany, Federal Geol. Survey, scale 1:250,000, 4 sheets.

Centre National de la Recherche Scientifique and Consiglio Nazionale delle Recherche (Italy), 1971, Geological map of the Danakil depression (northern Afar-Ethiopia): Paris, Géotechnip, 78, la Celle-Saint Cloud, scale, 1:500,000.

Fallon, N. L., I. G. Gass, R. W. Girdler, and A. S. Laughton, eds., 1970, A discussion of the structure and evolution of the Red Sea and the nature of the Red Sea, Gulf of Aden and Ethiopia rift junction: Royal Soc. London Philos. Trans., v. 267, 417 p.

Freund, R., 1965, A model of the structural development of Israel and adjacent areas since Upper Cretaceous times: Geol. Mag., v. 102, p. 189-205.

Girdler, R. W., 1966, The role of translational and rotational movements in the formation of the Red Sea and Gulf of Aden: Canada Geol. Survey Paper 66-14, p. 65-77.

Holwerda, J. G., and R. W. Hutchinson, 1968, Potash-bearing evaporites in the Danakil area, Ethiopia: Econ. Geology, v. 63, p. 124-150.

Hurley, P. M., 1968, The confirmation of continental drift: Sci. American, v. 220, p. 52-64.

International Upper Mantle Project, 1966, The world rift system; report of symposium, Ottawa, Canada, September 4, 5, 1965: Canada Geol. Survey Paper 66-14, 471 p.

Laughton, A. S., 1966, The Gulf of Aden, in relation to the Red Sea and the Afar depression of Ethiopia, *in* The world rift system: Canada Geol. Survey Paper 66-14, p. 78-97.

Lowman, P. D., Jr., 1964, A review of photography of the earth from sounding rockets and satellites: NASA Tech. Note D-1868, 25 p.

——— 1967, Geologic applications of orbital photography: NASA Tech. Note D-4155, 37 p.

——— and H. A. Tiedemann, 1971, Terrain photography from Gemini spacecraft; final report: Greenbelt, Maryland, Goddard Space Flight Center, X-644-71-15, 75 p.

Mohr, P. A., 1962, The geology of Ethiopia: Addis Ababa Univ. Press, 268 p.

——— 1967a, The Ethiopian rift system: Addis Ababa Univ. Geophys. Observ. Bull., no. 11, 65 p.

——— 1967b, Major volcano-tectonic lineament in the Ethiopian rift system: Nature, v. 213, p. 664-665.

——— 1968, Transcurrent faulting in the Ethiopian rift system: Nature, v. 218, p. 938-941.

——— 1970, The Afar triple junction and sea-floor spreading: Jour. Geophys. Research, v. 75, p. 7340-7352.

Peter, G., and O. E. DeWald, 1969, Geophysical reconnaissance in the Gulf of Tadjura: Geol. Soc. America Bull., v. 80, p. 2313-2316.

Phillips, J. D., and D. A. Ross, 1970, Continuous seismic reflexion profiles in the Red Sea: Royal Soc. London Philos. Trans., v. 267, p. 143-152.

Tazieff, H., 1970, The Afar triangle: Sci. American, v. 222, p. 32-40.

——— 1971, Sur la tectonique de l'Afar Central: Acad. Sci. Comptes Rendus, v. 272, p. 1055-1058.

——— G. Marinelli, F. Barberi, and J. Varet, 1970, Géologie de l'Afar septentrional: Bull. Volcanol., v. 33, p. 1039-1072.

——— and J. Varet, 1969, Signification tectonique et magmatique de l'Afar septentrional (Ethiopie): Rev. Géographie Phys. et de Géologie Dynamique, v. 11, p. 429-450.

Varet, J., 1971, Sur l'activité récente de l'Erta' Ale (Dankalie, Ethiopie): Acad. Sci. Comptes Rendus, v. 272, p. 1964-1967.

Reprinted by permission of the Bureau Recherches
Géologiques et Miniàres from P. Teleki and C. Weber,
eds., *Remote Sensing for Geologic Mapping;
Télédétection Appliquée á la Cartographie
Géologique*: Documents BRGM 82, 1984, p. 299-309.

REMOTE DETECTION OF GEOBOTANICAL ANOMALIES ASSOCIATED WITH HYDROCARBON MICROSEEPAGE USING THEMATIC MAPPER SIMULATOR (TMS) AND AIRBORNE IMAGING SPECTROMETER (AIS) DATA*

Barrett N. ROCK

ABSTRACT

As part of the continuing study of the NASA/Geosat Test Case Site at Lost River, West Virgina, Gulf Research and Development conducted an extensive soil gas survey of the site during the summer of 1983. This survey has identified an order-of-magnitude methane and ethane anomaly that precisely coincides with the linear maple anomaly reported previously by this author. This and other maple anomalies were previously suggested to be indicative of anaerobic soil conditions associated with methane microseepage. In <u>vitro</u> studies reported by Parrish and by Leone and co-workers support the view that anomalous distributions of tree species tolerant of anaerobic soil conditions may be useful indicators of methane microseepage in heavily vegetated areas having deciduous forest cover.

Remote sensing systems which allow discrimination and mapping of tree species and/or species associations will provide the exploration community with a means of identifying vegetation distributional anomalies indicative of microseepage. The success of the Lost River geobotanical investigation suggests that the airborne Thematic Mapper Simulator (TMS) is adequate for this purpose providing data are collected at the proper time of year (peak fall foliage display in the Lost River study). High-spectral resolution sensor systems currently under development at the Jet Propulsion Laboratory, such as the Airborne Imaging Spectrometer (AIS) and the Airborne Visible/Infrared Imaging Spectrometer (AVIRIS), may also provide this discriminatory capability at various times during the growing season. Direct identification of vegetation types may also be possible using such high-spectral resolution systems.

Details of the study and interpretation of TMS and AIS data sets collected from Lost River, West Virginia are presented, in addition to a brief review of the supervised vegetation classification approach to vegetation mapping used at Lost River. Preliminary study of AIS data suggests that contiguous high-spectral resolution data from a very limited portion of the spectrum (1.2 - 1.5µm) provide a greater discriminatory capability than do broad-band sensors such as the TMS covering a wider spectral range (0.45 - 2.35µm).

Jet Propulsion Laboratory, California Institute of Technology, Pasadena, California 91109.

* This research was carried out at the Jet Propulsion Laboratory, California Institute of Technology, under contract with the National Aeronautics and Space Administration.

Manuscript received Feb. 3, 1984   revised July 10, 1984.

RÉSUMÉ

Dans le cadre d'une étude continue sur le site expérimental NASA/Geosat sur Lost River (Virginie occidentale), Gulf Research and Development a réalisé une expérimentation sur les gaz du sol sur le site durant l'été 1983. Cette étude a mis en évidence une anomalie de méthane et éthane qui coïncide précisément avec une anomalie linéaire dans la végétation (érables), décrite antérieurement par l'auteur. Il avait été précisé précédemment que cette anomalie, comme d'autres, pouvait indiquer des conditions de sol anaérobies associées avec des micro-fuites de méthane. Des études in vitro faites par Parrish, Leone et collab. appuient l'hypothèse que les distributions anormales des espèces arborescentes tolérant des conditions de sol anaérobies peuvent être des indicateurs utiles de micro-fuites de méthane dans des zones de végétation dense d'arbres à feuilles caduques.

Les systèmes de télédétection qui permettent de différencier et cartographier les espèces végétales ou les associations fournissent au responsable de l'exploration des moyens d'identifier la distribution des anomalies de végétation relatives à des micro-fuites. Le succès des investigations géobotaniques sur Lost River montre que le Thematic Mapper Simulator (TMS) convient à cet objectif en fournissant des données recueillies à un moment propice de l'année (pic du feuillage d'automne à Lost River). Des systèmes de détecteur à haute résolution spectrale, actuellement en développement au Jet Propulsion Laboratory, tels que le spectromètre imageant aéroporté (AIS) et le spectromètre imageant aéroporté visible/infra rouge (AVIRIS), fournissent également une discrimination de la saison de croissance. Une identification directe des types de végétation peut être également possible en utilisant des systèmes de haute résolution spectrale.

On présente ici les détails de l'étude et de l'interprétation des données TMS et AIS recueillies sur Lost River, ainsi qu'une brève description de la méthode de classification de la végétation utilisée pour la cartographie de végétation à Lost River. L'étude préliminaire des données de l'AIS laisse penser que les données de haute résolution spectrale provenant d'une portion très limitée du spectre (1,2 à 1,5 μm) fournissent une meilleure possibilité de discrimination que des détecteurs à large bande, tel que le TMS qui couvre une gamme plus vaste (0,45 à 2,35 μm).

300

INTRODUCTION

Highly accurate supervised vegetation classification images of heavily forested terrain in the Ridge and Valley Province of the Appalachian basin of West Virginia, U.S.A., have been produced from simulated Landsat 4 Thematic Mapper (TM) data as part of the joint NASA/Geosat Test Case Study (Abrams, et al., 1984). Details of the vegetation mapping procedures and production of the supervised classification images are provided by Rock (1982).

The Lost River, West Virginia area (Figure 1) was selected for study because of the occurrence of a natural gas (methane) field located at depth in a Lower Devonian Oriskany Sandstone reservoir (Cardwell, 1974; Crow, 1981). Such a heavily forested site occurring over a source of potential microseepage of methane provides an ideal setting in which to identify vegetation response to methane-rich soil gas conditions, using remote sensing systems such as the TMS and AIS.

The Lost River site is forested (approximately 80%), representing a deciduous-evergreen (oak-hickory-pine) cover type. The non-forested areas are under cultivation (lawns, pastures and field crops such as corn and hay), and thus, surface reflectance data are totally dominated by various types of vegetation.

It was assumed that any potential botanical anomalies associated with methane microseepage would be either distributional or physiological (see Goetz, et al., 1983, for a discussion of various vegetation responses to geochemical stress.) Such anomalies may be due either to the presence of the gas itself or to alteration of the soil environment by the gas. In order to recognize vegetational anomalies, it was necessary to determine the normal physiological state and distribution patterns characteristic of the various physiographic sites within the study area.

To facilitate discrimination and identification of species and species complexes (communities) for this mapping exercise, TMS data were acquired during peak fall foliage display periods. Interpretation and ground checking of the supervised classification images led to the identification of subtle distributional vegetation anomalies. Detailed soil gas analysis has located methane through butane soil gas anomalies within the three vegetational anomalies tested. Gas sample measurements from within one of the vegetation anomalies represents the highest gas values detected over the entire gas field.

METHODS

An initial phase of the study involved use of preliminary field vegetation analysis (Rock, 1982) to establish a total of 35 training areas to be used in a supervised classification of airborne TMS (NASA NS-001) multispectral data acquired at peak fall foliage display (October 1980). The supervised vegetation classification image (Figure 2) was produced using a hybrid parallellepiped-Bayesian maximum likelihood classifier (Rock, 1982). The raw NS-001 band data used in the classification were: band 2 (0.52 - 0.60μm), band 3 (0.63 - 0.69μm), band 4 (0.76 - 0.90μm), band 5 (1.00 - 1.30μm), band 6 (1.55 - 1.75μm), and band 7 (2.08 - 2.35μm). These data were acquired at an altitude which delivered an average 15 m pixel at nadir.

301

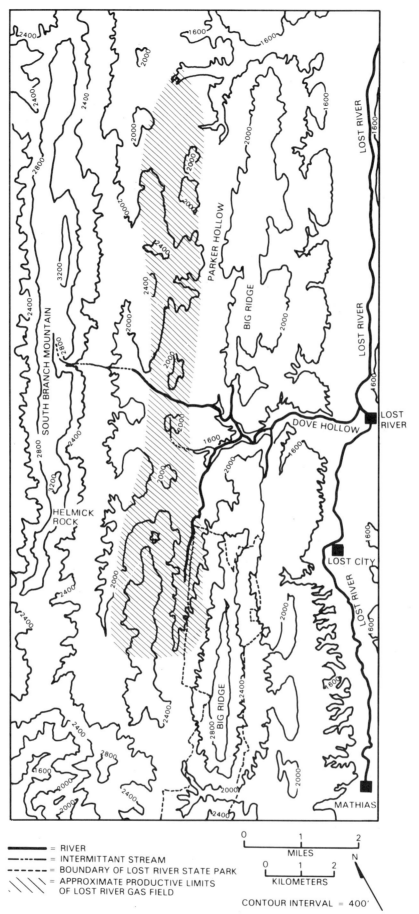

Figure 1. Diagrammatic representation of topographic and location data for the Lost
River field site. Images seen in Figure 2 cover approximately the same
area. Note the location of South Branch Mountain and Big Ridge for the
purposes of locating features in Figure 2.

302

214

Figure 2. NS-001 images of the Lost River, West Virginia (U.S.A.) study area. Image
on the left is a false color infrared (CIR) mimic of an aerial photograph
produced by assigning the colors blue, green and red to bands 2, 3 and 4,
respectively. This image is included for the purpose of orientation and
location of ground features. The image on the right is one of the
supervised vegetation classification images produced using a band and
band-ratio combination of 6, 3/5, 4/5, 5 and 3/6. The vegetation classes
recognized (but too small to be seen clearly in this copy): evergreen,
chestnut oak, chestnut oak shaded, scarlet oak, other oak, maple, field 1,
field 2, soil and scrub, respectively.

AIS data have been acquired over the field site during three stages of the growing cycle: late May, mid-August, and late October (foliage peak), 1983. For each overflight, data were acquired in 64 contiguous spectral channels, each 9.6 nm wide, spanning the 1.2 -1.8μm region of the spectrum at an altitude delivering data at approximately 10 m IFOV (see Rock; 1983, Vane, et al., 1983, for description of AIS and its botanical applications). During each of the flights, the NS-001 acquired data simultaneously with AIS. False color infrared (CIR) and panchromatic black and white film data were also collected for each flightline.

A soil gas analysis of the entire Lost River study area was conducted during July and August, 1983, by Gulf Research and Development Company, Houston, Texas. The method used employed a probe inserted into a three foot hole and an evacuated bottle to sample soil gases located at the bottom of the hole. Soil gas analyses were then conducted using a field portable flame ionization gas chromatography unit using techniques developed by Gulf.

RESULTS

Two supervised vegetation classification images were produced, one utilizing data from bands 2, 3, and 4 and band ratios of 4/5 and 5/7, and the other utilizing bands and band ratios of 6, 3/5, 4/5, 5 and 3/6. Both of the supervised vegetation classification images were based on nine vegetation classes. Determination of the most useful bands for vegetation discrimination was made by an evaluation of inter-band correlation matrices, principal component analyses, and stepwise discriminant analyses. In order of decreasing utility for vegetation discrimination, the four most useful bands were: band 6 (1.55 - 1.75μm), band 3 (0.63 - 0.69μm), band 5 (1.00 - 1.30μm) and band 4 (0.76 - 0.90μm). It should be noted that NS-001, band 5 (1.00 - 1.30μm) is not included on the Landsat-5 TM.

The selection of bands for production of the first supervised vegetation classification image (2, 3, 4, 4/5 and 5/7) was based on the comparison of a number of initial images (a band 2, 3, 4 color composite image; a 4/3, 4/5, 5/7 color ratio composite image; and a band 2/3 ratio color slice) with the preliminary field vegetation maps. The selection of bands for production of the second supervised vegetation classification image (6, 3/5, 4/5, 5 and 3/6 [shown in this report as Figure 1]) was determined by use of a stepwise linear discriminant analysis program.

A comparison of assignment of vegetation classes on each of the images with actual ground conditions shows both classifications to be highly accurate. During the summer of 1981, both of the classification images were field checked. For the Figure 2 image, a total of 289 field checks were made (using a modified 0.1 acre circular plot analysis method as suggested by Lindsey, et al., 1958) of selected broadleaf forest sites, and of these, 256 were correctly classified (88.6%) once the variations noted in class descriptions are considered (Rock, 1982). Correct assignment of the two field classes, based on comparison of images with aerial photography, is considered to be approximately 70% accurate. Based on field evaluations, the chief source of error for both classification images is the result of errors in training area assignments.

As a result of study of the classification images produced, subtle distributional anomalies were recognized (Rock, 1982). Figure 3 illustrates two of the anomalies in the form of concentrations of maples occurring at low elevation sites where normal plant community succession would be expected to result in oak-hickory climax communities. These and other maple anomalies were identified as a result of image analysis, followed by detailed field work conducted in July, 1981.

The soil gas survey conducted in 1983 identified an order-of-magnitude methane and ethane (with measurable amounts of propane and butane) anomaly which precisely coincides with the linear red maple anomaly identified on the NS-001 vegetation classification images. A comparison of Figures 3 and 4 illustrates this strong correlation. Comparison of methane concentration values in Figure 4 with the summary methane values for the entire study area (Figure 5) clearly indicates the anomalous nature of the gas values found within the linear maple site.

Preliminary study of AIS data acquired from the Lost River study site, as well as from the Virgilina district of North Caroline (a site having similar types of vegetation cover), suggests that AIS is superior to the NS-001 for the purpose of vegetation discrimination. As reported previously (Rock, 1983), the high-spectral resolution provided by AIS, although from only a limited spectral region, detects some of the spectral features associated with the near-infrared (NIR) plateau and shortwave-infrared (SWIR) that appear to be characteristic of individual species (Goetz, et al., 1983). Only a portion of this spectral region is detected by the Landsat TM. AIS data in the $1.2 - 1.5\mu m$ region indicate systematic variations associated with different types of plants and plant communities.

DISCUSSION

As a result of the Lost River botanical studies (both ground assessment and production of the classification images), the major broadleaf deciduous and needle-leaf evergreen vegetation types have been well documented in terms of their repeated association with specific topographic, lithologic, and aspect/slope features characterizing this portion of the Appalachian Mountains. When compared with the various types of forest cover occurring in West Virginia (Core, 1966), the woody vegetation found at the test site represents a typical oak-hickory-pine forest, characteristic of the Ridge and Valley Province of the state. The percent cover represented by each class, as derived from pixel assignments from both images, were compared (Rock, 1982) with data presented by Sturm (1977) for the Lost River State Park and were found to correlate very well with Sturm's assessment of vegetation cover in the area.

Laboratory studies (Parrish and Rock, 1983; Gilman, et al., 1981), in which woody plant species have been grown in contact with methane under controlled conditions, indicate that certain species are likely to be more tolerant of such conditions than others. Leone, et al. (1977) suggest that the influence of methane is indirect, generating an anaerobic soil condition. As red maples are commonly found growing along streambanks in which soils are likely to be saturated with water at least part of a growing season (creating an anaerobic soil condition), it is reasonable to consider them as candidates for growth in methane-rich soils.

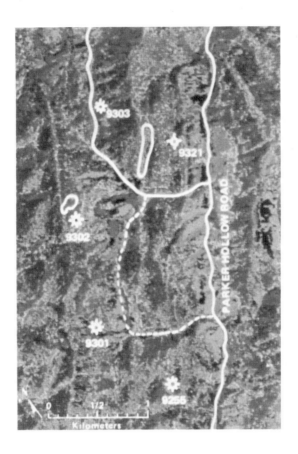

Figure 3. An enlargement of a small portion of the supervised classification image shown in Figure 2. The enlargement contains two of the maple anomalies discussed in the text, a circular anomaly and a linear anomaly. The color code assigned for Figure 2 is also appropriate for this enlargement.

306

218

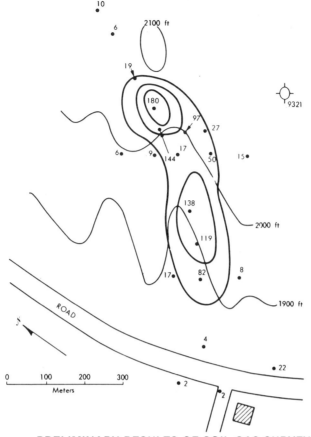

GULF SOIL GAS SURVEY
JULY 1983
(MATT MATTHEWS AND VICTOR JONES)

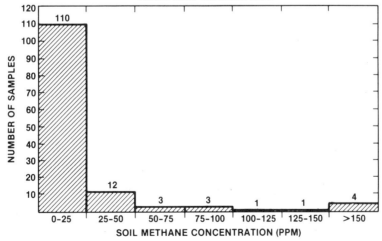

PRELIMINARY RESULTS OF SOIL GAS SURVEY
CONDUCTED AT THE LOST RIVER, WV TEST SITE, BASED
ON GAS SAMPLE ANALYSES FOR 20% OF THE TOTAL
SAMPLES COLLECTED

Figure 4. Results of the soil gas analysis conducted by Gulf Research and Development Company at the linear maple anomaly. Numbers represent the amount of
methane (ppm) contained within the probe sample taken at each sample
site. Light lines represent topographic contours and cultural features.
Heavy lines represent inferred concentration gradient for methane gas at
the site. The image of the linear maple anomaly and the lowest gradient
contour correspond very closely. The symbol labelled 9321 indicates a dry
gas well considered to represent the eastern boundary of the gas field.

Figure 5. Comparison of gas sample analysis for the Lost River, West Virginia study
site. Vertical bars represent the number of samples per methane concentration class. This comparison indicates the anomalous nature of the
methane concentrations from the linear maple anomaly (Figure 4).

Maples have been shown to require a different type of mycorrhizal fungus than oaks (Parrish and Rock, 1983; Abrams, et al., 1984) which suggests a possible explanation for maple's tolerance of the anaerobic soil conditions. Maples require a form of fungal association known as Endomycorrhizal which consist of lower (i.e., more primitive) forms of fungi, whereas, oaks require an Ectomycorrhizal association which is a higher form of fungus. Metabolic differences between the two forms of required fungal associations may allow those plants possessing Endomycorrhizal fungi to be at a competitive advantage in tolerating anaerobic soil conditions. If this is the case, use of this approach need not be site nor species specific. Remote discrimination, identification, and mapping of species known to require Endomycorrhizal associations, and thus more likely to be tolerant of anaerobic soil conditions and considered to be anomalously distributed, may prove to be a valuable exploration tool. High-spectral resolution remote sensing systems such as AIS, with its improved discriminatory capability, will prove most useful in this approach.

CONCLUSIONS

The techniques employed at the Lost River test site for vegetation discrimination and mapping have been very successful, and highly accurate supervised vegetation classification images have been produced (approximately 90% accurate assessment of seven forest classes and 70% accurate assessment of two field classes). The 15 m IFOV and the bandwidth selection provided by the airborne NS-001 TMS are well suited for vegetation mapping of heterogeneous deciduous forest cover. Multispectral data acquired during October foliage display proved valuable in allowing successful species discrimination and mapping. Based on data from this study, NS-001 band 2 (0.52 - 0.60μm, band 3 (0.63 - 0.69μm, band 4 (0.76 - 0.90μm, band 5 (1.00 - 1.30μm, band 6 (1.55 - 1.75μm, band 7 (2.08 - 2.35μm) are useful in providing accurate vegetation discrimination, classification, and mapping of eastern deciduous forest species and species associations. Anomalous distribution of maples detected on the supervised classification images have been shown to correlate very well with soil gas anomalies.

Preliminary analysis of AIS data acquired over this and other field sites suggests that high-spectral resolution imaging systems may provide even better vegetation discrimination and mapping capability when compared with the NS-001. Results of the Gulf soil gas analysis clearly indicate the value of vegetation discrimination, classification, and mapping in energy exploration activities carried out in heavily vegetated terrain.

308

REFERENCES

Abrams, M.J., Conel, J.E. and Lang, H.R., 1984, <u>The Joint NASA/Geosat Test Case Project Final Report</u>, Pasadena, California, JPL Technical Report, in press.

Cardwell, D.H., 1974, <u>Oriskany and Huntersville gas fields of West Virginia</u>, West Virginia Geological and Economic Survey, Mineral Resources Series, No. 5.

Core, E.L., 1966, <u>Vegetation of West Virginia</u>, Parsons, West Virginia, McClain Printing Co.

Crow, P., 1981, <u>Gas strikes spark play in eastern overthrust</u>, Oil & Gas Journal, Vol. 79, No. 17, pp. 109-113.

Gilman, E.F., Leone, I.A. and Flower, F.B., 1981, <u>The adaptability of 19 woody species in vegetating a former sanitary landfill</u>, Forest Science, Vol. 27:1, pp. 13-18.

Goetz, A.F.H., Rock, B.N. and Rowan, L.C., 1983, <u>Remote sensing for exploration: an overview</u>, Economic Geology, Vol. 78:4, pp. 573-590.

Leone, I.A., Flower, F.B., Arthur, J.J. and Gilman, E.F., 1977, <u>Damage to woody species by anaerobic landfill gases</u>, Journal of Arboriculture, Vol. 3:12, pp. 221-225.

Lindsey, A.A., Barton, J.D. and Miles, S.R., 1958, <u>Field efficiencies of forest sampling methods</u>, Ecology, Vol. 39, pp. 428-444.

Parrish, J.B. and Rock, B.N., 1983, <u>The effect of soil methane on in vitro growth and vigor of Quercus prinus seedlings</u>, Botanical Society of America, Vol. 70:5, Part 2, p. 52.

Rock, B..N., 1982, <u>Mapping of deciduous forest cover using simulated Landsat-D TM data</u>, International Geoscience and Remote Sensing Symposium (IGARSS '82), Vol. 1, WP-5, pp. 5.1-5.4.

Rock, B.N., 1983, <u>Preliminary Airborne Imaging Spectrometer vegetation data</u>, International Geoscience and Remote Sensing Symposium (IGARSS '83), Vol. 2, FA-1, pp. 5.1-5.4.

Sturm, R.L., 1977, <u>Comparison of forest cover types in seven environmentally diverse areas in West Virginia</u>, Morgantown, West Virginia, unpublished M.S. Thesis, West Virginia University.

Vane, G., Goetz, A.F.H. and Wellman, J., 1983, <u>Airborne Imaging Spectrometer: a new tool for remote sensing</u>, Internationl Geoscience and Remote Sensing Symposium (IGARSS '83), Vol. 2, FA-4, pp. 6.1-6.5.

309

THERMAL INFRARED IMAGERY

Reprinted by permission of the publisher from "Geologic mapping using thermal images" by M. J. Abrams, A. B. Hahle, F. D. Palluconi, and J. P. Schieldge, *Remote Sensing of Environment*, v. 16, no. 1, p. 13-33. Copyright 1984 by Elsevier Science Publishing.

REMOTE SENSING OF ENVIRONMENT 16:13–33 (1984)   13

Geologic Mapping Using Thermal Images

MICHAEL J. ABRAMS, ANNE B. KAHLE, FRANK D. PALLUCONI, AND JOHN P. SCHIELDGE

Jet Propulsion Laboratory, California Institute of Technology, Pasadena, CA 91109

In the past, remote sensing from aircraft and satellite for geologic mapping concentrated on the visible and reflective infrared parts of the spectrum, because of the availability of Landsat and aircraft multispectral scanners operating in this spectral range. With the launch of the Heat Capacity Mapping Mission (HCMM) satellite, regional thermal image data also became available. We have examined the HCMM data for geologic information over two desert areas in southern California, the Trona area and the Pisgah area. Three techniques were used for displaying and combining thermal data and visible and near infrared, including color additive composites, principal components, and calculation of thermal inertia images. Use of the color additive composite image was simplest and allowed for simultaneous display of both thermal and reflectance properties. Thermal data were found to provide additional geologic information, unavailable from Landsat data or from aircraft visible and near-infrared data alone. The addition of these data relating to thermal properties allowed separation of rock types with differing thermal properties but with similar reflectance characteristics.

Introduction

Remote sensing by aircraft and satellite multispectral scanners has proven to be a valuable tool for geologic mapping (Rowan et al., 1974; Goetz and Rowan, 1981). The most common methods involve observations of the surface in the reflective visible and near-infrared (VNIR) spectral regions. With the launch of the first Landsat satellite in 1972, synoptic, global multispectral data have been available for geologic applications. These data, however, have limitations because (1) VNIR spectral images do not yield unique compositional information and (2) surface coatings on exposed materials can mask their true composition. Thermal radiance data can augment the surface reflectance data and provide additional remote sensing information related to material composition.

Early workers used aircraft-acquired, broad-band thermal images to solve several problems not amenable to solution using reflectance data alone (Bakker et al., 1978; Pohn et al., 1974; Sabins, 1969; Vincent and Thomson, 1972; Watson, 1975). Separation of certain rock types and delineation of folds and faults were based on differences in thermal properties of materials imaged. The thermal data were either used directly by converting digital or analog signals to images or used to compute thermal inertia (TI), a body property of materials related to density, conductivity, and specific heat (Kahle, 1977; Kahle et al., 1976; Price, 1977; Pratt and Ellyett, 1979). These studies indicated the value of thermal data, but also revealed the problems of limited coverage and availability of thermal data acquired by aircraft. On 26 April 1978, the Heat Capacity Mapping Mission (HCMM) thermal satellite was launched to acquire thermal data appropriate for reconnaissance geologic mapping. The satellite measured broad band reflective

0034-4257/84/$3.00

VNIR (0.5–1.1 μm) and thermal infrared (10.5–12.5 μm) radiance data during midday and thermal infrared radiance during the night. The nominal ground resolution at nadir was 600 m, and the swath width of the scan was approximately 700 km (HCMM, 1980).

Several reports (Kahle et al., 1981; Watson, 1982; Watson et al., 1981, 1982) have recently appeared describing the application of HCMM data to thermal inertia calculations for delineation of regional lineaments and rock types. This paper presents the results of two detailed analyses of HCMM data for geologic mapping, in the areas of Pisgah, CA and Trona, CA.

At Pisgah, we first investigated methods for displaying and interpreting HCMM data, and compared HCMM data to Landsat data. A simple new technique is presented for combining visible and near IR data with thermal data. We also investigated aircraft multispectral VNIR and broad band thermal IR data for Pisgah having much higher spatial resolution. Both thermal inertia and principal component images were interpreted.

Because the spatial resolution of HCMM data was found to be inadequate for geologic problems in a small area like Pisgah, we then investigated the utility of HCMM data for lithologic discrimination at a regional scale, examining HCMM data for the Trona 1° × 2° area.

Pisgah Study Area

The Pisgah test site is located in the Mojave Desert region of southeastern California in the vicinity of latitude 34°45′N, longitude 116°15′W (Fig. 1). The topography is variable: alluvial fans and dry lake beds surround rugged moun-

FIGURE 1. Location of Trona and Pisgah study areas in California.

tains with relief in excess of 700 m. Vegetation is either sparse or absent owing to the minimal annual precipitation.

A simplified geologic map of the area, redrawn from the California Division of Mines and Geology (1969) map is shown in Fig. 2. Rocks exposed in the area are igneous intrusives and extrusives, and sediments derived from these sources. Sedimentary rocks are dominated by fanglomerates and fan gravels derived from adjacent mountains and deposited as alluvial fans. Alkaline clay occurring in the dry lake beds and windblown sand deposits form surficial sediments of non-volcanic origin.

Structures in the area are dominated by northwest trending right-lateral strike-slip faults with some normal displacement, and minor normal faults of various trends (Fig. 2).

Several HCMM data sets consisting of day VNIR and thermal IR data, and night IR data acquired within 12 or 36 h of the

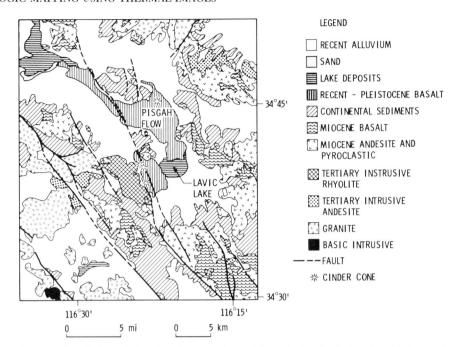

FIGURE 2. Geology of Pisgah study area. Modified from the California Division of Mines and Geology (1969) map.

daytime, were acquired over the Pisgah site. The best day–night data set was selected on the basis of sharpness of features in the images. These are shown in Figs. 3(a)–3(c).

The HCMM VNIR image [Fig. 3(a)] allows separation of a few units. The darkest areas are basalt outcrops, playas are the lightest, medium grays correspond to the remaining intrusive rocks. Very little discrimination is possible. The HCMM daytime IR image is shown as Fig. 3(b). The cold areas (dark) correspond to Lavic Lake playa, and outcrops of pyroclastics, andesite, some basalts, dacite, biotite granite, and granite. The area of aa at the Pisgah flow is colder than the pahoehoe, though the flow is not separable from the surrounding alluvial fans. The warmest areas are those corresponding to older fan deposits (north central part of the image). The HCMM nighttime IR image is shown as Fig. 3(c).

The warmest areas generally outline bedrock outcrops; alluvial areas are colder; Lavic Lake playa is quite cold. Very little lithologic discrimination is possible based on this image.

The various day–night image sets were coregistered, and then registered to a topographic base map. Registration was accomplished by identifying common tie points between the data sets, then "rubber-sheet" stretching one image to another using a fifth-order polynomial surface fit to the distortion field.

Thermal inertia images were produced from several of the day–night HCMM data sets, using models and techniques developed earlier (Kahle, 1977). However, we found that a more informative image product was a simple-to-construct color additive composite (CAC) of the three measured data sets (Kahle et al., 1981). One such image is shown in Fig. 4 (left). This image is a combination of the day

FIGURE 3.   HCMM images of Pisgah study area; data obtained 6–7 July 1978: (a) Visible image; (b) day infrared image; (c) night infrared image.

IR, night IR, and visible data, displayed as blue, red, and green, respectively. All three components were complemented, so areas that are cold or have low albedo appear as strongly colored in the image. This particular display was selected as providing the best visual separation of units in the scene after examination of many other possible combinations. This CAC was found to be superior to the TI images because, while the TI images contain more quantitative information on a single physical property of the surface material (TI), the VNIR properties have been removed. The CAC retains and displays information on both thermal and reflective properties.

For comparison, a Landsat scene of the same area was computer processed; of several enhancements produced, the best, a principal components transformed image, is shown in Fig. 4 (right). Interpretation maps of each are presented in Fig. 5. The maps were produced using standard photointerpretation techniques to draw boundaries around distinctive areas. Comparison with the detailed published geologic map (Dibblee, 1966) allowed rock type names to be assigned to the interpreted and outlined units. Note that these rock names differ somewhat from those given in Fig. 2 which were taken from the less detailed published map (California Division of Mines and Geology, 1969) of the entire $1° \times 2°$ area.

The HCMM CAC image presents a wide variety of colors for bedrock areas, whereas most of these areas are similarly colored on the Landsat image. Many of the rocks are spectrally quite flat in the Landsat wavelength region, and so cannot be separated. These same rocks, however, can be separated using HCMM data due to different thermal properties.

On the HCMM image CAC (Fig. 4) pahoehoe basalts are green. The aa flows at Pisgah and west of Pisgah are yellow, and easily separable. Felsic rocks are generally orange; the biotite quartz monzonites in the south are somewhat redder. The dacite west of Pisgah is the same orange color; compositionally this rock is similar to the granites. The other extrusive rocks are orange-yellow in color.

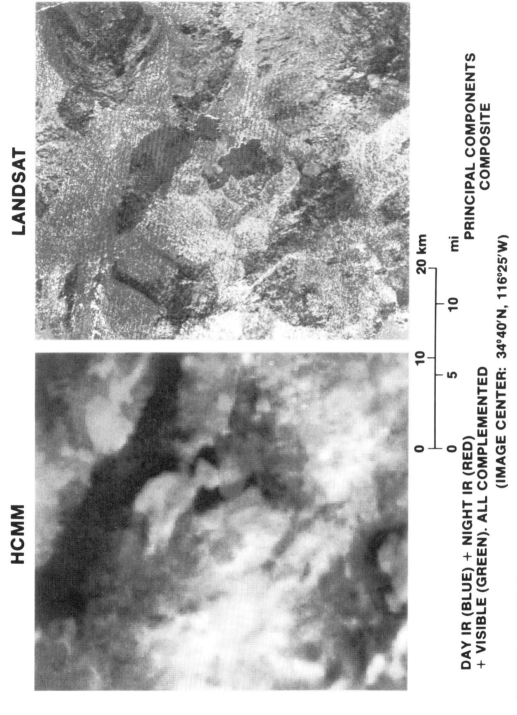

LANDSAT

HCMM

PRINCIPAL COMPONENTS COMPOSITE

DAY IR (BLUE) + NIGHT IR (RED) + VISIBLE (GREEN). ALL COMPLEMENTED

(IMAGE CENTER: 34°40'N, 116°25'W)

20 km 10

10 5 0 0 mi

FIGURE 4. HCMM and Landsat color composites of Pisgah test area. HCMM image displays day IR, night IR, and visible data as blue, red, and green, respectively; in addition, all three components are negatives. Landsat image is a principal components composite.

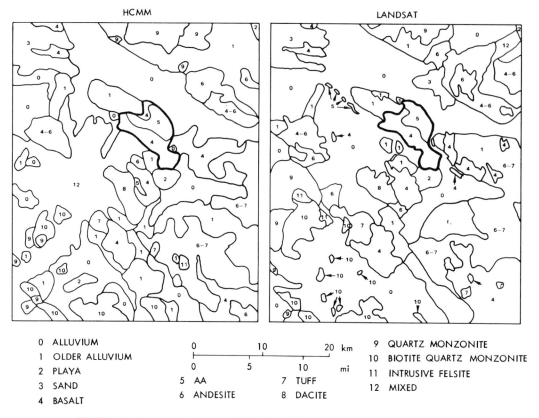

FIGURE 5.   Interpretation maps of HCMM and Landsat images appearing in Fig. 4.

The sand dunes in the northwestern corner are red (cold at night, warm in the day, high albedo) as is Lavic Lake playa. The varied colors of the alluvial areas reflect the variation in their source rock composition.

On the Landsat image very few separations can be made between the various rock types. HCMM images show lithologic changes more effectively. On the other hand, a major advantage of Landsat images is their higher spatial resolution which permits mapping of smaller features. The spatial resolution of the HCMM images was found to be quite unsatisfactory.

The value of greatly increased spatial resolution provided by aircraft scanners

was examined using day–night data acquired on 13–14 July 1977, with NASA's Bendix Modular Multispectral Scanner (MMS). This instrument acquired data simultaneously in 10 channels between 0.38 and 1.06 μm, and one channel between 8.05 and 13.7 μm near noon. The same area was reflown before sunrise, when only thermal data were obtained. A thermal inertia image (Fig. 6) was produced from the visible and near infrared data, the daytime thermal data, and the nighttime thermal data using the modeling method developed by Kahle (Watson et al., 1982). A crude interpretation map was produced from this image (Fig. 7).

Some thermal inertia differences associated with different rock types are de-

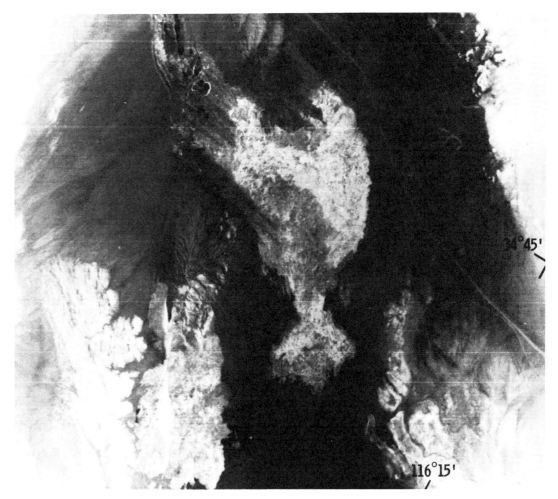

FIGURE 6. Thermal inertia image of Pisgah study area derived from aircraft data. Areas of high thermal inertia are displayed as white; areas of low thermal inertia appear dark.

picted by varying gray tones in the image. The Pisgah aa flows are clearly separable from the pahoehoe flows. The area of low TI of Lavic Lake playa in the south central part of the image is also readily apparent. A major characteristic of the terrain, discernible due to the higher resolution, is geomorphology and drainage texture. With the addition of this information, several units can be separated. The older fans, for example, have a dissected appearance due to high drainage density. The dacite porphyry appears distinctive due to linear features (dikes) cutting it.

Flow features are visible on the Pisgah flows. Alluvial fans appear fairly homogeneous with parallel to braided drainage patterns. This type of information, vital to photointerpreters, is not available from low resolution HCMM data.

The north-trending linear feature in the southwest part of the image (dashed line indicated by arrows on Fig. 7) is interpreted to be an unmapped extension of a mapped fault. This feature can be distinguished readily on the TI image. It seems to be expressed as a narrow valley or series of low areas covered with al-

231

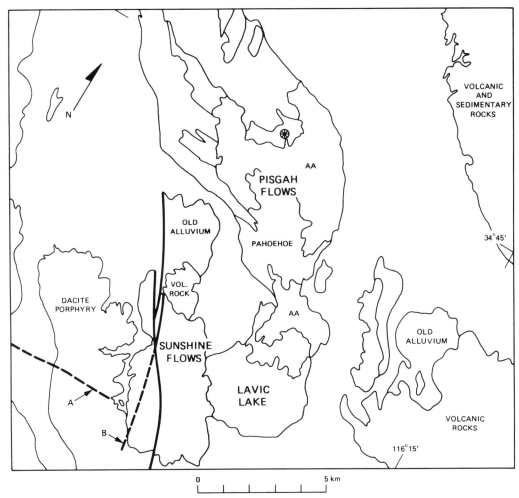

FIGURE 7.   Interpretation map of thermal inertia image (Fig. 6). See Table 1 for explanation of symbols. Heavy lines are mapped faults; dashed heavy lines delineate previously unmapped features, interpreted to be faults.

luvium, which have a lower thermal inertia than the surrounding bedrock. Another linear feature which may be an unmapped fault is indicated by a dashed line and arrow B. Future field work will be necessary to determine the actual nature of these features.

The aircraft scanner data were also used to evaluate the geologic information which could be extracted using both visible and near-infrared multispectral data and thermal data together. Due to the large number of wavelength variables available,

principal components analysis was used to process the aircraft data.

Principal components analysis (PCA) has been used in statistical processing of data for a considerable time (Hotelling, 1933) and is well illustrated in many texts (Davis, 1973). The method produces new variables (components) which are uncorrelated with each other and which are linear combinations of the original variables. The new axes for the variables are computed so that the first axis or component contains the greatest proportion of the variance, followed by later components which contain successively less of the scene variance. In this fashion, it is often possible to ignore the later components because they contain little or no useful information to the analyst. Geometrically, a PCA can be viewed as a rotation and translation of the original coordinate axes to new axes passing through the data. PCA is also a method for reducing the dimensionality of data by removing intervariable correlation and producing new, orthogonal axes, usually less than the number of original data axes.

Twelve variables were used for the PCA analysis (Table 1): nine channels of visible and near IR data (MMS channel 7 was not used due to a limit of 12 variables in the computer program), the daytime thermal data, the nighttime thermal data, and the derived thermal inertia values.

Twelve components (eigenvectors) were output from the program, with weightings and percent of scene variance shown in Table 2. It can be seen that the first component accounts for almost 95% of the scene variance, and the first three components account for more than 99%.

The 12 eigenvector pictures (V1–V12) are shown in Fig. 8. VI is essentially an average of the visible and near-infrared channels, or albedo. The basalts are the darkest materials in the scene, the playa is the brightest. V2 is basically the difference between thermal inertia and daytime infrared (DIR). Bright areas are those with high thermal inertias and low daytime temperatures; these include the volcanic and sedimentary rocks in the northeast corner, and the dacite porphyry in the southwest corner.

V3 is heavily weighted on channels 1, 2, DIR, and night infrared (NIR). Materials that have low reflectances and low temperatures in these channels will be bright on the image. These include playas, alluvial fans in the northern part of the area, and the dacite prophyry. It is surprising to have a low-order component weighted on both reflectance channels and thermal channels. In previous analyses using PC transformations, the two data types fell almost entirely into separate components. Probably the presence of large areas of very dark rocks caused there to be a high correlation between the thermal and reflectance channels. V4 is positively weighted on channel 10 and NIR, negatively weighted on channel 3. On the V4 image of the Pisgah flows, aa has higher values than pahoehoe. The small white dots on the pahoehoe surfaces are pockets of aeolian sand. The volcanic and sedimentary rocks on the eastern side of the area are all bright. V5 is strongly weighted on the NIR. Bright areas are those warmer at night. The dacite porphyry and hydrothermally leached andesite porphyry in the upper right corner are two prominent areas which stand out. There is also good separation of the pahoehoe and aa flows. V9 is weighted positively on DIR and TI, negatively on NIR. Some topographic features are visi-

TABLE 1 Variables Used for Principal Components Analysis

VARIABLE	CHANNEL OR VALUE
1	MMS 1, 0.38–0.44 μm
2	MMS 2, 0.44–0.49 μm
3	MMS 3, 0.49–0.54 μm
4	MMS 4, 0.54–0.58 μm
5	MMS 5, 0.58–0.62 μm
6	MMS 6, 0.62–0.66 μm
7	MMS 8, 0.70–0.74 μm
8	MMS 9, 0.76–0.86 μm
9	MMS 10, 0.97–1.06 μm
10	day infrared, 8.05–13.7 μm
11	night infrared, 8.05–13.7 μm
12	thermal inertia

TABLE 2 Loadings for Eigenvectors MMS Channel or Value[a]

Component	1	2	3	4	5	6	8	9	10	DIR	NIR	TI	% OF VARIANCE
VI	0.30	0.36	0.33	0.28	0.28	0.31	0.36	0.36	0.38	0.01	-0.04	-0.16	94.8
V2	0.19	0.08	0.02	0.0	0.0	0.0	0.0	0.03	0.07	-0.64	0.11	0.73	3.0
V3	-0.72	-0.32	0.0	0.13	0.14	0.13	0.13	0.17	0.24	-0.32	-0.33	-0.04	1.4
V4	0.01	-0.29	-0.37	0.28	-0.13	0.0	0.17	0.26	0.57	0.25	0.42	0.13	0.4
V5	-0.40	0.03	0.28	0.29	0.22	0.14	0.02	-0.04	-0.28	0.28	0.63	0.27	0.2
V6	0.11	-0.16	-0.33	-0.18	0.13	0.35	0.52	0.23	-0.59	-0.06	0.01	0.01	0.1
V7	0.15	-0.17	-0.21	0.06	0.34	0.44	-0.02	-0.71	0.22	-0.11	0.08	-0.11	<0.1
V8	0.36	-0.61	-0.02	0.47	0.22	-0.06	-0.35	0.30	-0.08	0.06	-0.05	0.03	<0.1
V9	-0.02	0.04	-0.03	-0.01	0.07	0.14	0.04	-0.10	0.06	0.57	-0.54	0.58	<0.1
V10	0.24	-0.14	0.40	-0.41	-0.11	0.22	0.34	-0.60	0.33	-0.02	-0.02	-0.02	<0.1
V11	0.17	0.07	-0.29	0.58	-0.52	-0.15	0.46	-0.25	0.09	-0.01	0.0	0.0	<0.1
V12	0.11	-0.01	0.02	-0.13	0.45	-0.76	0.45	-0.01	-0.02	0.0	0.0	0.0	<0.1

[a] Largest weightings underlined.

ble due to solar heating anisotropies, particularly in the dacite porphyry. V10, V11, and V12 have very little information and contain mostly noise.

Based on examination of the individual black and white component pictures, several triplets were selected to produce color composites. One of those is shown in Fig. 9. This image displays V3, V6, and V7 as blue, green, and red, respectively.

The amount of geologic information displayed in this image greatly surpasses that interpreted from the HCMM data for the same area described earlier. In fact, more lithologic separations can be made from an analysis of this image than are shown on the detailed 15' published geologic map of the area (Dibblee, 1966). The Pisgah flows (see sketch map, Fig. 7) are clearly separable into two subunits, corresponding to aa and pahoehoe; Pisgah Crater, composed of cinders, is a different color than the surrounding flows. The geologic map does not distinguish the two flow units. The separation of units in the volcanic rocks (lower right corner) and volcanic and sedimentary rocks (upper right corner) is similar to and locally more detailed than the geologic map. The varied colors of Lavic Lake (a clay playa) reflect differences in composition and moisture.

The ability to portray these factors has important implications for engineering geology and hydrology. Finally, the vivid portrayal of alluvial fans, ranging in age from Recent to Pleistocene, would provide a valuable mapping tool for Quaternary geomorphologists.

The data sets discussed in this section clearly illustrate the improvement in lithologic discrimination that can be achieved using thermal infrared data combined with visible and near-infrared data. However, it was also clear that HCMM did not have sufficient spatial resolution to be very useful at the scale of the geologic features we were examining at Pisgah. Therefore, we chose to investigate the utility of HCMM data over a larger area —the Trona region—at a scale more suitable to the HCMM resolution.

Trona Study Area

The Trona study area is in southeastern California (Fig. 1) and lies between latitudes 35° and 36°N, and between longitudes 116° and 118°W, about 110 × 180 km. The regional geology is displayed on the Trona Sheet of the Geologic Map of California series (California Division of Mines and Geology, 1963) at a scale of

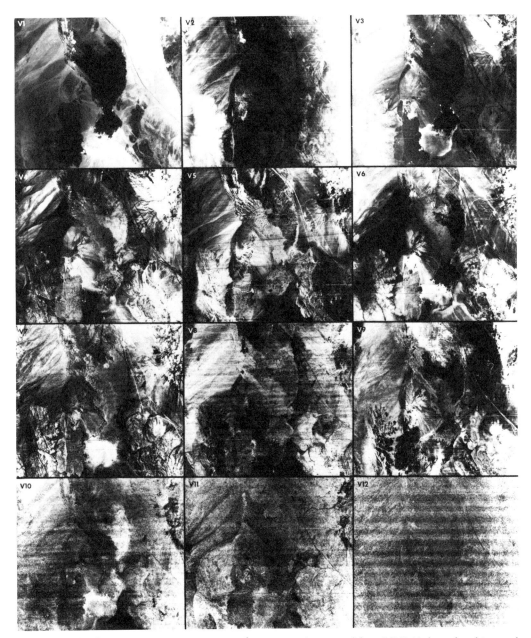

FIGURE 8. Twelve eigenvector pictures (principal components) computed from MMS 11-channel multispectral scanner data and thermal inertia data.

1:250,000. The image data were examined at this same scale.

The majority of the area is in the Mojave Desert region, bounded on the north by the Garlock fault; the northwestern corner falls in the Tehachapi Mountain region, while the northeastern part lies in the southern Death Valley region. The Mojave region is characterized by isolated mountain masses surrounded by pediments, but large parts consist of alluvium, within which there are several

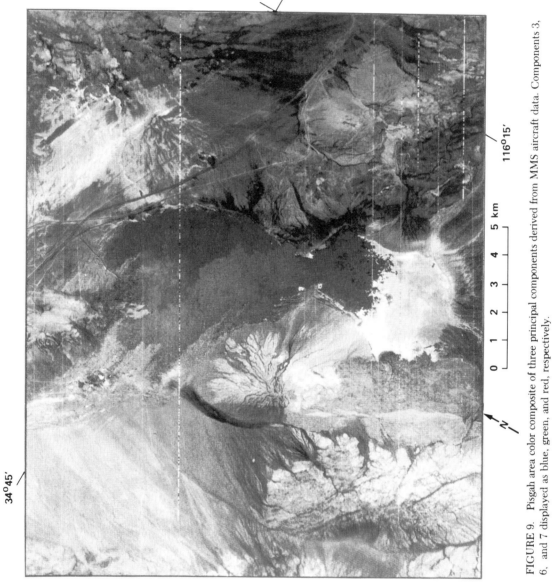

34°45'

116°15'

0 1 2 3 4 5 km

FIGURE 9.   Pisgah area color composite of three principal components derived from MMS aircraft data. Components 3, 6, and 7 displayed as blue, green, and red, respectively.

large and small playas. The Tehachapi Mountains region is characterized by a relatively flat upland surface which is often considered to be the southern extension of the Sierra Nevada. The Death Valley region is characterized by north-trending mountain ranges and deep, relatively narrow, alluviated valleys, many of which have playas at their lowest elevations.

The geology of the Trona area is complex, reflecting a long history of tectonism. For this report, only the distribution of rock types was considered. A summary of lithologies exposed in the area is shown in Table 3; a simplified geologic map is found in Fig. 10. The state geologic map depicts units based on age and, in some cases, composition. Quaternary units are separated into dune sand, playa lake deposits, salt deposits, and alluvium. Marine sediments, on the other hand, are mapped on the basis of age, rather than rock type; this unit therefore includes limestone,

dolomite, sandstone, shale, chert, wacke, and quartzite. In some cases, volcanic rocks are subdivided by lithology; in others they are undifferentiated. The units chosen for the simplified map in this paper group similar general rock types, regardless of age; for example, the metamorphic rock unit includes Mesozoic and Precambrian rocks which are gneisses, schists, metasediments, and metavolcanics. The unit "granite" consists primarily of granite but includes Mesozoic intrusive rocks ranging from intermediate to acidic in composition.

HCMM data from multiple dates were processed for the Trona area. The day–night image pairs were registered to a topographic base map. The day–night data sets were examined in a preliminary way on an interactive computer processing system to choose the best images and enhancement methods for lithologic evaluation. As at Pisgah, the best HCMM image enhancement was found to be a

TABLE 3   Geologic Units of Trona Area

Unit	Ages	Description
Alluvium	Quaternary	alluvial fan deposits
Salt	Quaternary	salt deposits on playas
Lake sediments	Quaternary	playa deposits; mainly clay, silt, and salts
Sand	Quaternary	dune sand
Nonmarine sediments	Tertiary	older alluvium, sandstone fanglomerate, conglomerate
Hypabyssal intrusives	Tertiary	plutonic and intrusive rocks of rhyolitic, andesitic, and basaltic composition
Volcanics	Cenozoic, Mesozoic	basalt, andesite, rhyolite extrusives
Basic intrusives	Mesozoic	diorite, gabbro, and amphibolite
Granite	Mesozoic	intrusive rocks ranging from intermediate to acidic compositions
Metamorphic	Mesozoic, Precambrian	Metasediments, metavolcanics, schists, gneisses
Marine sediments	Mesozoic, Paleozoic	sandstone, shale, quartzite, limestone, dolomite

26

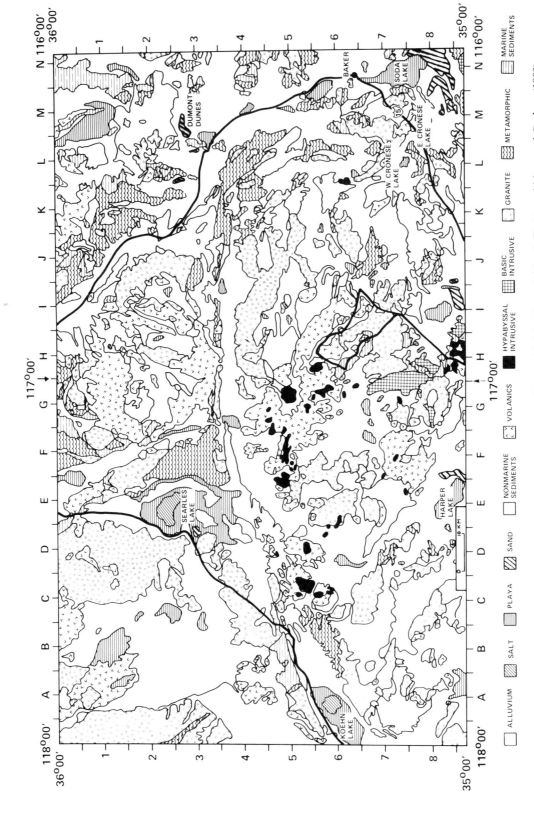

FIGURE 10. Simplified geologic map of the Trona 2° sheet, California. Modified from California Division of Mines and Geology (1963).

238

color additive composite (CAC) of the three variables, displaying positive or negative images of the variables in blue, green, and red light. After many combinations were examined, the 30–31 May 1978, day–night pair was selected for intensive study. A CAC was produced (Fig. 11) by displaying the visible channel as red, the daytime infrared channel as blue, and the complement (negative) of the nighttime infrared channel as green. The data were also contrast enhanced prior to production of the composite image. The colors on the image are due to combinations of physical properties: materials which have high albedo, high daytime temperature and/or low nighttime temperatures will have strong red, blue, and/or green color contributions, respectively. Playas, for example, have high albedo and low TI; they will be displayed with large red (high albedo), large green (cold at night), and low blue (cold during the day) color contributions, which will result in their appearing yellow or orange.

A color print of the HCMM composite image was produced at a scale of 1:250,000, and a positive color transparency of the geologic map was overlayed onto the HCMM image. This facilitated systematic comparison of the color variations on the HCMM image compared to the lithologic units portrayed on the map. Of particular interest were those units that had consistent colors on the image, and those that showed heterogeneities or other discrepancies. Those areas which seemed anomalous were delineated, and field investigations were undertaken in May 1982 to examine the cause of their behavior.

Playas, or dry lakes, are common in the Trona region. On the CAC HCMM image (Fig. 11) the majority are displayed in yellow-orange. Examples include Coyote Lake (I–8), Silver Lake (M.5–5.5) (note M.5 denotes halfway between M and N), Amargosa Lake (M–1), and West Cronese Lake (L–7.5). From the color, one can infer that the surfaces have high albedos (red component), low nighttime temperatures (green component), and low daytime temperatures (absence of blue). This is consistent with a composition of light-colored clay/silt, which has a low TI. Several playas, however, appear distinctly different, in whole or in part. These include Soda Lake (N–7.5), Harper Lake (E–8.5), Searles Lake (E–2.5), Koehn Lake (A–6), and East Cronese Lake (L.5–7.5).

Soda Lake has a green crescent-shaped area at its south end (N–8). When this feature was examined in the field, a large corresponding area of very dense, 1-m-high dry bushes was found. The green color results from low albedo, low daytime temperature, and low nighttime temperature. Even though the ground under the vegetation is bright playa material, the bushes effectively shadow the ground and control the area's reflectance and thermal properties. The large green area at the west end of Koehn Lake (A–6) was similarly observed to be highly vegetated from cultivation. The same color feature at the west end of Harper Lake (E–8) is assumed to be due to the same cause.

Searles Lake (E–2.5) appears on the CAC as a panoply of colors. There is an outside ring of orange-yellow color, similar to the other clay/silt playas. The core of the playa is red on the CAC, and this area corresponds on the geologic map to a large elliptical area of salt deposits. There is also a smaller red area at the south end of the lake. A black patch and several blue-lavender areas can also be seen on the CAC. Examination of the playa in the

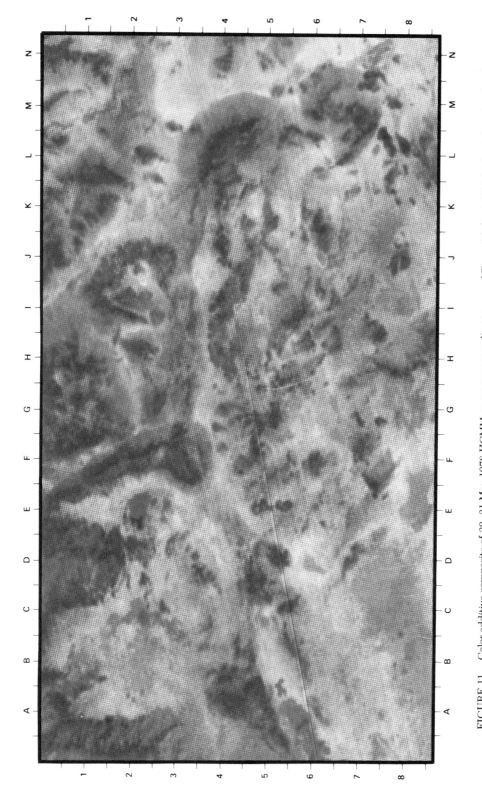

FIGURE 11. Color additive composite of 30–31 May 1978 HCMM scene corresponding to area of Trona 2° sheet. Visible, day infrared and night infrared channels displayed as red, blue, and green, respectively. The night infrared (green) channel is a negative: cold areas have large green contributions on the color image.

field revealed that the black area corresponds to the lowest part of the playa, where standing water is often found. Black is produced by low albedo (no red), low daytime temperatures (no blue), and high nighttime temperatures (no green). These characteristics are consistent with the presence of standing water. Red color is produced by high albedo, low daytime temperatures, and high nighttime temperatures. The composition of the playa in this area is dominantly salt and silt. The material was found to be very bright, and moist just below the surface. The combination of higher density and moisture results in higher TI, producing the thermal behavior seen on the CAC. The blue-lavender areas have moderate albedo, high daytime and nighttime temperatures. These areas in the field were composed of a light brown silt/clay surface with salt underneath and are inferred to have been less moist.

East Cronese Lake (L.5–7.5) is dark red on the CAC. This is inferred to be due to lower albedo and higher nighttime temperature, than with playas having about the same daytime temperature as West Cronese Lake (L–7.5). In the field, both playas were dry; however, the surfaces of the two playas were distinctly different. The western playa was pale yellowish-brown with large polygonal cracking; the surface was undulating but smooth. The eastern playa was darker brown, with polygonal cracks extending to 30 cm; the surface had a popcorn texture with a frosting of salt. The field observations are inconsistent with the appearance on the CAC. Examination of the rainfall data for Death Valley for 1978 indicates that a wet winter was followed by precipitation in May. In addition, the 15′ topographic map of the area shows

that one of the drainages of the Mojave River is into East Cronese Lake. Water is prevented from flowing to West Cronese Lake by a drainage divide due to a spur of the Cronese Mountains. These two facts suggest that, at the time of the HCMM data acquisition, standing water was present on East Cronese Lake. The color on the CAC further suggests that the water was quite shallow; the thermal behavior is characteristic of water, but the moderate albedo suggests penetration and reflection of visible light from the lake bottom.

A similar procedure was used to evaluate areas of rock outcrop. In the CAC image, rock outcrops were found to be displayed in a variety of colors: dark red, red, red-blue, dark blue, blue, dark green, green, dark orange, etc. Similarly colored areas were outlined from the CAC, then the outline map was superimposed on the geologic map to identify the units. Most of the red areas corresponded to granitic rocks (J–2, K.5–8, H.5–7, F–3, for example); a few corresponded to marine sediments (N–4, M.5–1.5, for example). However, the comparison indicated that some areas mapped as granitic rocks appeared dark blue on the CAC (M–7, H.5–7.5); these areas were targeted for visiting in the field to determine the composition and characteristics of the materials responsible for the CAC colors.

Volcanic rocks appeared on the CAC as dark blue, blue, dark blue-green (F–7.5, E–5, H–3, A.5–1, for example) and dark orange (F–5). Basic intrusive rocks were dark green or blue (G.5–7); hypabyssal intrusive rocks were dark green or blue also (D.5–5, G.5–6). Dune sands were not discernible on the HCMM image; nonmarine sediments were many different colors, reflecting the variety of compositions of the source rocks. Metamorphic rocks were dark blue or green (M.5–7.5, J–6.5, for example). Large green areas on the CAC correspond to mountain ranges with high elevations, such as the Avawatz Mountains (L–4.5) and the Argus Range (D–1). The green color indicates moderate to low albedo, cold day temperatures, and cold nighttime temperatures. This is consistent with the effects of lower temperatures as a function of increasing elevation, and the presence of substantial vegetation cover. The HCMM data were not corrected for the first effect, and the presence of vegetation would mask the underlying bedrock.

Field observations of granitic rocks at many locations revealed that those appearing red on the CAC were light-colored, intermediate to acidic rocks; dark-colored "granites" (diorite, gabbro, etc.) appeared dark blue or red on the CAC. The same daytime and nighttime temperature signatures were combining with variable albedos (red component) to produce the different colors. Therefore, a useful feature of the HCMM data was the ability to subdivide mapped granitic rocks into mafic and felsic subunits.

Field examination of volcanic rocks indicated that blue areas on the CAC corresponded to outcrops of basalt; green and dark orange areas were pyroclastics, andesites, rhyolites, and tuffs. Basalt has very low albedo, high daytime temperatures, and relatively high nighttime temperatures due to its high thermal inertia; these characteristics account for the blue color. The other volcanic rocks have higher albedos (rhyolites, K.5–.5, orange area) or lower thermal inertias (green areas). Therefore, the HCMM data

provide some compositional information related to physical characteristics of the rock types.

A sharp boundary on the HCMM data was seen in the Owlshead Mountains (J–2.5), juxtaposing a red area to the north with a dark red-blue area to the south. The geologic map shows a contact between granitic rocks to the north and metamorphic rocks to the south. The HCMM boundary, however, is displaced 6 km to the north relative to the geologic map. In the field, the HCMM boundary was located and was related to a sharp change in rock albedo. Access to the area was too difficult to examine the rock types firsthand; either the geologic map depicts the position of the boundary incorrectly or a less acidic facies of the intrusives at the south end produces a similar appearance to the metamorphic rocks on the CAC, so that the HCMM boundary is separating felsic granitic rocks from intermediate intrusive rocks and metamorphic rocks.

Summary and Conclusions

Our primary goal was to investigate the utility of HCMM data for geologic applications in some arid and semiarid regions of the western United States. Our original intent was to study the geology in terms of the thermal inertia of the surface materials as derived from the diurnal temperature variation and albedo. However, the computation of thermal inertia uses the VNIR data strictly to define the albedo effect upon the diurnal heating of the surface. It is possible to retain the albedo information along with the thermal information by use of the simple-to-produce color additive composite (CAC) of day IR, night IR, and day VNIR.

HCMM data in the CAC format were compared with processed Landsat MSS data for the Pisgah test area. Although the Landsat data had higher spatial resolution (80 m), the lack of spectral differences between exposed surface materials in the Landsat wavelength region greatly limited the ability to separate various rock types. Most of the bedrock materials in the area are volcanic or intrusive rocks of intermediate composition, with low albedos, but differing thermal properties. The HCMM thermal data were more effective at producing separations of compositionally different rock types.

Aircraft multispectral scanner data, acquired over the Pisgah study area, were examined to assess the effect of increased spatial resolution and to evaluate the utility of combining multispectral visible and thermal data for interpretation. A thermal inertia image produced from the data differentiated some rocks based on differences in thermal properties. A previously mapped fault appeared on the TI image to extend further than originally mapped. This feature was not easily seen on other images of the area.

The technique of using principal components analysis was found to be an effective method of combining disparate data sets (multispectral reflectance and thermal data) after coregistering the data sets. For the Pisgah study area, more lithologic separations were evident on the principal components composite than appeared on the detailed published geologic map.

The HCMM CAC data were used to produce an image for a $1° \times 2°$ area covered by the Trona, CA, geologic map. The color picture was enlarged to a scale of 1:250,000, and a photographic transparency of the geologic map was overlayed onto the HCMM image to evaluate the geologic separations interpretable from the HCMM image.

Lithologic variations were revealed by the combination of thermal and reflectance differences. Playas were found to be displayed in a variety of colors. In all cases, the colors could be related to physical characteristics related to the playas' composition, moisture content, presence of standing water, and vegetation cover. Granitic rocks were depicted in two colors. Field work revealed that this division corresponded to differences in composition: felsic rocks could be distinguished from mafic rocks. This distinction was not portrayed on the state geologic map. Volcanic rocks were displayed in a variety of colors, dependent on the albedo and thermal inertias of the materials. Basalts, for example, were distinguished by low albedo, high daytime temperature and high nighttime temperature; less dense pyroclastics and tuffs were distinguishable by their higher albedo and lower daytime and nighttime temperatures. Unconsolidated to consolidated Tertiary nonmarine sediments were not separable on the HCMM images from recent alluvial fans, due to lack of contrasting physical properties. Similarly, areas of dune sands were not distinguishable, though in this case the coarse HCMM resolution (500 m) may have played a significant factor, as the dunes are quite small.

At both test areas, the spatial resolution of HCMM data was the major limiting factor in the usefulness of the data. Future thermal infrared satellite sensors need to provide spatial resolution comparable to that of the Landsat MSS or TM data.

This increased spatial resolution would be much more important for geologic studies than increased thermal resolution.

This study of HCMM data has confirmed that thermal data add a new dimension to geologic remote sensing. This study has examined only the broad band thermal infrared data which leads to information about the thermal properties of the surface materials. Multispectral thermal infrared data provide a different type of information which is also useful. Both types of information could, in principle, be acquired from a single multispectral thermal infrared sensor. Future satellite missions should be undertaken to allow the continued development of the thermal infrared data for geology.

Computer processing of HCMM and aircraft data was done by Ron Alley of JPL's Image Processing Laboratory. Alan Gillespie offered useful suggestions and criticisms which improved this manuscript. Work was done at the Jet Propulsion Laboratory, California Institute of Technology, through contracts with the National Aeronautics and Space Administration.

References

Bakker, P., Church, D., Feuchtwanger, T., Grootenboer, J., Lee, C., Longshaw, T., and Viljoen, R. (1978), *Mining Mag.*: 398–413.

California Division of Mines and Geology (1963), Geologic Map of California, Trona Sheet, Sacramento, CA.

California Division of Mines and Geology (1969), Geologic Map of California, San Bernardino Sheet, Sacramento, CA.

Davis, J. (1973), *Statistics and Data Analysis in Geology*, Wiley, New York.

Dibblee, T. (1966), Geologic Map of the Lavic Quadrangle, San Bernardino County, CA, *Misc. Geol. Invest.* Map I-472, U.S. Geological Survey, Washington, DC.

Goetz, A. F. H., and Rowan, L. C. (1981), Geologic remote sensing, *Science* 211: 781–791.

Heat Capacity Mapping Mission (HCMM) Data Users Handbook (1980), Goddard Space Flight Center, NASA, Greenbelt, MD, 120 pp.

Hotelling, M. (1933), Analysis of a complex of statistical variables into principal components, *J. Educ. Psych.* 24:417–441.

Kahle, A. B., Gillespie, A. R., and Goetz, A. F. H. (1976), Thermal inertia imaging: a new geologic mapping tool, *Geophys. Res. Lett.* 3:26–28.

Kahle, A. B. (1977), A simple thermal model of the Earth's surface for geologic mapping by remote sensing, *J. Geophys. Res.* 82: 1673–1680.

Kahle, A. B., Schieldge, J. P., Abrams, M. J., Alley, R. E., and LeVine, C. J. (1981), Geologic applications of thermal inertia imaging using HCMM data, JPL publication 81-55, NASA/Jet Propulsion Laboratory, Pasadena, CA.

Pohn, H. A., Offield, T. W., and Watson, K. (1974), Thermal inertia mapping from satellite—discrimination of geologic units in Oman, *J. Res. U.S. Geol. Surv.* 2:147–158.

Pratt, D., and Ellyett, C. (1979), The thermal inertia approach to mapping of soil moisture and geology, *Remote Sens. Environ.* 8:151–168.

Price, J. C. (1977), Thermal inertia mapping: a new view of the Earth, *J. Geophys. Res.* 18:2582–2590.

Rowan, L. C., Wetlaufer, P. H., Goetz, A. F. H., Billingsley, F. C., and Stewart, J. H. (1974), Discrimination of rock types and altered areas in Nevada by the use of

ERTS images, Prof. Paper 883, U.S. Geological Survey, Washington, DC.

Sabins, F. F. (1969), Thermal infrared imagery and its application to structural mapping in southern California, *Geol. Soc. Am. Bull.* 80:397–404.

Vincent, R. K., and Thomson, F. (1972), Rock-type discrimination from ratioed infrared scanner images of Pisgah Crater, California, *Science* 175: 986–988.

Watson, K. (1975), Geologic applications of thermal infrared images, *Proc. IEEE* 63:128–137.

Watson, K. (1982), Regional thermal-inertia mapping from an experimental satellite, *Geophysics* 47:1681–1687.

Watson, K., Hummer-Miller, S., and Offield, T. W. (1981), Geologic applications of thermal-inertia mapping from satellite, Final Report, HCMM Investigation S-40256-B, U.S. Geological Survey, Denver, CO, July, 72 pp.

Watson, K., Hummer-Miller, S., and Sawatzky, D. L. (1982), Registration of Heat Capacity Mapping Mission thermal satellite images, *Photogramm. Eng. Remote Sens.* 48:263–267.

FLOYD F. SABINS, JR.   *Chevron Oil Field Research Company, P. O. Box 446, La Habra, California 90631*

Thermal Infrared Imagery and Its Application to Structural Mapping in Southern California

Abstract: Thermal infrared imagery is obtained by airborne scanning devices that detect thermal radiation from the earth's surface and record it as an image in which bright tones represent relatively warm temperatures. Scanners sensitive to wavelengths between 8 and 14 microns span the radiant power peak of the earth at 9.7 microns and coincide with an atmospheric "window."

An example of 8 to 14 micron nighttime infrared imagery from the Imperial Valley, California, is interpreted and compared with aerial photographs of the same area. In this monotonous-appearing desert terrain, the imagery exhibits greater contrast and geologic detail than the photography. On the imagery, deformed Tertiary sedimentary bedrock (relatively cool) is distinguished from Holocene windblown sand cover (relatively warm).

Of especial geologic interest is a faulted plunging anticline in flat terrain. It is obscure both on aerial photographs and to a ground observer. On nighttime infrared imagery, however, the fold is clearly shown by the outcrop configuration of the individual siltstone and sandstone strata comprising the structure. Apparently the radiometric temperature differences between strata are sufficient to outline the fold on the imagery. The obscure expression of the fold on aerial photographs may be due to insufficient contrast in light reflectance between the different strata.

CONTENTS

INTRODUCTION

"Remote sensing" is the general term for obtaining airborne imagery of the earth. Formerly this was confined to aerial photography, but new technology makes it possible to obtain imagery by use of energy from other regions of the electromagnetic spectrum, such as the thermal infrared and the microwave. The relationship of these wavelengths to those of light are illustrated on Figure 1. General discussions of all aspects of remote sensing have been published recently by Colwell (1968), and Badgley and others (1967).

Geologists are becoming increasingly interested in this new technology, both for their own research and for instructing students. Well-documented examples are still relatively scarce, however, because of the newness of the methods and security restrictions on the

Geological Society of America Bulletin, v. 80, p. 397–404, 5 figs., 2 pls., March 1969

397

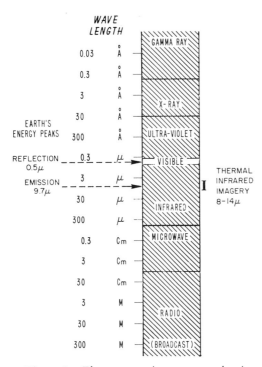

Figure 1. Electromagnetic spectrum, showing spectral region imaged on Plate 1B.

imagery. The purpose of this paper is to review the technology of thermal infrared imaging and present a previously unpublished example of imagery that provides more geologic information than existing aerial photography. A previous paper by this author (Sabins, 1967) demonstrated that thermal infrared imagery can be used to interpret lithology and structure. Additional geologic applications are described in the papers by Cantrell (1964), Fischer and others (1964), Lattman (1963), and Wallace and Moxham (1967).

ACKNOWLEDGMENTS

The writer is indebted to Chevron Oil Field Research Company for supporting the research that went into this paper and for permitting its publication. Mr. T. R. Ory and the HRB-Singer Company furnished the imagery. On field trips to the area, the writer has benefited from discussions with members of his Graduate Seminar in Remote Sensing at the University of Southern California Geology Department. The following individuals have also provided valuable help in the field: Professor R. J. P. Lyon of Stanford University, Professor Joseph Lintz of the University of Nevada, and Eugene Borax of the Union Oil Company. The manuscript was read by W. A. Fischer, C. F. Everett, T. R. Ory, and E. Borax. Figures were drafted by J. C. Keeser and T. G. Naves.

PRINCIPLES OF THERMAL INFRARED IMAGERY

Radiant energy from the sun is absorbed and reflected by the earth, which has a maximum reflected energy peak at a wavelength of 0.5 microns, corresponding to green light. In addition, the earth radiates, or emits, absorbed energy back to the atmosphere. At the average ambient surface temperature of approximately 300° K, the peak emission of the earth is at a wavelength near 9.7 microns. As is shown on Figure 1, this emission peak is within the infrared region, which extends from wavelengths of about 0.7 to 1000 microns. The emission peak coincides with an atmospheric "window" between 8 and 14 microns, so-called because radiation at these wavelengths is only slightly attenuated in its passage through the atmosphere. Another "window," between 3 and 5 microns, is also of interest to earth scientists.

The very short wavelength portion of the infrared spectrum, between 0.7 and 0.9 microns, can be "imaged" directly on infrared sensitive film, and is referred to as "photographic infrared." The longer wavelength radiation is called "thermal infrared," and cannot be imaged by photographic methods. Geologic interpretation of thermal infrared imagery is the subject of this paper.

Infrared Detectors and Scanners

Thermal infrared radiation can be measured by photodetectors fabricated from single crystals of semiconductor materials. In the region between 8 and 14 microns, germanium doped with copper or mercury is an excellent photoconductive detector; that is, its electrical conductivity changes as a function of the infrared radiation striking it.

The spot size of a detector system is rather small; a typical value is 3 milliradians. This means that at a height of 1000 ft the detector receives infrared radiation from a ground area, or spot size, of approximately 3 ft by 3 ft. An airborne scanning device, such as the one shown diagramatically on Figure 2, is used to convert the detector response into an image. A rotating mirror scans the earth's surface at right angles to the aircraft's flight direction and focuses the radiant infrared energy onto the detector. The

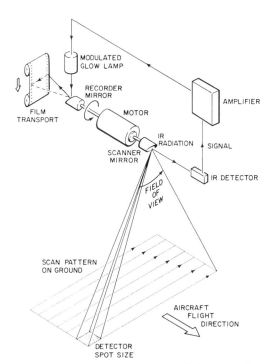

Figure 2. Schematic diagram of thermal infrared scanner.

detector converts the infrared energy into an electric signal, which is amplified and then modulates the brightness of a small glow lamp. A recorder mirror, rotating synchronously with the scanner mirror, sweeps the modulated light signal across a strip of recording film. One sweep across the film thus corresponds to one scan line on the ground. As the aircraft moves across the terrain, the film is advanced so that successive scan lines form an image, such as the one illustrated on Plate 1B. For the sake of simplicity, the diagram on Figure 2 omits the liquid helium or other cooling system used to increase the sensitivity of the detector. Also omitted are the additional mirrors that focus the infrared radiation onto the detector.

Radiant temperature sensitivity of infrared scanners varies with detector size and optical configuration, but may be on the order of 0.5° C. Quantitative temperatures are not recorded; rather, the gray scale of the imagery records the relative amount of radiation emitted by one spot on the ground as compared with adjacent spots. Since the technique is passive (there is no energy transmitted to the ground), it is restricted to observing the thermal properties of the upper few microns of the

earth's surface and cannot be evaluated in terms of penetration power.

The mirrors in a typical scanner system rotate at 6000 rpm. Aircraft ground speed is on the order of 150 to 180 mph, and the flying height is generally less than 10,000 ft above the terrain. Greater heights are used for regional investigations with smaller scale imagery, and lower heights (less than 5000 ft above terrain) are used for detailed investigations. Depending on the make and model of scanner, the angular field of view ranges from 73° to 120°. At a height of 1000 ft above terrain, the ground width, normal to flight direction, of a 120° scanner is 3464 ft; at 5000 ft above terrain, the width is 17,320 ft.

Image Distortion

Scanner imagery has a characteristic distortion, which is explained on Figure 3. The scanner operates at a constant angular rate, so the detector spot size (or "look element") at either edge of the field of view is larger than in the center, directly beneath the aircraft (Figure 3A). The imagery is recorded on the film, however, at a constant linear rate, so that each "look element" is recorded in an equal space on the film trace. This compresses the edges of the imagery, resulting in the distortions illustrated on Figure 3B. The "sigmoid" distortion of straight roads trending diagonally across the flight path is particularly characteristic of scanner imagery. To minimize the effects of distortion, interpreters generally work with the central two-thirds of the image

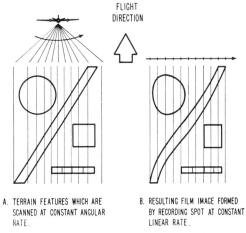

A. TERRAIN FEATURES WHICH ARE SCANNED AT CONSTANT ANGULAR RATE.

B. RESULTING FILM IMAGE FORMED BY RECORDING SPOT AT CONSTANT LINEAR RATE.

Figure 3. Distortion characteristics of scanner imagery.

strip. Ideally, adjacent flight lines are spaced to allow at least one-third overlap on the edges.

For more details of infrared physics and technology, see Simon (1966), or Wolfe (1965).

INTERPRETATION OF INFRARED IMAGERY

General

Mr. T. R. Ory of the HRB-Singer Company provided a set of thermal infrared scanner imagery of the Imperial Valley, California, which was flown by his organization in October, 1963. A small-scale mosaic of the imagery was published by Williams and Ory (1967). This imagery records wavelengths from 8 to 14 microns and was flown under nearly ideal conditions. Flights were made in the predawn hours to image radiant energy without the complicating effects of reflected solar energy.

An enlarged portion of the imagery and a corresponding aerial photograph are illustrated on Plate 1, and a location map is shown on Figure 4. The area is 10 miles northwest of El Centro, in secs. 25 and 36, T. 14 S., R. 12 E., and adjacent secs. 19 and 30, T. 14 S., R. 13 E., of Imperial County. The area is covered by the U.S. Geological Survey topographic map of the Brawley NW quadrangle.

The imagery (Pl. 1B) is enlarged from the original, which was recorded on 70 mm film. The flight line from which this was taken was oriented NNW–SSE. Light areas on the imagery represent relatively warm areas on the ground, and dark images are relatively cool. Variations in emissivity, as well as temperature, will influence the radiometric temperature imaged by the scanner, but for this interpretive report, "warmer" and "cooler" are adequate terms. Moisture content has a strong influence on rock and soil temperatures, but no precipitation had occurred in this arid area for some time prior to the flight (T. R. Ory, personal commun.).

Figure 5 is an interpretation of the infrared imagery (Pl. 1B). Note the characteristic scanner distortion of Imler Road, which is actually straight. The road, which images warm, is surfaced with hard-packed sand. The faint trace along the power line on the imagery is actually a seldom-used access road. At the south edge, the Fillaree irrigation canal images warmer than the adjacent ground. Because of its higher thermal capacity, water does not cool as rapidly as land after sundown, so is actually warmer in the predawn hours. The thermal variations in the cultivated areas reflect different agricultural practices. The very warm imaging (light) field north of the canal was probably flooded with water at the time to leach salts from the soil. The very cool imaging (dark) fields had probably been recently irrigated, but had no standing water, so that evaporative cooling of the damp ground caused the cold pattern. Fields with intermediate temperatures probably had intermediate moisture content.

Geologic Interpretation

Aside from the irrigated fields, the area of Plate 1 is desert with very sparse vegetation. The sea-level contour crosses the northwest corner, and most of the area is below sea level. Within the five square miles covered, total relief is less than 100 ft. The most prominent topographic features are sand dunes up to 30 ft high, which are stabilized by mesquite bushes. These appear dark on the aerial photo and bright (warm) on the thermal imagery. Vegetation and sand typically appear warm on this type of nighttime imagery.

Bedrock in the area consists of nonresistant, brownish-gray, lacustrine siltstone with thin interbeds of well-cemented, moderately brown sandstone. Flaggy pieces of the sandstone litter the surface where it crops out. Light-colored, nodular, thin layers within the siltstone help define bedding trends in this otherwise monotonous sequence of Tertiary age. Some gastropods and thin-shelled pelecypods were the only fossils noted. Much of the area is covered by a sheet of windblown sand, which supports most of the sparse vegetation. Numerous exposures of Tertiary bedrock indicate that the sand rarely exceeds a few feet in thickness. The El Centro sheet of the geologic map of California, by the California Division of Mines and Geology, shows that this area was covered by ancient Lake Coahuilla during Quaternary time. Except for some large boulders coated with travertine deposits, any Coahuilla deposits have been reworked by the wind.

On the imagery the bedrock outcrops are dark (cool), and the windblown sand is shown by lighter shades of gray (warm). The very warm Y-shaped image in the west-central part of the area is a thick accumulation of windblown sand lodged against an earthen levee.

The major geologic feature is the east-plunging fold pattern in the center of the imagery. Field work showed that this is an

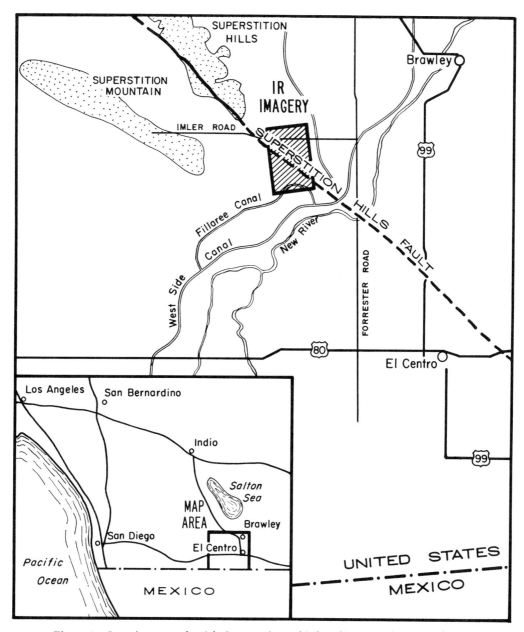

Figure 4. Location map of aerial photography and infrared imagery shown on Plate 1.

anticline in the Tertiary siltstone bedrock and has an axial length of about three-quarters of a mile and a width of one-quarter mile. Had the imagery not indicated its presence, we could have walked across the anticline in the field without recognizing it, for there were no conspicuous lithologic or topographic patterns. The plunge of the anticline was defined by walking out the light-colored nodular beds and the resistant sandstones. Structural attitudes are obscure in the siltstone, but dips up to 45° were measured in the sandstones, and the dip reversal across the fold axis was located. A broad, low ridge that has up to 30 ft of relief coincides with the general area of the anticline.

On Figure 5 the anticline is shown as solid bedrock, but actually there are numerous thin

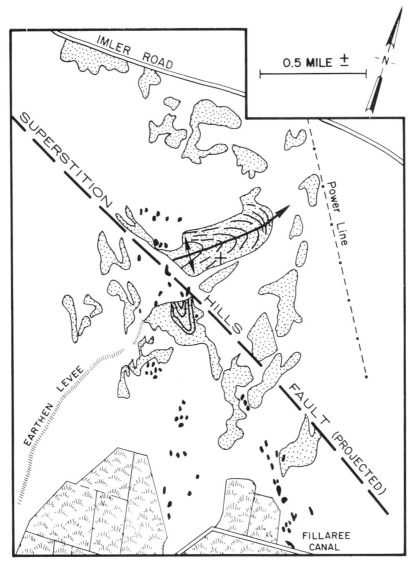

OUTCROPS OF DEFORMED TERTIARY LAKE DEPOSITS, SHOWING TRENDS OF BEDDING.

RECENT WIND BLOWN SAND COVER.

CULTIVATED AREAS.

STABILIZED SAND DUNES AND TUFA-COATED BOULDERS.

BRASS CAP MARKING SE CORNER SEC. 25, T.14S, R12E.

Figure 5.   Geologic interpretation of infrared scanner imagery shown on Plate 1B.

patches of windblown sand, which account for the gray tones on the imagery. The dark-imaging core of the anticline consists of silt-stone. The alternating light and dark pattern outlining the limbs of the fold appears to correlate with outcrops of sandstone (warm) and siltstone (cool), respectively. This was confirmed, although four years later (March, 1968), by predawn measurements with portable radiometers. The sandstones generally measured 2° F warmer than the siltstones. The sand dunes, vegetation, and windblown sand were warmest, as indicated on the imagery. This relationship agrees qualitatively with a similar survey made in the Indio Hills, north of the Salton Sea (Sabins, 1957).

The west end of the anticline is abruptly truncated along a line coincident with the southeastward projection of the Superstition Hills fault, as shown on the El Centro geologic map sheet. The inferred trace of the fault is covered with windblown sand and is not evident on the imagery, but siltstone outcrops in the immediate vicinity are strongly deformed, suggesting drag folding.

South of the anticline there is a small arcuate pattern of alternating dark and light bands. Field inspection showed this to be an exposure of gently dipping siltstone and sandstone bedrock with an outcrop pattern generally resembling that of the imagery. The pattern is more pronounced on the imagery than on the ground.

COMPARISON OF IMAGERY AND AERIAL PHOTOGRAPHY

The most recent large-scale aerial photography of the area was flown in May, 1953, for the U.S. Geological Survey and predates the imagery by ten years. The imagery itself was not available to us until 1967. Despite the elapsed time, it was possible to correlate features on the photography, on the imagery, and in the field with a high degree of confidence. This is probably due to the combination of low-relief terrain in an arid, uniform climate, with little human activity, so that changes are minimal.

The most striking and obvious difference is the greater detail and contrast on the imagery than on the photography. One might blame incorrect exposure or processing for the lack of contrast on the aerial photography. Inspection in the field, however, shows that the photography is a faithful reproduction of the appearance of the terrain. Dark-green mesquite bushes photograph black, and light-colored windblown sand photographs white, so the full dynamic range of the film was utilized. Under stereoscopic viewing, individual clumps of vegetation and minute details of drainage can be resolved, so the photography is representative of the state of the art.

Through the courtesy of Norman Stanley of Growers Aerial Service, El Centro, the writer made an aerial reconnaissance over the anticline and vicinity in July, 1968. This flight confirmed the generally monotonous appearance of the terrain and the obscurity of the anticline. High-angle oblique photographs were taken with Kodachrome and infrared Ektachrome films, using a variety of filters in an attempt to enhance the expression of the anticline. These photographs were not significantly better than the black and white photographs.

One possible explanation for the greater contrast on the thermal infrared imagery is that in this area the range of nighttime thermal emittance variations is greater than the range of daytime visible reflectance variations. This explanation could be evaluated by making nighttime radiometer measurements in the area and comparing them with corresponding daytime photometer measurements. We have taken some preliminary measurements and hope to continue the investigation.

On the aerial photographs the windblown sand is lighter in tone than the bedrock outcrops. The same relationship occurs on the imagery, but the tonal contrast between the warmer sand and cooler outcrop is more pronounced. The stabilized sand dunes are dark on the photography because of the associated vegetation, but are very bright on the imagery. Our nighttime radiometric measurements of the dunes showed that both the vegetation and the sand were relatively warm, which accounts for their bright appearance on the imagery. Elsewhere in the area, vegetation is too sparse to affect the imagery.

The enhanced appearance of the anticline on the imagery is not due to poor reproduction of the aerial photography, for both illustrations were reproduced with maximum care. In fact, the aerial photography is essentially a contact print, whereas the imagery has been enlarged by a factor of two from a contact print, with attendant loss of resolution. Most geologists with whom I have consulted agree that during a normal stereoscopic study of the aerial photography, they would not have noticed the anticline. After examining the infrared imagery, however, they were able to locate the anticline on the photography.

CONCLUSIONS

Thermal infrared scanning provides imagery of the earth from electromagnetic energy at the wavelengths of the radiant power peak of the earth. This imagery provides valuable information about the thermal characteristics of the surface of the earth. In the example described here, this thermal imagery has greater contrast and resolution than conventional aerial photography. Interpretation of imagery from other areas, such as the Indio Hills, has also shown its geologic value, particularly in arid and semi-arid terrain, for: (1) Discriminating between outcrop and cover; (2) recognizing subtle structures; (3) distinguishing variations in surface moisture content.

This does not imply that thermal infrared imagery should replace aerial photography, but rather that it is a new and useful way of looking at the surface of the earth.

REFERENCES CITED

Badgley, P. C., Childs, L., and Vest, W. L., 1967, The application of remote sensing instruments in earth resources surveys: Geophysics, v. 32, p. 583–601.

Cantrell, J. L., 1964, Infrared geology: Photogramm. Eng., v. 30, p. 916–922.

Colwell, R. N., 1968, Remote sensing of natural resources: Sci. American, v. 217, p. 54–69.

Fischer, W. A., Moxham, R. M., Polycn, F., and Landis, G. H., 1964, Infrared surveys of Hawaiian volcanoes: Science, v. 146, p. 733–742.

Lattman, L. H., 1963, Geologic interpretation of airborne infrared imagery: Photogramm. Eng., v. 29, p. 83–87.

Sabins, F. F., 1967, Infrared imagery and geologic aspects: Photogramm. Eng., v. 33, p. 743–750.

Simon, Ivan, 1966, Infrared radiation: New York, D. van Nostrand Co., 119 p.

Wallace, R. E., and Moxham, R. M., 1967, Use of infrared imagery in study of the San Andreas fault system, California: U.S. Geol. Survey Prof. Paper 575-D, p. D147–D156.

Williams, R. S., and Ory, T. R., 1967, Infrared imagery mosaics for geological investigation: Photogramm. Eng., v. 33, p. 1377–1380.

Wolfe, W. L., *Editor*, 1965, Handbook of military infrared technology: Washington, U.S. Govt. Printing Office, 906 p.

MANUSCRIPT RECEIVED BY THE SOCIETY JULY 18, 1968
REVISED MANUSCRIPT RECEIVED SEPTEMBER 5, 1968

Reprinted by permission from *Economic Geology*, v.
78 (1983), p. 573-590.

ECONOMIC GEOLOGY

AND THE

BULLETIN OF THE SOCIETY OF ECONOMIC GEOLOGISTS

| VOL. 78 | JUNE–JULY, 1983 | No. 4 |

Remote Sensing for Exploration: An Overview

ALEXANDER F. H. GOETZ, BARRETT N. ROCK,

Jet Propulsion Laboratory, California Institute of Technology, Pasadena, California 91109

AND LAWRENCE C. ROWAN

U. S. Geological Survey, Reston, Virginia 22092

Abstract

A review of remote sensing as applicable to resource exploration is presented based on the
sixteen papers making up the special issue of this journal. Background material on the measurement techniques and their physical basis as applied to rocks, soils, and vegetation is covered.
Special emphasis is given to new developments in high spectral resolution remote sensing for
mineralogic and vegetation mapping.

Introduction

THIS volume of *Economic Geology*, edited by the
authors of this paper, contains a collection of sixteen
papers addressing questions on the use of remote-sensing data acquired on the ground, from aircraft, and
from satellites for making maps that are useful as
exploration tools. This paper seeks to familiarize the
reader with the results discussed in this volume, fill
the gaps not covered, and discuss the future potential
of remote sensing in the exploration process.

The present collection of papers is unique in that
almost half of the contributions deal with questions
of vegetation cover and the use of the spectral properties of plants to identify conditions present in the
soil. This heavy emphasis on vegetation analysis is a
recent and welcome trend because, if the vegetation-covered areas of the earth are ignored, only approximately 30 percent of the land surface would be open
to exploration by remote sensing. For the reader not
familiar with basic botanical concepts and remote-sensing techniques a section on concepts and principles is included.

The use of vegetation as an indicator of potentially
important sites of mineralization or hydrocarbon occurrence requires a multidisciplinary approach involving a number of subdisciplines within the sciences
of botany and geology. Such an approach, especially
when coupled with remote sensing, demands well-coordinated efforts by exploration geologists, soil scientists, and field-oriented botanists, all of whom
should be familiar with remote sensing. For those

readers interested in a more detailed treatment of the
procedures employed in both geobotanical and biogeochemical investigations, the works of Brooks (1972),
Cannon (1960), Chikishev (1965), and Kovalevskii
(1979) are highly recommended (the latter two are
available in English translations).

Past to Present

Historically, remote-sensing techniques are based
on aerial photography. Since 1858, when the first
aerial photograph was made from a balloon, aerial
photographs have been used for map making and the
description of terrain. With the advent of other kinds
of sensors, such as scanners and imaging radar, the
more general term remote sensing has become widely
accepted. Remote sensing can be defined as measurement of an object from afar without physical contact. Therefore, gravity and magnetic measurements
can be considered remote sensing, but the accepted
definition is based on the use of reflected or emitted
electromagnetic energy. The electromagnetic spectrum available for remote sensing extends from the
ultraviolet at 0.4 μm to the microwave extending to
50 cm, as shown in Figure 1.

Aerial photography

Aerial photography is very useful in the study of
both vegetated and nonvegetated surfaces, owing
largely to the high spatial resolution provided and the
availability of stereo. Even today, aerial photography
is one of the most important tools used in geologic

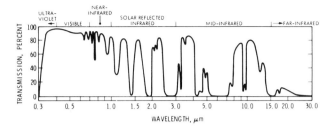

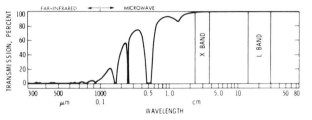

FIG. 1. Generalized absorption spectrum of the atmosphere at the zenith with the named spectral regions outlined (from Goetz and Rowan, 1981).

mapping and exploration. Photointerpretation reveals bedding, surface morphology, and structure as well as provides targets for further ground exploration and detailed mapping. Color aerial photography provides another dimension of information over black and white photography, because inherent properties of materials, such as their reflectance characteristics, can be used for identification and discrimination among similar-appearing rocks and soils.

True color aerial photographs are appealing because they present a photographic image which depicts surface features as the human eye sees them. In the case of vegetation, however, most visible wavelengths are strongly absorbed as a result of photosynthetic activity; only yellow and green wavelengths (Fig. 2) are weakly reflected. Thus as a remote-sensing device, true color aerial photographs record only a limited amount of data over heavily vegetated terrain.

False color infrared sensitive film produces photographic images (referred to variously as false color infrared, color infrared, or simply infrared photos) which record a small part of the highly reflective portion of the reflectance spectrum (0.75–0.9 μm), characteristic of vegetation, that is known as the near-infrared plateau (Fig. 2). In recording a photographic image on infrared sensitive film, a yellow filter is used to remove reflected blue from the target being photographed so that the blue, green, and red sensitive layers of emulsion in the film are available to record reflected green, red, and near-infrared radiation, respectively. Since for vegetation the most reflective portion of the spectrum is the infrared, recorded in the red sensitive emulsion layer, actively growing,

healthy vegetation appears bright red in false color infrared photographs, while unhealthy vegetation appears blue to gray.

Since little variation is seen in the visible portion of the spectrum for various types of green foliage (Fig. 2), true color aerial photography is of only limited value in geobotanical investigations involving discrimination and mapping of subtle vegetation differences. Color infrared aerial photography takes advantage of the considerable differences noted in the near-infrared reflectances characterizing various vegetation types and therefore is of much greater value in conducting such mapping exercises.

The most serious shortcoming of color infrared film is that it is sensitive to near-infrared radiation only out to approximately 0.9 μm. As will be seen below, much useful information resides in the near-infrared and shortwave infrared beyond 0.9 μm, information relating to the moisture content and cellular structure of the leaf. This information, when combined with pigment information from the visible region, may be used to infer the state of health of vegetation as well as provide additional data for discrimination and mapping. Both kinds of data are very useful in geobotanical assessments.

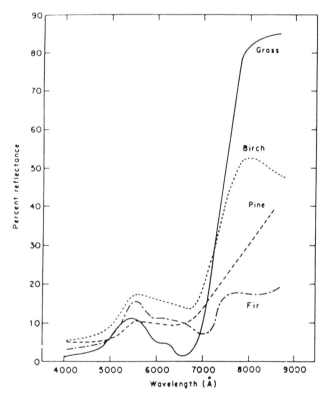

FIG. 2. Spectral reflectance curves for various types of foliage. The spectral region covered is that to which infrared films are sensitive (approximately 0.4–0.9 μm). True color film records reflected radiation from 0.4–0.65 μm (from Brooks, 1972).

Satellite remote sensing

Modern remote sensing began in 1972 with the launch of Landsat 1 (EROS, 1979). The great advantage of using images obtained from satellite altitude is in the synoptic view of large areas of the earth's surface, views that were not available from aerial photography, even from aerial photography mosaics. Not only does Landsat afford a view of a large area of the earth's surface (34,000 km²), but the angles subtended by the imaging instrument are small enough (11° for Landsat 1–3 and 15° for Landsat 4) that the entire image appears uniformly illuminated. This property is especially important when mosaics of images are produced. In addition to the narrow angles subtended, Landsat operates in a sun synchronous orbit which means that for the same time of year at a given latitude, the surface illumination is consistent worldwide. This property makes it possible to extrapolate results from one area to another, even though widely separated.

The Landsat multispectral scanner (MSS) collects image data by scanning detectors across the track underneath the spacecraft and recording the brightness values of terrain elements in four spectral bands. The data are recovered in digital form and therefore can be manipulated by computer before the final image is formed. The MSS exposed a significant portion of the geologic community to the concept of multispectral imaging as well as digital image processing. The result was a significant turn away from the use of standard photointerpretative techniques on black and white or color aerial photography to an expanded use of the spectral properties of surface materials for mapping. The MSS fostered research in digital image processing techniques and an expanded use of the spectrum for surface material identification (Goetz and Rowan, 1981). In addition, spectral research supported the incorporation of other infrared spectral channels on the recently launched Landsat 4 thematic mapper (Engel and Weinstein, 1982).

Landsat data have been widely disseminated in the geologic community because of an enlightened policy of NASA and the Department of Interior to make satellite data available to anyone at the cost of reproduction.

Concepts and Principles

Spectral data collection

The spectral properties of materials can be used for identification purposes, and continuous laboratory spectra extending out to 200 μm have long been used for analysis. Some spectral reflectance information can be derived from Landsat MSS images. The MSS acquired data in four broad spectral bands or chan-

nels: band 4 in the green (0.5–0.6 μm), band 5 in the red (0.6–0.7 μm), and bands 6 and 7 in the near infrared (0.7–0.8 and 0.8–1.1 μm, respectively). The MSS was designed primarily for agricultural applications and not for geologic purposes, although much valuable geologic information has been derived from it. The major drawbacks of using Landsat MSS images for geologic purposes have resulted from the lack of stereo, coarse spectral resolution, and limited spectral coverage (0.5–1.1 μm) which does not extend into regions of the spectrum of most use in characterizing the spectral properties of geologic surface materials. The Landsat 4 thematic mapper images will be of greater value for geologic remote sensing because of the extended spectral coverage (band 5 at 1.55–1.75 μm and band 7 at 2.08–2.35 μm). See Figure 3 for a comparison of band placement and coverage between Landsat MSS and thematic mapper systems. Although the thematic mapper images will contain more information than the MSS images, there is a growing need for high spectral resolution data for direct identification of surface materials. The papers in this volume deal almost exclusively with measurements and analyses of data in the 0.4- to 2.5-μm region, characterized by reflected solar radiation, and several of these are concerned with interpretation of high-resolution data.

Spectral remote sensing can be used to derive information about surface materials based on their reflectance behavior, as seen in Figure 4. The slopes of the spectral reflectance curves, as well as the position of the absorption features, are diagnostic of the surface material. However, in the wavelength region under discussion remote-sensing data are acquired

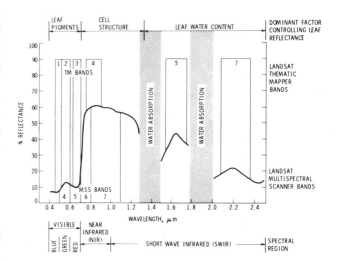

FIG. 3. Typical vegetation reflectance curve acquired in the field. Landsat MSS and thematic mapper (TM) bands are indicated. Gaps in the spectral curve at 1.4 and 1.9 μm are due to atmospheric water absorption (data were not gathered at these points).

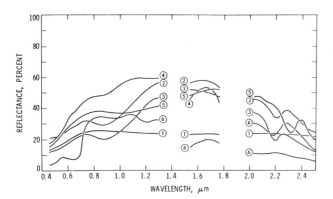

FIG. 4. Field-acquired reflectance spectra. 1, unaltered tuff fragments and soil; 2, argillized andesite fragments; 3, silicified dacite; 4, opaline tuff; 5, tan marble; 6, ponderosa pine. The gaps at 1.4 and 1.9 μm are the result of atmospheric water absorption (from Goetz and Rowan, 1981).

only from the upper micrometers of the surface because of the generally high opacity of natural materials in this region.

Spectral properties of rocks and minerals

The spectral reflectance and spectral emittance characteristics of rocks and minerals in different wavelength regions are the result of different physical and chemical properties. The spectral reflectance of minerals in the visible and near infrared short of 1 μm is determined by the presence or absence of transition metals. Charge transfer bands, which are the result of exchange of electrons between neighboring metal ions, create absorptions in the wavelength region short of 0.4 μm, and the wings of these bands extend into the visible portion of the spectrum. Charge transfer bands are the source of the general rise in reflectance toward longer wavelengths, as seen in Figure 4. Electronic transitions in transition elements, resulting from changes in energy level in d-shell electrons within the crystal field of the mineral results in absorption features such as those seen in spectrum 2 and 3 in Figure 4 (Hunt, 1977). Iron, because of its ubiquitous distribution on the earth's surface, is the most common cause of the charge transfer and electronic transition absorption features seen in rock and soil reflectance spectra. The position of the absorption minimum for the Fe^{+3} electronic transition can be influenced by the substitution of aluminum for iron, as shown by Buckingham and Sommer (1983). The absorption band between 0.85 and 0.92 μm, as well as the slope of the reflectance curve in the visible region, is used to identify the presence of limonite in Landsat MSS images (Rowan et al., 1974; Rowan et al., 1977).

The short wavelength infrared region between 1 and 2.5 μm contains both broad band and narrow

band spectral features that are diagnostic of the composition of minerals. Absorption bands for both bound and unbound water at 1.4 and 1.9 μm can be used for diagnostic purposes in laboratory measurements, but these bands (Fig. 4) coincide with strong atmospheric water bands, making this region problematical for remote sensing. The strong fundamental OH vibration at 2.74 μm influences the behavior of hydroxl-bearing minerals. Clays, in particular, exhibit decreasing spectral reflectance beyond 1.6 μm, and this broad band behavior has been used to identify clay-rich areas associated with hydrothermal alteration zones by using images taken in the region around 1.6 μm and in the 2.1- to 2.4-μm region; alunite, pyrophyllite, and muscovite also reduce the reflectance in this region (Podwysocki et al., 1983; Abrams et al., 1983; Rowan and Kahle, 1982; Prost, 1980).

The use of ratio images made in 0.2-μm-wide bands centered at 1.6 and 2.2 μm was shown by Abrams et al. (1977) to be highly effective in mapping clays and alunite in the Cuprite, Nevada, mining district. This study was the basis for the selection of the seventh band (2.08–2.35 μm) to be incorporated in the Landsat 4 thematic mapper.

Overtone bending-stretching vibration for layered silicates for Al-OH and Mg-OH create narrow spectral absorption bands within the 2.1- to 2.4-μm region. Figure 5 shows examples of laboratory spectra for some common minerals (Goetz and Rowan, 1981). The exploitation of these diagnostic features has just begun in the last few years with the advent of remote-sensing instruments that have sufficient spectral resolution. Collins et al. (1983) have developed an aircraft spectroradiometer with which they have demonstrated the use of the overtone bands for direct identification of clay minerals. From orbit, using a 10-channel radiometer, Goetz et al. (1982) have made the first direct identification of kaolinite and carbonate rocks.

The exploitation of the diagnostic spectral characteristics in the reflective infrared portion of the spectrum is possible not only with aircraft and spacecraft data but also with field instruments. Whitney et al. (1983) and Gladwell et al. (1983) describe measurements with a portable ratio-determining radiometer utilizing a simultaneous measurement in two spectral channels for the field identification of carbonate and hydroxyl-bearing minerals. The field geologist can use the instrument as another pair of eyes sensitive to longer wavelengths.

Although not covered in any papers in this volume, the midinfrared region between 8 and 12 μm is gaining importance for remote geologic mapping (Kahle and Rowan, 1980). In this region it is possible to distinguish between most silicate and nonsilicate rocks and to discriminate among different silicate rocks.

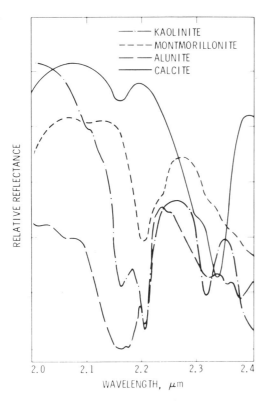

FIG. 5. High-resolution laboratory reflectance spectra in the region from 2.0 to 2.4 μm of common minerals typically associated with hydrothermal alteration.

Silicates exhibit features in spectral emittance spectra that are the result of fundamental Si-O vibrations. These features, called reststrahlen bands, result in regions of metallic-like reflection. The position of the bands is dependent on the extent of the interconnection of the Si-O tetrahedra comprising the crystal lattice (Lyon, 1965). Quartz has an emittance minimum of 8.5 μm, while olivine exhibits a broad minimum centered at approximately 10 μm. In the East Tintic, Utah, mining district, Kahle and Rowan (1980) showed that a variety of silicate rocks such as quartzite, quartz monzonite, monzonite, quartz latite, and latite were separable on a digitally processed multispectral thermal infrared scanner image taken from an aircraft. These lithologic units were not separable on thematic mapper simulator images (Rowan and Kahle, 1982). Construction of a new airborne, 6-channel thermal-infrared multispectral scanner with increased sensitivity was completed recently, and preliminary results substantiate the findings at East Tintic and show that similar separations among silicate rock types can be made at other locations (A. B. Kahle and A. F. H. Goetz, written commun.).

The use of microwave measurements in resource exploration has been confined almost exclusively to imaging radar systems in the 3- and 25-cm wave-length regions, also known as X and L bands, respectively. The radar backscatter is controlled mainly by the slope of the terrain and the surface roughness as scaled to the wavelength of the transmitted signal (Elachi, 1980). The variations in dielectric constants of rocks are too small to have a significant effect on the returned signal, and therefore the composition of rocks cannot be determined by a direct means using imaging radar. However, the roughness of the surface, manifested in the weathering or jointing patterns, may be indicative of rock type in a given climatic environment (Daily et al., 1978). Recently a shuttle-imaging radar (SIR-A) experiment (Elachi et al., 1982) has demonstrated the usefulness of microwave imaging for structure mapping in heavily vegetated areas, as a complement to Landsat in all environments and for subsurface mapping in hyperarid areas (McCauley et al., 1982).

Spectral properties of vegetation

A typical reflectance spectral curve for an actively photosynthetic plant is shown in Figure 3. This spectral curve was produced using data acquired with the Jet Propulsion Laboratory's portable field reflectance spectrometer (Goetz et al., 1975). The absorption features centered at approximately 0.480 and 0.680 μm are the result of strong chlorophyll absorption[1] in the leaf involving electronic transitions in the chlorophyll molecule centered around the magnesium component of the photoactive site. The weak reflectance feature at 0.52 to 0.60 μm indicates the portion of the visible which is not strongly absorbed, resulting in the green appearance of plants to the human eye. The high reflectance feature centered at approximately 1.0 μm is referred to as the near-infrared plateau and is characteristic of healthy leaf tissue. The sharp rise in the curve between the chlorophyll absorption feature at 0.68 μm and the near-infrared plateau is referred to as the red edge by Barber and coworkers (Horler et al., 1980, 1983), and the slope of the red edge has been related to chlorophyll concentrations in the leaf (Horler et al., 1983). Figure 6 shows the changes that occur in the spectral curve of leaves during the growing season, which in turn reflect changes in the chlorophyll concentrations as well as buildup of additional pigments such as tannins within leaves. Note that both the position and slope of the red edge change during the gradual change in the leaf from active photosynthesis to total senescence. Such information concerning phenologic stage and/or state of health, as well as taxonomic placement, is a most important aspect

[1] Absorption in the 0.480-μm region is actually the result of electron transitions in carotenoid pigments (carotenes and xanthophylls) which function as accessory pigments in the photosynthetic process in addition to the chlorophylls.

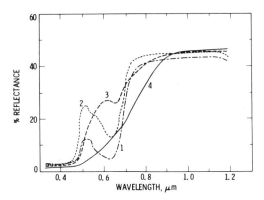

FIG. 6. Reflectance spectra for a healthy beech leaf (1) and beech leaves in progressive phases of senescence (2–4) (from Knipling, 1969).

of detection of stress in vegetation (Labovitz et al., 1983) and in geobotanical investigations (Rock, 1982).

According to the work of Collins et al. (1981) and Horler et al. (1982), the spectral region most affected by geochemical stress lies between 0.55 and 0.75 μm, with symptomatic shifts in position and slope of the red edge being most diagnostic.

Collins and coworkers (1981; Chang and Collins, 1983; Collins et al., 1983; Milton et al., 1983) note a blue shift, consisting of a 0.007- to 0.010-μm shift of the chlorophyll shoulder or red edge to slightly shorter wavelengths (0.665–0.675 μm) in plants influenced by geochemical stress (see Chang and Collins, 1983, fig. 4, p. 727). In the study of phenologic changes (ontogenic or maturation stages) of crop plants such as corn, wheat, and sorghum, Collins (1978) has demonstrated a red shift (0.007–0.010 μm shift of the red edge to longer wavelengths, 0.690–0.700 μm) associated with the conversion from vegetative growth to reproductive growth (heading and flowering). The red shift may also be useful in separation of crop types (Chang and Collins, 1983, fig. 4, p. 727). In the case of the blue shift (mineral-induced stress), the shift may be related to subtle changes in the cellular environment (Collins et al., 1983), while the red shift may be related to a decrease in chlorophyll production.

The red shift-blue shift can only be resolved using a high spectral resolution instrument. The instrumentation used by Collins (1978; Collins et al., 1981; Chang and Collins, 1983) is a 512-channel spectroradiometer which makes measurements with 0.0014-μm-wide bands, collecting data in the 0.400- to 1.100-μm spectral region. Collins (1978) also notes that the red shift can be measured with 0.010-μm-wide spectral bands centered at 0.745 and 0.785 μm. The red shift-blue shift is not seen in data provided by the MSS or thematic mapper systems or in color infrared photography.

The reflectance levels at 1.6 and 2.2 μm (Fig. 7) provide an accurate indication of leaf water content (Rohde and Olson, 1971; Tucker, 1980), while similar information might be obtained from leaf temperature. A rapidly transpiring leaf will have a high water content, assuming a readily available soil water source, and will be cooler than ambient temperature, owing to evaporative cooling of the inner surfaces of the leaf, while nontranspiring leaves will be warmer. Spectral data from the reflective infrared and thermal infrared should prove to be most useful in detection of early water stress in vegetation.

As indicated in the discussion of features of a typical reflectance curve, the near-infrared plateau was cited as characteristic of healthy leaf tissue. Subtle absorption features (Fig. 8) do occur in the near-infrared plateau region of the spectrum (0.75–1.3 μm), and these features may be related to both cellular arrangement within the leaf and hydration state (Gates, 1970; Gausman et al., 1977, 1978). Cellular arrangement within the leaf is genetically controlled and thus is of taxonomic significance (Esau, 1977). The near-infrared plateau region of the reflectance curve of vegetation may prove to be useful in the same way that the 2.0- to 2.5-μm region has for mineral identification. High spectral-spatial resolution remote-sensing systems capable of detecting such taxonomically significant spectral data will provide tremendous potential for vegetation discrimination, identification, mapping and, thus, geobotanical investigation in the future.

The phenologic cycle of a given species is also under strong genetic control, even under adverse environmental conditions (Baker et al., 1982). Identification of plant community composition at the genus

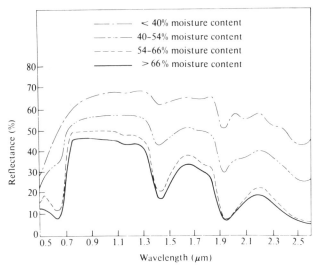

FIG. 7. Spectral curves representing leaves of various moisture contents. Reflectance data for corn leaves (after Hoffer and Johannsen, 1969).

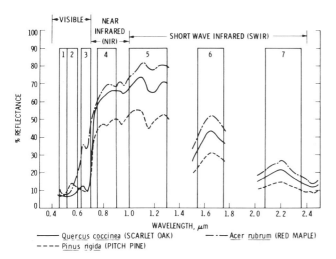

FIG. 8. Reflectance spectra acquired using the Jet Propulsion Laboratory's portable field reflectance spectrometer during the fall foliage peak at Lost River, West Virginia. Data acquired October 19 to 21, 1981. The numbered bands refer to NS-001 thematic mapper simulator bands.

and species level will be possible using remote-sensing systems that acquire data on the slope and position of the red edge, on subtle absorption features of the near-infrared plateau and reflective and immissive effects associated with leaf water content. In addition to determining the state of health of community members, such a remote-sensing capability will allow for discrimination of secondary regrowth communities from climax communities, which are of considerable interest in geobotanical investigations.

Botanical aspects of geochemical stress

The use of remote-sensing data to assess vegetation response to geochemically induced soil anomalies requires an understanding of how vegetation is likely to respond to given geochemical conditions and how that response will be recorded in the remotely sensed data. Although it is not the authors' intent to present the reader with detailed data on plant physiology, mineral nutrition, and vegetation remote sensing, a limited exposure will prove useful in interpreting remote-sensing data acquired over vegetated field sites and identifying geobotanical features of significance.

Geobotanical anomalies may be expressed in a number of ways: (1) anomalous distribution of species and/or plant communities, (2) stunted growth and/ or decreased percent ground cover, (3) alteration of leaf pigment complexes and/or physiologic processes which produce a yellowing or chlorosis and/or altered transpiration rate, and (4) an anomalous phenologic cycle which causes early foliage change or senescence in the fall, alteration of flowering periods, and/or late leaf flush in the spring.

Some of these anomalies are subtle (1), and/or seasonal (4) in their expression and require developing a sound understanding of what is normal for the site in question. Other responses are more obvious (2, 3), both on the ground and in remotely sensed data. Since all plant community members undergo environmentally triggered, genetically controlled phenologic cycle changes such as variations in pigment concentrations, physiological processes, and growth rates, it becomes essential that remote-sensing data be acquired for a given site several times during the year with emphasis placed on the growing period, from leaf flush in the spring to fall senescence. Such multitemporal data sets are absolutely necessary for the normal condition to be determined accurately, allowing the anomalies to be identified.

In most cases, geochemical anomalies result in an alteration in the availability of soil nutrients required by plants growing at these sites. (Table 1 lists those nutrients required by plants for normal growth.) Such alterations will also influence the availability of water to the plants, so that geochemically stressed vegetation may show water stress symptoms.

In many cases, high concentrations of certain heavy metals in the soil such as Ca^{+2}, Mg^{+2}, Cu^{+2}, and other bivalent cations will actually generate deficiency symptoms in plants. Although this sounds somewhat contradictory—as a result of competitive inhibition in the root zone, in which one ion competes with other similar ions for carrier sites as well as for transport and utilization within the leaf—apparent toxicity symptoms may in fact be deficiency symptoms. Table 2 lists common deficiency symptoms resulting from inadequate amounts of given nutrients being made available for plant uptake.

Competitive inhibition is seen, even at low level concentrations, between pairs of cations such as Ca^{+2} and Mg^{+2}, Mn^{+2} and Ca^{+2}, and possibly Fe^{+2} and Ca^{+2}. Chlorosis[2] seen in leaves will result in plants growing in soils having high levels of calcium carbonate (called lime-induced chlorosis), probably as a result of competitive inhibition between Ca^{+2} and Fe^{+2} at carrier sites in root cell membranes. Similar chlorotic symptoms resulting from magnesium and manganese deficiencies may also result from high levels of Ca^{+2} in the soil.

It is not only the high concentrations of calcium and associated competitive inhibition that induces deficiency symptoms in plants growing in calcareous soils. The effect of soil pH on availability of required nutrients is well known (Fig. 9), and the high pH associated with calcareous soils strongly limits the

---

[2] The term chlorosis refers to the yellowing of leaves which results from decreasing chlorophyll content in the mesophyll cells; iron deficiency leads to chlorosis because iron is required for the synthesis of chlorophyll within these cells.

TABLE 1. Nutrients Essential for Normal Growth and Development in Vascular Plants

Element	Form in which absorbed	Some functions
Macronutrients		
Carbon	CO_2	Carbohydrates, both structural (cellulose of the cell wall) and functional (energy sources such as starch and glucose), proteins, fats, etc.
Hydrogen	HOH	Also in carbohydrates, proteins, fats, etc.; functional in energy transfer mechanisms; osmotic and ionic balance
Oxygen	O_2	Also in carbohydrates, proteins, fats; required in free form for aerobic metabolic processes
Nitrogen	NO_3^-m or NH_4^+	Amino acids, proteins, nucleotides, nucleic acids, chlorophyll, and coenzymes
Potassium	K^+	Enzymes, amino acids, and protein synthesis; activator of many enzymes; opening and closing of stomata
Calcium	Ca^{+2}	Calcium of cell walls; cell membrane permeability; enzyme cofactor
Phosphorus	H_2PO_4 or HPO_4^{-2}	Formation of high-energy phosphate compounds; nucleic acids; phosphorylation of sugars; several essential coenzymes; phospholipids
Magnesium	Mg^{+2}	Part of the chlorophyll molecule; activator of many enzymes
Sulfur	SO_4^{-2}	Some amino acids and proteins; coenzyme A
Micronutrients		
Iron	Fe^{+2}	Chlorophyll synthesis, cytochromes, and ferredoxin
Chlorine	Cl^-	Osmosis and ionic balance; probably essential in photosynthesis in the reactions in which oxygen is produced
Copper	Cu^{+2}	Activator of some enzymes
Manganese	Mn^{+2}	Activator of some enzymes
Zinc	Zn^{+2}	Activator of many enzymes
Molybdenum	MoO_4^{-2}	Nitrogen metabolism
Boron	BO^{-3} or $B_4O_7^{-2}$ (borate or tetraborate)	Influences Ca^{+2} utilization; functions unknown

Modified from Raven et al. (1976)

availability of nutrients such as iron, phosphorus, manganese, boron, copper, and zinc.

A complete syndrome of stress-related symptoms is well known for nutrient deficiencies and edaphic (soil) conditions, both at the gross, macroscopic level (note Table 2) and at the microscopic, anatomical level in stems and leaves (Hewitt, 1963). When macroscopic stress-related deficiency symptoms are pronounced, they are expressed in the canopy and are so characteristic that they may even be utilized by photointerpretors in assessing the exact nature of forest damage as recorded in remotely sensed infrared photography, either black and white or false color (Murtha, 1972).

In addition to the adverse growth conditions caused by the high calcium concentrations and associated pH characteristic of calcareous soils, calcium deficiencies caused by low calcium concentrations in residual soils

TABLE 2. Nutrient Deficiency Symptoms

Symptoms	Deficient element
Older leaves affected first	
Effects mostly generalized over whole plant; lower leaves dry up and die	
Plants light green; lower leaves yellow, drying to brown; stalks become short and slender	Nitrogen
Plants dark green; often red or purple colors appear; lower leaves yellow, drying to dark green; stalks become short and slender	Phosphorus
Effects mostly localized; mottling or chlorosis; lower leaves do not dry up but become mottled or chlorotic; leaf margins cupped or tucked	
Leaves mottled or chlorotic, sometimes reddened; necrotic spots; stalks slender	Magnesium
Mottled or chlorotic leaves; necrotic spots small and between veins or near leaf tips and margins; stalks slender	Potassium
Necrotic spots large and general, eventually involving veins; leaves thick; stalks short; rosetting of leaves	Zinc
Young leaves affected first	
Terminal buds die; distortion and necrosis of young leaves (terminal die-back)	
Young leaves hooked, then die back at tips and margins	Calcium
Young leaves light green at bases; die back from base; leaves twisted	Boron
Terminal buds remain alive but chlorotic or wilted, without necrotic spots	
Young leaves wilted; without chlorosis; stem tip weak	Copper
Young leaves not wilted; chlorosis occurs	
Small necrotic spots; veins remain green	Manganese
No necrotic spots	
Veins remain green	Iron
Veins become chlorotic	Sulfur

Modified from Bidwell (1979)

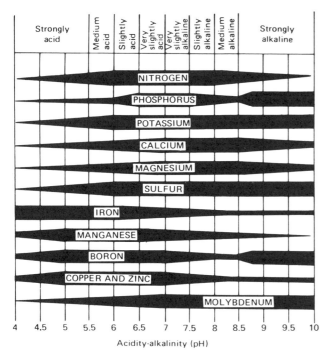

FIG. 9. The effect of soil pH on availability of nutrients to plants. The thickness of the horizontal band represents the relative solubility and, thus, the availability of the nutrient. Taken from Bidwell (1979).

derived from the weathering of serpentinized rocks lead to quite different adverse growth conditions. The role of calcium in plant nutrition has only recently been recognized as relating to membrane permeability (Bidwell, 1979; Epstein, 1972) and thus having a pronounced influence on the uptake and transport of both water and associated ions. Calcium has been shown to be absolutely essential for unimpaired ion transport into cells (Epstein, 1961), and the lack of calcium results in loss of the differential permeability characteristic of the normal cell membrane system. In other words, the cellular membranes become leaky as a result of low calcium concentrations, and the plant's capacity for normal nutrient uptake and transport as well as retention of ions is lost (Epstein, 1972).

As a result of loss of selective permeability, low calcium concentrations in the soil will lead to increased leaf, stem, and root concentrations of a wide variety of ions. High concentrations of zinc, copper, manganese, iron, aluminum, and molybdenum have been reported in leaves and roots of crop plants grown in low calcium soils (Wallace, 1971). It has also been reported that plants are more susceptible to heavy metal "toxicities," which may actually be deficiencies, as a result of low calcium levels in the soil (Frolich et al., 1966; Wallace et al., 1966). Table 1 lists those nutrients required by healthy, green plants for normal growth. The absence or decreased availability

of any one of these required nutrients will result in characteristic deficiency symptoms (Table 2). A partial list of general plant responses to specific geochemical stress conditions is given in Table 3.

Applications

Landsat MSS images continue to be the most widely used data set in remote-sensing studies of mineral deposits because of their demonstrated applicability, worldwide availability, and the more limited areal coverage of data acquired using more advanced systems. The most common application is still for delineating regional structural features that might influ-

TABLE 3. Some General Plant Responses to Geochemical Stress

Edaphic condition	Possible effect
Ca^{+2} in high concentrations	Competative inhibition of transport of required cations such as Fe^{+2}, Mg^{+2}, and Mn^{+2}; deficiency symptoms[1] such as lime-induced chlorosis may result
Ca^{+2} absent or in low concentrations	Cellular membranes become leaky, differential permeability lost and ability to absorb water impeded under certain conditions; deficiency symptoms[1] and/or stunting may result
Cu^{+2}, Zn^{+2}, Mg^{+2}, etc., in high concentrations	Competative inhibition of transport of required cations may result; effects similar to high Ca^{+2} concentrations possible; may be coupled with Ca^{+2} deficiency symptoms[1] also
SO_4^{-2} in high concentrations	Cellular membrane permeability affected; rate of cation uptake impeded; deficiency symptoms[1] could result
Low soil pH (acid soils)	Heavy metals such as Al^{+2}, Cu^{+2}, Zn^{+2}, and Mn^{+2} become soluble in soil solution (thus will be available to plants growing in such soils); may be linked with Ca^{+2} deficiency since many calcium-deficient soils are acid; required nutrients may become limited owing to increased insolubility below pH 6.0[2]
High soil pH (basic soils)	Heavy metals required for growth (Fe^{+2}, Zn^{+2}, Cu^{+2}, and Mn^{+2}) may become limited owing to increased insolubility above pH 7.0[2]

As noted in the text, patterns of uptake and thus response varies among plant species, the plant parts affected, and the differential metals absorbed. The responses cited above are common but not universal

[1] Note Table 2
[2] Note Figure 9

ence the emplacement of intrusive bodies and associated mineral deposits, but important advances have been made in lithologic and geobotanical studies. In reasonably well-exposed areas, many lithologic units are distinguishable in MSS images on the basis of their spectral properties, and anomalously limonitic rocks can be identified. Although lithologic units are not mapped directly in vegetated areas, some can be distinguished by studying the areal density and community associations of plants (Raines et al., 1978; Raines and Canney, 1980; Milton, 1983). In a few areas, anomalous vegetation patterns detected in MSS images are related to metal anomalies in the soil (Bolviken et al., 1977; Schwaller and Tkack, 1980).

The results reported in this volume and at recent geological remote-sensing symposia clearly indicate a high level of interest in measurements made in other wavelength regions and with high spectral resolution. This level of interest was stimulated by recognition of the limitations imposed by MSS spectral bands for mapping and defining the mineral content of lithologic units, especially hydrothermally altered rocks, and for studying botanical phenomena related to mineralization. Although higher spatial resolution is also needed, emphasis has been placed on spectral considerations. Consequently, most of the papers in this volume are concerned with lithologic and geobotanical studies of mineral deposits using thematic mapper spectral bands and high spectral resolution airborne measurements. The limited number of structural studies reported here (Prost, 1983) are not representative of the worldwide use of remote sensing in mineral exploration but are consistent with the trend toward spectral studies. The interested reader is referred to the publications of Goetz and Rowan (1981), Rowan and Wetlaufer (1981), Sabins (1978), and Gold (1980) for discussion of structural applications. Two important wavelength regions, the midinfrared and microwave, are not represented by papers in this volume but were discussed earlier in this paper.

Lithologic studies

As noted earlier, the geological remote-sensing community argued successfully for inclusion of a band centered near 2.2 μm in the thematic mapper, because analysis of laboratory, field, and aircraft measurements had demonstrated the effectiveness of such a band, coupled with one near 1.6 μm, for detecting the presence of OH-bearing minerals, a characteristic of many hydrothermally altered rocks. This capability provides a basis for distinguishing between limonitic hydrothermally altered rocks and most limonitic igneous rocks in which the limonite is the product of oxidation of ferromagnesian and opaque minerals. In addition, highly leached, nonlimonitic

altered rocks can be detected where OH-bearing minerals are present (Abrams et al., 1977; Rowan et al., 1977; Rowan and Kahle, 1982). Because the MSS lacks spectral bands beyond 1.1 μm, these distinctions cannot be made consistently in these images. Since the time of the design and construction of the thematic mapper, numerous studies, particularly those of the NASA-Geosat test case program, have documented the value of the 1.6- and 2.2-μm spectral bands for mineral exploration. A further testimony to the value of these images, which have been acquired from aircraft, is a survey of roughly 15,000 square miles by a consortium of mineral and petroleum companies.

In this volume, papers by Abrams et al. (1983) and Podwysocki et al. (1983) illustrate the processing, analysis, and interpretation of thematic mapper simulator images for mapping porphyry copper deposits in Arizona and a multimetal deposit at Marysvale, Utah, respectively. In Arizona, the Helvetia-Rosemont, Silver Bell, and Safford deposits were studied using Landsat MSS as well as thematic mapper simulator images. In each area, the simulator images processed to enhance spectral contrasts proved to be extremely useful for mapping mineralogic differences between the altered and unaltered rocks and among the altered rocks. Structural information increased in such images in concert with improved lithologic information. In spite of the lower spatial resolution and limited spectral bands, Landsat MSS images yielded valuable regional structural and some lithologic information, mainly the distribution of limonitic rock and soil.

At Safford, Abrams et al. (1983) mapped the three principal types of hydrothermally altered rocks in a thematic mapper simulator image in which a canonical transformation had been applied to the radiance data (see Abrams et al., 1983, fig. 11, p. 603). Propylitic alteration is widespread and characterized by the presence of epidote and chlorite. Biotization is also common, especially in Tertiary porphyritic andesites. Quartz-sericite alteration occurs in silicic to intermediate intrusive rocks, especially in dikes along northeast-oriented shear zones which were important in the localization of mineralization.

In the upper part of figure 11 in Abrams et al. (1983, p. 603), the canonically transformed image is compared with a color aerial photograph obtained simultaneously with the thematic mapper simulator image, and in the lower part, the resulting alteration map is compared with a map prepared using conventional field and laboratory methods. Abrams et al. (1983) noted that the blue areas in the map derived from the simulator image represent propylitized rocks, red areas mark quartz-sericite alteration, and yellow areas are the most intensely altered with abundant pyrite, secondary silica, and the largest amount of copper-oxide minerals. These distinctions were based on color differences in the simulator image which are related to spectral reflectance differences among the mineral assemblages in altered rocks.

The Marysvale, Utah, mining district is dominated by Tertiary volcanic and related intrusive rocks that range from intermediate to silicic compositions. Intermediate volcanic rocks were intruded by quartz monzonite stocks about 23 m.y. ago, which resulted in convecting hydrothermal cells in the adjacent wall rocks (Cunningham and Steven, 1979; Podwysocki et al., 1983). Altered rocks containing alunite and kaolinite are exposed in the cores of these cells. Silicic volcanic rocks erupted between 21 and 14 m.y. ago and water-laid sediments cover large parts of the area.

The altered rocks are characterized by absorption features centered between 2.17 and 2.22 μm owing to the Al-OH bending and stretching vibrations that occur in alunite, kaolinite, and montmorillonite (see Podwysocki et al., 1983, fig. 3, p. 679). These absorption features are absent or weakly expressed in the unaltered rock spectra. However, diagenetically zeolitized tuff has intense absorption features at 1.4 and 1.9 μm owing to constituent water, which results in low reflectance in the 2.0- and 2.5-μm region (fig. 2 in Podwysocki et al., 1983). Some of the unaltered rocks are limonitic and locally the altered rocks lack limonite staining.

The color ratio composite image shown in figure 4 in Podwysocki et al. (1983, p. 683) was constructed to take advantage of the spectral differences between the hydrothermally altered and unaltered rock, and to distinguish between vegetation and altered rocks, which are spectrally similar to the broad 1.6- and 2.2-μm bands. Altered rocks are shown in yellow where limonite is lacking and white where they are limonitic. Limonitic rocks lacking significant OH-bearing minerals are light blue. The typically spectrally flat unaltered volcanic rocks are green. Vegetation varies from magenta in coniferous cover to dark blue for sagebrush to red or orange in grassy areas (see Podwysocki et al., 1983, fig. 4, p. 683).

Comparison of the areal distribution of altered rocks as interpreted from the color ratio composite image presented in Podwysocki et al. (1983, fig 4, p. 683) with the detailed field map generally shows excellent agreement. However, several areas now known to consist of altered rocks were not mapped in the field (C). One of these is the uranium-producing Central mining area (B), where the alteration is confined to fracture zones in the volcanic and intrusive rocks. On the other hand, one area of highly silicified hematic rocks is not portrayed in the yellow and white colors typical of the hydrothermally altered rocks, because the content of OH-bearing minerals is low.

The zeolitic tuff (E) is yellow owing to the low reflectance at 2.2 μm relative to 1.6 μm.

Because of the sensitivity of the 1.6 μm/2.2 μm ratio to the presence of OH-bearing minerals, Podwysocki and colleagues, using an approach described by Ashley and Abrams (1980), prepared a color-coded map of the areal variation of this ratio. The highest ratio values coincide with the central parts of the paleohydrothermal cells and along faults in the Central mining area. The level of detail was notably better than on the color ratio composite image.

As these samples indicate, thematic mapper images should prove to be very useful for mapping hydrothermally altered rocks, but the lack of specific mineralogic determinations in these broad bands will lead to several important ambiguities. For example, altered rocks and carbonate rocks, which typically have an absorption band near 2.33 μm, are not likely to be separable spectrally in thematic mapper images. In the Marysvale district, high spectral resolution airborne measurements were used for identifying the dominant OH-bearing minerals along several flight lines (Podwysocki et al., 1982). These data, collected using the Mark II system, consist of radiance reflected from 20-m^2 areas which are recorded in 64 spectral channels in the 2.0- to 2.5-μm region (Collins et al., 1981; Podwysocki et al., 1982). Figure 10 shows examples of airborne spectra for kaolinite, alunite, jarosite, and clay plus gypsum. Analysis of measurements taken along flight lines crossing the altered areas resulted in identification of mineralogic variations that characterize different levels of alteration intensity.

In the Oatman mining district in northwestern Arizona, Marsh and McKeon (1983) used a combination of high spectral resolution field measurements and Mark II airborne measurements for studying subtle alteration mineralogical variations. The Oatman district consists of Tertiary intermediate and silicic volcanic rocks that have been intruded by two epizonal plutons, the Times and Moss Porphyries (Clifton et al., 1980; Marsh and McKeon, 1983). Pre-Tertiary rocks are Precambrian schists, gneiss, and granite.

Four alteration assemblages are present in the Oatman district: (1) pervasive argillic alteration which is characterized by the presence of alunite and sericite, (2) phyllic (illite) alteration, (3) silicic alteration which is characterized by the introduction of quartz along fractures, and (4) propylitic alteration where chlorite, calcite, and epidote formed along veins. The association of the orebodies, mainly precious metal epithermal veins, with the altered rocks in categories 2 through 4 is clear, but the significance of the pervasive argillic alteration is less certain (Thorson, 1971; Clifton et al., 1980; Marsh and McKeon, 1983).

High spectral resolution measurements of spectral

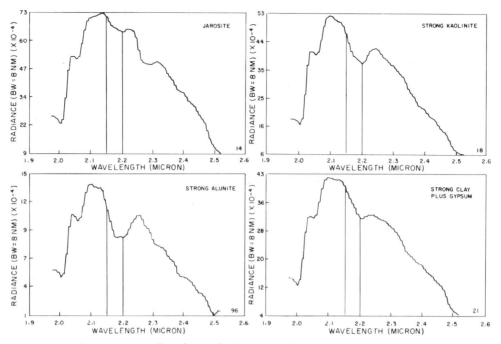

FIG. 10. Airborne spectra collected over the Marysvale, Utah, area with the spectroradiometer used by W. Collins. The spectra are for kaolinite, alunite, jarosite, and clay plus gypsum (Podwysocki et al., 1982).

reflectance were made in the field for areas within the alteration types, and the resulting absorption features were analyzed using the results of X-ray diffraction determinations. Marsh and McKeon show that characteristic spectra for the units reflect their mineralogic assemblages. These results indicate that in situ measurements can be used for rapidly separating and mineralogically characterizing visibly similar altered rocks, such as the argillic and phyllic alteration at Oatman. Compared with conventional procedures, a very substantial savings in time is indicated, because most of the spectral reflectance analysis can be conducted in the field by referring to reflectance spectra for individual alteration minerals and mixtures.

The reflectance spectra recorded in the field and related mineralogic results also provided a critical basis for interpretation of the high spectral resolution airborne measurements. Four approaches were taken for analysis of the airborne measurements, including comparative visual analysis of spectra such as those shown in figure 6 in Marsh and McKeon (1983, p. 625). The residual image map (see Marsh and McKeon, 1983, fig. 7, p. 626) was especially effective because it uses a color graphical display to show both the major spectral anomalies and the small spectral features in a two-dimensional image for the survey area.

Analysis of shuttle multispectral infrared radiometer measurements made during the November 1981 flight of Columbia have shown that some minerals, including kaolinite, alunite, calcite (dolomite), and probably montmorillonite can be identified using as

few as five spectral channels (Goetz et al., 1982), where these are the dominant minerals with absorption features in the 2.0- to 2.5-μm region. However, identification of other such minerals and the components of mixtures of OH- and/or CO_3-bearing minerals will require a larger number of spectral channels in the 2.0- to 2.5-μm region, probably at least 10 to 12 (W. Collins, pers. commun., 1982), or as many as 25 if the region were to be covered with 0.02 μm channels equally spaced between 2.0 and 2.5 μm.

An important development associated with the design of the shuttle multispectral infrared radiometer experiment in 1976 was the portable ratio-determining radiometer (Whitney et al., 1983). Although this instrument was developed as an aid in the field for interpretation of shuttle radiometer measurements, it has attained wider usage because of the flexibility provided by interchangeable filters. Ten channels, five positioned in each of two parallel optical trains, allow 25 different ratios to be measured without changing filters. The radiometer head and battery pack are separate and weigh about 2 kg each. Individual channels or ratios can be recorded either by visual display on a screen or continuously on an analog recorder (Whitney et al., 1983).

Whitney and others show that for kaolinite, montmorillonite, alunite, calcite, gypsum, illite, serpentine, and chlorite only three ratios are needed to achieve 100 percent correct identification of 19 samples. However, the accuracy of identification decreased when these three ratios were analyzed for a larger set of samples (56), mainly because of the presence of mixtures of OH- and CO_3-bearing minerals and impurities. When all 25 ratios were analyzed, 51 of the 56 samples were correctly identified; 13 ratios were used, since no improvement occurred with the addition of other ratios. Again, misclassifications were due to overlapping spectral features resulting from mixtures and to impurities.

The portable ratio-determining radiometer appears to have several field applications such as indicating, from trials made either on the ground or from helicopters, the presence of OH- and CO_3-bearing minerals in ore samples, mines, and outcrops. One such application described in this volume deals with the characterization of lithologic units in the Kerr Addison mine, Ontario, Canada, according to spectral reflectance measurements made on samples from the 3850 level (Gladwell et al., 1983).

Interest in the spectral absorption features in the 2.0- to 2.5-μm region has overshadowed advances made in the 0.4- to 1.1-μm region. The positions of spectral absorption features owing to electronic transitions in Fe^{+3} ions also vary depending on composition and crystal structure (Hunt et al., 1971; Hunt and Ashley, 1979; Buckingham and Sommer, 1983), and these spectral variations can be used to map the distributions of Fe^{+3} minerals (Peters, 1983). Further research will undoubtedly permit additional mineralogic determinations to be made in the visible and short-wavelength near-infrared region.

Geobotanical studies

The vegetation-oriented papers in this volume represent geobotanical applications of remotely sensed data sets of a variety of types. High spectral resolution remote-sensing data from both aircraft and laboratory experiments are used by Collins and his coworkers (Chang and Collins, 1983; Collins et al., 1983; and Milton et al., 1983) to locate and enhance the so-called blue shift in the red edge portion of the vegetation reflectance spectrum (cf. spectral properties of vegetation) associated with geochemical stress. Milton (1983) describes how broader band spectral data acquired by aircraft may be effectively correlated with field-acquired spectral data. Darch and Barber (1983) address the question of developing maximum use of the broad band spectral data available from the Landsat MSS for interpretation and utilization of vegetation cover.

Several of the papers in the present volume also provide detailed comparisons between biogeochemical analyses, actual chemical analyses of leaf material, and spectral properties (Labovitz et al., 1983; Darch and Barber, 1983; and Chang and Collins, 1983). Several of the papers (Darch and Barber, 1983; Labovitz et al., 1983; and Segal, 1983) emphasize the need to understand, and thus appreciate, the multitemporal variability inherent in vegetation spectral data in making use of such data in exploration approaches.

Of the three papers dealing with the interpretation of high spectral resolution data (Collins et al., 1983; Chang and Collins, 1983; and Milton et al., 1983), each presents very important points to be considered. It is suggested that Collins et al. (1983) be read first since it serves to introduce the reader to Collin's airborne spectroradiometer, exploration approach, and use of waveform analysis to extract information on the position of the chlorophyll absorption red edge. In addition, this paper presents the reader with a general overview of geobotanical-biogeochemical investigations conducted to date. The areas studied, Cotter Basin, Montana, and Spirit Lake, Washington, are characterized by coniferous forest species, and the consistent demonstration of a blue shift in conifer forest canopies growing over sites of known mineralization suggests that the phenomenon is a real pigment-related response to geochemical stress.

It is interesting to note (Collins et al., 1983) that the blue shift was not detected over iron sulfide zones whereas it is detected consistently over lead and copper sulfide zones. These data are consistent with the competitive inhibition model of selective ion uptake in the root zone for explaining heavy metal toxicity and may well provide information useful in interpreting the true botanical nature of the blue shift.

The second paper dealing with the high spectral resolution provided by Collin's airborne spectroradiometer (Milton et al., 1983) presents similar data for mixed deciduous forest canopies from mineralized sites associated with the Carolina slate belt at Pilot Mountain, North Carolina. The fact that the blue shift phenomenon can be demonstrated in both broadleaf forest species and coniferous species strongly suggests a relatively universal pigment response on the part of green plants to specific types of geochemical stress.

The laboratory studies conducted by Chang and Collins (1983) tend to underscore the point that the blue shift represents a universal response to certain but not all forms of geochemical stress. Specifically, they found a pronounced shift in greenhouse-grown sorghum in response to high concentrations of various salts of zinc, copper, and manganese but not to similar concentrations of other heavy metals. It is interesting to note that high concentrations of copper and zinc alone generate stress, while similar concentrations of copper and zinc, in combination with iron and lead, do not generate stress. This is again a strong suggestion that competitive inhibition is the mechanism functioning at the root zone in regard to the uptake of bivalent cations. This paper represents the first verification of the blue shift phenomenon in response to controlled in vitro experiments.

One word of caution needs to be raised concerning interpretation of Chang and Collin's (1983) laboratory data. In most cases only a single sorghum plant was exposed to a given chemical treatment. There is much danger in drawing broad implications from such a limited data set. Individual variability among sorghum plants is not taken into account, and it is suggested that a future study be conducted, involving a larger number of plants exposed to each given treatment. A large number of control plants—those not exposed to heavy metal treatments—should also be maintained in order to allow determination of individual specimen variation under normal growth conditions. Perhaps plants subjected to water stress conditions could also be included in order to determine if water stress alone could generate a blue shift response.

The spectral data presented in the study by Labovitz et al. (1983) were acquired with a portable radiometer in spectral regions duplicating the thematic mapper bands 3 (0.63–0.69 μm), 4 (0.76–0.90 μm), and 5 (1.55–1.75 μm). They found that reflectance values increase in bands 3 and 5 when leaves from mineralized sites were compared with leaves from nonmineralized sites, suggesting that variations are due to decreases in chlorophyll, resulting in increased reflectance in band 3 and leaf water content characterized by increased reflectance in band 5.

The fact that no change was detected in reflectance values in band 4 suggests either that little or no change in cellular structure occurs within the leaf in response to mineralization or that the change, if it does occur, is not detectable with a relatively broad band sensor such as the thematic mapper. These data do suggest, however, that the thematic mapper will be a more powerful tool than the MSS in detecting subtle changes in chlorophyll pigment and leaf water content variations in response to geochemical stress.

Labovitz et al. (1983) cite data suggesting that certain trace elements such as Zn and Mn are found in concentrations in the leaves of white oak that are roughly proportional to the soil concentrations, while other elements such as Cu and Pb did not concentrate in leaves in proportion to soil concentrations. These data generally agree with findings by Wickland (1982), which indicate that plant populations from heavy metal-contaminated mine sites—derelict Cu, Zn, and Pb mine sites from the Carolina slate belt, North Carolina—may or may not have significantly higher concentrations of metals in their leaf tissues than do plants from control sites. Wickland states that patterns of uptake and amounts of metal concentrated vary among plant species, between tissues sampled, and for each metal. Groups of species with similar uptake patterns were recognized (Wickland, 1982). Evergreen trees appear to have slightly lower con-

centrations of metals in their leaves and twigs than do deciduous trees; however, in terms of relative amounts of leaf and twig metal concentrations, deciduous trees tend to have higher concentrations of Cu and Zn in their leaves, whereas evergreen trees have just the reverse, with the higher metal concentrations in their twigs. The variation in the patterns of uptake and the concentrations of specific metals noted by Labovitz et al. (1983) and Wickland (1982) may well be the results of species variation and serve to underscore Wickland's view that uptake patterns for a given metal are highly species specific.

From the standpoint of remote sensing it is significant to note that Labovitz et al. (1983) have noted spectral variations in leaf reflectance that are directly related to relative amounts of metals in the soils but not to amounts in the leaves. From these data, it would appear that the soil concentrations of metals can produce their effects on leaf chlorophyll and water concentrations without actually being concentrated in the leaves, suggesting that in some cases geobotanical remote-sensing observations may be a more accurate indication of soil geochemistry than biogeochemical analyses.

As noted by Labovitz et al. (1983), decreases in chlorophyll content and leaf moisture represent a normal phenologic change in temperate deciduous species known as senescence, and that detection of early senescence in a forest canopy may be an important indication of a geochemical anomaly. The work of Darch and Barber (1983) clearly demonstrates the need for consideration of the phenologic cycle in considering both biogeochemical analysis as well as geobotanical remote sensing.

Data from Darch and Barber (1983) suggest that concentrations of copper, zinc, and molybdenum are correlated with leaf emergence and flowering in some species, while concentrations of nickel, lead, chromium, cobalt, iron, aluminum, boron, and manganese increase to a peak immediately before leaf senescence. These data strongly suggest that seasonal changes in plant heavy metal cycling, and thus stress, affect the reflectance spectra of the vegetation since heavy metals interfere, via competitive inhibition, with the plant's uptake of iron and other essential nutrients. Such changes are detectable remotely, even using the coarse spectral resolution of the Landsat MSS. Darch and Barber found that early spring (April and May) Landsat MSS images provided the best opportunity for locating geobotanical anomalies associated with copper stress.

Emphasis cannot be placed too heavily on the need for multitemporal data sets in detecting and interpreting geobotanical anomalies associated with heavy metal stress. The work of Labovitz et al. (1983) and Darch and Barber (1983) certainly suggests that success in recognizing heavy metal-stressed vegetation in various data sets may depend heavily on the season when the data were acquired. Selecting imagery to be used in exploration of sites with even moderate vegetation cover, on the basis of percent cloud cover and image quality, neglects the consideration of phenology and heavy metal cycling within the plant—features which may have a pronounced influence on the spectral properties being remotely detected.

Segal (1983) offers a method of separating the spectral signatures of vegetation, limonitic hydrothermally altered rocks, and unaltered, nonlimonitic rocks. The standard color ratio composite consisting of MSS band ratios 4/5, 5/6, and 6/7 and used by many workers (Rowan et al., 1974, 1977) to map limonitic rocks in arid and semiarid areas is cited as ineffective in heavily vegetated areas, and Segal suggests the use of a compound MSS ratio—(4/5)/(6/7)—as a means of separation.

In summary, the geobotanical papers included in this volume suggest a number of major points. Increasing spectral resolution has been shown to provide increasingly specific data concerning the nature of the geochemically induced stress. MSS data allow for identification of certain seasonally pronounced geobotanical anomalies, thematic mapper data allow for recognition of changes in both pigments and leaf water content, and high spectral resolution data suggest specific types of heavy metal-induced stress. Multitemporal data sets are clearly shown to be of great value in terms of both detection and interpretation of geobotanical anomalies noted on imagery. The utility of gathering ground spectral data for aiding in the interpretation of airborne or spaceborne remotely sensed data sets is demonstrated as especially valuable in the study of sites influenced by vegetation. Finally, the science of geobotanical remote sensing is beginning to approach the degree of sophistication which characterizes geologic remote sensing, and vegetation ground noise is actually becoming a useful signal for the exploration geologist.

Future

The development of remote-sensing techniques and applications for resource exploration is now at a crossroads. The initial decade of space remote sensing with the Landsat MSS has shown that multispectral images taken from the vantage point of space have found their place in the modern geologist's toolbox. Albeit a fair amount of equipment is required to extract all the information from the data, but then ion microprobes are not inexpensive either. The most important effect that Landsat has had on the geologic community is to whet its appetite for more powerful tools and techniques just now coming over the horizon.

This volume holds some of the first papers outlining results from the next generation of data acquisition techniques and machine-processing methods. These techniques will influence the types of problems to be addressed in the future by remote sensing, and as evidenced by the emphasis in this volume, geobotanical studies will receive new impetus from the directions now being taken by spectral technique development.

Technique development

The thrust in technique development is split into two complementary areas, data acquisition and data analysis. Interpretation will be considered in the next section.

In data acquisition, the first steps toward acquiring high spectral resolution data that can be used for direct identification of materials based on their spectral characteristics have been made by Collins (1978), Collins et al. (1981 and 1983), and Goetz et al. (1982). From aircraft Collins has acquired spectroradiometer data in 512 channels in the 0.4- to 1.0-μm region and 64 channels in the 1.0- to 2.5-μm or 2.0- to 2.5-μm region. Goetz et al. (1982) acquired data from shuttle orbit with a radiometer covering the 0.5- to 2.35-μm region in 10 channels. The 2.1- to 2.35-μm region contained 5 channels; 3 were 0.02 μm wide and centered at 2.17, 2.20, and 2.22 μm. With both instruments, direct identification of hydroxyl-bearing minerals and carbonates were made using the diagnostic reflectance spectra. With Collins's instrument, in the visible and near-infrared region, vegetation spectral reflectance anomalies that are only detectable with high spectral resolution measurements were obtained in known mineralized areas.

The next obvious step in data acquisition is the creation of images in adjacent narrow spectral bands covering the entire reflective portion of the spectrum. In this case each picture element would have associated with it an entire spectrum, potentially allowing direct identification of many surface materials. The concept of an imaging spectrometer has been discussed by Wellman and Goetz (1980). Such an instrument has been built and tested recently (Goetz and Vane, 1982), and the feasibility of simultaneously acquiring image data in 128 spectral bands was demonstrated. Given sufficient resources, such an instrument could be making measurements from orbit within five years and sooner from aircraft. The implications for data handling and analysis of data with such high dimensionality are enormous.

Entirely new data analysis methods will have to be developed to extract and present the relevant information. Collins et al. (1983) use a method of dimensionality reduction involving fitting spectral curves with Chebyshev polynomials and plotting ratios of coefficients relevant to the identification problem at hand. With 512-channel coverage of the visible and near-infrared, Collins et al. (1983) have found that the ratio of the 9th and 5th coefficients indicates the presence of the blue shift associated with metal-stressed trees. The reduction in dimensionality by a factor of 50 was an important step in making the analysis of a very large data set manageable.

Pattern recognition techniques such as those applied by Collins et al. (1983) and discussed above, rather than statistical analyses of data sets as presently practiced (Swain and Davis, 1978) on Landsat data, will be necessary to extract information from high spectral resolution image data.

The compilation and display of geophysical and remote-sensing data sets for ease of interpretation, as demonstrated by Guinness et al. (1983), will play a major role in exploration research using diverse data sets. Digital aeromagnetic, Bouguer gravity, and airborne gamma-ray data have been superimposed on Landsat image data. Correlations between magnetic and Bouguer anomalies within a riftlike feature in southern Missouri have been shown. These techniques will receive much more attention in the future.

Interpretation

The technique development described above will lead to significant challenges in the area of data interpretation. The ability to map distributions of single minerals or mineral suites should have a profound effect on the development of models for the use of surface mineralogic information in exploration. Marsh and McKeon (1983), Abrams et al. (1983), and Podwysocki et al. (1983) touch on the potential use of such information. Similarly, the development of models for the use of geobotanical data will be affected significantly by the newfound ability to map subtle changes in chlorophyll absorption and water content in leaves, and therefore potential species and maturation and stress effects. A large portion of the earth's surface, not previously studied with remote-sensing techniques, will be amenable to investigation by these new methods.

The development of imaging spectrometers will increase significantly the sampling density for high spectral resolution data. Instead of the present contouring from individual spectrophotometer profiles (Marsh and McKeon, 1983), images will be available that will allow the identification of individual outcrops and tree stands.

Techniques for displaying coregistered data sets now exist (Guinness et al., 1983; Rebillard and Evans, 1983). The interpretation of such data, in particular the combination of diverse data types such as radiometric, potential field, and spectral reflectance, will require new insight and new models to understand the coupling between the information obtained at different scales and different depths. This development will prove to be a significant advance in the use of remote-sensing data for resource exploration.

Acknowledgments

The research described in this paper was carried out by the Jet Propulsion Laboratory, California Institute of Technology, under contract with the National Aeronautics and Space Administration.

February 28, 1983

REFERENCES

Abrams, M. J., Ashley, R. P., Rowan, L. C., Goetz, A. F. H., and Kahle, A. B., 1977, Mapping of hydrothermal alteration in the Cuprite mining district, Nevada, using aircraft scanner images for the spectral region 0.46 to 2.36 μm: Geology, v. 5, p. 713–718.

Abrams, M. J., Brown, D., Lepley, L., and Sadowski, R., 1983, Remote sensing for copper exploration in southern Arizona: Econ. Geol., v. 78, p. 591–604.

Ashley, R. P., and Abrams, M. J., 1980, Alteration mapping using multispectral images—Cuprite mining district, Esmeralda County, Nevada: U. S. Geol. Survey Open-File Rept. 80-367, 17 p.

Baker, G. A., Rundel, P. W., and Parsons, D. J., 1982, Comparative phenology and growth in three chaparral shrubs: Bot. Gaz., v. 143, p. 94–100.

Bidwell, R. G. S., 1979, Plant physiology, 2nd ed.: New York, MacMillan, 726 p.

Bolviken, B., Honey, F., Levine, S. R., Lyon, R. J. P., and Prelat, A., 1977, Detection of naturally heavy-metal-poisoned areas by Landsat 1 digital data: Jour. Geochem. Explor., v. 8, p. 457–471.

Brooks, R. R., 1972, Geobotany and biogeochemistry in mineral exploration: New York, Harper and Row, 290 p.

Buckingham and Sommer, 1983, Mineralogical characterization of rock surfaces formed by hydrothermal alteration and weathering: Application to remote sensing: Econ. Geol., v. 78, p. 664–674.

Cannon, H. L., 1960, Botanical prospecting for ore deposits: Science, v. 132, p. 591–598.

Chang, S. H., and Collins, W., 1983, Confirmation of the airborne biogeophysical mineral exploration technique using laboratory methods: Econ. Geol., v. 78, p. 723–736.

Chikishev, A. G., 1965, Plant indicators of soils, rocks, and subsurface waters [trans.]: New York, Consultants Bur., 210 p.

Clifton, C. G., Buchanan, L. J., and Durning, W. P., 1980, Exploration procedure and controls of mineralization in the Oatman mining district, Oatman, Arizona: New York, Soc. Mining Engineers AIME, Preprint 80-143, 17 p.

Collins, W., 1978, Remote sensing of crop type and maturity: Photogramm. Eng. Remote Sensing, v. 44, p. 43–55.

Collins, W., Chang, S. H., and Kuo, J. T., 1981, Detection of hidden mineral deposits by airborne spectral analysis of forest canopies: NASA Contract NSG-5222, Final Rept., 61 p.

Collins, W., Chang, S. H., Raines, G., Canney, F., and Ashley, R., 1983, Airborne biogeochemical mapping of hidden mineral deposits: Econ. Geol., v. 78, p. 737–749.

Cunningham, C. G., and Steven, T. A., 1979, Uranium in the central mining area, Marysvale district, west-central Utah: U. S. Geol. Survey Misc. Inv. Ser. Map I-1177, 1:24,000.

Daily, M., Elachi, C., Farr, T., and Schaber, G., 1978, Discrimination of geologic units in Death Valley using dual frequency and polarization imaging radar data: Geophys. Research Letters, v. 5, p. 889–892.

Darch, J. P., and Barber, J., 1983, Multitemporal remote sensing of a geobotanical anomaly: Econ. Geol., v. 78, p. 770–782.

Elachi, C., 1980, Spaceborne imaging radar: Geologic and oceanographic applications: Science, v. 209, p. 1073–1082.

Elachi, C., Brown, W. E., Cimino, J. B., et al., 1982, Shuttle imaging radar experiment: Science, v. 218, p. 996–1003.

Engel, J. L., and Weinstein, O., 1982, The Thematic Mapper—An overview *in* Internat. Geoscience Remote Sensing Symposium, Munich, 1982, Digest: New York, Inst. Electrical Electronics Engineers, v. 1, p. WP1.1–1.7.

Epstein, E., 1961, The essential role of calcium in selective cation transport by plant cells: Plant Physiology., v. 36, p. 437–444.

—— 1972, Mineral nutrition of plants: Principles and perspectives: New York, Wiley, 412 p.

EROS (Earth Resources Observational Satellite Data Center), 1979, Landsat data users handbook 1979: Sioux Falls, South Dakota, U. S. Geol. Survey.

Esau, K., 1977, Anatomy of seed plants, 2nd ed.: New York, Wiley, 550 p.

Frolich, E. F., Wallace, A., and Lunt, O. R., 1966, Plant toxicity resulting from solutions of single salt cations and their amelioration by calcium, *in* Wallace, A., ed., Current topics in plant nutrition: Los Angeles, Univ. California Press, p. 120–126.

Gates, D. M., 1970, Physical and physiological properties of plants, *in* Remote sensing with special reference to agriculture and forestry: Washington, Natl. Acad. Sci., p. 224–252.

Gausman, H. W., Escobar, D. E., and Knipling, E. B., 1977, Relation of *Peperomia obtusifolia's* anomalous leaf reflectance to its leaf anatomy: Photogramm. Eng. Remote Sensing, v. 43, p. 1183–1185.

Gausman, H. W., Escobar, D. E., Everitt, J. H., Richardson, A. J., and Rodriguez, R. R., 1978, Distinguishing succulent plants from crop and woody plants: Photogramm. Eng. Remote Sensing, v. 44, p. 487–491.

Gladwell, D. R., Lett, R. E., and Lawrence, P., 1983, Application of reflectance spectrometry to mineral exploration using portable radiometers: ECON. GEOL., v. 78, p. 699–710.

Goetz, A. F. H., and Rowan, L. C., 1981, Geologic remote sensing: Science, v. 211, p. 781–791.

Goetz, A. F. H., and Vane, G., 1982, High spectral resolution imaging from aircraft [abs.]: Internat. Symposium Remote Sensing Environment, Fort Worth, 1982, Abstracts, p. 37.

Goetz, A. F. H., Billingsley, F. C., Gillespie, A. R., Abrams, M. J., Squires, R. L., Shoemaker, E. M., Lucchitta, I., and Elston, D. P., 1975, Application of ERTS images and image processing to regional geologic problems and geologic mapping in northern Arizona: Jet Propulsion Laboratory, Pasadena, California, Tech. Rept. 32-2597, 188 p.

Goetz, A. F. H., Rowan, L. C., and Kingston, M. J., 1982, Mineral identification from orbit: Initial results from the shuttle multispectral infrared radiometer: Science, v. 218, p. 1020–1024.

Gold, D. P., 1980, Structural geology, *in* Siegal, B. S., and Gillespie, A. R., eds., Remote sensing in geology: New York, Wiley, p. 419–483.

Guinness, E. A., Arvidson, R. E., Leff, C. E., Edwards, M. H., and Bindschadler, D. L., 1983, Digital image processing application to analysis of geophysical and geochemical data for southern Missouri: ECON. GEOL., v. 78, p. 654–663.

Hewitt, E. J., 1963, The essential nutrient elements: Requirements and interactions in plants, *in* Stewart, F. C., ed., Plant physiology—a treatise: New York, Academic Press, v. 3, p. 137–360.

Hoffer, R. M., and Johannsen, C. J., 1969, Ecological potentials in spectral signature analysis, *in* Johnson, P. L., ed., Remote sensing in ecology: Athens, Univ. Georgia Press, p. 1–16.

Horler, D. N. H., Barber, J., and Barringer, A. R., 1980, Effects of heavy metals on the absorbance and reflectance spectra of plants: Internat. Jour. Remote Sensing, v. 1, p. 121–136.

Horler, D. N. H., Dockray, M., Barber, J., and Barringer, A. R., 1983, Red edge measurements for remotely sensing plant chlorophyll content: Comm. on Space Research Symposium on Remote Sensing and Mineral Exploration, Ottawa 1982, Proc., in press.

Hunt, G. R., 1977, Spectral signatures of particulate minerals in the visible and near infrared: Geophysics, v. 42, p. 501–513.

Hunt, G. R., and Ashley, R. P., 1979, Spectra of altered rocks in the visible and near infrared: ECON. GEOL., v. 74, p. 1613–1629.

Hunt, G. R., Salisbury, J. W., and Lenhoff, D. J., 1971, Visible and near-infrared spectra of minerals and rocks. III. Oxides and hydroxides: Modern Geology, v. 2, p. 195–205.

Kahle, A. B., and Rowan, L. C., 1980, Evaluation of multispectral middle infrared aircraft images for lithologic mapping in the east Tintic Mountains, Utah: Geology, v. 8, p. 234–239.

Knipling, E. B., 1969, Leaf reflectance and image formation on color infrared film, *in* Johnson, P. L., ed., Remote sensing in ecology: Athens, Univ. Georgia Press, p. 17–29.

Kovalevskii, A. L., 1979, Biogeochemical exploration for mineral deposits (trans. of 1974 text): Washington, U. S. Dept. Interior, 136 p.

Labovitz, M. L., Masuoka, E. J., Bell, R., Siegrist, A. W., and Nelson, R. F., 1983, The application of remote sensing to geobotanical exploration for metal sulfides: ECON. GEOL., v. 78, p. 750–760.

Lyon, R. J. P., 1965, Analysis of rocks by spectral infrared emission (8 to 25 microns): ECON. GEOL., v. 60, p. 715–736.

Marsh, S. E., and McKeon, J. B., 1983, Integrated analysis of high resolution field and airborne spectroradiometer data for alteration mapping: ECON. GEOL., v. 78, p. 618–632.

McCauley, J. F., Schaber, G. C., Breed, C. S., Grolier, M. J., Haynes, C. V., Issawi, B., Elachi, C., and Blom, R., 1982, Subsurface valleys and geoarcheology of the eastern Sahara revealed by shuttle radar: Science, v. 218, p. 1004–1019.

Milton N. M., 1983, Use of reflectance spectra of native plant species for interpreting airborne multispectral scanner data in the east Tintic Mountains, Utah: ECON. GEOL., v. 78, p. 761–769.

Milton, N. M., Collins, W., Chang, S.-H., and Schmidt, R. G., 1983, Remote detection of metal anomalies on Pilot Mountain, Randolph County, North Carolina: ECON. GEOL., v. 78, p. 605–617.

Murtha, P. A., 1972, A guide to air photo interpretation of forest damage in Canada: Ottawa, Canadian Forestry Service Pub. 1292, p. 12–35.

Peters, D. C., 1983, Using airborne multispectral scanner to map alteration related to roll-front uranium migration: ECON. GEOL., v. 78, p. 641–653.

Podwysocki, M. H., Segal, D. B., Collins, W. E., and Chang, S. H., 1982, Mapping the distribution and mineralogy of hydrothermally altered rocks by using airborne multispectral scanner and spectroradiometer data, Marysvale, Utah [abs.]: Internat. Symposium Remote Sensing Environment, Fort Worth, 1982, Abstracts, p. 18.

Podwysocki, N. H., Segal, D. B., and Abrams, M. J., 1983, Use of multispectral scanner images for assessment of hydrothermal alteration in the Marysvale, Utah, mining area: ECON. GEOL., v. 78, p. 675–687.

Prost, G. L., 1980, Alteration mapping with airborne multispectral scanners: ECON. GEOL., v. 75, p. 894–906.

—— 1983, Mineral exploration with Skylab photography: ECON. GEOL., v. 78, p. 633–640.

Raines, G. L., and Canney, F. C., 1980, Vegetation and geology, *in* Siegal, B. S., and Gillespie, A. R., eds., Remote sensing in geology: New York, Wiley, p. 365–380.

Raines, G. L., Offield, T. W., and Santos, E. S., 1978, Remote-sensing and subsurface definition of facies and structure related to uranium deposits, Powder River Basin, Wyoming: ECON. GEOL., v. 73, p. 1706–1723.

Raven, P. H., Evert, R. F., and Curtis, H., 1976, Biology of plants, 3rd ed.: New York, Worth, 685 p.

Rebillard, P., and Evans, D. L., 1983, Analysis of coregistered Landsat Seasat and SIR-A images of varied terrain types: Geophysics Research Letters, v. 10, p. 277–280.

Rock, B. N., 1982, Mapping of deciduous forest cover using simulated Landsat D TM data *in* Internat. Geoscience Remote Sensing Symposium Munich, 1982, Digest: New York, Inst. Electrical Electronics Engineers, p. WP5 3.1–3.5.

Rohde, W. G., and Olson, C. E., Jr., 1971, Estimating foliar moisture content from infrared reflectance data, *in* Third biennial workshop, color aerial photography in the plant sciences: Falls Church, Virginia, Am. Soc. Photogrammetry, p. 144–164.

Rowan, L. C., and Kahle, A. B., 1982, Evaluation of 0.46- to 2.36-μm multispectral scanner images of the East Tintic mining district, Utah, for mapping hydrothermally altered rocks: Econ. Geol., v. 77, p. 441–452.

Rowan, L. C., and Wetlaufer, P. H., 1981, Relation between regional lineament systems and structural zones in Nevada: Am. Assoc. Petroleum Geologists Bull., v. 65, p. 1414–1432.

Rowan, L. C., Wetlaufer, P. H., Goetz, A. F. H., Billingsley, F. C., and Stewart, J. H., 1974, Discrimination of rock types and detection of hydrothermally altered areas in south-central Nevada by use of computer-enhanced ERTS images: U. S. Geol. Survey Prof. Paper 883, 35 p.

Rowan, L. C., Goetz, A. G. H., and Ashley, R. P., 1977, Discrimination of hydrothermally altered and unaltered rocks in visible and near-infrared multispectral images: Geophysics, v. 42, p. 522–535.

Sabins, F. F., Jr., 1978, Remote sensing—principles and interpretation: San Francisco, W. H. Freeman, 476 p.

Schwaller, M. R., and Tkack, S. J., 1980, Premature leaf senescence as an indicator in geobotanical prospecting with remote sensing techniques: Internat. Symposium Remote Sensing Environment, 14th, San Jose, Costa Rica, Proc., p. 347–358.

Segal, B. D., 1983, Use of Landsat multispectral scanner data for the definition of limonitic exposures in heavily vegetated areas: Econ. Geol., v. 78, p. 711–722.

Swain, P. H., and Davis, S. M., eds., 1978, Remote sensing: The quantitative approach: New York, McGraw Hill, 396 p.

Thorson, J. P., 1971, Igneous petrology of the Oatman district, Mojave County, Arizona: Unpub. Ph.D. dissert., Univ. California, Santa Barbara, 189 p.

Tucker, C. J., 1980, Remote sensing of leaf water content in the near infrared: Remote Sensing Environment, v. 10, p. 23–32.

Wallace, A., 1971, Regulation of the micronutrient status of plants by chelating agents and other factors: Univ. California, Los Angeles 34P51-33, Los Angeles, Arthur Wallace, p. 307.

Wallace, A., Frolich, E., and Lunt, O. R., 1966, Calcium requirements of higher plants: Nature, v. 209, p. 634.

Wellman, J. B., and Goetz, A. F. H., 1980, Experiments in infrared multispectral mapping of earth resources: Am. Inst. Aeronautics Astronautics Conf., Sensor Systems for the 80's, Colorado Springs, 1980, Proc., p. 163–174.

Whitney, C. G., Abrams, M. J., and Goetz, A. F. H., 1983, Mineral discrimination using a portable ratio-determining radiometer: Econ. Geol., v. 78, p. 688–698.

Wickland, D. E., 1982, Patterns of vegetation response to heavy metal stress [abs.]: Internat. Symposium Remote Sensing Environment, Fort Worth, 1982, Abstracts, p. 126–127.

RADAR

Reprinted by permission of the Papua New Guinea
Chamber of Mines and Petroleum from G. J. Carman
and Z. Carman, eds., *Petroleum Exploration in Papua
New Guinea*, 1990, p. 319-336.

Petroleum Exploration in Papua New Guinea: Proceedings of
the First PNG Petroleum Convention, Port Moresby, 12–14th
February 1990, Carman, G.J. and Z., (Eds).

A Structural Interpretation of the Onshore Eastern Papuan Fold Belt, Based on Remote Sensing and Fieldwork

Fons Dekker[1], Hugh Balkwill[1], Alan Slater[1], Robert Herner[2],
Wim Kampschuur[3]

Abstract

Over the Eastern Papuan Fold Belt, in the Gulf Province of Papua New Guinea, aerial photographs, Landsat Multispectral Scanner (MSS) imagery and Synthetic Aperture Radar (SAR) data are available to interpret the hydrocarbon prospectivity of the onshore area.

Analysis of the available remote sensing sources in the study area revealed a more detailed picture than displayed before on any geological map. SAR provided the first comprehensive regional mosaic of the area. Because of the low, artificial 'sun angle' it showed many structural elements that had not previously been appreciated.

There is a distinct difference in structural style between the northern segment of the study area and the southern segment. The northern segment shows open folding with widely separated anticlines embedded in featureless valleys. Structures are discontinuous across the area. The southern segment is tightly folded, with few clearly recognizable anticlines and synclines, but structural components can easily be traced for tens of miles.

Recent fieldwork supported the structural interpretation from the remote sensing data which suggested that most, if not all, of the anticlines in the northern segment are overturned.

The combination of remote sensing and fieldwork proved invaluable in better understanding the fold belt tectonics and aided considerably in the decision making process regarding a drilling location.

Introduction

The study area, near Kerema in the Eastern Papuan Fold Belt is located on the northern margin of the Australian Craton (Fig. 1). The Tertiary collision with the Bismarck and Solomon Plates created a series of generally northwest-trending, narrow tectonic domains. These range from the undeformed Fly Platform in the SW through the folded and thrusted Papuan Fold Belt to the metamorphics of the Owen Stanley Metamorphic Complex and the oceanic allochthons northeast of the Ramu Markham Fault Zone (Slater et al, 1988).

In the past, exploration emphasis was placed on some of the major surface anticlines as seismic could not easily be obtained due to the tropical rain forest cover and deep weathering. Wells drilled onshore in the study area (Kariava-1, Upoia-1, Rarako Creek-1) were located on these surface anticlines; they never reached the prospective section beneath the Aure Beds (Fig. 2).

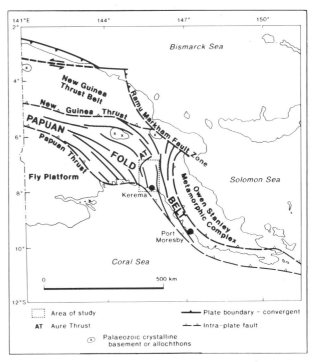

Figure 1. Location map with tectonic domains and major structural trends, after Rogerson et al. (1987).

[1]Petro-Canada Inc, Calgary, CANADA
[2]MARS Associates, Inc., Phoenix, U.S.A.
[3]International Tectostrat Geoconsultants bv, Amsterdam, THE NETHERLANDS

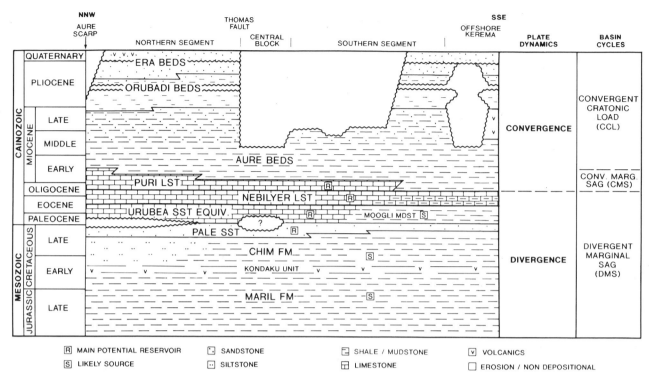

Figure 2. Generalized stratigraphy of the study area.

In recent years, significant hydrocarbon exploration activity took place in the study area which has numerous oil and gas seeps (Fig. 3). Much of the activity was directed at the investigation of the structural style and the importance of surface anticlines as indicators of structuring at reservoir depth.

Various field parties recently visited a variety of locations, from the Aure Scarp in the northwest to the Kerema coast in the southeast (Fig. 3). Traditionally, aerial photographs were used for location and interpretation, but presently modern remote sensing techniques are applied in the study area. Landsat Multispectral Scanner (MSS — Fig. 4) and Synthetic Aperture Radar (SAR — Fig. 5) were obtained and interpreted together with the aerial photographs, and the interpretations were tested in the field. This approach produced a regional interpretation while visiting only a limited area in the field.

Stratigraphy

The stratigraphy is summarized in Figure 2. It represents deposition in a distal setting on Australian continental crust during middle Mesozoic to Palaeogene (divergent marginal sag cycle) and deposition in a foreland basin during the Neogene (convergent cratonic load cycle).

Lithological control for the Cretaceous to Palaeogene section is established at the Aure Scarp, where a 4450 m section exposes Neocomian to Mio-Pliocene beds (Carman, 1986, 1987); the section includes

950 m of Paleocene to Oligocene carbonates and up to 200 m of Campanian sandstone (Pale Sandstone) which form potential reservoirs to the south and in the offshore Kerema area (Fig. 3).

Neogene clastics form the surface exposures over most of the area. Athough the terrain is covered with jungle canopy the difference between the Aure Beds, Orubadi Beds and Era Beds can be recognized by their radar signatures. The slope, relief and resistance to weathering of these lithological packages affects the morphology of the terrain which is imaged with SAR (Fig. 5). Unconformities are postulated to exist over the area at base Quaternary and mid-Pliocene levels based on interpretation of the offshore seismic (Fig. 2).

The Aure Beds (generally Tm and Tma on the geologic maps, Figs. 6 and 7) comprise interbedded lithic sandstone and mudstone which are interpreted to be supra-fan and distal fan turbidites (Sari, 1985) deposited in middle to upper bathyal environments. They are highly indurated and form the prominent belt of resistant ridge topography over most of the Southern Segment, all of the Central Block and the individual anticlines of the Northern Segment (Figs. 3 and 5). They are most indurated in the Central Block where they may represent the lower part of the Aure Beds, as indicated by biostratigraphic analyses (N9–N13).

The overlying Orubadi Beds (Tmup and Tmu on the geologic map) consist of upper bathyal mudstones with neritic sands at the top. They crop out in

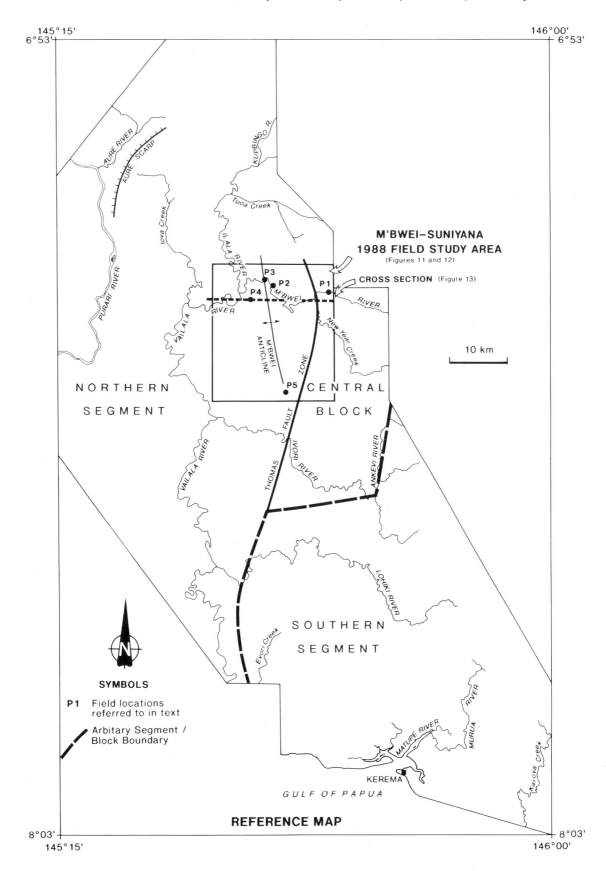

Figure 3. Reference map for the study area.

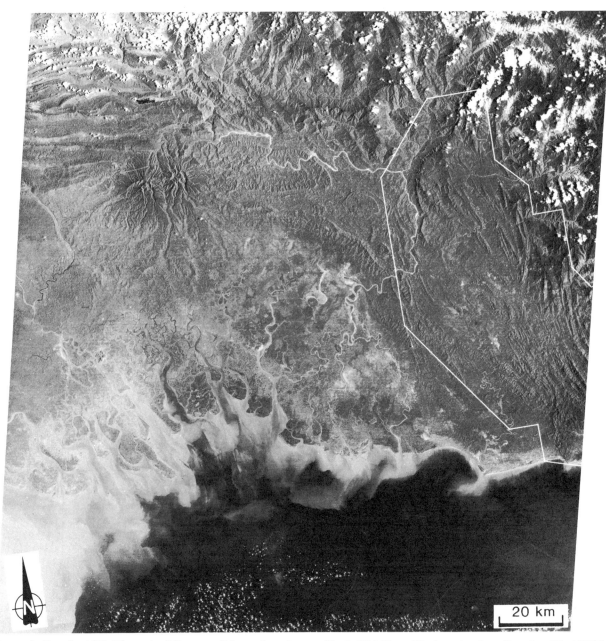

Figure 4. Landsat Multispectral Scanner false colour image. Path 104, Row 65. Acquired on October 30, 1980 by Landsat 2. Processed by Earth Satellite Corporation. Study area is outlined.

the Northern Segment in synclines, where they form non-resistant terrain.

The Orubadi is overlain by the Plio-Pleistocene Era Beds (TQe on the geologic map). This mainly sandstone and mudstone sequence with minor conglomerate was deposited in shallow-marine, transitional and non-marine conditions. The unit crops out along the western edge of the area, where the coarser clastics form a resistant irregular terrain. Quaternary alluvium (Q), a flat surface on the SAR, is easily identified in the valleys and coastal areas while stratified volcanics (also Q) form sheet-like terrain in the northeast, and obscure the topography.

Remote Sensing Data Base

For this study the authors had access to aerial photographs which are generally available for Papua New Guinea, Landsat MSS and SAR. Landsat Thematic Mapper (TM) and SPOT data were not available due to a variety of technical (data relay) and quality (cloud cover) problems.

Since the launch of the first Landsat satellite in 1972, there has been a more or less continuous recording over Papua New Guinea with a 16–18 day cycle and over the last few years, with both Landsats 4 and 5 recording simultaneously, every 8 days. Despite the good coverage there are not too many cloud free scenes over the study area because of its relief, which ranges from sea level to about 3000m. Clouds are almost always present near the higher elevations at the time the Landsat satellites pass over the area around 9:30 am.

The heavy forest cover in the study area masks most of the lithological information otherwise obtainable through the use of a variety of spectral bands: the false color images are more or less completely red. The high sun angle also hampers the Landsat images of Papua New Guinea. When the satellites pass over the area, the sun has already risen 45–50 degrees above the horizon. This obscures the subtle structures even though larger trends are clearly visible (Fig. 4).

The Landsat MSS imagery provided a reasonable regional picture for the Eastern Papuan Fold Belt, but details are not discernible. One full Landsat image covers an area of 185 × 185 km (34,000 km²), this provides a truly regional perspective even though resolution is limited. For MSS the pixel (picture element) size is 79 × 79 m and since several pixels are needed to define an object, the resolution is in the order of several hundred metres.

The image in Figure 4, with the Purari River in the centre of the picture, was enhanced for geological purposes through the Geopic program (trademark of Earth Satellite Corporation). Imagery at 100,000 scale was produced using a variety of enhancement techniques (contrast stretch, principal component, low pass filter) and directional filters (north, northeast, east and southeast). These directional filters helped to highlight groups of lineaments which otherwise were not clearly visible. Other enhancements were largely hampered by the dense rain forest cover.

SPOT imagery provides 20 × 20 m resolution in MSS mode and 10 × 10 m resolution in the panchromatic mode. The high resolution is a great improvement over the Landsat system but the limitations are largely the same: subdued structural information due to high sun angle, inability to penetrate cloud cover and lack of lithological definition due to continuous canopy. Unfortunately, no cloud free imagery had been obtained over the study area since the launch of the SPOT system in 1986 and the time of this study in 1988.

Radar

Radar is a useful tool in mapping heavily forested regions such as eastern Papua New Guinea. Radar does not penetrate tree canopy, but images the surface created by the continuous canopy. Tree canopy follows the structuring of underlying lithology. Various lithologies have their own specific ways of structuring, like bedding of clastics and karstification of carbonates. The low angle radar beam, transmitted from the radar instrument, skims over the canopy, which mimics the ground surface, and reveals a picture similar to the underlying terrain. The structuring on the image then allows the experienced interpreter to map the lithology.

For detailed information on the technical aspects of radar, its use in Papua New Guinea and radar survey designs, the reader is referred to Ellis and Pruett (1986) who documented the radar application over the Iagifu area of the Papuan Fold Belt. Petro-Canada used the same instrument and a similar design for its 1987 survey. Dekker (1989) described the use of radar as an exploration tool in tropical rain forest areas and made comparisons with other remote sensing devices.

Radar Survey Over The Study Area

The project area encompasses approximately 5,600 km² (Fig. 5). Radar imagery was acquired on 10 June 1987 in the wide swath (45 km) mode using a north look direction. The north look was selected based on the orientation of the fold belt, fault patterns and other lineaments known from the study area. An east or west look would have been a viable alternative, although shadowing might have become too heavy in the tightly folded Southern Segment (Fig. 3). Based upon the aircraft altitude of 27,000 feet, the near-range depression angle (measured from the horizon) is about 26° and the far-range angle 7°. Line spacings were positioned to provide 55–60 per cent side lap between adjacent strips for construction of a mosaic and for stereo interpretation. Each strip image was digitally processed into analog film negatives at a scale of 1:250,000.

The existing 1:250,000 scale and 1:100,000 scale topographic maps in Papua were used for mosaic control during assembly, after reconciling control points to a master UTM grid. The topographic base maps in Papua New Guinea (produced by the Royal Australian Survey Corps) have proved to be unusually good compared to similar map series published in Southeast Asia. Because of the good side lap between adjacent strip images, relatively uniform illumination across ther entire mosaic area was achieved (average depression angle of 17°) and terrain elements are shown in their true perspective. Images are best viewed with the shadows towards the observer.

Stratigraphic Interpretation

Figure 6 represents a montage of the currently published geological maps of the area. The seam at longitude 145°30′E is indicative of the correlation problems from map sheet to map sheet. The radar interpretation maps have been subdivided into stratigraphic (Fig. 7) and structural (Fig. 8) maps to improve legibility.

Figure 7 shows the distribution of the stratigraphic units. The Upper Oligocene to Upper Miocene Aure Beds (Fig. 2) clearly dominate the outcrop in the map area. Biostratigraphic analyses from many locations

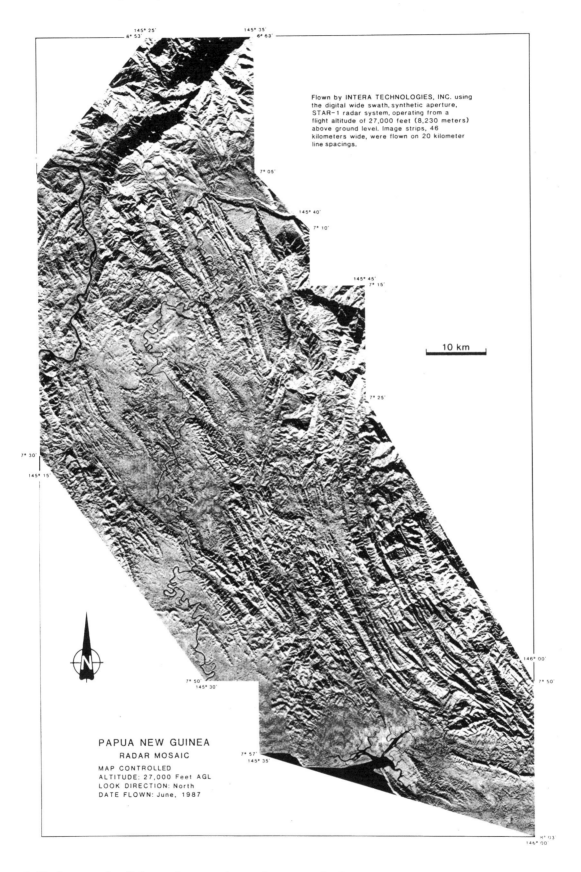

Flown by INTERA TECHNOLOGIES, INC. using the digital wide swath, synthetic aperture, STAR-1 radar system, operating from a flight altitude of 27,000 feet (8,230 meters) above ground level. Image strips, 46 kilometers wide, were flown on 20 kilometer line spacings.

10 km

PAPUA NEW GUINEA

RADAR MOSAIC

MAP CONTROLLED
ALTITUDE: 27,000 Feet AGL
LOOK DIRECTION: North
DATE FLOWN: June, 1987

Figure 5. Radar mosaic of the study area. Area of coverage is the same as in Figure 3.

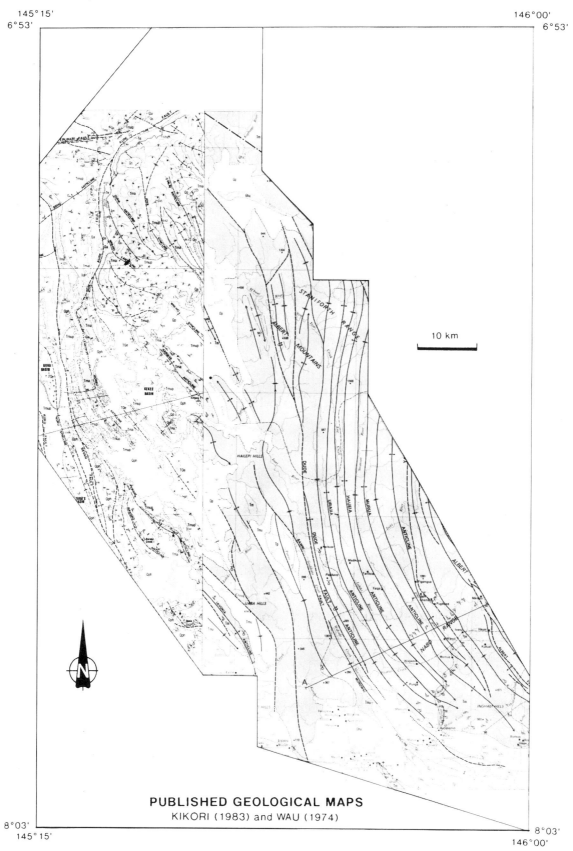

PUBLISHED GEOLOGICAL MAPS
KIKORI (1983) and WAU (1974)

Figure 6. Published geological maps of the study area. Wau map sheet (east) compiled by Dow et al. (1974) and Kikori map sheet (west) compiled by Pieters (1983).

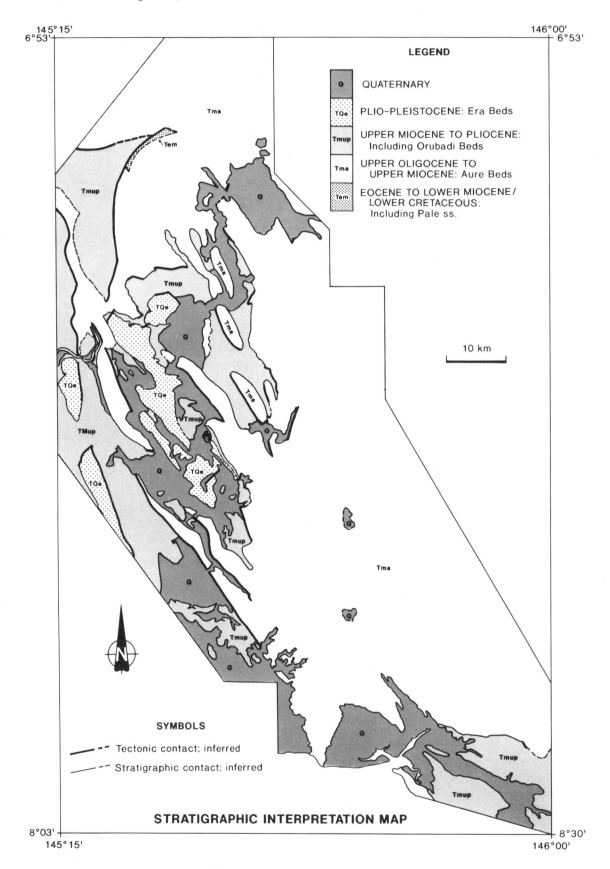

Figure 7. Stratigraphic interpretation map.

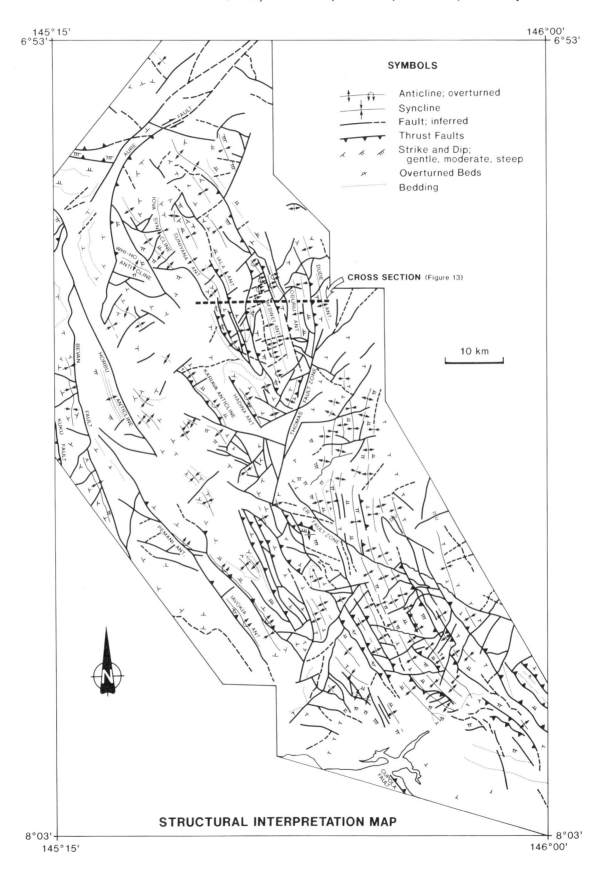

Figure 8. Structural interpretation map. Strike and dip symbols are representative of many more measurements.

suggest that the base of the Aure Beds is only reached in the Aure Scarp and nowhere else in the study area, although the number of analyses on the eastern side is limited. The estimated stratigraphic thickness of the Aure Beds is about 2500 m (Slater et al, 1988). This is based on seismic interpretation offshore Kerema and on field measurements near Kerema (Sari, 1985). On the dip slope of the Aure Scarp the Aure Beds are estimated to be at least 1800 m thick (Carman, 1986).

Exposures of older rocks are present at the Aure Scarp in the northwest (Fig. 3) where Campanian Pale Sandstone crops out below a thick section of Palaeogene limestones (Carman, 1987). In the Northern Segment (Fig. 3), Aure Beds within anticlinal crests are surrounded by Plio-Pleistocene and younger valley fill including volcanics (Q in the northeast). The Southern Segment is completely dominated by Aure Beds and only near the Kerema coast are younger formations present. The coastal zone itself is formed by a thick unit of Aure Beds (Sari, 1985; Petro-Canada, 1989). Many of the contacts between formations are along older-over-younger thrust faults.

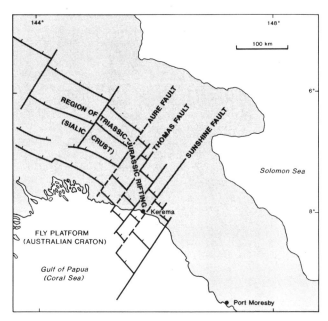

Figure 9. Basement structures due to Mesozoic rifting.

Regional Structural Interpretation

The regional structural style of the project area is different from that in the Papuan Fold Belt to the northwest. A change in trend is the most obvious disparity viewed on the radar and satellite imagery (Figs. 4 and 5). From the Irian Jaya border to Mount Murray (lat. 6°45′S, long. 144°E) the fold belt trend is approximately northwest–southeast (Fig. 1). Due east of Mt. Murray an east–west trend is assumed. Where the Purari and Aure Rivers meet (Fig. 3), the trend becomes north-northwest to south-southeast (Figs. 6 and 8). This latter trend is known as the Eastern Papuan Fold Belt.

Within the Eastern Papuan Fold Belt two major structural segments are identified (in addition to several minor ones). To the north are anticlines, mostly asymmetrical, commonly bounded by thrust faults (Fig. 8) . Near the coast there are sub-parallel tight folds and closely-spaced faults. Both of these areas are part of an allochthonous sheet that is bounded by the Aure Fault to the north and the Bevan or Bevan-Pemani Fault to the west (Fig. 8). A southern boundary fault, which may lie near the coast, will be discussed later.

The boundary between the two segments is a north-northeast to south-southwest trending fracture zone, referred to here as the Thomas Fault Zone. Together with the Aure and Sunshine fault zones, this line is suggested to represent a transfer fault, developed during Mesozoic extensional rifting and reactivated (with opposed slip) during the Tertiary convergence (Fig. 9).

Aure Beds in the study area display only a few massive sandstones and a lack of limestones. This thick, mechanically incompetent succession was deformed as a system of asymmetric folds, with the fore-limbs commonly broken by moderately to steeply dipping reverse faults. The faults are suggested to gather at a sub-horizontal basal detachment zone, probably lying within ductile upper Mesozoic shales (Fig. 13). Most of the faults are east dipping but west dipping back thrusts occur, as seen in the field and on offshore seismic (Fig. 8).

Structural analysis of the area, both from radar (Fig. 10A) and from field measurements (Fig. 10B), using the Palaeo-Stress Analysis System (trademark of International Tectostrat Geoconsultants bv), indicates that the direction of maximum compressional stress was from east-northeast to west-southwest (M'Bwei Phase, Fig. 10A). Conjugate fault zones developed along north-northeast to south-southwest trends with dextral motion and along west-northwest to east-southeast trends with sinistral motion. Consequently, the boundary between the Northern and Southern Segments, the Thomas Fault Zone (Fig. 3), is interpreted to be dextral.

The difference in the structural style between the segments may be related to the depth to basement. It is apparent from stratigraphic relationships that the Northern Segment is downthrown along the boundary fault relative to both the Central Block and to the Southern Segment (Fig. 3) since younger beds crop out in the Northern Segment (Fig. 7).

Along the Kerema coast, beds from the Southern Segment turn sharply to the southeast, suggesting sinistral drag (synthetic conjugate faulting). Fieldwork along the coastline (Petro-Canada, 1989) revealed very strong deformation in the Aure Beds, with abun-

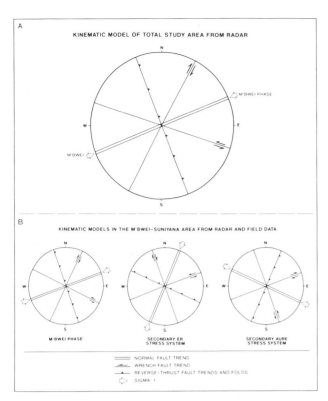

Figure 10. Kinematic models calculated through the Palaeo-Stress Analysis System; A: Based on statistical analysis of all lineaments from radar interpretation of total study area. B: M'Bwei–Suniyana area based on radar interpretation and numerous field measurements. The M'Bwei Phase stress system created the secondary stress systems of Aure and Er.

dant growth faults, both normal and reverse. Structurally, the onshore Aure Beds occur hundreds to perhaps thousands of metres above the Aure Beds in the immediately adjacent offshore area. All of this leads to the conclusion that a major northeast-dipping thrust fault (the newly named Cupola Fault) is located just offshore with the Southern Segment thrusted over the offshore part (Fig. 8). Gas seeps have been reported offshore, a short distance from the coastline at Cape Cupola, and may be due to leakage along the Cupola Fault.

Detailed mapping shows that anticlines are commonly cut by transverse faults, and are therefore not continuous for long distances. En echelon folds are far more common than regionally continuous ones (cf. Figs. 6 and 8). Specific elements emphasized in Figure 8 are an unnamed large fault parallel and close to the Aure Fault; the Thomas Fault Zone; the Er Fault Zone and the Cupola Fault. These faults may be related to the rejuvenation of Mesozoic divergent faults in which the Aure and Thomas faults were Mesozoic transfer faults and the Er and Cupola faults were originally extension faults (Fig. 9).

The lineaments bounding the antiformal structures are interpreted as steeply dipping reverse faults. Low-angle thrusts are very difficult to discern on remote sensing images, especially if the area of interest is covered by rain forests.

Structural Interpretation of the M'Bwei–Suniyana Area

The 1988 Field Party
Detailed structural/stratigraphic fieldwork was done in the M'Bwei–Suniyana area in 1988 by Petro-Canada Inc. (Fig. 3). During the field program, a roughly east–west traverse from the Central Block at P1 (Fig. 11) to the Suniyana Anticline in the west was completed along the M'Bwei River. By combining the use of a helicopter and an inflatable boat, the crew could visit almost every outcrop on both sides of the river. The field study was designed to elucidate the nature of detailed structures — including possible overturned anticlines interpreted from radar — in order to predict the geometry of subsurface objectives in areas of potential drilling.

M'Bwei Anticline
The M'Bwei Anticline has long been a target of exploration interest (Fig. 3). This structure is shown on the Wau map sheet (Fig. 6) as a simple, unlabelled anticline flanked by a companion syncline to the east and is visible on the Landsat image (Fig. 4). Both radar and aerial photograph (Fig. 12) displays of almost the same region are provided here to allow for comparison between the two techniques. The images are orientated with the radar shadows towards the reader to enhance viewing (Fig. 11).

The axis of the M'Bwei Anticline trends north-northwest to south-southwest; the directions of dip are recognizable from the asymmetry of slopes and outcrop "V's". Detailed imagery analysis reveals such features as truncated beds along both the northern and southern noses, bedding offset along strike, perceptible change in dip (including overturned beds), and possible repetition of beds.

The eastern flank (backlimb) of the M'Bwei Anticline (Figs. 11 and 13), which dips east-northeast, has two erosional characteristics: the southern portion possesses a smooth gently- to moderately-dipping slope developed upon a relatively resistant unit, and the northern part is scalloped, and incised by numerous streams. The asymmetry of slope is sufficient criteria by which to determine the southern section's dip. To the north the dip is derived through the analysis of outcrop V's, as seen along stream valleys. Cause for the differences in slope characteristics may in part be due to a proximity to local base level (M'Bwei and Vailala Rivers), change in dip magnitude (dip is slightly steeper to the north), and northeast–southwest and northwest–southeast trending faults.

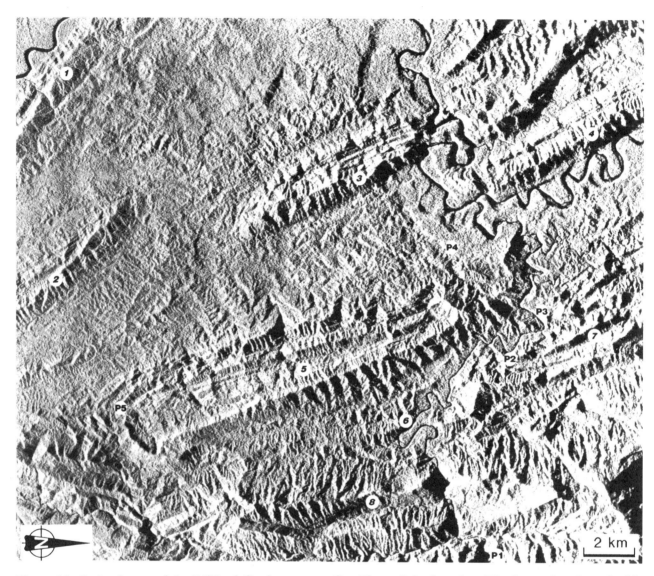

Figure 11. Radar image of the M'Bwei–Suniyana area. See Figure 3 for location. The image is oriented with north arrow horizontal to place radar shadows towards the bottom of the picture since this provides better spatial viewing. Anticlines: 1 = Kariava; 2 = Hadina; 3 = Suniyana; 4 = Iala; 5 = M'Bwei; 6 = M'Bwei River; 7 = Crocodile; 8 = Oburu. P1–P5: locations mentioned in text.

The M'Bwei Anticline's western flank (forelimb) is more complex. Dip along the northern two-thirds is generally west to west-southwest while to the south it steepens. Beds are locally overturned. This was interpreted from radar imagery and confirmed in the field (Petro-Canada, 1988), both on the northside (Location P4, Fig. 11) and the southside (P5) of the M'Bwei Anticline.

At P4, the M'Bwei Thrust crops out over an area of about 1–2 km^2. The fault is subhorizontal to slightly east dipping (14 degrees), and is undulatory (Fig. 13). Above the thrust plane, the Aure Beds dip very gently to the northwest (10–12 degrees). Stratigraphically lower, the dips steepen to vertical towards the axial plane of the M'Bwei Anticline. Recognition of low-

angle faults such as the M'Bwei Thrust has importance for regional structural geometry of the study area.

The deepest part of the west flank of the M'Bwei Anticline is exposed over a map distance of at least one kilometre. The beds are overturned, with dips ranging from near vertical to 73 degrees east. Overturned beds between Location P4 and the axis of the anticline dip eastward, but having west-facing, fining-upward, Bouma-sequence graded beds (Petro-Canada, 1988).

East-northeast to west-southwest trending tear faults have offset the axial trace of the M'Bwei Anticline (Fig. 8). The tightness of the anticline may change drastically across such tear faults. Most of

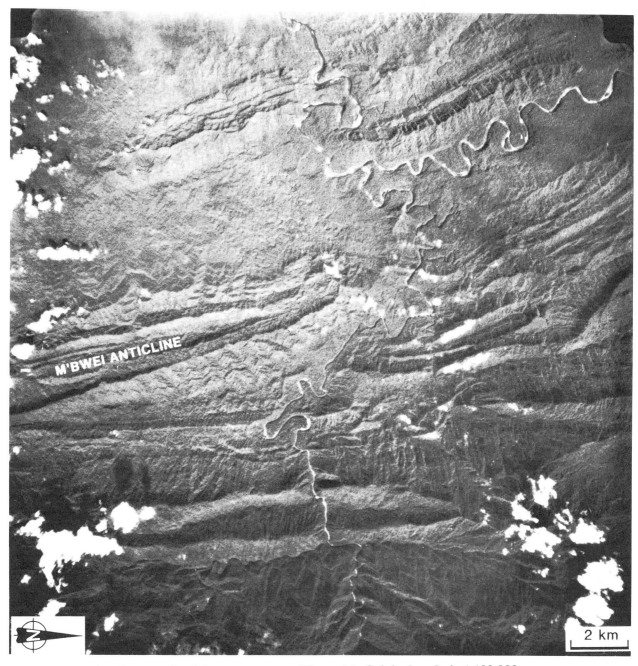

Figure 12. Aerial photograph of the same area as Figure 11. Original scale is 1:100,000.

these faults are suspected as possessing a lateral sense of displacement, although a degree of vertical offset can also be inferred from slickensides. To the north, radar imagery suggests clockwise rotation among blocks indicating right-lateral motion. To the south the same geometrics are assumed although morphologic features are less distinct. The southern nose of the M'Bwei Anticline has been cut off by the north-northeast to south-southwest trending Thomas Fault Zone, which belongs to the dextral wrench fault system of the M'Bwei phase (Fig. 10).

As the M'Bwei Anticline itself is offset by faults, there is not a continuous bedded sequence that can be traced around the structure. However, the correlation of beds between opposite flanks is not believed to be markedly disparate. The impression from the imagery is that there is not a one-on-one correlation between flanks, although many of the beds probably match. The various field surveys could not establish a one-on-one lithological correlation between both flanks because of the lack of unique markers in this heterogeneous, thick turbidite sequence, and the poor, discontinuous exposure between the creeks.

In this area, several smaller north-northeast and

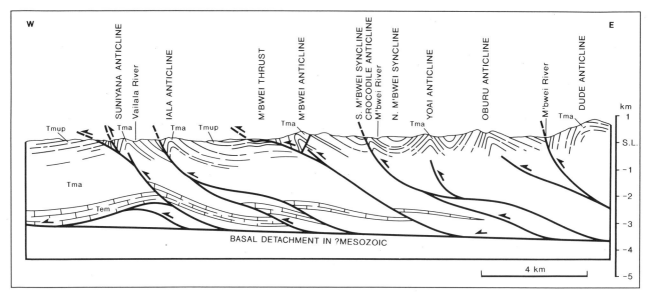

Figure 13. East–west cross-section: schematic model based on field measurements and offshore seismic interpretation. See Figure 3 for line of section and Figure 7 for stratigraphic abbreviations.

west-northwest trending anticlinal and synclinal structures are superimposed on the northwest-trending M'Bwei Anticline. The smaller structures are thought to be related to stresses developed perpendicular to the main wrench fault directions (Aure and Er secondary stress systems in Figure 10B — Aure developed perpendicular to the Aure and Thomas fault zones; Er perpendicular to the Er Fault Zone) when these wrench faults were (re)activated by the M'Bwei Phase stress system (Fig. 10A) due to the Tertiary collision between the Australian Craton and the Pacific plates.

The Central Block

At P1, in the oldest Aure Beds analysed in this area (N9–N13), the turbidite sediments consist of complete Bouma sequences (Fig. 14). Bed thicknesses vary considerably both from sequence to sequence and along bedding planes.

Faulting and fracturing is abundant along the M'Bwei River, especially when approaching the Thomas Fault Zone (Fig. 15). An array of northwest and northeast striking faults show apparent offsets of up to several metres vertically but slickensides indicate a significant lateral component. Analysis of the slickensides provided an extensive database from which the directions and relative ages of fault movement could be calculated. From this information the diagrams for the M'Bwei Phase and for the secondary Aure and Er stress systems were constructed (Fig. 10).

An example of fracturing is shown in Figures 15A and B. Quartz and calcite veins infill the fractures. A simple shear pattern with an en echelon, rhombohedral shaped fracture pattern can be observed, limited on its southern edge by a N60E (150/90) trending main dextral wrench fault (part of secondary Aure

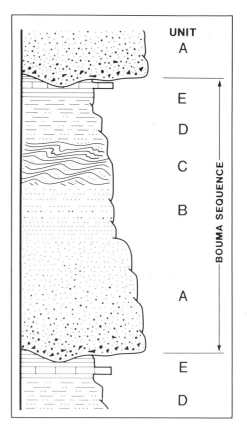

Figure 14. Bouma sequences occurring at Location P1. A: coarse massive or graded sandstone, B: medium grained, parallel laminated sandstone, C: rippled sandstone/siltstone, D: laminated siltstone and mud, E: mud from turbidity current and black hemipelagic mud, often with a thin limestone bed and sometimes carbonaceous material.

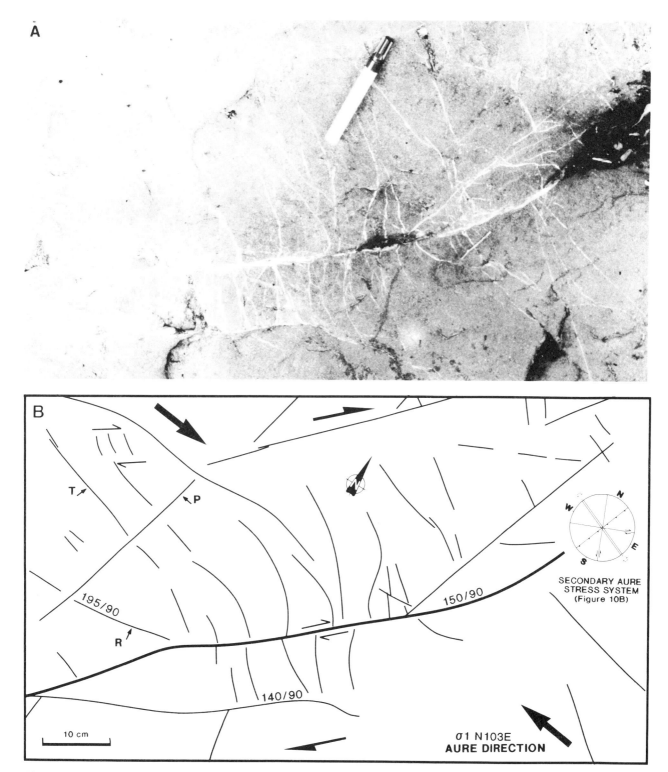

Figure 15. Fracture system at Location P1 in the Central Block. See Figure 11 for location. A: Fractures are filled with calcite and quartz. B: Interpretation of A. Measurement 150/90 means dip is 90° in direction N150E (strike is N60E). P and R: P and R shears; T: tensional fault with clockwise rotation.

stress system, Fig. 15B). The maximum compressional Aure stress at this location is interpreted to be N103E directed. The rhombohedral shaped fractures are the P and R shears. These are faults which originate in an early stage of the wrench evolution before the main dextral wrench fault comes into existence.

The Northern Segment
Crossing the Thomas Fault Zone, from P1 to the M'Bwei area (Fig. 11), reveals a distinct difference in rock character. Central Block rocks are extremely indurated whereas the Northern Segment sediments are more friable. Although the composition of sands appears to be similar, the Bouma sequences are incomplete. Along the M'Bwei River, just west of the Thomas Fault Zone, there are several massive sand and conglomerate units up to ten metres thick, rising sometimes vertically out of the forest. Further west the character changes and often only the upper units of the Bouma sequence are present. This suggests an increase in flow distance from east to west.

The field party investigated all exposed anticlines along the M'Bwei River and found that all had overturned west flanks where deeper parts of the anticline were exposed. The level of exposure is crucial as illustrated by the M'Bwei Anticline where at P4 and P5 the overturning is seen in the field to be in the core of the anticline, whereas at P3 only a very gently dipping anticlinal crest is exposed with a flat top and dips of a few degrees towards the northeast and southwest on each side of the axial plane (Fig. 11).

Overturning on the Crocodile Anticline
The small Crocodile Anticline is located at P2, east of the M'Bwei Anticline (Fig. 11). The west flank (Fig. 16) consists mostly of the black shale with thin limey sandstone beds dipping about 50° towards the east. Ripples and parallel laminations are the only visible sedimentological structures in the silty, limey sandstone beds and they occur upside down relative to exposures upstream.

About 400m downstream of Loc. P2 (towards the northwest), a massive vertical sandstone bed occurs with the same strike (N5W) as the beds at P2 (Fig. 17). Bedding between P2 and this location is continuously east-dipping. Coarse grained, pebbly sandstone cuts into finely laminated, limely siltstone (Fig. 18). The stratigraphic top is at the top of Figure 18 which is the west side in Figure 17. Therefore the stratigraphic top is west and the east-dipping strata infer overturning for the section in Figure 16.

The field party confirmed that the western limbs of the Crocodile, M'Bwei, Iala and Suniyana anticlines are all overturned along the M'Bwei River (Fig. 11).

Overturning and its Consequences
Radar interpretation indicates that several other anticlines are overturned (Kariava and Hadina, Fig. 11). Additionally, published maps show overturning in parts of the Oburu and Dude Anticlines (Figs. 6 and 11); and the Australasian Petroleum Company Pty Ltd. reported overturned measurements in several anticlines in the study area on their 1948 maps (Kariava, Suniyana — Fig. 11, and Iavokia — Fig. 8) (APC, 1948). We therefore propose that most, if not all, of the anticlines in this part of the Eastern Papuan Fold Belt are overturned at deeper levels.

Based on outcrop geology, and on seismic data from offshore (Slater et al, 1988), a schematic cross-section was constructed through the area of detailed investigation (Fig. 13). Taking into account the stratigraphic thickness of about 2500m for the Aure Beds, we conclude that in many cases the surface anticlines will be unrelated to structuring at pre-Aure reservoir levels.

Conclusions
Interpretation of radar images indicated the possibility of previously unmapped structural complexities in the Eastern Papuan Fold Belt. Ground checks confirmed the existence of northeasterly overturned beds in the internal parts of forelimbs of asymmetric anticlines; additionally, northeastward-striking transfer faults have cut and displaced the noses of some plunging anticlines. Both of these fabrics can have critical importance for interpreting the subsurface structural style, and assessing the prospectivity, of some large, possibly oil-bearing anticlines. We interpret some large structural elements to be expressions in the Cenozoic cover rocks of deeply buried, basement-rooted, Mesozoic-age features: northwestward striking Cupola and Er faults may be rooted in basement faults that acted in extension during Mesozoic rifting; Aure and Thomas faults are suggested to be superimposed on reactivated transfer faults in basement.

Acknowledgements
The authors would like to thank the managements of Petro-Canada Inc., Petroleum Corporation of New Zealand (Exploration) Ltd. and Austin Oil N.L. for permission to publish this paper.

References
Australasian Petroleum Company Pty. Ltd. (APC), 1948, Topographical and Photogeological maps: Maropo East, Paku West, Vailala East.

Carman, G.J., 1986, Report on the M'Bwei–Aure Scarp Geological Survey PPL30. Southeastern Oil & Gas Pty. Ltd., open file report.

Carman, G.J., 1987, The stratigraphy of the Aure Scarp, Papua New Guinea: Petroleum Exploration Society of Australia, Jrnl. No. 11, p. 26–35.

Dekker, F., 1989, Radar as an exploration tool in tropical rainforest areas: ARPEL (Association for Reciprocal Assistance of Latin American State Oil Companies) Conference, Calgary, Alberta, June 1989, Petro-Canada Internal Report, (unpublished).

Figure 16. Overturned shaley Aure Beds in west flank of Crocodile Anticline at Location P2, looking north (Fig. 11). Strike is N5W, bedding dips 50° towards the east. Units C to E of Bouma sequence.

Figure 17. Vertical Aure Beds in west flank of Crocodile Anticline 400 m northwest (downstream) of P2 (Fig. 11).

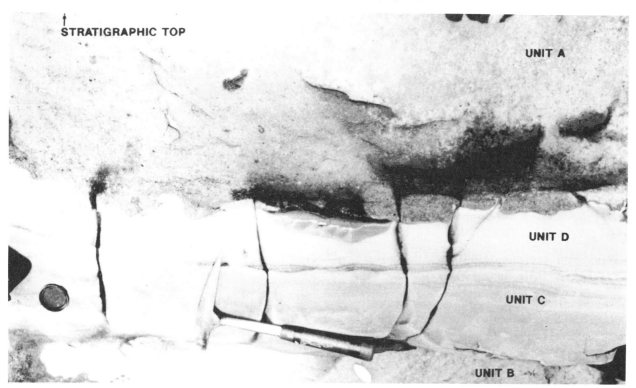

Figure 18. Details of centre area in Figure 17. Coarse pebbly sand of unit A of the Bouma sequence cuts into limey siltstone to micritic limestone of unit D.

Dow, D.B., Smith, J.A.J., and Page, R.W., (compilers), 1974, Wau, Papua New Guinea, 1:250,000 Geological Series — Map Sheet SB/55–14. International Index: Department of Minerals and Energy, Australia and Geological Survey of Papua New Guinea.

Ellis, J.M., and Pruett, F.D., 1986, Application of Synthetic Aperture Radar (SAR) to southern Papua New Guinea Fold Belt exploration: in Proceedings of the Fifth Thematic Conference on Remote Sensing for Exploration Geology; Reno, Nevada; p. 15–34. Environmental Research Institute of Michigan (ERIM).

Petro-Canada Inc., 1988, Field Study, Kerema Region: Proprietary Report, (unpublished).

Petro-Canada Inc., 1989, Field Study, Kerema Coastline: Proprietary Report, (unpublished).

Pieters, P.E., (compiler), 1983, Kikori, Papua New Guinea, 1:250,000 Geological Series Map Sheet SB/55–13. International Index: Department of Minerals and Energy, Australia and Geological Survey of Papua New Guinea.

Rogerson, R., Hilyard, D., Francis, G., and Finlayson, E., 1987, The foreland thrust belt of Papua New Guinea: in Proceedings of the Pacific Rim Congress 87, Gold Coast, Australia; p. 579–583. Australian Institute of Mining and Metallurgy.

Sari, J.K., 1985, Stratigraphy and sedimentology of a Neogene sequence in the Kerema district, Gulf Province, Papua New Guinea: M.Sc. Thesis, Sydney University, (unpublished).

Slater, A., Balkwill, H.R., and Fong, G.U., 1988, Seismic evidence for structural style in the offshore Kerema area, Papua New Guinea: Application to petroleum exploration: 7th Offshore South East Asia Conference Preprints p. 141–149.

Reprinted from *Radar Geology: An Assessment*: Jet
Propulsion Laboratory Publication 80-61, 1980, p.
457-501.

RADAR, AN OPTIMUM REMOTE-SENSING TOOL FOR DETAILED PLATE
TECTONIC ANALYSIS AND ITS APPLICATION TO HYDROCARBON
EXPLORATION (AN EXAMPLE IN IRIAN JAYA, INDONESIA.)

Claude M. Froidevaux
Phillips Petroleum Company
Bartlesville, Oklahoma   74004

ABSTRACT

In many parts of the planet, a general dynamic pattern of plate
motion is accepted, but the geologic details of plate boundaries are
not examined closely enough to be fitted into the scheme.

Radar images, because of their intermediate scale, geomorphic
definition, and synoptic quality, appear to be optimum tools for detailed testing of the plate tectonics model. At the same time, the
critical elements for the four-dimensional analysis of local or regional hydrocarbon history so essential in improving our predictabil-
ity in exploration are provided. The method is the best available in
areas of prevailing cloud and vegetal cover.

Radar was a useful tool in deciphering the evolution of the
Salawati area, a province of oil-producing Miocene reefs located at
the western extremity of Irian Jaya, Indonesia (Figures 1 and 2).
The Salawati area lies in the complex area of interaction of four
major crustal plates: the Pacific oceanic plate on the north, the
Australian continental plate on the southeast, the Asian continental
plate on the west, and the Indian oceanic plate on the southwest.

Geometric, geomorphic, and structural information derived from
the examination of radar imagery and combined with geologic and geophysical evidences strongly indicates that Salawati Island was
attached to the Irian Jaya mainland during the time of Miocene lower-
Pliocene reef development, and that it was separated in middle Pliocene to Pleistocene time, opening the Sele Strait rift zone. The
island moved 17.5 km southwestward after an initial counterclockwise
rotation of 13°.

The rift zone is subsequent to the creation of the large left-
lateral Sorong fault zone that is part of the transitional area separating the westward-moving Pacific plate from the relatively stable
Australian plate. The motion was triggered during a widespread magmatic intrusion of the Sorong fault zone, when the basalt infiltrated
a right-lateral fault system in the area of the present Sele Strait.

457

Rifting along three parallel major left-lateral strike-slip faults can be traced from the Sele Strait to the southern part of Salawati Island. The amount of relative displacement increases from the southeast fault to the northwest fault. These faults later become the site of important down-to-the northwest normal faulting to accommodate the subsidence resulting from the load of Pliocene-Pleistocene deposits derived from the high northern basaltic mountains.

Pliocene-Pleistocene diastrophism thus has defined several zones of varied structural character: the Sele Strait, the Irian Jaya mainland, Salawati Island, and their respective surroundings. An understanding of the dynamic character of the area was essential to successfully study hydrocarbon migration and accumulation:

(1) When the paleogeography is restored to a predrift position, the distribution of oil-bearing Miocene reefs is different from the present arrangement.

(2) The changing nature of structural unit boundaries implies that a given fault could have acted both as a barrier or an avenue for hydrocarbon migration at different times.

I. INTRODUCTION

The purpose of this presentation is to illustrate the important and useful role played by radar imagery in a comprehensive study of a prospective hydrocarbon-bearing province located within a complex area of crustal plate interaction.

II. POTENTIAL OF RADAR IMAGERY

The most useful ability of radar sensing is to emphasize morphology. This is illustrated by the example shown on Figures 3 to 5. Figure 3 is an enlargement of a portion of the radar mosaic represented on Figure 6. It shows the obvious presence of fish traps around elongated islets in the form of small white arrows. One of these structures is seen on Figure 4.

Figure 5 is a reproduction of an aerial photograph of the same area, showing a great abundance of detail hidden to radar sensing. Corals can be identified under several feet of water, the depth of the sea could be estimated from the change in color, and different species of trees are recognizable along the beach. But the fish traps remain very hard to detect. The same difficulty subsists even when comparing the original aerial photograph (scale 1:20,000) with the original radar mosaic (scale 1:100,000), which is 25 times smaller in area.

The morphologic characters enhanced by radar sensing are:

(1) Shape. Radar imagery expresses clearly and objectively the outline of the land. Relationships of shapes provided the original intuition of the thesis developed in this study: mainly, that Salawati Island appears to have been detached by rifting from the mainland.

458

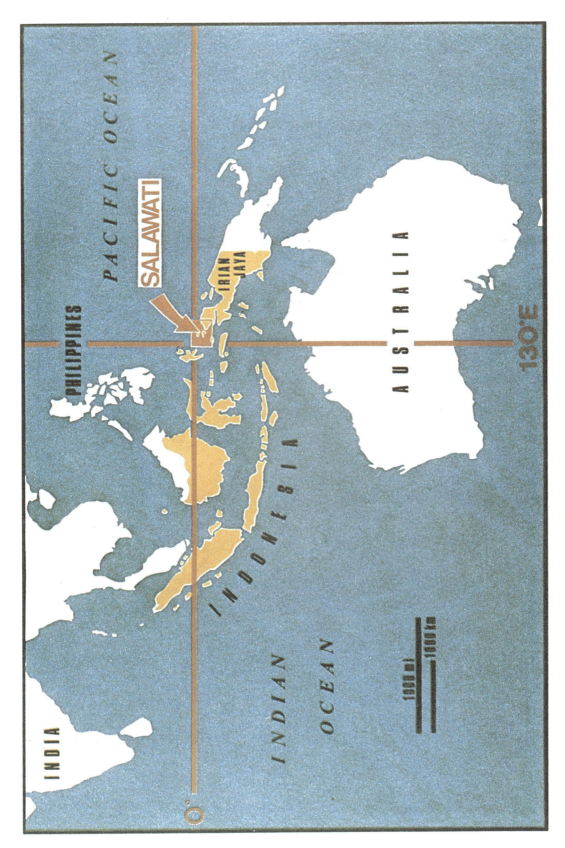

Figure 1. Location. The red rectangle represents the example area.

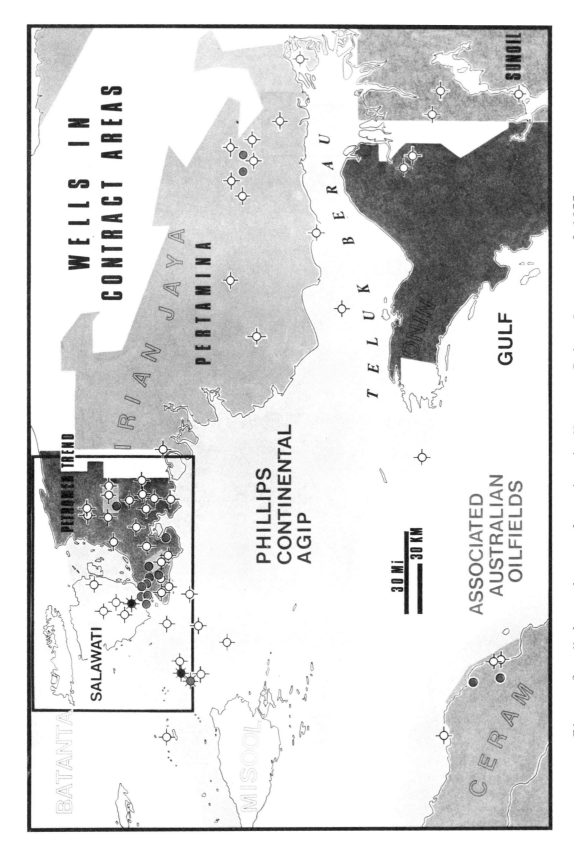

Figure 2. Hydrocarbon exploration in Western Irian Jaya, as of 1977.
Salawati area is in rectangle.

460

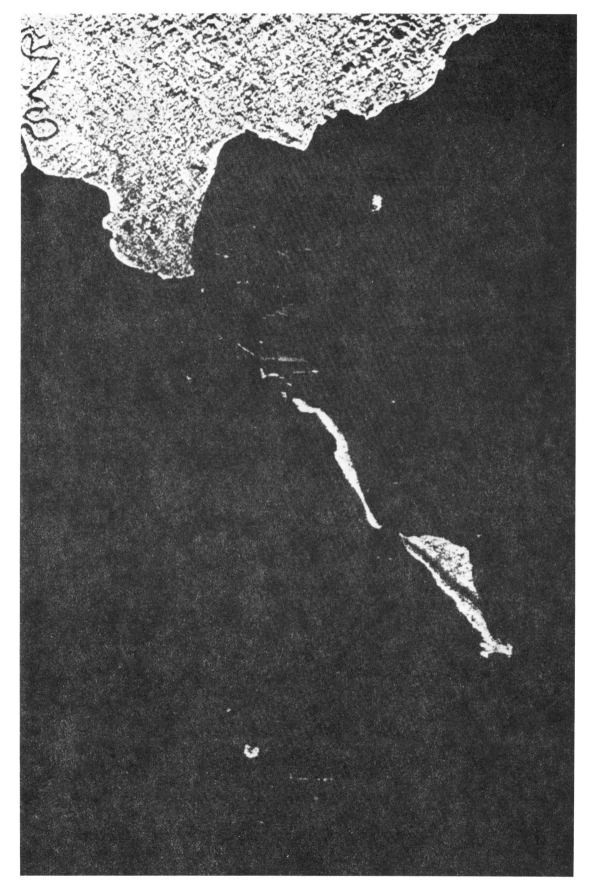

Figure 3. Enlargement of the upper center part of radar mosaic, Figure 6, showing the sharp signature of fish traps as white arrows. South is on top.

461

Figure 4. Oblique aerial view of the same fish traps as shown in Figure 3. View is to the South.

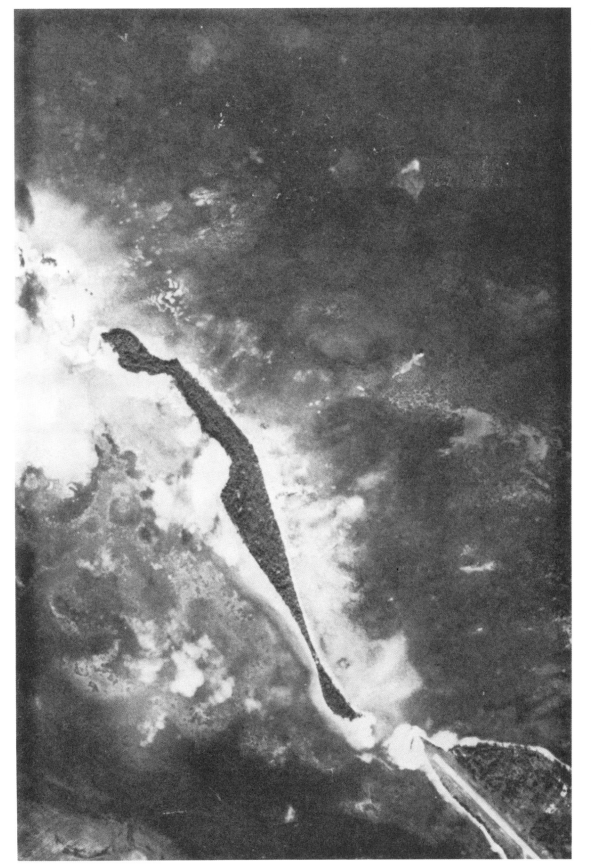

Figure 5. Aerial photograph of the elongated island shown in Figures 3 and 4. Fish traps are very hard to detect, although a much greater amount of detail is available on this picture than on the radar image.

Figure 6. Radar mosaic of the Salawati area, from a 1974 survey by Aero Service - Goodyear, flown in north-south strips with synthetic aperture. White bar represents 10 km.

(2) Linears. Radar imagery enhances a large amount of linear features, such as coastline segments, textural change boundaries, and thin, contiguous shadow and bright zones, which cannot be seen easily in the field, but which can nevertheless be identified by analogy as faults, fractures and/or eroded fold ridges when compared to the relatively rare ground data. Radar linears can be assembled into comprehensive structural maps (two-dimensional), which can serve as excellent bases for three-dimensional geometric analysis substantiated by subsurface data.

(3) Patterns. Arrangements of linears in patterns are clues to tectonic analysis (four-dimensional), because their geometric properties reflect the orientation of the stresses, and indirectly the forces, exerted on the deformed body of rock. The coexistence of different patterns in the same rock and at the same location implies successive tectonic events, and therefore calls for the analyst to decipher the geologic history of an area.

(4) Texture. Texture is mostly useful in two-dimensional analysis (as is tone) for the making of geologic maps emphasizing lithology rather than structure. However, in some cases, a change in texture can be the expression of a fault (see Figure 7).

The following is an example of the application of Radar data analysis to detail plate tectonic study related to hydrocarbon exploration.

III. GENERAL TECTONIC SETTING OF SALAWATI AREA

The oil-producing Salawati area is located at the western extremity of the island of New Guinea. The main elements of the general tectonic framework are represented in Figure 8.

A.   Sorong Fault Zone

The dominant tectonic feature of the Salawati area is the Sorong fault zone that comprises the northern part of the island. The fault is part of a large global transcurrent zone that can be traced from eastern Papua New Guinea to the vicinity of Celebes (Sulawesi). It separates the westward-moving Pacific oceanic plate from the relatively stable Australian continental plate. As a result, the Sorong fault is a left-lateral strike-slip fault zone. The exact amount of relative motion along this zone is unknown: Some workers (Hamilton, 1973) believe that it may be as much as 1000 km. This may be excessive, but the displacement probably is of the order of tens of miles at least. The fault zone is 8 to 13 km wide in the Salawati area, where mixtures of rocks of all kinds have been recognized in a disordered assemblage.

B.   Tarera-Aiduna Fault Zone

This is a large-scale tectonic feature similar to the Sorong fault, but of lesser extent. It can be traced from the southern coast of Irian Jaya to possibly the south flank of Ceram Island. Left lateral motion was recognized early by Dutch geologists southeast of the Lengguru foldbelt. It was confirmed later, offshore on the west, by the study of seismic records. The horizontal displacement on this zone may be of the order of 60 km at the longitude of the

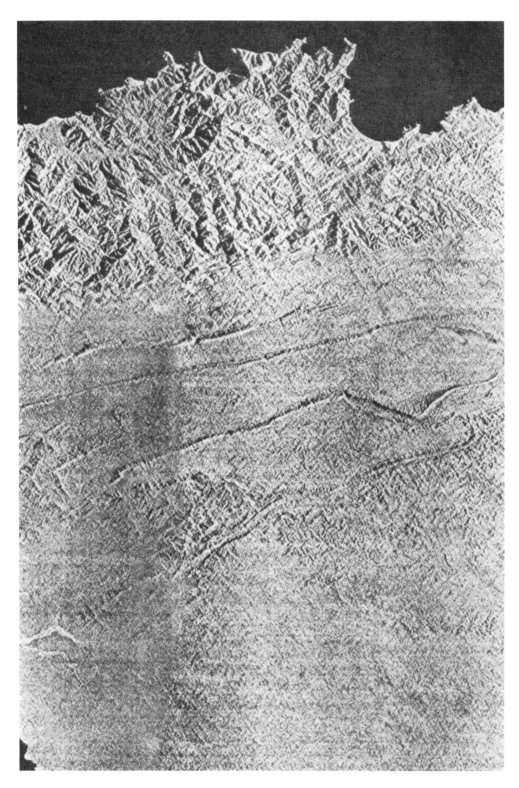

Figure 7. Radar image of the northern part of the Salawati Island showing:
(1) the southern border of the Soron fault zone as to textural change
from the coarse basalt signature on the north to the fine grain of the
clastic sediments on the south, (2) ridges of eroded gravity folds that
were derived from the northern elevated area. (Enlarged from radar
mosaic of Figure 6).

466

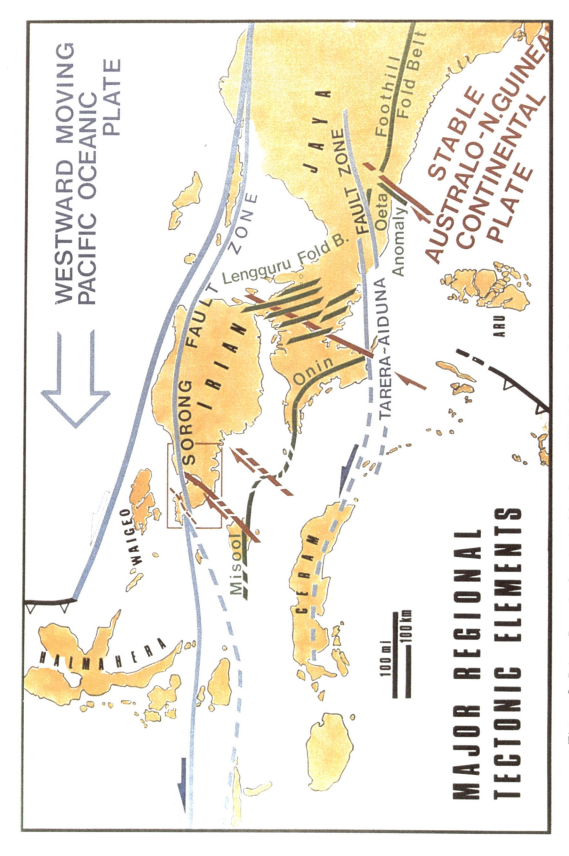

Figure 8. Irian Jaya is dominated by large E-W, left-lateral fault zones (blue) and associated right lateral NE-SW faults (red). Salawati area is in red triangle.

303

Salawati area, according to field observations made in southern Ceram on shifted drainage and offset outcrops of metemorphic rocks (Tjia, 1973).

C. Oblique Zones of Right-Lateral Motion

The two previously mentioned left-lateral fault zones delimit a broad area in which secondary large-scale features have been generated by coupling effect. They are the high-angle right-lateral motion zones expected in a left-lateral wrench system. From east to west these features are the Oeta anomaly, Lengguru foldbelt, Misool-Onin anticlinorium flexure, and Sel Strait-Misool fault.

1. Oeta Anomaly. This zone was recognized as a longitudinal high in early New Guinea gravimetric exploration (Visser and Hermes, 1962). The Oeta feature is a folded structure on the west side of a steep west-dipping, right-lateral fault that borders the anomaly. The fold is the result of upthrust drag along the fault which is the probable cause of the anomaly. The effect of both the fold and fault can be seen at the intersection of the Oeta anomaly with the foothill foldbelt: the other east-west trending fold system (related to the time of southward subduction of the Pacific plate) was bent right laterally by the fault. The Triassic anticlinal core at the intersection assumed a domal shape as a result of superposition of the younger Oeta upthrust fold, which is related to the time of westward motion of the Pacific plate (Figure 8).

2. Lengguru Foldbelt. This is a right-lateral bend of the generally east-west-trending New Guinea mountain system. The folds have been reoriented into a typical right-lateral en echelon pattern. The direction of faulting is outlined by a long and narrow bay.

3. Misool-Onin Anticlinorium Flexure. The eastern end of Misool Island is a mirror image of the western end of the Onin Peninsula. Both terminate as four narrow plunging anticlines, strongly indicating that they belong to the same folded ridge. However, the opposite plunging directions are not oriented exactly toward each other and suggest, rather, that the ridge is displaced right-laterally somewhere under the sea.

4. Sele Strait-Misool Fault. The Sele Strait longitudinal axis is aligned with a right-lateral fault that shaped the straight east coast of Misool Island, which can be seen on radar imagery offsetting the prominent ridge of Eocene limestone of the north flank of the anticlinorium (Figure 9).

IV. STRUCTURAL ANALYSIS BASED ON RADAR IMAGERY

A. Radar Survey

The study area was surveyed in 1974 by Goodyear-Aero Service in north-south flight strips, using a synthetic aperture radar. A copy of a semicontrolled mosaic is represented on Figure 6. The geomorphologic emphasis of this radar picture can be appreciated by comparison with the best false color Landsat image of the same area (Figure 10), which mainly reflects healthy vegetation (red) and cloud cover, two characters of little use for our study.

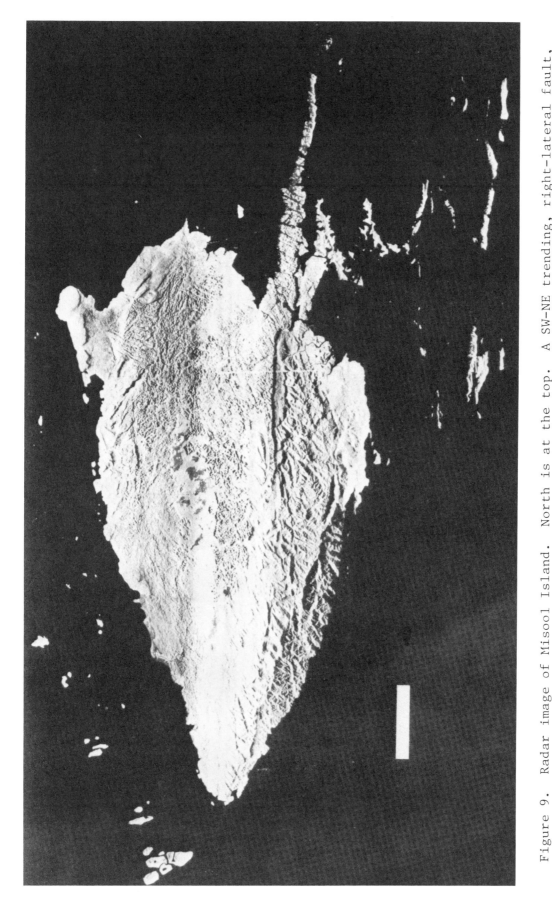

Figure 9. Radar image of Misool Island. North is at the top. A SW-NE trending, right-lateral fault, visibly offsetting the Eocene limestone on the east coast, is in line with the Sele Strait that separates Salawati Island from the mainland (see Figures 8 and 14 for location). The coexistence of elevated Pleistocene reefs in the northwestern islets, with drowned karst in the southeast (expressed by jagged coastlines) suggests a recent east southeastward tilting of Misool Island (see Recent Tectonics). White bar represents 10 km.

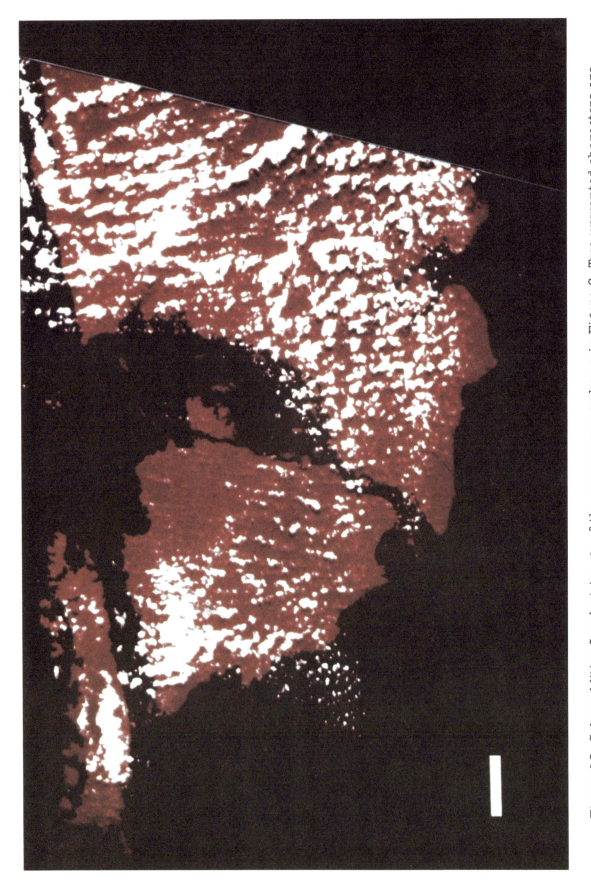

Figure 10. Color additive Landsat image of the same area as shown in Figure 6. Two unwanted characters are emphasized: A healthy vegetation (red) and cloud cover. White bar represents 10 km.

B. Shape

A glance at the radar mosaic of the Salawati area (Figure 6), which is the best cartographic unbiased expression of its land shapes, suggests that Salawati Island was detached from the Irian Jaya mainland and drifted away, opening the gap of the Sele Strait. Furthermore, if the island is placed in its proposed original position, the northern elevated area will fit perfectly between the mountainous Batanta Island and the high coastal ridge of the mainland. The Sele Strait will be closed almost completely by its islets. Warir Island (see Figure 11) is, however, too big to be enclosed completely between the two present shores, and the northwest corner of Salawati corresponds almost exactly to the eastern bay of Batanta, although the closing southern part of the bay seems to be in the way (Figure 12). These three spatial problems can be solved when considering data from surface geology (see below).

C. Linears and Patterns

Linears and patterns derived from radar imagery are represented on Figure 13.

1. _Sorong Fault System._ The southern edge of the fault zone (blue line) roughly parallels the southern foot of the basaltic mountains of the mainland and Salawti. On the west, a modification of the current interpretation is proposed. The fault trace can be carried from the Kofiau Island (Figure 14) area to the deep channel located just south of Batanta Island, instead of being traced directly to Salawati. This interpretation is supported by both seismic and geomorphologic data (see below). The Batanta part of the fault trace is believed to end near the eastern extremity of the island, at the contact of the basalt and clastic rocks, by an offsetting fault. The basic geometric elements of the Sorong fault zone are an approximate east-west direction and left-lateral strike-slip motion, with a dynamic compartment on the north side (related to the Pacific palte), and a relatively static compartment on the south side (related to the Australian-New Guinean plate).

The direction of the fault must be measured on the fixed mainland block, where it is about N80 to 85°E. (The trace of the fault on Salawati assumed a more southwesterly direction after rotation of the island.)

2. _Sele Strait Fault System._ The original direction of the fault (red line) is probably roughly parallel with the stable east side of the strait, making an angle of about 60° with the Sorong fault, which corresponds to the theoretical direction of an antithetic set of strike-slip faults in a left-lateral wrench system. This possibility is corroborated by the Sele Strait gap being aligned with the right-lateral fault of eastern Misool (Figure 14).

a. _Strain Ellipsoid._ The hypothesis that the Sele Strait was initially a right-lateral system can be checked by a study of its internal geometric properties, which can be measured directly on radar imagery (Figure 11). The geometric elements are plotted on Figure 13. A right-lateral fault should have developed a conjugate set with approximately the same angular relations as the one initiated by the Sorong fault, because the rock material is the same. This is exactly what can be seen in the group of islands on the south side of the Sele Strait near the mainland (green lines). The trapezoidal shape of several islets is modeled after a conjugate set of fractures related to the main

471

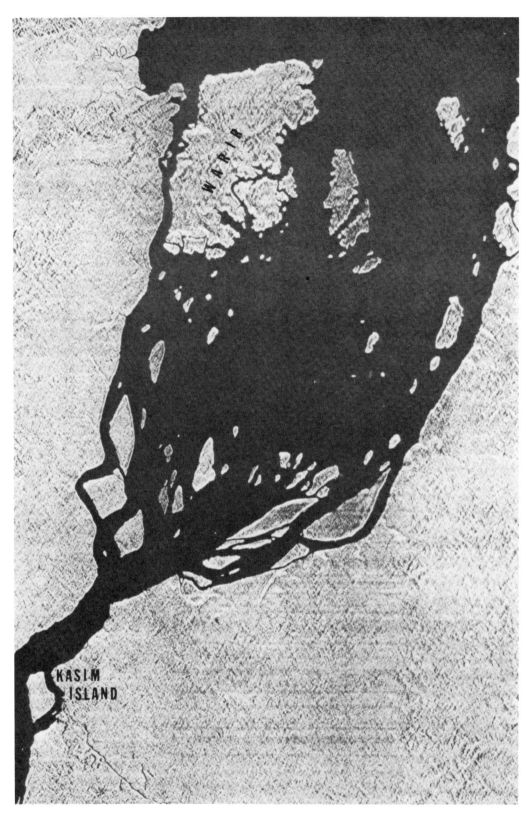

Figure 11. Enlargement of a portion of the radar mosaic of Figure 6, showing the sharply defined shapes of the islets in Sele Strait, used in structural analysis for the construction of strain ellipsoids (see Figures 13 and 24). North is on top.

472

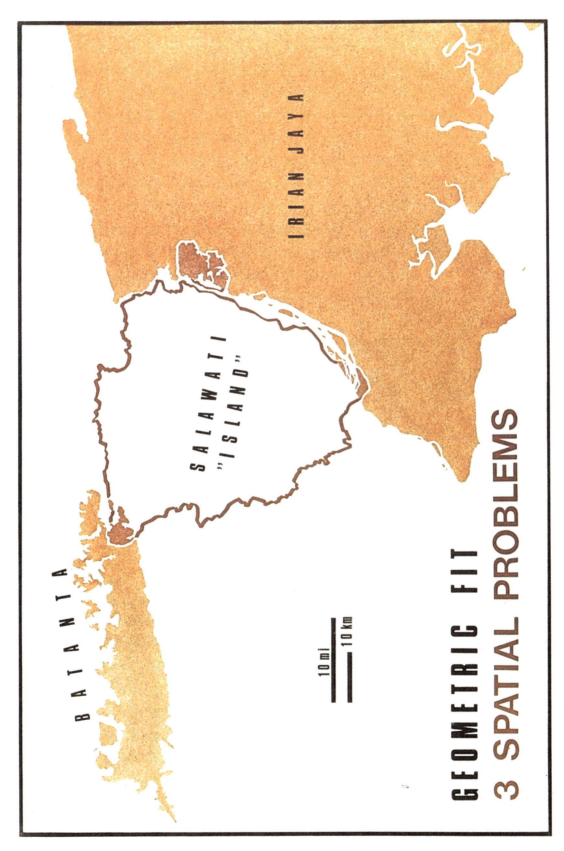

Figure 12. When the Salawati block is placed back in its alleged original position, three areas overlap the surrounding lands (red). These spatial problems can be solved when considering geological data. (see Figure 15.)

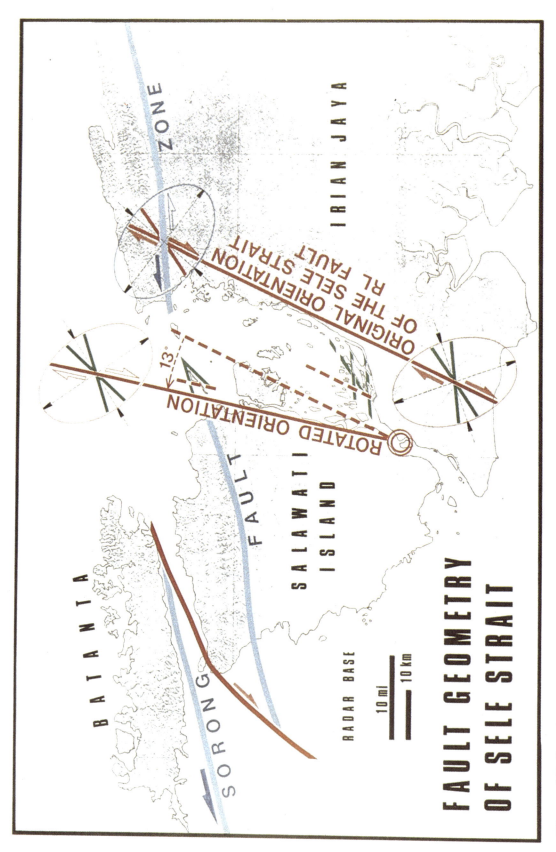

Figure 13. Rectilinear west coast of strait appears to be rotated 13 degrees CCW with respect to east coast. Orientation of east coast corresponds to that of antithetic right-lateral fault in conjugate set (red) defined by left-lateral system of Sorong fault (blue). Orientation of fracture-controlled shape (green) of Sele Strait islets corresponds to the expected conjugate set of the right-lateral system of the strait. Furthermore, it is also rotated 13 degrees CCW on the Salawati side (upper center).

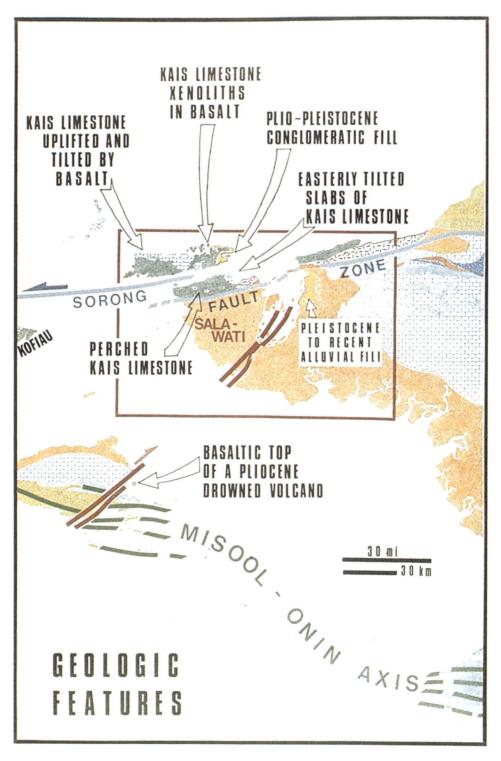

Figure 14. The right-lateral fault of Misool (red) is in line with the Sele Strait, that separates Salawati Island from the mainland. Red rectangle corresponds to Salawati area of Figures 10 and 11. Dark green equals basalt, red equals granite, patterned white equals tectonic brecia, brown and yellow equals Plio-Pleistocene, patterned blue equals Eocene-Oligocene, pale green equals Mesozoic, dark blue equals Paleozoic.

direction of right-lateral faulting of the Sele Strait. The ellipsoid thus defined has the same geometric properties as the one along the Sorong fault, but in reverse, because here the main axis of wrenching is right-lateral. The high-angle conjugate set amounts also to 60°.

b. Rotation. On the west side of the Sele Strait, the conjugate set of fractures is the same, but it appears to have been rotated counterclockwise. The east coast of Salawati Island can be taken as a reference for the rotated direction of the Sele Strait right-lateral fault. The strain ellipsoid is outlined at Warir Island, where the north coast is parallel with the high-angle conjugate set, also making an angle of 60° with the narrow straight channel that separates Warir from Salawati. The airstrip islet of Jefman (Figures 3 to 5), which underwent the greatest amount of rotation, is shaped according to the elements of the strain ellipsoid, rotated 13° counterclockwise.

The counterclockwise rotation indicated by the study of the strain ellipsoid can be checked by other means. A reference line can be chosen and traced between two points on Salawati Island. Its measured direction can be compared to the direction of the same line in the initial prerotated position of the block. It is preferable to pick two points that are easily recognizable such as the southwest tip and northeast corner of the island. The method shows a counterclockwise angle of rotation of 13° also.

c. Translation. Rotation alone can account for a large part of the opening of Sele Strait but is not sufficient to explain the present position of Salawati Island. A comparison of the position of the island after its 13° counterclockwise rotation with its present location indicates a subsequent southwestward translation of the block of 17.5 km. This translation was measured by comparing pairs of displaced points picked on the island in their position before and after translation.

It should be noted that the "translation" is in the strict sense also a rotation because any displacement on the spherical surface of the earth can only be described as such. In the present case, the arc of 17.5 km corresponds to a rotation of less than 10 minutes along a great circle.

V. TESTING OF RIFT HYPOTHESIS

A. Data from Surface Geology

A look at the geology (Figure 14) confirms the alleged fit of the Salawati block in its former mainland frame, and even allows for a better match (Figure 15). The distribution of the geologic elements becomes simpler. All the mountains appear as a continuous basaltic ridge (characterized by a coarse texture on radar imagery, Figure 6) parallel and contiguous to the Sorong fault. Some granite outcrops in northeast Salawati match similar exposures near the northwest corner of the mainland.

The spatial problems raised by the northwest corner of Salawati and the Island of Warir are easily explained. The islet in the bay of eastern Batanta and the promontories closing the bay on its south side are made of an accumulation of coarse Pliocene to Pleistocene conglomerates containing mostly limestone and basaltic boulders. Once this young mass of rock is removed, it leaves a larger gap that can accommodate the northwest corner of Salawati almost perfectly.

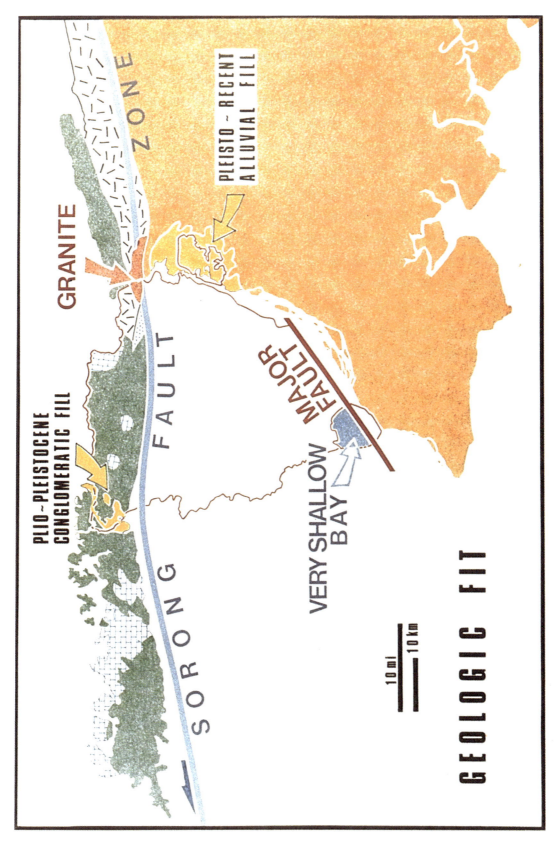

Figure 15. The NW and SE geometric overlaps coincide with areas of deposits younger than the time of rifting. The southern tip of Salawati can be slipped along a known major fault to provide a perfect geometric fit. Good geologic fit is further emphasized by the linear arrangement of the basalt (green) and the matching of granite outcrops (red).

313

Just across the Sele Strait from Warir Island, the mainland rivers have filled a gap on the coast with Pleistocene alluvial deposits. The outline of the coastal alluvial plain corresponds to the shape of the drifted island.

If the south tip of Salawati is moved southwestward along a known major fault (Figure 15), so that its present shallow bay shore becomes aligned with the west coast of Salawati, the block closes the Sele gap with the best possible fit. In that position, the present south tip of Salawati Island would adjoin the north end of Kasim Island (Figure 11).

B.   More Data from Radar

In western Irian Jaya, the eastern exposures of Miocene limestone (Figure 14) clearly display on radar imagery a conjugate set of fractures, enhanced by karstic solution (Figure 16). The two systems intersect at approximately 60°, symmetrically with respect to a meridian. This arrangement is typical of a north-south compressional stress field, and thus suggests its relation to an ancestral tectonic model developed at the time of the birth of New Guinea, when the Pacific plate was being subducted southward under the Australian continental plate.

The stream pattern in the lower land of Pleistocene to recent exposures west of the Miocene limestone outcrops geometrically resembles the fracture system (Figure 17). The coincidence of the old fracture system with the present drainage pattern suggests that western Irian Jaya has remained tectonically quiet for a long time, probably since its creation, and supports the interpretation that the area was a relatively fixed block when Salawati was rifting away.

On Salawati, where no limestone rocks crop out, there is no clear surface expression of the fracture pattern. The most obvious geomorphic features related to structure in Salawati are anticlinal ridges carved in Pliocene-Pleistocene sandstones that were folded at a later stage of the deformation. The folds seem to have been formed by gravity sliding from the northern high mountains. They are confined to and parallel with the mountain front, asymmetric to the south, partly thrusted, shallow, and abut against the first major normal fault in the central part of the island. Their signature on radar imagery is outstanding (Figures 6 and 7).

C.   Data from Seismic

Figure 18 represents the fault patterns interpreted from seismic data.

The subsurface fault pattern of western Irian Jaya (Redmond and Koesoemadinata, 1977) presents the same geometric arrangement as the pattern exposed at the surface on radar imagery (Figure 16). The fractures are distributed in the same two prevailing directions 30° apart from a meridian. The same properties apply to the area south offshore. This general compatibility is an expected condition for a structurally stable area where no drastic tectonic event occurred to disturb the original configuration.

The faults mapped at the level of the granite basement in Salawati Island show a pattern slightly different from the mainland, with more right-angle fault intersections. This angular relation may be a reflection (and is typical) of the original tensional fractures developed during the cooling period of an intrusion, especially since the characteristic is confined largely to the older faults.

478

Figure 16. Enlargement of the SE portion of the radar mosaic of Figure 6, showing fracture and river patterns in the mainland, used in the construction of NE quadrant of Figure 1/. North is on top.

479

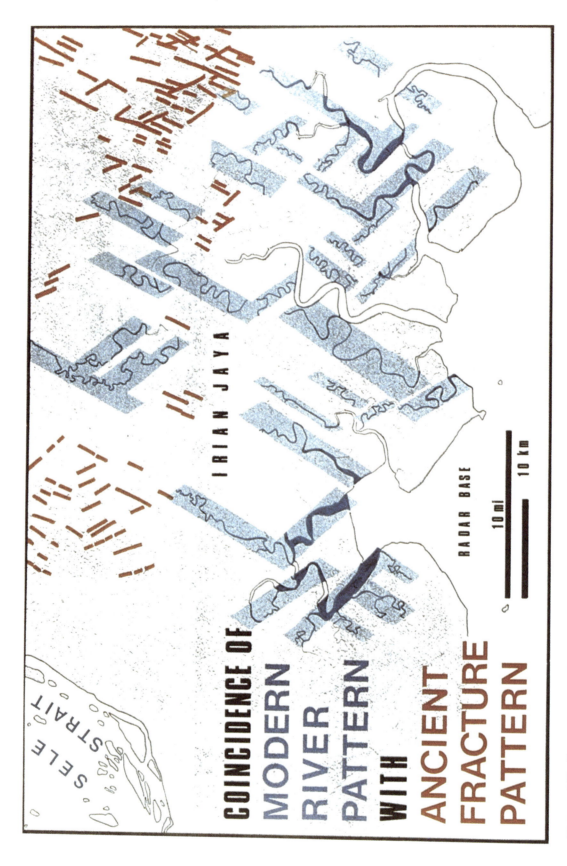

Figure 17. Coincidence of major drainage pattern with ancient fracture pattern in the mainland suggests that this area remained tectonically stable during the rifting of Salawati Island. Radar patterns are shown on Figure 16. North is on top.

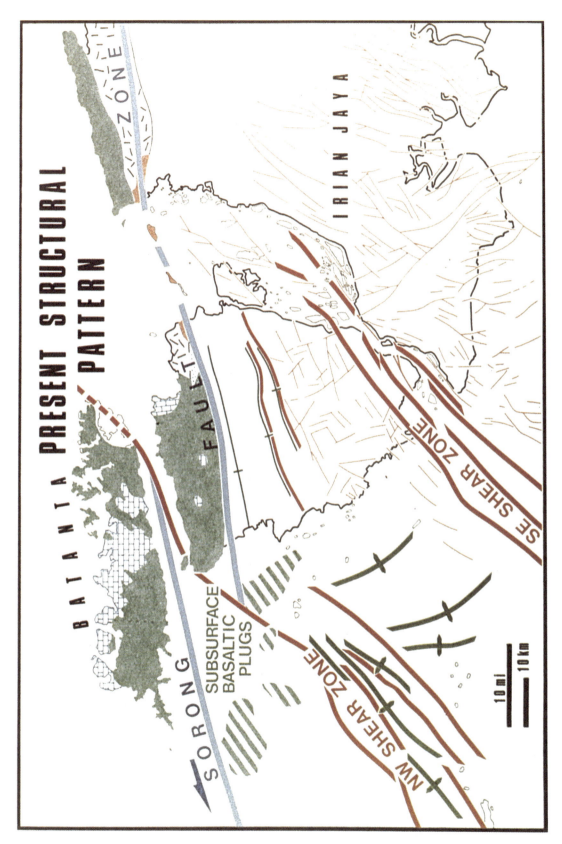

Figure 18. Fault patterns derived from seismic data. Two sheer zones (red) bound the moving island. Expected compressional folds are present between them in front of the block. Other folds are located within the NW sheer zone and are oriented in contradiction to the compression due to the block motion. Their possible origin is explained on Figure 19.

Offshore Salawati Island, from the southern entrance of the Sele Strait toward the direction of eastern Misool Island, a series of seismic discontinuities can be mapped as a set of parallel, southwest trending faults. They are interpreted as the southeast left-lateral shear zone that bounded the Salawati block during rifting. No evidence of compression is present along this zone outside Sele Strait.

Offshore, northwest of Salawati, a major zone of disturbance extends from the eastern end of Batanta Island, southwestward through the bordering strait and beyond to the vicinity of Misool Island. This zone, called the northwest shear zone, originated during the drift as a right-lateral system with a maximum relative displacement of about 30 km near Batanta (larger than on the southeast shear zone because of the additional greater effect of the initial counterclockwise rotation of the Salawati block).

The area offshore Salawati Island on the southwest, between the two shear zones described previously, was the logical place for compressional strain to have developed under the effect of the drifting block. Gravity and seismic work suggests that the edge of the underlying central Salawati granite mass corresponds approximately to the west coast of Salawati. The expected effect of a moving rigid block is to induce folding and/or thrusting in the softer material in front of it, perpendicular to its motion. The detected pattern, however, is not so simple. Only a few monoclinal flexures and large but low-amplitude folds, with axes approximately parallel with the front edge of Salawati Island, can be interpreted in the relatively undisturbed offshore area bounded by the two shear zones. The alignment of islets parallel with the west coast of Salawati suggests a geomorphic relation with this deformation. These folds seem more intense toward the north. This phenomenon may be caused by the increasing amount of displacement in the north compared to that in the south owing to the initial counterclockwise rotation and distance away from the center of rotation at Kasim Island. Most of the compressional strain can be seen clearly along the bounding northwest shear zone. The faults themselves indicate evidence of thrusting between slightly folded sheets. This apparently contradictory orientation of the folds can be explained by the combined effect of the shear zone and the push exerted by the moving Salawati block. Folds may have been initiated along the shear zone as drag folds which, at an incipient stage, would have been oriented more nearly diagonal to the shear zone, and also more nearly parallel with the western edge of the Salawati block (Figure 19). Once started, the folds would have defined the weak area, where further strain could develop. Then, under continuous coupling effect, they would progressively arrange themselves more nearly parallel with the shear zone, finally to assume their present orientation.

The stress field also was complicated by vertical intrusions near the northern outside edge of the shear zone. They are probably of volcanic nature, as suggested by their proximity to the basaltic source. These intrusions may have overshadowed in part the effect of compression related to the southwestward movement of Salawati Island by introducing a southeast stress component into the deformation. Thus, tectonic features detected in the extended vicinity of Salawati Island appear to be in harmony with the hypothesis of rifting.

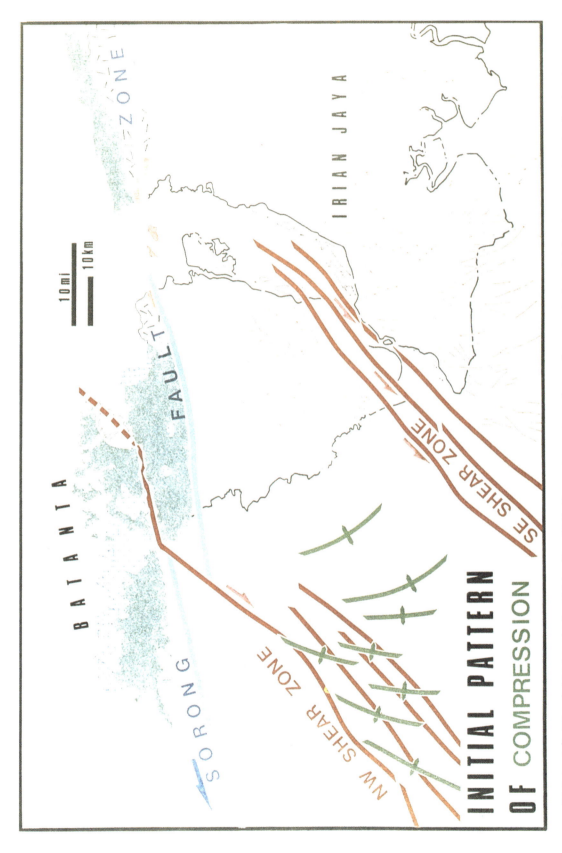

Figure 19. The folds located within the NW sheer zone may have been echelon folds at an earlier tectonic stage, oriented more in harmony with the compressional direction of the moving block. Further sheering stretched them parallel to the sheer zone.

319

Four areas of high positive Bouguer anomalies (Figure 20) were recognized by early Dutch surveys (Visser and Hermes, 1962). In decreasing order of magnitude, they are (1) a bulging nose in the northern part of Sele Strait plunging southward along the axis of the strait, (2) a high centered around the Kalamono oil field, (3) the Sele dome centered around the Sele-Walio oil fields, and (4) the significantly lower Salawati dome centered near the middle part of Salawati Island.

The largest anomaly may be the expression of the basalt intrusion into the strait from the Sorong fault zone. The other three highs seems to be related to the shape of the granitic basement.

E.   Data from Submarine Morphology

The morphology of the sea bottom illustrated on Figure 21 provides several elements in good harmony with the geometry of our rift model:

(1)   A 300-ft (90 m) break in slope coinciding with the northwest shear zone detected by seismic.

(2)   Two parallel furrows in the Sele Strait corresponding to fault traces of the southeast shear zone.

(3)   A deep underwater channel bordering the edge of a very shallow shelf in line with the trace of the major fault of the strait.

(4)   A relatively shallow bottom in the northern part of the Sele Strait, which can best be explained by basaltic invasion. If it were not for the basalt filling, one would expect the strait to be deeper in the north than in the south, because of a larger amount of rift displacement.

VI.   CHRONOLOGY OF TERTIARY TECTONIC EVENTS

The coherence and consistency of geometric properties and dynamic relation in the fracture pattern of the Salawati area, as seen on radar imagery, suggest that the present structural framework was initiated by the major westward motion of the Pacific plate, expressed locally by the left-lateral Sorong fault zone. Successive events are relatively easy to trace from the time of incipient motion along the Sorong fault zone. Previous events are more difficult to decipher because their effects have been obliterated by the younger large-scale diastrophism. A stratigraphic sketch summarizing the geologic evolution of the Salawati area before the phase of major normal faulting is represented on Figure 22.

A.   Pre-Miocene Tertiary

We know from well records and radioactive dating that a granite was emplaced probably in Late Cretaceous time, in the Salawati area and that its sedimentary cover was uplifted and eroded until almost mature peneplanation of the granite surface before the invasion of the middle Miocene sea.

484

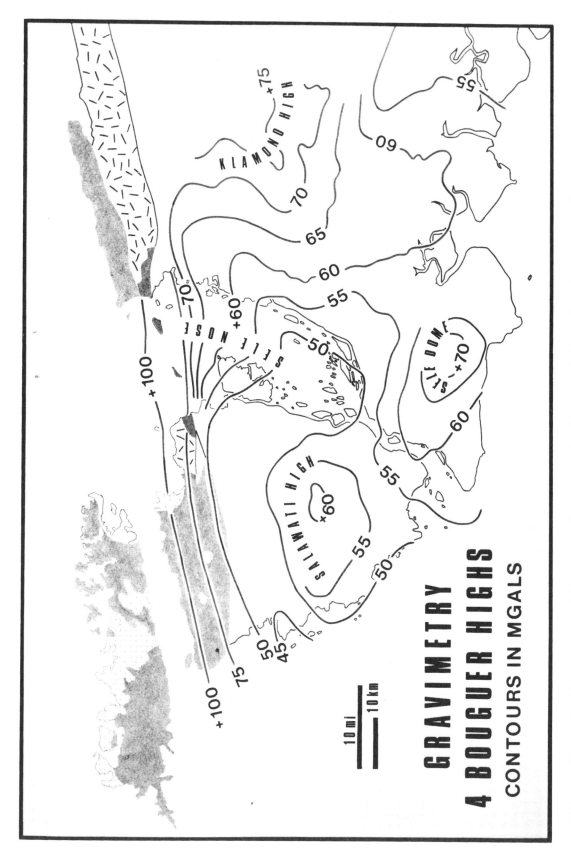

Figure 20. The largest positive gravity anomaly is probably related to an infiltration of basalt in the northern part of Sele Strait. The other anomalies reflect the shape of the granitic basement.

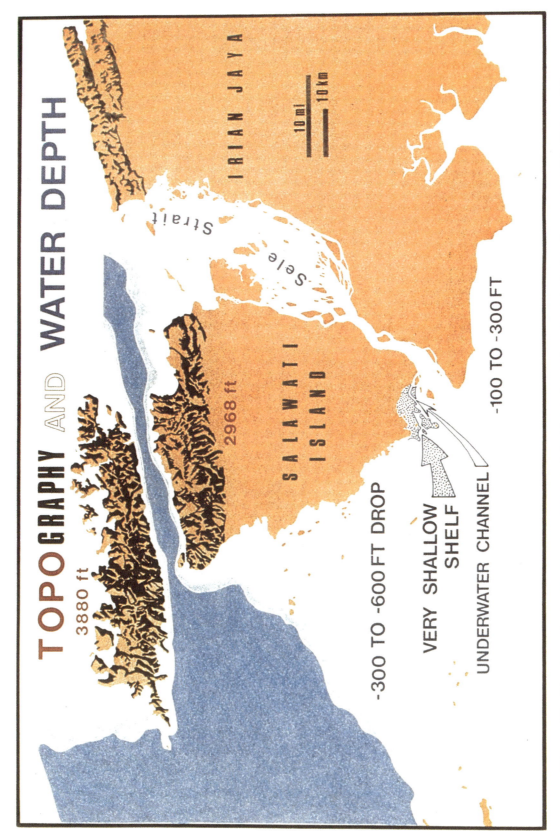

Figure 21. The two sheer zones are expressed in the submarine morphology. Compare with Figure 18.

322

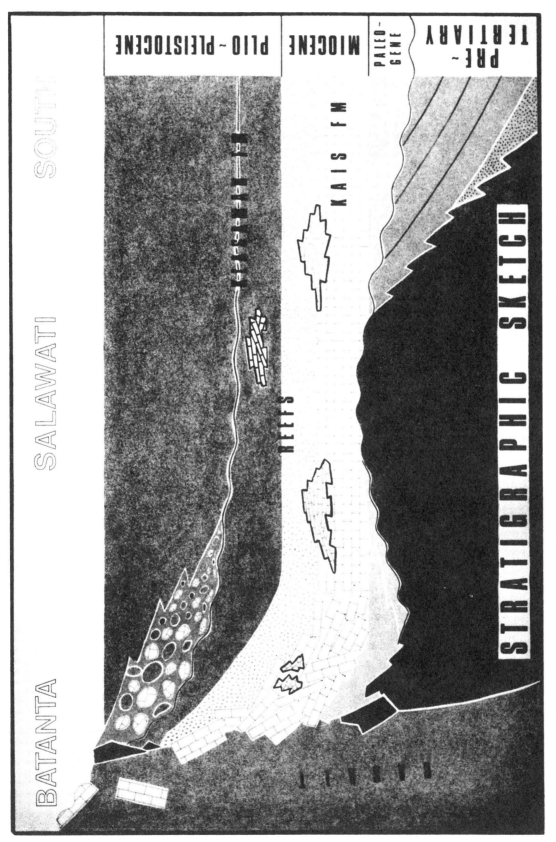

Figure 22. The emplacement of the Late Cretaceous granite was followed by 65 m.y. period of peneplanation ending with a Miocene marine invasion, which initiated reef development. During Middle Pliocene (after the formation of the last reefs), basalt erupted along the Sorong fault zone, creating a high mountain ridge in the north, that became the source of voluminous Plio-Pleistocene clastics.

The early Tertiary (Paleocene to Oligocene) apparently was locally a time of continental environment. The lower Eocene of Misool Island is characterized by wood-bearing near-shore sandstones. The Pliocene offshore of Onin (Figure 14) is unconformable on the Cretaceous. Thus a landmass is outlined stratigraphically from an area north of Misool to the Onin Peninsula area. It included Salawati and probably the western part of the mainland as far as the northern and eastern edge of the Miocene limestone outcrop area where a continuous lower Tertiary section is exposed. The present expression of this ancient crustal high may be the positive Bouguer anomalies of the Salawati, Sele dome, and Klamono areas (Figure 20). If the Salawati block is placed in its initial, pre-Pliocene position, the continuity of the anomaly is more striking (Figure 28).

Volcanic activity occurred in very early Eocene time as testified by K-Ar dating of a basalt from a well located east of Misool Island (Figure 2).

B.   Miocene

At the beginning of Miocene time, the area became the site of widespread marine transgression, initiating the phase of carbonate deposition and reef development. The sea extended northward beyond the present line of the Sorong fault zone, which was formed later. The configuration of the shallow crystalline and metamorphic (?) basement certainly was determinant in the distribution of reefal activity.

The fracture pattern or erosional shape of the Miocene basement surface remains unknown. From structural analysis most of the present framework appears to be the result of Pliocene diastrophism.

C.   Post-Miocene

Since the inception of the Pliocene strike-slip motion along the Sorong fault, the tectonic evolution of the Salawati area is easier to deduce. The following succession of events is believed to have taken place:

1.   Formation of Sorong Fault. The Pacific oceanic plate had begun its westward motion, after a Miocene phase of southward subduction, during which the backbone of New Guinea already was roughly outlined.

2.   Right-Lateral Wrenching in Sele Strait Area. A conjugate set of strike-slip faults was developed concurrently. The direction of tensional stress expressed by the geometric properties of the set (as seen in the present geomorphology of the Sele Strait islands) is oriented across the axis of the strait (Figure 24).

3.   Basaltic Intrusion Along Sorong Fault Zone. The Salawati block was still a part of the Irian Jaya mainland, closing the gap of the present strait and defining a continuous zone of volcanic activity (Figure 15). The amount of left-lateral motion along the Sorong fault may already have been as much as 100 km, the present distance between the Miocene reefs of Kofiau Island (located on the westward moving compartment) and Salawati. Such a displacement would bring the Batanta Kais reefs (Miocene: Figure 14) approximately 30 km east of the northwest corner of mainland New Guinea. It is not unreasonable to imagine that the Salawati bioherms were thriving at the edge of a landmass represented by ancient Irian Jaya, facing an open sea on the west, somewhat as at present,

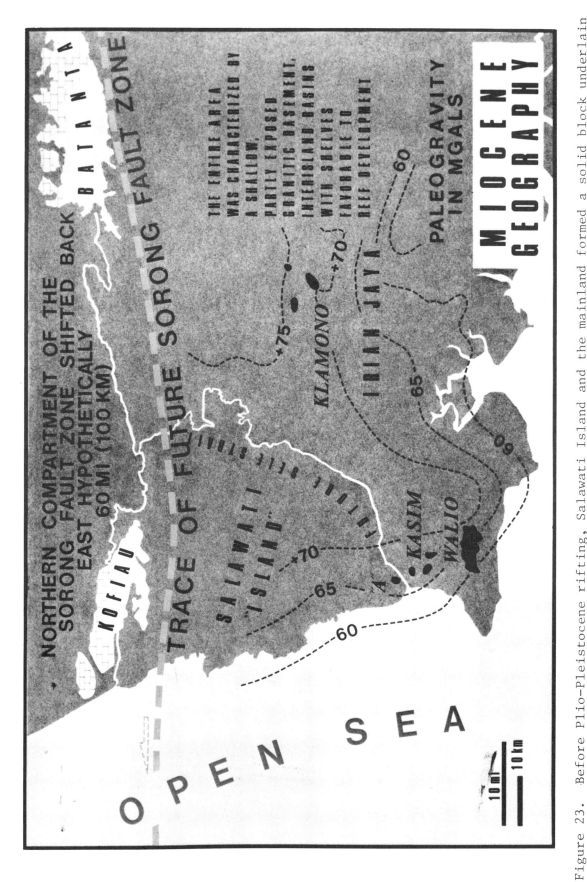

Figure 23. Before Plio-Pleistocene rifting, Salawati Island and the mainland formed a solid block underlain by granitic basement, outlined here by reconstructed gravity contours (compare with Figure 20). The "A" oil field appears to have been located north of Kasim oil field, instead of SW as today (compare with Figure 28), outlining a north-south reef trend facing an open sea to the west. Later effect of left-lateral Sorong fault zone has been removed by sliding Batanta and Kofiau islands back east about 100 km.

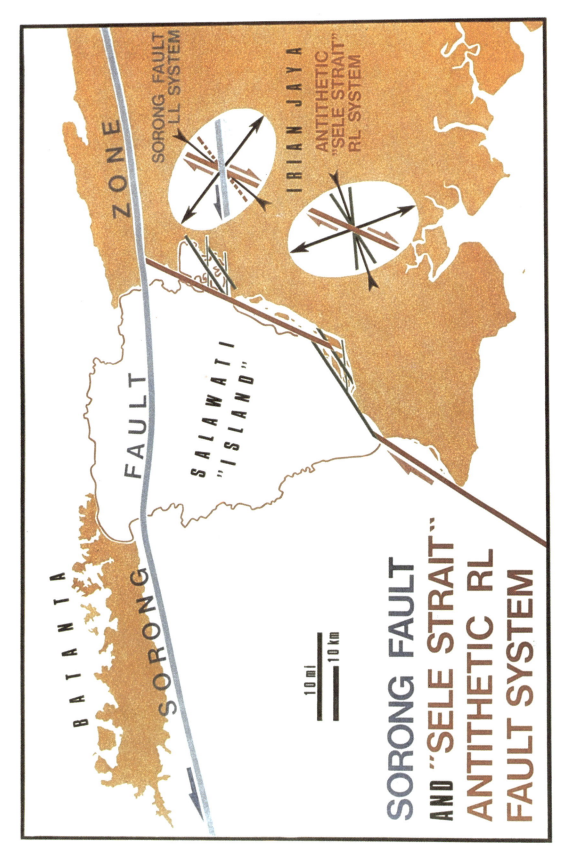

Figure 24. Reconstruction of strain ellipsoid characteristics before rifting based on the examination of radar imagery (see Figures 6, 11, 13).

because the large island of New Guinea was already largely defined during Miocene time. The rising basalt lifted the Miocene carbonate cover now tilted on the north coast of Batanta. The basalt partly digested the limestone, now floating as xenoliths, and filled fractures. Isolated blocks of Kais limestone are perched in the Salawati Mountains. The north Salawati mountain ridge became the source of the Klasaman Formation sediments. The downward decreases in volcanic content and increase in calcareous character of the Klasaman correlate with the progressive erosion of a carbonate-covered basaltic mass. At an early stage, most of the mountain tops probably were still carved in uplifted Miocene limestone.

    4.   Opening of Sele Strait by Westward Rifting of Salawati Island. The basaltic intrusion probably triggered the process already outlined by the direction of extension of the local stress field. The rifting proceeded in two phases: first rotation, then translation.

        a.   Initial 13° Counterclockwise Rotation (Figure 25). The rotation of the Salawati block may have been caused mainly by the wedging effect of the basaltic intrusion from the Sorong fault zone. For that reason, it may have been a rapid event of near-volcanic speed with the magma ripping through the right-lateral fracture system of Sele Strait.

    The presence of a southward-bulging gravimetric high in the northern shallow part of the Sele Strait strongly suggests that the basalt partly invaded the strait (Figure 20). The possibility of a basement wedge rather than an intrusive body as a cause of the gravity anomaly is unlikely, for it would act in opposite direction to the stress field of the general fault system.

    This suggestion is supported further by the presence of a small basaltic dike in a fracture of the Upper Cretaceous limestone, parallel with and near the right-lateral fault zone of eastern Misool. The exposed dike is only a few miles long but it is in line, through intermediate karstic channels, with an islet on the north, which is believed to be the energent top of a drowned Pliocene volcano (Figure 14). The basalt in a nearby well originated during a very early Eocene volcanic event (K-Ar dating) unrelated to the Pliocene intrusion.

    Figure 25 indicates a large bay on the north side of the basaltic ridge at the intersection of the right-lateral Sele fault. This bay is interpreted to be a gap left by the part of the Sorong fault basalt that was flushed into the Sele Strait, as if the basalt first rose along the Sorong fault to a certain height, and then collapsed by leaking into the Sele Strait. The present west coast of the bay is characterized by slabs of Kais limestone steeply tilted east-southeast (Hermes, 1959), as though they fell toward the opening gap. The rotation also caused the basaltic prism to break right-laterally in the area now occupied by the eastern extremity of Batanta Island. At this stage, the Salawati block became detached from Batanta and Irian Jaya and ready to move freely.

    In the initial phase of rifting, the conjugate set of the original right-lateral fault system of the Sele Strait became the site of tensional faulting to accommodate the loss of volume. Horsts and grabens were created in the rift zone. These now are well expressed on radar imagery by the geomorphology of the straight — the islands are horsts separated by drowned grabens. The pivot point of

491

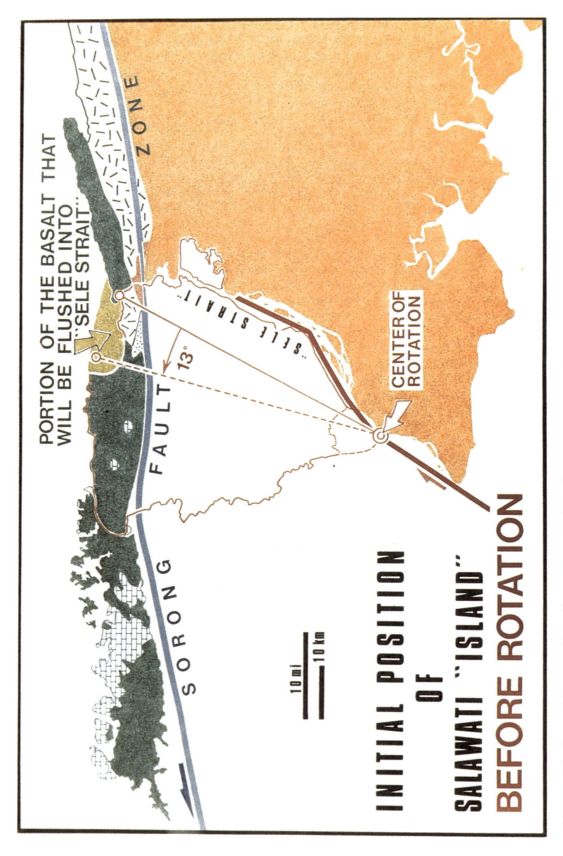

Figure 25. Geometric considerations suggest that the center of rotation of the Salawati block was located near the present Kasim Island (see Figure 28). A gap (pale green) in the northern basaltic ridge suggests that the basalt withdrew when infiltrating the Sele Strait right-lateral fault, initiating Salawati rotation by wedging effect.

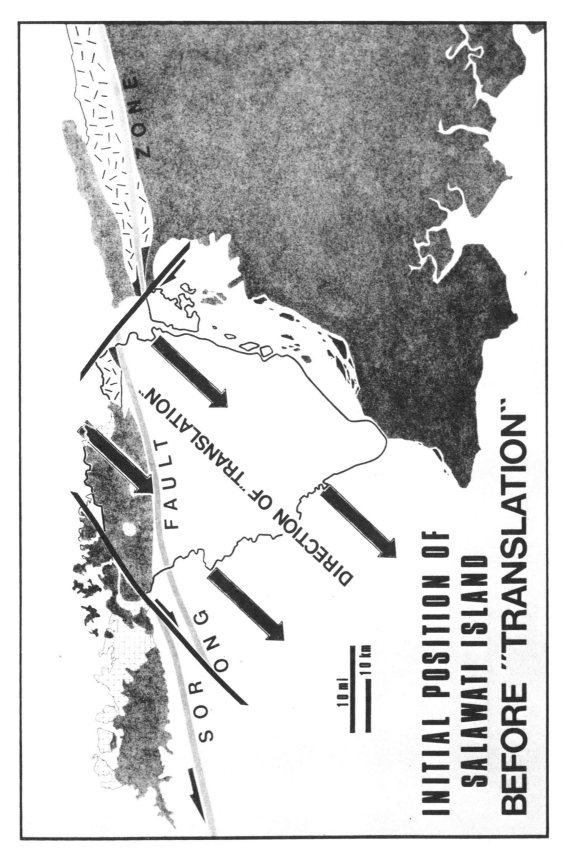

Figure 26. After a 13° CCW rotation, the Salawati block was in a proper position to move freely to the SW.

493

329

Figure 27. Intermediate stage of translation showing the separation of the southern tip of Salawati from the main block, and the creation of Kasim Island as a broken corner of the mainland, at the former center of rotation.

the rotating block was located near the southern tip of Salawati previously located just north of Kasim Island (Figure 25). The density of islands (horsts) is
greater in the southern part than in the northern part of the Sele Strait, as
expected near the center of rotation where less lateral displacement occurred.
The fact that the western end of the Pliocene landmass broke off instead of
being bent supports the interpretation of a generalized crystalline basement in
the area. It is also significant that the area is the extremity of the world's
second largest island.

b. <u>Final 17.5-km Southwestward Translation</u>. After its 13° counterclockwise rotation, the Salawati block was in a favorable position to assume a
southwestward motion (Figure 23). Its east side became aligned with the present
narrow part of the Sele Strait, along which it was able to slip left laterally
under the continuing push exerted on its northeast corner by the progressively
cooling basaltic wedge. When the basaltic wedge had hardened, the block continued its left-lateral slip motion by the slow westward push of the northern
compartment of the Sorong fault. Without the initial rotation, this linear
displacement would have been impeded by the resistance offered by the Sele dome
(Figure 20).

The new left-lateral friction zone developed generally within the original
zone of right-lateral motion of the Misool-Sele fault system. However, the slippage was exerted on two main subparallel fault planes, about 3 km apart, roughly
aligned with the new general southwest direction of motion and extending into
the Sele Strait, where they are expressed by two furrows on the sea bottom
(Figure 21). They were the Salawati main fault and the original right-lateral
fault of Sele Strait. The Salawati main fault is expressed at the surface by
the straight coast northwest of Kasim Island and extends southwest to the
parallel edge of a very shallow bay (Figure 21) bordered by a deep underwater
channel; on the northeast, the fault corresponds to one of the sea-bottom
furrows of Sele Strait. The original right-lateral fault of Sele Strait is
represented at the surface by the narrow and deepest zone (120+ m) of the entire
strait.

The two fault planes defined a sliver of transitional left-lateral motion
that absorbed most of the strain due to the friction of the moving Salawati
block along the fixed mainland. This intermediate narrow block comprises the
southern tip of Salawati Island and neighboring islets. A third minor plane of
friction is located between Kasim Island and the mainland.

The total amount of slippage (17.5 km) is distributed along these three
planes of left-lateral motion in southern Salawati. In the strait itself, most
of the slippage occurred along the extension of the plane of the main Salawati
fault. The first effect of southwest translation was the breakoff of the tip
of Salawati from its former location at the south edge of the shallow bay. The
extremity of the block remained at the center of rotation, which was the main
point of resistance (the least affected by rifting). The slippage started along
the main Salawati fault plane, probably as soon as rotation was carried far
enough to orient the block in a position favorable to lateral displacement. The
south tip later moved left-laterally along the Sele Strait fault, slipping away
from the center of rotation, and finally tearing off a corner of the mainland
block, thus creating Kasim Island. Kasim Island slid along the third minor
strike-slip fault (Figure 27). The amount of relative left-lateral motion
decreased from 8.5 km at the northwest fault (main Salawati fault), to 8 km at

the middle fault, and 1 km at the southeast fault (Kasim Island). The northwest block moved 17.5 km; the middle sliver, 9 km; and Kasim Island, 1 km. Thus it appears that the total left-lateral motion of 17.5 km was distributed on these three fault planes. The mainland remained immobile. No horizontal slippage seems to have taken place along the northernmost major fault of Salawati, which moved with the central block. Figure 28 shows the present position of Salawati.

c. Rate of Separation. The rate of separation after the initial phase of rapid rotation can be estimated. It is known that the motion started in Pliocene time, therefore not earlier than 5.5 m.y. ago, and probably later because it reasonably can be assumed that rifting occurred after the quiet period of carbonate deposition that ended with the development of Lower Pliocene Klasaman reefs. The subsequent thick clastic Pliocene section was derived from the northern basaltic mountains that were formed shortly before the rifting of Sele Strait. Consideration of the Klasman rate of sedimentation (to be discussed later) indirectly suggests that separation started about 2.7 m.y. ago. If it is still active today, the average velocity of southwestward motion is therefore of the order of 7 mm/year.

5. Major Normal Faulting in Salawati Island. A basement structural map indicates a general downdropping of northern compartments along the southwest-northeast fault system during a later phase of deformation. The balance of the amount of upward and downward motion recorded along the faults located between the southern and northernmost wells is comparable to the present difference in elevation between the top of a Miocene reef on the south and the top of a Pliocene reef on the north. Because during their lifetime both reefs were practically at sea level, it can be concluded that the downward motion occurred after the development of the Klasaman reefs, sometime during the Pliocene. The phase of large-scale normal faulting probably was initiated by rapid deposition of the Klasaman Formation, supplied by the high mountain ridge on the north. The marked northward thickening of the Pliocene section indicates that the source was in that area.

There is a possibility that normal faulting started during the phase of southwestward movement of Salawati away from the mainland. However, it could not have been before sufficient erosion of the basalt mountains had supplied the required load for the subsidence of the downfaulted blocks. The rate of sedimentation was fast enough to cause adjustment by normal faulting along the previously outlined zones of shear. The northernmost measured total thickness of Pliocene-Pleistocene section amounts to 3200 m and corresponds to a time span of 5.5 m.y., representing a fast average rate of sedimentation of 56 cm/1000 years. This value is a minimum because the computation includes the lower fifth of the section (up to the top of the Klasaman reef), which corresponds to a time of quiet and slow deposition favorable to reef development. If we assume, for example, that the slow rate was about 2.5 times slower than the later fast rate, the rate of clastic deposition was 93 cm/1000 years. This rate is also a minimum because it is not even measured near the mountain source.

6. Recent Tectonics. The map of earthquake epicenters by Hamilton (1974) shows three zones of activity in western Irian Jaya (Figure 8): (a) on the north, the Sorong fault zone; (b) on the south, the Ceram area, as a possible western extension of the Tarera-Aiduna fault zone; and (c) on the east, the right-lateral fault zone underlying the Lengguru foldbelt.

496

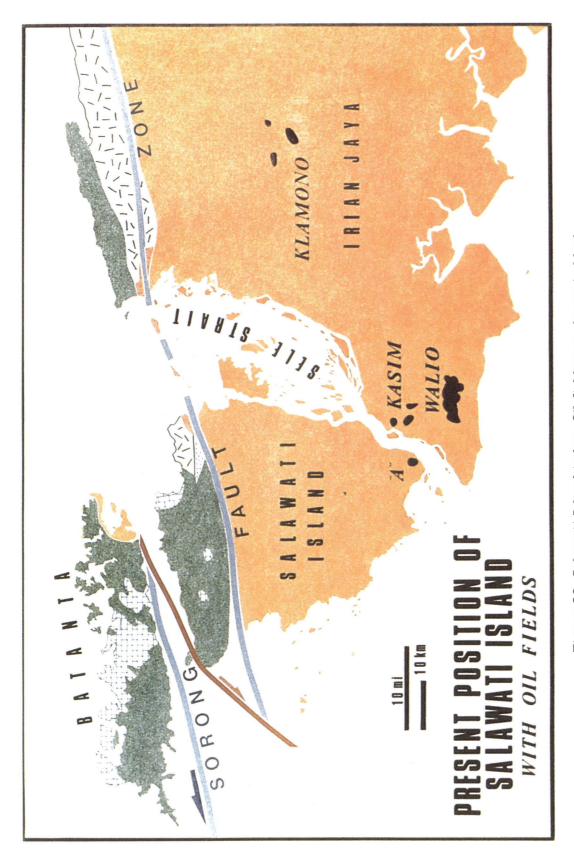

PRESENT POSITION OF
SALAWATI ISLAND
WITH OIL FIELDS

Figure 28. Salawati Island today. Oil fields are shown in black.

333

The presence of the these zones suggests that the Salawati area is still under the effect of a general east-west left-lateral fault system.

Recent tectonic unrest is evidenced at Misool Island by several geomorphic features (Figure 9). An east-southeastward tilting of the island is shown by elevated Pleistocene reefs and meanders encased within Pliocene rocks on the northwest, and drowned karst on the east (Froidevaux, 1975). The southwestward movement of Salawati Island relative to Irian Jaya may be the force behind the tilt. Since the beginning of oil exploration in the Salawati area, at least two earth tremors have been reported by the rig and seismic crews. The island still may be moving westward away from Irian Jaya together with, but separate from, the Pacific oceanic plate.

VII. ECONOMIC IMPLICATIONS

Knowlege of the paleogeography at the time of reservoir and source rocks development, and of the successive diastrophic phases that affected them, is essential to successful hydrocarbon exploration. It is not possible, for proprietary reasons, to expose in detail all the economic implications of the proposed model, but two guiding conclusions can be offered:

A.   Miocene Landscape

Figure 23 is a representation of the Salawati area before the time of Pliocene diastrophism. Salawati Island the Irian Jaya mainland formed one solid block that probably extended northward slightly beyond the present trace of the Sorong fault, toward the latitude of the Waigeo archipelago (Figure 8). The Miocene sea transgressed over the area and the carbonate development was initiated. The landscape was characterized by a shallow granitic basement, partly exposed as low-relief islands with shelves favorable to reef growth and separated by small basins.

Kais reefs are exposed on Batanta Island and appear offshore on the north on correlatable seismic lines. Some are perched in the Salawati Mountains. Bioherms of the same age are exposed in Kofiau Island on the west and in the Waigeo Island group.

The Kais carbonate rocks on the north side of the Sorong fault must be shifted back eastward to restore the Miocene picture properly. This suggests that the edge of the old Miocene Irian Jaya shelf extended farther north than today, and that it was facing an open sea on the west. On the south, the edge of the landmass was located approximately along a line joining Misool Island and the Onin Peninsula.

Once the Salawati block and its southern sliver are placed in their initial Irian Jaya frame, the Miocene position of the reefs shows that the old coastal area of reef development was oriented more nearly north-south, suggesting an open basin on the west and southwest. The reefs on the east side of Sele Strait were in the same position during the Miocene, being part of the fixed land. The bioherms on the intermediate fault sliver were farther northeast, so that they were partly north of, rather than west of, the Kasim field. The reefs of the central part of Salawati shifted even farther, having been at the longitude of the Walio field. They typically are arranged circularly around an old, erosion and/or structural basement high, that was probably one of the last islands to support reefal life in the central Salawati area in early Pliocene time just before the rapid terminal subsidence.

498

B. Prospective Quality of Structural Elements

Understanding the tectonic history of an area that has undergone drastic structural modification provides useful guidelines for the study of hydrocarbon migration and accumulation. Figure 29 illustrates in a simplified manner the distribution of structural units of basically different characters that were acquired during the Tertiary evolution of the Salawati area as a result of the rift and possibly drift event. The changing nature of the faults through time, as unit boundaries acting either as permeability barriers or avenues, can be integrated into a synthesis of hydrocarbon history.

VIII. CONCLUSION

The significant contributions made by the radar images in this plate tectonics study are tied initially to the ability of the radar to define a clear land-water interface. This was extremely important in that it enabled the author to closely and accurately examine shorelines of the islands involved. This initial examination provided a basis for a working hypothesis. Secondly, radar images provided, together with internal features, the geometric elements (linears and patterns) for structural analysis, from the construction of a comprehensive map (2-dimensional) to the reconstruction of a tectonic history (4-dimensional).

The intermediate scale of radar imagery provides the advantageous synoptic view of Landsat, while at the same time eliminating the unnecessary details of aerial photography. The tool suppresses the disadvantages by removing the effect of vegetal and cloud cover.

The combination of synoptic view and sharply defined patterns is particularly suited to detailed 4-dimensional analysis of mobilistic models, such as proposed by plate tectonics.

In our example, the radar geomorphic framework permitted the fitting of a great abundance of scattered, and seemingly unrelated subsurface data, into a coherent geologic story, which is always the key to sound economic exploration.

ACKNOWLEDGMENT

The author is grateful to the American Association of Petroleum Geologists for permission to reprint the parts of this article that previously appeared in its 1978 bulletin, Vol. 62, No. 7. He is also thankful to the management of Phillips Petroleum Company for authorization to publish this information.

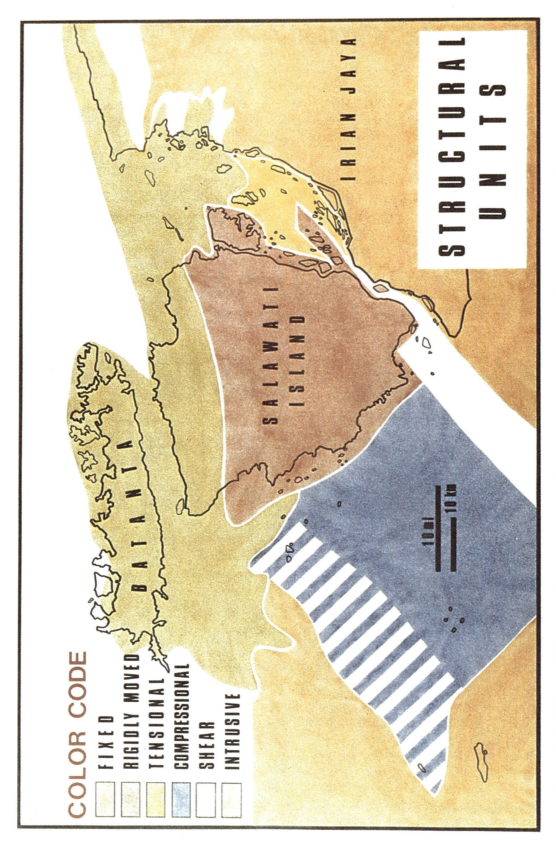

Figure 29. Rifting of Salawati Island created several structural units and boundaries of different and changing permeability characters important in deciphering hydrocarbon history.

REFERENCES

Froidevaux, C. M., 1975, Geology of Misool Island (Irian Jaya): Indonesian Petroleum Assoc. 3rd Ann. Conv., 1979, Proc., No. 3, pp. 189-196.

_____1978, Tertiary Tectonic History of Salawati Area, Irian Jaya, Indonesia: AAPG Bull., Vol. 62, No. 7, pp. 1127-1150, 14 Figs.

Hamilton, W., 1973, Tectonics of the Indonesian Region, in Regional Conference on the Geology of Southeast Asia, Proc.: Geol. Soc. Malaysia Bull., No. 6, p. 3-10.

_____1974, Earthquake Map of the Indonesian Region: U. S. Geol. Survey Misc. Inv. Ser. I-875-C, seismic map, scale 1:5,000,000.

Hermes, J. J., 1959, Geology of the Radja Ampat Islands and North Coast of the West Vogelkop: Nederlansche Nieuw Guinee Petroleum Maatschappij (NNGPM) Geol. Rept. 477.

Redmond, J. L,, and R. P. Koesoemadinata, 1977, Walio Oil Field and the Miocene Carbonates of Salawati Basin, Irian Jaya, Indonesia: Indonesian Petroleum Assoc. 5th Ann. Conv., 1976, Proc., Vol. 1, pp. 41-57.

Tjia, H. D., 1973, Displacement Patterns of Strike-Slip Faults in Malaysia-Indonesia-Philippines: Geol. en Mijnbouw, Vol. 52, pp. 21-30.

Vincelette, R. R., and R. A. Soeparjadi, 1976, Oil-Bearing Reefs in Salawati Basin of Irian Jaya, Indonesia: AAPG Bull., Vol. 60. pp. 1448-1462.

Visser, W. A., and J. J. Hermes, 1962, Geological Results of the Exploration for Oil in Netherlands New Guinea: Koninkl. Nederlands Geol. Mijnbouwkundig, Genoot. Seol. Ser., Vol. 20, 265 p.

THE GABON BASIN: ITS REGIONAL SETTING WITH

RESPECT TO ONSHORE BASEMENT TECTONIC ELEMENTS AS

INTERPRETED FROM SIDE-LOOKING AIRBORNE RADAR IMAGERY

Robert R. Herner

MARS Associates, Inc.
Phoenix, Arizona

ABSTRACT

Side-looking airborne radar imagery
has been acquired of the entire country of
Gabon, West Africa. This high-resolution
data provides geologists with a synoptic
view by which to assess the regional geology
of the country. Major structures, viewed in
the Precambrian basement terrain, are
similar in trend to those reported lying in
the subsurface of the hydrocarbon-bearing
Gabon Basin. Geomorphic phenomena, such as
the antecedent Ogooué River, incised into a
vast relic peneplane surface, are also
related to the evolution of the basin.

A preliminary interpretation of the
radar data postulates a gross relationship
between basement structures trending NNW and
NE, deltaic-shallow marine sediments
deposited at the mouth of the ancestral
Ogooué River, and the occurrence of
hydrocarbons. This relationship may
possibly be duplicated to the south in the
Congo/Cabinda and Cuanza basins, both of
which lie proximal to major west-flowing
drainage systems.

1. INTRODUCTION

Under the supervision of the Service Géologique du Gabon, an extensive
program was undertaken in 1971 to assess the distribution and reserves of the
nation's primary non-renewable natural resources (petroleum, uranium,
manganese), and to locate and evaluate ores significant in quantity to be of
potential economic value. Accordingly, a subsequent program followed in 1975
emphasizing the incorporation of field reconnaissance with remotely-sensed
data. A priority, within the latter study, was to compile detailed geologic
maps of the region between Libreville and Franceville that border the path of
the Transgabonese railway, excluding the well-studied coastal plain portion.
Remotely-sensed data were sparse during this study, and useable, cloud-free,
Landsat imagery had not been acquired. Moreover, available aerial photographs
often lacked sharpness and clarity due to the attenuating effects of the
dense, humid, tropical atmosphere.

---

*Presented at the Fourth Thematic Conference: "Remote Sensing for Exploration
Geology", San Francisco, California, April 1-4, 1985.

31

In order to alleviate the paucity of available high quality remotely-sensed data, and to further probe the country's remote regions heretofore relatively unexplored, a new natural resources exploration and development program was established during the late 1970's under the auspices of the Ministry of Mines and Hydrocarbons. Research proposed under this new program included (1) additional field geologic mapping (including stream sediment and heavy mineral geochemistry of selected regions), (2) radiometric and magnetic surveys of the entire country (excluding the coastal plain), and (3) a side-looking airborne radar survey of the entire country (267,000 km2).

The radar survey was of primary importance since cloud-free images of large areas could be acquired rapidly during either daytime or nightime hours regardless of weather conditions. In addition, the radar mosaics could be used as base maps for field studies. Furthermore, 1:50,000-scale enlargements of the original radar mosaics would be beneficial to the coordination and navigational planning of the pending airborne geophysical survey.

Radar Acquisition

The radar survey of Gabon was flown during December 1981 by Aeroservice Corporation employing a Goodyear GEMS 1000, X-band, synthetic aperture radar system. All flight lines were flown in either a north or south direction with a sidelap of approximately fifteen percent and a consistent east-look (illumination) direction. Depression angles were thirty-two degrees for the near range portion of the image strips and ten degrees for the far range portion. Reflected electromagnetic radiation was received through the system's antenna and recorded on film in the form of a photographic hologram. These data were then processed through an optical correlator and reproduced onto photographic film as image strips. The imagery was acquired at a scale of 1:400,000 and subsequently enlarged to a scale of 1:200,000 for mosaicking. The resultant mosaics correspond to the existing 1:200,000-scale topographic map series of Gabon. There are thirty-two 1:200,000-scale mosaic sheets, and one, of the entire country, at a scale of 1:1,000,000 (Fig. 1).

Radar Interpretation Techniques

The geologic and geomorphic interpretation of radar imagery is, in many respects, similar to photogeologic interpretation; however, special considerations must be made which pertain to the properties of reflected microwave radiation, the peculiarities of the "radar shadow", and artifacts created during the mosaicking process. Scene elements also must be properly assessed with respect to their shape, pattern, tone, and image distortion, such as layover. Therefore, it is essential that photogeologic interpretation techniques be modified when applied to radar imagery.

The interpretation of the radar imagery was performed using both image strip-prints and positive transparent mosaics. The use of the strip-prints was essential in spite of the fact that all the data were registered to the mosaics, because the radar strip-prints are second-generation contact prints from original negatives, whereas the mosaics are fourth-generation, cosmetically improved, photographic products. Moreover, features that are artifacts of the mosaicking process, such as match-lines, will not be erroneously interpreted if the strip-prints are used for interpreting.

Certain optical-mechanical devices were used as aids when interpreting the positive-transparent radar mosaics. Light tables equipped with variable-intensity rheostats were used, often in conjunction with color-density filters (translucent color film), to provide greater image contrast.

32

Figure 1.   Radar mosaic of Gabon, West Africa.

33

2. GEOLOGICAL SUMMARY OF THE GABON BASIN

The Gabon Basin lies along the offshore and coastal plain portion of western Gabon and is composed of a thick sedimentary sequence that attains a maximum thickness of 16,000 to 18,000 meters. Sedimentation into the basin originated during late Jurassic or early Cretaceous time in concert with the breaking apart of the African and South American continents, and was structurally controlled by a system of three major NNW-trending hinge zones that are characteristic of horst and graben tectonics (Fig. 2). Hinge zones 1 and 2, to the east, are faults, whereas zone 3, to the west, is a flexure, the Atlantic hinge belt (Brink, 1974). An eastern and western sub-basin were formed as a consequence of a shift in the position of the hinge zones. The depocenter gradually migrated to the west toward the relatively stationary Anguille high. A salt layer deposited during late Aptian time separates primarily continental facies (typical of the eastern sub-basin) from predominantly marine facies (more common to the west).

The stratigraphic sequence that lies above the pre-Mesozoic basement is referred to as the pre-Cocobeach and Cocobeach series, named after a small coastal village in northwestern Gabon. Sediments in this series attain a combined thickness of between 7,000 and 9,000 meters in the N'Toum Graben, in the eastern sub-basin, and in the Atlantic basin, in the western sub-basin. Deposition is related to the degree of activity along the hinge zones where paleobreaks in relief were formed. West of the hinge zones coarse clastics were deposited in alluvial fans, with sediment size decreasing rapidly to the west.

The pre-Cocobeach sediments have been studied by geologists almost exclusively from wells and outcrops in the eastern sub-basin. The lower limit of the series consists of conglomerates, sandstones, and shales which overlie pre-Mesozoic basement. The upper limit is defined as the top of the massive N'Dombo sandstone. This massive sandstone was probably derived from braided-stream deposits and ranges in thickness, south to north, from 150 meters to 400 meters. Its porosity is very high, ranking it among the best potential reservoir rocks in the entire Gabon Basin.

The Cocobeach series is typically divided into lower, middle, and upper units. The lower unit consists of alternating sandstones and shales with occasional intercalations of siltstone and limestone. The environment of deposition for this unit is probably similar to the N'Dombo sandstone, except that finer-grained clastics are more common. The middle Cocobeach, deposited west of hinge zone 1, consists of over 50 percent sandstone which alternates with greenish, brown, gray, and black shales. Much of this unit was probably deposited in an alluvial fan/lacustrine environment where coarser debris settled near the hinge zone and fine-grained sediments were laid down further west. During this time the Lambarene and Gamba horsts were apparently uplifted as evidenced by the remains of a few remnants of middle Cocobeach found capping the Gamba Horst. It is estimated that approximately 1,000 to 2,000 meters of middle Cocobeach sediments were eroded prior to the Gamba transgression. The maximum amount of upper Cocobeach sedimentation occurred west of the Lambaréné Horst along hinge zone 2. Here thicknesses are estimated to range between 1,500 and 2,500 meters. The uppermost section of the upper Cocobeach is the Gamba Formation, marking the initial transgression of the sea into the basin. It was deposited in a belt 40 to 60 kilometers wide, east of hinge zone 2, and unconformably overlies middle, lower, and pre-Cocobeach sediments, and basement (Brink, 1974).

A salt sequence, overlying the Gamba Formation, was deposited in late Aptian time along the western margin of Africa. It extends north beneath the Douala Basin to the Cameroon slope anomaly, and south beneath the Gabon,

34

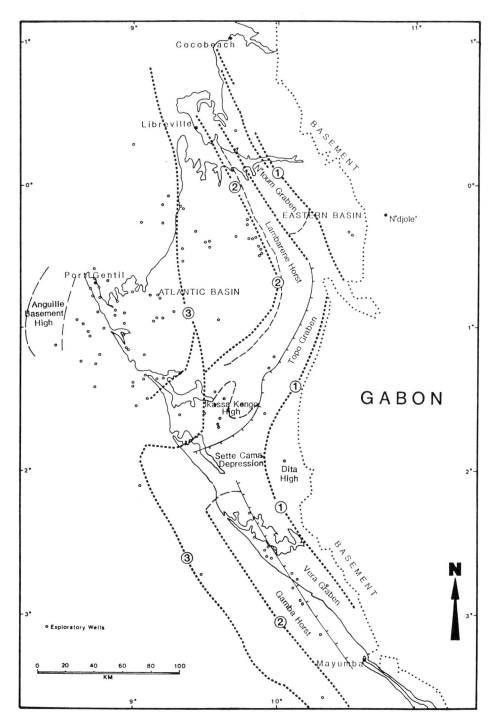

Figure 2. Structural elements of the Gabon Basin (after Brink, 1974).

35

343

Congo/Cabinda, and Cuanza (Angola) basins, a distance of 1,500 kilometers. The width may reach 200 kilometers. This wide distribution of salts indicates that the above basins were interconnected at the time of deposition (de Ruiter, 1979). The average thickness of the salt sequence is approximately 300 meters; however, due to strong halokinesis, thicknesses have reached 900 meters in the Congo Basin (Belmonte et al., 1965). The sea, from which the salts were precipitated, apparently entered the basin from the south and was probably restricted by the Anguille high to the west.

The appearance of the uppermost Aptian-Albian Madiéla Formation signals the beginning of open marine conditions in the Gabon Basin. Its carbonate petrology is very distinct east of hinge zone 3 and is easily correlated on electric logs; however, west of the hinge zone the limestones grade into dolomites and are often anhydritic and typically laden with a high percentage of sand. The depocenter for this formation lies between the Lambaréné Horst and hinge zone 3.

Stratigraphically above the Madiéla Formation are a series of reddish sandstones that are intercalated with varicolored marls and shales. This unit is sometimes referred to in the literature as the "Series Rouges", though the name N'Komi Formation is proposed by Brink (1974). West of hinge zone 3 the N'Komi Formation grades laterally into siltstones and shales and is known as the Cap Lopez Formation.

The laterally equivalent Azilé and Sibang formations, deposited on opposite sides of the Atlantic hinge belt (hinge zone 3) were the next stratigraphic units laid down. The Sibang Formation comprises the near-shore facies and is comprised of carbonates, whereas the Azilé Formation was deposited in a deep marine environment. The deposition of these units was followed by an erosional period, prior to the deposition of the transgressive Pointe Clairette Formation.

The Pointe Clairette Formation is comprised of a transgressive and regressive sequence grouped together as a single unit. Following the basal transgressive phase, the sea almost completely withdrew from the platform east of hinge zone 3; during this time several hundred meters of continental-type sandstones, conglomerates, and shales were deposited. West of the hinge line, shallow-marine sandstones were deposited which grade laterally seaward into shales. Several sedimentation cycles are distinguished on electric logs, separated by unconformities, which mark the closing of Pointe Clairette time.

The composition of Tertiary stratigraphic units in the Gabon Basin is generally transformed from thick sequences of shales, intercalated with thin carbonates, to interbedded shales and sandstones, and finally, sandstones. The Anguille high, once prominent in the western portion of the basin, subsided by Miocene time and no longer influenced deposition. As a result, sedimentation was generally west of the continental shelf. Stratigraphic nomenclature pertaining to the Tertiary units varies; see, for example, Belmonte (1965), Reyre (1966), Brink (1974), and Vidal et al. (1975).

3. PETROLEUM OCCURRENCE

Traces of hydrocarbons have been discovered essentially throughout the entire stratigraphic section of the Gabon Basin, though production is limited primarily to the upper Aptian Gamba Formation (upper Cocobeach) and the Senoian Pointe Clairette Formation (including the Port Gentil and Batanga formations). Apart from the Gamba Formation, no significant hydrocarbon discoveries have been made in the pre-Cocobeach or Cocobeach; however, petroleum has been discovered in Cocobeach-equivalent beds in the Congo/Cabinda Basin (Brink, 1974) and in the Reconcavo Basin in Brazil (Cohen,

1985). A lack of permeability and porosity below a depth of 2,000 to 2,500 meters is reported in the Cocobeach series which renders the rocks to be poor hydrocarbon reservoirs (Brink, 1974). Moreover, many potential reservoir rocks east of the Atlantic hinge belt lack impermeable shale cap rocks.

The majority of hydrocarbons produced in the Gabon Basin are from lithologies post-salt in age, and accumulation is almost always related to structures resultant from salt tectonism. The magnitude of these structures ranges from broad, slightly updomed, pillows, to relatively narrow piercement domes, that may reach several kilometers in height. An increase in the frequency of salt domes exists, from west to east, where gently domed structures were formed beneath great thicknesses of overburden, to steep piercement domes, occurring where overburden is less (Brink, 1974).

Deposition of the lower post-salt formations was influenced by the Atlantic hinge belt. To the west, hydrocarbon source beds were deposited mostly in deep marine environments whereas reservoir beds were typically deposited on the shelf to the east. As with the Cocobeach sediments, accumulations of hydrocarbons may be restricted in the east because reservoir beds often lack a shale cap.

Most of the large producing fields lie west of the Atlantic hinge belt in sediments deposited in estuarine and deltaic environments which were subsequently subjected to salt tectonism. The offshore Grondin field, consisting of either deltaic or shallow marine platform facies sediments, lies above a gently-arched salt structure, for example (Vidal, 1980); however, Seiglie and Baker (1984), on the basis of foraminiferal faunas, do not support the shallow marine platform environment hypothesis.

4. INTERPRETATION RESULTS

Coastal Plain

During the course of the radargeologic interpretation of Gabon, special emphasis was not given to the coastal plain region (50,000 km2), in spite of its hydrocarbon potential. If anything, greater emphasis was given to the interpretation of basement regions, such as the Francevillien, where economically important mineral zones are known to exist.

Upon a cursory glance of the radar imagery, the coastal plain appears to be relatively featureless with respect to the basement terrain. Only pre-Cocobeach sediments, such as the N'Dombo sandstone, that crop out in the eastern sub-basin appear morphologically and structurally distinct (Fig. 3). Throughout most of the coastal plain stratigraphic and structural data usually are discernible only after thorough examination of the imagery.

Stratigraphic differentiation between lithologies in the coastal plain is primarily dependent upon detecting subtle textural variations and/or delineating east-facing erosional scarps, indicative of lithologic contacts. Changes in the drainage density typically reflect lithologic differences. The finely dissected beds of the slightly metamorphosed Noya complex and the overlying, undissected, upper Paleozoic N'Khom and Agoula series serve as an example (Fig. 4). Textural differences are also apparent between the smooth-textured Azilé Formation and the moderately dissected Cap Lopez Formation. The most conspicuous formation, the upper Senonian to Paleocene "Formation des Cirques du Grand Bam-Bam", is best displayed north of the Ogooué River. It is deeply incised, tonally distinct due to a change in vegetation cover, with its upper and lower limits distinguished by prominent east-facing scarps.

37

Figure 3. Faulted beds of the resistant N'Dombo Sandstone
and non-resistant M'Vone Series. Basement rocks of the
Monts de Cristal appear in the upper right-hand corner.

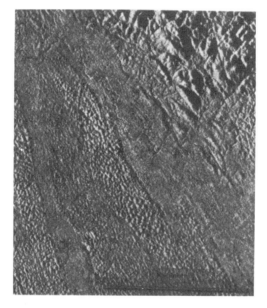

Figure 4. Differential erosion between the Noya complex
(coarse texture) and Agoula and N'Khom series (smooth
texture). In the upper right-hand corner are the Monts
de Cristal.

38

Structurally, the coastal plain is a broad, gentle, west-dipping, cuestaform that is divided by the Lambaréné Horst (Fig . 5, where basement rocks are exposed at the surface, into an eastern sub-basin (10,000 km2) and a western sub-basin (40,000 km2). Though lineaments are readily interpreted throughout, visible offsets, indicative of faulting, are not apparent except for disturbed pre-Cocobeach sediments lying near the northeastern margin (Fig 3). To the south, west of the village of Fougamou, lie a series of structurally controlled features (Fig. 6). These features have been penetrated in drill holes but, at present, the results are not available in the literature. Finally, among the most apparent structures in the coastal plain are numerous mound-shaped salt domes. These are most common north of the Ogooué River, and west of the Lambaréné Horst (Fig. 7). Often the structures have collapsed due to solution and appear as circular lakes.

Basement Terrain

Numerous structural features are readily interpreted on the radar imagery in the basement terrain adjacent to the coastal plain. Foremost among these structures are the NNW-trending Nyanga Syncline and L'Ikoye Fault. Their NNW trend is essentially the same as the hinge zones which lie buried beneath the coastal plain. Additional structures, including longitudinal shears, conjugate shears, and east-vergent thrust faults have also been interpreted. These interpretations suggest that the country's western region is structurally more complex than presently depicted on published geologic maps.

East-vergent thrust faults are interpreted along both the eastern perimeter of the Ogooué terrain, where metasediments of the Ogooué series are truncated against the western edge of the Booué Basin (Herner, 1984), and due west of the Nyanga Syncline (Fig. 8). Evidence for thrust faulting in the eastern Ogooué terrain has recently been discovered in the field (P. Molina, personal communication), whereas the thrust faults interpreted west of the Nyanga Syncline are shown on unpublished geologic maps compiled by J.P. Bassot (personal communication).

The major NNW-trending faults are here interpreted as having originated as lateral shears, based on the configuration of the Ndjolé Basin and from the geometrical alignment of conjugate faults (Fig. 9). Secondary, prominent, NE-SW-trending faults are also interpreted in Gabon, particularly to the north.

The tectonic setting of equatorial West Africa changes abruptly in Equatorial Guinea, where the NNW structural trend, dominant in Gabon, is truncated by the NE-SW trend (Fig. 10). The latter trend parallels the Cameroon slope anomaly and the Cameroon volcanics line. The most paramount feature among this NE-SW trend is a large graben, interpreted as reactivated along a former basement trend, possibly in conjunction with the Cameroon volcanics event. This trend corresponds to the northern limit of the Gabon Basin.

The southern limit of the Gabon Basin is formed by the NE-SW-trending Mayumba spur. This feature, delineated primarily by drill-hole and geophysical data (Brink, 1974), is not surfically expressed on the radar imagery of Gabon. However, it is possible that there may be surficial expression of this trend in the Republic of Congo, where radar imagery is not available. A fragmentation of the Nyanga Syncline toward the southern border may bear some relationship with this structure, but this cannot be substantiated by radar evidence.

An additional NE-SW-trending fault, considered here to be of major significance, is interpreted in the vicinity of the village of Fougamou (herein referred to as the Fougamou Fault). Northeasterly, from the coastal

39

Figure 5. The NNW-trending Lambaréné Horst.

Figure 6. NNW-trending structures west of the Nyanga Syncline, in the coastal plain.

40

Figure 7. Salt domes in the coastal plain.

Figure 8. NNW-trending folds truncated by a
N-S-trending east-vergent thrust fault.

41

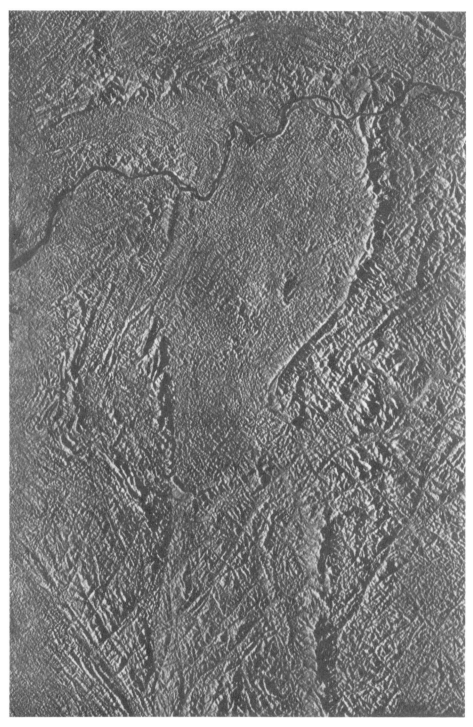

Figure 9. The Ndjolé Basin. Note the L'Ikoye
Fault in the lower left-hand corner.

42

Figure 10. Radar mosaic of western Equatorial Guinea.

43

plain, it cuts a deep linear slice through the Massif de Koumounabwai, dextrally offsets the Ngounié River, continuing until it is finally truncated by the L'Ikoye Fault (Fig. 11). When extrapolated into the coastal plain, this fault lines up with the offset Ikassa Kongo High and the Sette Cama Depression.

5. CONCLUSIONS

Numerous faults and lineaments, developed in the basement terrain, have been interpreted from the radar imagery of Gabon and assessed with regard to the Gabon Basin. The dominant structural trend is NNW, becoming more diffuse west to east. Additional subordinate NNE and NE trends are also interpreted as well as east-vergent thrust faults.

By analyzing the structural configuration of the Ndjole Basin and the azimuth of conjugate faults lying within the Monts de Cristal region, the sense of motion along the dominant NNW trend is determined to be sinistral. Though the absolute age of displacement cannot be accurately discerned from the radar imagery, it is apparent, from cross-cutting fault and fold relationships, that movement began no earlier than latest Precambrian.

There is strong correlation between the basement hinge zones, which controlled sedimentation into the Gabon Basin, and the NNW-trending faults interpreted from radar. The hinge zones are hypothesized as being basement faults that were tensionally reactivated during the breakup of the African and South American continents. At the same time, the NE trend was probably reactivated as lateral (transform?) faults with a vertical component. This system of NE-trending faults appears to be typified by the Fougamou Fault when extended from the basement into the coastal plain. The Fougamou Fault was also apparently influenced by the course of the ancestral Ogooué River. It is here suggested that the Cocobeach series was deposited by a series of pre-Ogooué rivers possibly time-equivalent to the Kretacic or Moorland planation cycles of King (1983). The deposition of the late Albian salt, and to a lesser degree the Madiéla Limestone, was prior to a period of renewed uplift during which time the course of the Ogooué River became entrenched. Water gaps, cut by the Ogooué River through very resistant quartzites near Booué and through the Ndjolé Basin, are interpreted as antecedent.

In summary, the broad tectonic framework of western Gabon is interpretable from basement terrain features viewed on radar images. These features represent geologic structures which were reactivted as a consequence of the breakup of the African and South American continents. Furthermore, they provide important clues which are helpful in modelling the structural nature and genesis of the Gabon Basin. From the radar data it is concluded that:

(1) The major structural trend in western Gabon, exemplified by the Nyanga Syncline and L'Ikoye Fault, is north-northwest.

(2) There is a second prominent trend, exemplified by the fault trends in Equatorial Guinea, the Fougamou Fault, and the Mayumba spur, to the northeast.

(3) The major component of the original faults was lateral.

(4) The NNW trend was tensionally reactivated during the breakup of the African and South American continents. The hinge zones in the Gabon Basin are such faults.

44

Figure 11. The NE-trending Fougamou Fault. Note the right-lateral offset of the Ngounié River.

45

(5)   The crust flooring the Gabon Basin may have been slightly attenuated by pull-apart action along the NNW-trending lateral faults.

(6)   The latitudinal offset along the Fougamou Fault affected the depositional setting between the northern and southern portions of the Gabon Basin.

(7)   Vast peneplanation surfaces were likely formed during pre-Cocobeach/Cocobeach time.

(8)   Renewed uplift in eastern Gabon caused the rejuvenation and incisement of the Ogooué River.

(9)   Sediments derived from the Ogooué River were originally deposited in the northern portion of the Gabon Basin (north of the Fougamou Fault).

(10) The structural framework of the (northern) Gabon Basin and the relatively fixed course of the paleo-Ogooué River are parameters of equal importance as parameters conditional to hydrocarbon occurrence.

(11) The Congo/Cabinda and Cuanza Basins, lying to the south of Gabon, may have had an evolutionary history similar to the Gabon Basin, as each is fed by a large west-flowing river, the Zaire and Cuanza, respectively.

The radar imagery of Gabon contains a considerable amount of data. Its greatest benefit is the synoptic view provided, by which geologists may assess broad regional structural relationships. From this source of data, new ideas and hypotheses have been formulated in accordance with available, current literature. Continued assessment of these hypotheses, with respect to new and proprietary geological and geophysical data, will serve to further clarify the relationship between the sedimentological/structural evolution of a basin and hydrocarbon occurrence.

REFERENCES CITED

Belmonte, Y., P. Hirtz, and R. Wenger, 1965, The salt basins of the Gabon and Congo (Brazzaville), a tentative paleographic interpretation, in Salt basins around Africa: London, Inst. Petroleum, p. 55-74.

Brink, A.H., 1974, Petroleum Geology of Gabon Basin: AAPG Bulletin, v. 58, p. 216-235.

Cohen, C.R., 1985, Role of Rejuvenation in Hydrocarbon Accumulation and Structural Evolution of Reconcavo Basin, Northeastern Brazil: AAPG Bulletin, v. 69, p. 65-76.

De Ruiter, P.A.C., 1979, The Gabon and Congo Basins Salt Deposits: Economic Geology, v. 74, p. 419-431.

Herner, R.R., 1984, Geologic Interpretation of Radar Imagery of Gabon, West Africa, in Technical Papers, 1984 ASP-ACSM Fall Convention, p. 644-654.

46

Reyre, D., Y. Belmonte, F. Derumaux, and R. Wenger, 1966, Evolution geologique du bassin gabonais, in Bassins sedimentaires du littoral african; lere partie: littoral atlantique; Association des services geologiques africains, p. 171-191.

Seiglie, G.A., and M.B. Baker, 1984, Relative Sea-Level Changes During the Middle and Late Cretaceous from Zaire to Cameroon (Central West Africa), in J.S. Schlee, ed., Interregional Unconformities and Hydrocarbon Accumulation: AAPG Memoir 36, p. 81-88.

Vidal, J., 1980, Geology of Grondin Field, in M.T. Halbouty, ed., Giant Oil and Gas Fields of the Decade: 1968-1978: AAPG Memoir 30, p. 577-590.

Vidal, J., R. Joyes, and J. Van Veen, 1975, L'exploration petroliere au Gabon et au Congo, 9th World Petroleum Conference, Tokyo 1975, v. 3, p. 149-165.

47

Reprinted by permission of the International Institute
for Aerospace Survey and Earth Sciences (ITC) from
the *ITC Journal*, 1983-3, p. 223-231.

Spaceborne imaging radars, present and future

Bas N Koopmans*

ABSTRACT

The evolution of spaceborne imaging radar is described, from
the first Seasat (1978) to present-day SIR-A. SIR-A (1981), with
a 43° incidence angle, provides a much better configuration than
Seasat for land surveying. Comparative studies show that SIR-A
imagery is better able to show terrain morphologic features, as well
as lineament patterns and faults. The influence of surface roughness
on the radar backscatter makes terrain differentiation possible. Its
penetration capacity of loose sand cover in hyperarid areas, however, has provided the most exciting results in the SIR-A programme. SIR-B, SIR-C and programmes of various other space agencies
are also described.

During the late 1960s and the first half of the 1970s,
large parts of the world were covered by airborne side-
looking radar surveys. Most of these surveys were carried
out over Third World countries where the complete lack of
any type of map severely hampered development progress.
The "Radam" project in Brazil is the best known example
and the largest in areal extent (4.5 million km²). The
survey was later expanded to cover the entire Brazilian
territory of 8.5 million km². With other radar surveys over
the Colombian, Ecuadorian, Peruvian and Bolivian Amazon areas, the largest cartographic blank in the world (the
South American Amazon area) was coloured in–thanks to
the fast method of data acquisition (independent of weather conditions) and the synoptic view provided by radar
mosaics, which permit relatively rapid reconnaissance surveying.

With the launch of Seasat in 1978, spaceborne imaging radar for earth observation "came into its own". The
early highly promising results of the Seasat data caused a
shift of attention in the remote sensing scientific world
towards the microwave field. The SAR 580 campaign of
the European Space Agency during 1981-83 attests to this
change in emphasis.

In a recent paper entitled "Characteristics of the
second generation earth observation satellites", Hempenius *et al* [7] largely ignored the satellite programmes in
advanced stages of planning or even execution which will
carry microwave remote sensing packages onboard for
earth resource surveying, directed to either oceanographic
or land surveying applications. In this article, I would like
to draw *ITC Journal* readers' attention to some of the
spaceborne imaging radar programmes of this decade.

MICROWAVE IMAGING PROGRAMMES

One of the most striking developments in remote

sensing, apart from the French SPOT programme, is the
rapid advancement in spaceborne microwave surveying.
What Landsat and multispectral scanner investigations
did for the 1970s, microwave and imaging radar surveying
are doing for the 1980s. A growing stream of radar publications is emerging and several imaging radar satellite programmes are being executed or are in advanced stages of
planning (SIR-A, MRSE, SIR-B, SIR-C, ERS-1 and -2
(European), ERS (Japan), SIR- D, Radarsat). With this
development, there is also a partial shift of interest to
surveying the other 74 percent of the earth (see Mekel's
address in this issue) covered by sea and ice. A number of
radar satellite programmes will be directed to ocean and
sea ice surveying, surveillance of the maritime domain, the
study of oceanographic processes and monitoring of coastal dynamics. Information may also be obtained indirectly
about bottom configurations in shallow waters with SAR
and radar altimeters for ocean depth. The dynamic aspects
related to sea state, ice distribution and flow are worth
monitoring, and such information will be of economic
benefit to many.

Land surveying is also receiving full attention with
the shuttle imaging radar programme [8], but much basic
research is still required to understand fully the interaction
between microwave scattering and the different terrain
and vegetation parameters.

SEASAT

Seasat was launched in 1978 in a circular orbit with
an inclination of 108° and an average altitude of 795 km
above the earth. During Seasat's life of 106 days, 3200
minutes of side-looking radar data were collected, providing a wealth of information and a tremendous amount of
research data. Unfortunately, Seasat stopped functioning
at an early stage because of a general power failure. The
payload consisted of a radar altimeter (ALT), a scanning
multichannel microwave radiometer (SMMR), Seasat-A
scatterometer (SASS) and a synthetic aperture radar
(SAR). The imaging radar operated in the L- band (1,275
GHz). The radar look angle was 20° from nadir, with a
variation of only 6° over the 100 km wide swath. According to specifications, the spatial resolution was 25 m. The
extremely large data rate made onboard data storage
infeasible, so areas could be imaged only when the satellite
was within the radius of a receiving station for real-time
transmission. There were three stations in the United
States (Alaska, Florida and California), one in Canada
(Newfoundland) and one in Europe (England). This left
most of the Third World countries and their surrounding
sea areas outside its imaging capability (see Figure 1). The

* Geology Department, ITC

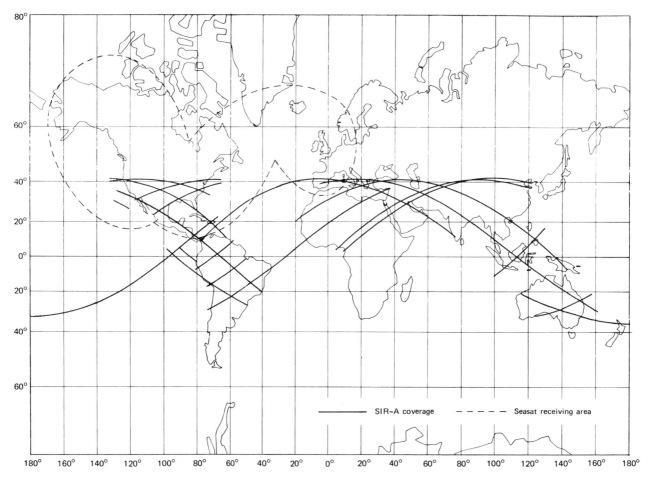

FIGURE 1 Coverage map of SIR-A radar imagery and Seasat receiving area

Seasat SAR raw data were optically as well as digitally correlated. The optical correlated imagery was effected by a number of sources of image degradation, such as tracking errors effecting azimuth resolution, misalignment in the optical correlator, film-drive "jitter" in the correlator and incomplete correction for range migration.

Pravdo *et al* [12], evaluating Seasat performance, described deterioration of spatial resolution for optically correlated imagery ranging up to 50 m in azimuth and 70 m in range. Most of these errors do not occur in the digitally processed imagery. The Interim Digital Processor (JPL's IDP), which processed most available Seasat data in the United States, achieved a resolution of 25 m in azimuth and range, whereas improved algorithms for research purposes have led to an azimuth resolution of 6 m.

The purposes of the different instruments in the Seasat payload were for ocean and coastal studies, as follows:

(1) Radar altimeter (ALT) to measure precise satellite heights above the sea surfacee (+/- 10 cm) for geoid determination and for determination of significant wave heights at sub-satellite points.

(2) Scanning multichannel microwave radiometer (SMMR) to measure ocean surface temperatures and surface wind speeds and to determine ice coverage and characteristics. The radiometer had a swath width of 600 km and a spatial resolution varying between 22 km for the higher frequencies and 100 km for the lower frequencies.

(3) Seasat-A scatterometer (SASS) to determine local wind vectors and to map wind fields on the basis of scattering coefficient measurements over wide ocean tracks. Four fan-shaped beams, each 500 km wide on both sides of the satellite ground track, allowed a spatial resolution in the order of 50 km.

(4) Synthetic aperture radar (SAR) to obtain radar imagery of ocean wave patterns, to study the water/land interaction in littoral zones and to study sea ice characteristics and its dynamic aspects.

Imagery obtained over land also made land application studies possible, although the principal parameters of the system were directed to optimize oceanographic surveying.

Figure 2 shows the surface expression of the bottom topography in a coastal area in the Netherlands (Waddensea). The tidal channels and bottom topography indirectly visible on the Seasat radar imagery are identical to the marine topographic map for the area. Because microwaves cannot penetrate the water surface, it must be the water surface expression (Bragg scattering of small gravity waves, currents influencing wave pattern, etc) which reflects the bottom topography. Sandbanks located at depths of 30 to 40 m in the North Sea are also expressed on

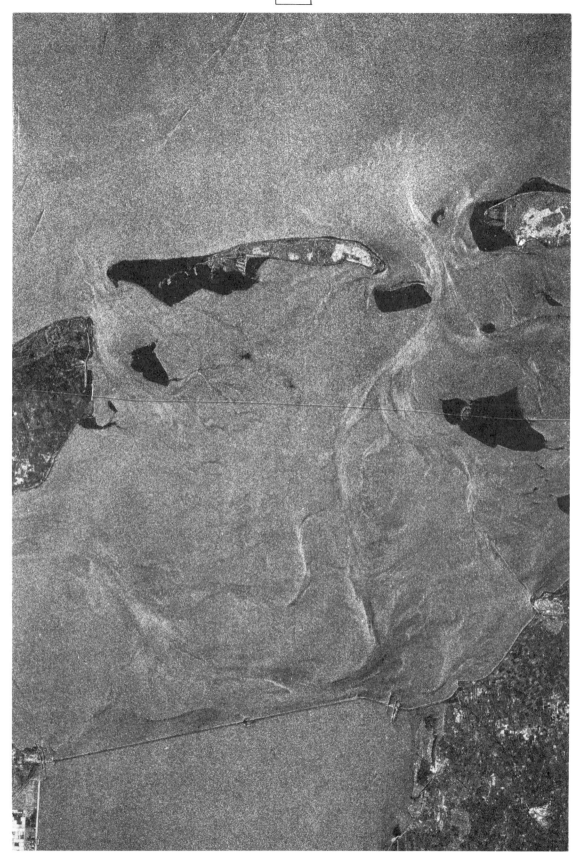

FIGURE 2   Seasat imagery of western Waddenzee, The Netherlands

The island of Vlieland is in the middle part of the picture. Backscatter from the water surface in the tidal area is clearly influenced by bottom topography and the tidal current, which effect the small scale gravity waves. Note the smooth return from the water surface south of the dike closing off the open sea and the wake in the top corner of several fishing boats

225

Seasat imagery. Further research is required to fully resolve the question of the exact origin of the differences in radar backscatter in relation to bottom topography.

The radar altimeter (ALT) has given indirect information on bottom configuration of much larger depth and smaller scale. This concerns the calculation of the geoid over large oceanic tracks. Oceanographic depth differences influence the gravitational field and consequently the average oceanographic water surface topography. The radial orbit precision for Seasat was in the order of +/- 70 cm, whereas the measuring precision for ALT was better than 10 cm [3]. The average oceanographic water surface topography could be measured within the amazing internal precision of approximately 1 m. This allowed Marsh *et al* [11] to construct mean water surface topographic maps with a 1 m contour interval precision. The influence of the bottom topography (small-scale geoidal features), such as ocean trenches, mid-ocean ridges, transform faults and guyots, is expressed on the water surface topography as anomalies in the order of several meters to tenths of meters (see Figure 3).

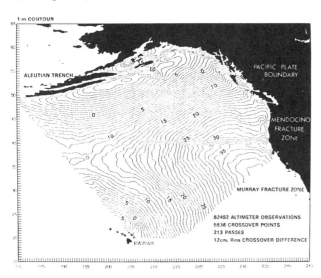

FIGURE 3 The average water surface topography of the northeastern Pacific calculated from Seasat altimeter measurements, after adjustments of track intersections (1 m contours), (from [11])

A recently published book on satellite microwave remote sensing [2] gives the results of the Seasat Users Research Group of Europe (SURGE). It appeared that sea surface conditions were more accurately revealed by Seasat than by weather charts based on ships' reports. The study of wave fields and large internal waves promises a better understanding of ocean surface dynamics and interaction with meteorological phenoma, permitting future sea-state forecasting and procedures for ship routing.

Despite its short functional life, Seasat more than justified its promise. Allan states "In the history of marine science and exploration its brief mission may yet come to be considered as one of the most important events since Challenger left port over a century ago to sail around the world on a three and one-half years voyage and lay the foundations for the systematic study of the sea."

SHUTTLE IMAGING RADAR (SIR-A)

America's shuttle imaging radar programme consists of a series of spaceborne radar experiments making use of the space shuttle. The programme started with the second shuttle mission (STS-2), launched on 12 November 1981, with a remote sensing payload consisting of an L-band synthetic aperture radar of 23.5 cm wavelength and horizontal polarization. During eight hours of sensor activity in the course of the mission, over 10 million km² were imaged (see Figure 1) [6]. The shuttle was launched in a near-circular orbit with a 38° inclination. The look angle of the radar was 47° (+/- 3°) off nadir towards the north so that the northern hemisphere was viewed to slightly higher latitudes (41°) than the southern hemisphere (36°). This orientation of the look direction is a disadvantage for humans who are accustomed to better perception of relief when shadows are oriented towards the observer. This means that to obtain a good impression of relief when viewing SIR-A imagery, it should be observed with south at the top. In the following articles dealing with SIR-A, all imagery is oriented according to the accompanying maps with north at the top. To obtain a better perception of relief, the Journal should be held upside down.

The swath width of the radar strips is approximately 50 km; spatial resolution was specified as 40 m in azimuth as well as range. The signal films were correlated only optically, so that the products were positive films on an approximate scale of 1:500,000.

Dr C Elachi, JPL's principal investigator for the SIR programme, kindly provided four strips of the SIR-A coverage for study. Based on this material, a comparative study programme was set up in ITC in cooperation with a number of ITC alumni. Results of these studies in Argentina, Brazil, Thailand, China, Turkey and Greece are included as individual articles in this issue. In all areas studied, a comparison was made with available Landsat imagery. In the Brazilian area, further comparison was made with airborne SAR and aerial photographs.

One of the great advantages of radar is that it functions independent of weather conditions. This is clearly shown in the tropical rainforest belt (*eg*, Thailand) where no cloudless Landsat frames were available for comparative studies (see Muenlek and Koopmans, this issue, Figure 5).

Most of these authors emphasize radar's enhancement of topographic relief and lineaments–compared with Landsat. Although this is undoubtedly true in general, it should be noted that the flight configuration for SIR-A was far from optimal for relief enhancement. The very steep look angle for Seasat (20° off nadir) resulted in a strong relief distortion. "Layover" and slope foreshortening played a very important role in this (see Figure 4), making structural interpretation a difficult task. (According to the *Manual of Remote Sensing*, Vol 2, layover is defined as the "displacement of the top of an elevated feature with respect to its base on the radar image. The peaks look like dip-slopes".) In high relief areas, deep valleys may have been completely obscured by the layover of the surrounding mountains. On the other hand, the effect of shadow is minimal because few slopes are steeper

FIGURE 4 Seasat imagery, Hodna range, Algeria
Strong layover and foreshortening combined with the adverse orientation of radar look direction with respect to the main structural trend make interpretation difficult

than the radar depression angle. The absence of radar shadow does not help to enhance the relief. The foreslopes, on the other hand, tend to be oriented roughly perpendicular to the incoming beam (slope angles of approximately 20°), causing a very high backscatter of the radar and in this way enhancing the relief. As shown in Figure 3, topographic slopes oriented towards the radar in hilly areas give a high radar return; even the small side slopes of the rill pattern in the flat semi-desert areas also give high responses, enhancing the drainage pattern.

The SIR-A sample in Figure 5A shows a tropical rainforest area in the Brazilian Amazon. With a 47° look angle, scarcely any layover occurs; slope foreshortening is also very much reduced compared with the Seasat configu-

ration. Because the average incidence angle is larger than 40°, not much shadowing develops at the backslope unless there is a sharp relief, nor is there much perpendicular orientation of slope with respect to the incoming radar beam, as occurred with Seasat. This results in neither relief enhancement by shadowing effect nor relief enhancement by high backscatter caused by 0° local incidence angle.

Comparing this image with an X-band airborne radar mosaic of the same area (see Figure 5B) it can be observed that relief is more enhanced by shadows. The radar look angle for the mosaic varies between 60° and 80°, giving a better relief impression than the 47° for the SIR-A. Pedreira (see the following article), comparing X-band airborne radar with SIR-A, concluded that X-band

FIGURE 5A SIR–A imagery (spaceborne) L–band

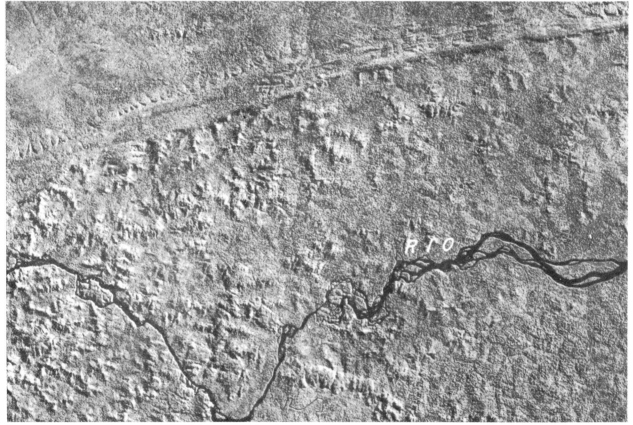

FIGURE 5B SAR radar GEMS 1000 (airborne) X–band

Rio Teles Pires, Mato Grosso, Brazil (approximate coordinates 56° 40' W, 9° 25' S). Northern part consists of sedimentary platform cover of the young Precambrian Benificente Group. The conspicuous strike ridges obliquely cross cut the imagery from the southern extension of the Cachimbo graben where the Benificente Group dips gently northwards. South of this line are the Uatuma Group and the older Xingu complex in which approximately circular granite intrusions are found (Teles Pires granite) which gave radiometric age determinations of 1589 (± 36) million years

228

was better for lineament determination because of its sharpness (better spatial resolution) and shadowing, but that SIR-A was better for dip determination–allowing a better structural interpretation–because of the lack of disturbing shadows. One of the objectives of the future SIR-B experiment is to determine the optimal look angle for interpretation purposes over different terrains under different climatic conditions.

Most of the authors in this issue agree on the use of SIR-A imagery for regional mapping purposes and emphasize the complementary information content of radar with respect to Landsat. Radar gives more information on pattern, texture and small scale surface roughness, whereas Landsat gives more spectral information and detail regarding vegetation. Lithologic differentiation is considered by some as poor with SIR-A; others, such as Pedreira, rate it as good, whereas Muenlek states that for southern Thailand a reasonable subdivision can be made on the basis of surface pattern expression, contrast and drainage pattern. The influence of surface roughness in unvegetated areas on the radar backscatter makes terrain differentiation possible–a facility not attainable with Landsat. SIR-A is considered good for structural information although detail is often difficult to observe. Aerial photographs in most cases give valuable additional detail.

For lineament interpretation and fault analysis, SIR-A gives excellent results, although care should be taken with the directional bias. The radar look direction is found to be strongly under-represented. Orthogonal or oblique look directions would be preferable to minimize this directional dependency.

The most exciting result of the SIR-A programme is the penetration capability of microwaves in loose sand cover in hyperarid areas. Breed *et al* [4] found buried stream channels and bedrock structures in the western desert of Egypt and Sudan under windblown loose sand cover several centimeters to more than a meter thick. This allowed them to map "radar valleys" not evident in the terrain or on Landsat. Figure 6 shows an area in the hyperarid zone of northeastern Mali–the westerly end of the Adrar des Iforas region. The most conspicuous structures are the round Younger Granites with ring complexes, the A Idoenyan in the south and the Timedjelajen complex in the north. An extensive field of dike swarms of acid as well as basic composition are related to these ring complexes, which cut the Lower Proterozoic granite basement. Particularly in the central valley part, many of these dikes are lightly covered by windblown sands. Their appearance on the SIR-A imagery is nevertheless quite clear and sharp and certainly much more detail can be extracted from the radar imagery than from Landsat.

The subject of radar penetration capability still merits further investigation–which is planned for future radar shuttle missions.

FIGURE 6 SIR-A imagery, Adrar des Iforas, Mali

0   5   10 km

Younger Granite ring complexes are conspicuous at the top and bottom of the imagery. A field of dike swarms cross cutting in different directions is clearly visible although they are partly covered by a thin layer of windblown sand (central part)

229

FUTURE SPACEBORNE IMAGING RADAR PROJECTS

THE SHUTTLE MISSIONS

SIR-B is planned for the 17th shuttle mission (STS-17) in August 1984. Like SIR-A, it will be an L-band synthetic aperture radar with horizontal polarization [9] The innovative aspects of the SIR-B shuttle experiment are the use of selectable radar look angles with a possibility of varying between 15 and 60°. This will allow the study of backscatter characteristics of surface materials and vegetation covers under different incidence angles during passes on subsequent days. Further insight into the penetration capability in surface materials (eg, hyperarid areas) related to incidence angles may be obtained. Studies in these fields will allow the determination of the optimal radar look configuration for certain survey aspects.

The multiple incidence angle configuration will also permit acquisition of stereo radar imagery and study of the optimal combination of angles for stereo image interpretation and radargrammetric measurements. Image correlation will be done in both optical and digital modes.

Radiometric calibration experiments will be carried out over a number of point targets (radar corner reflectors) as well as over extended targets. The Amazon rainforest area has been selected as the extended target because of its homogeneity in composition and texture.

An experiment with the "spotlight" mode configuration, making use of a controlled yaw maneuver of the shuttle during imaging, will allow a target area to remain within the radar beam for a longer period. This will increase the synthetic antenna length and consequently improve the azimuth resolution, allowing multiple look processing without degrading the spatial resolution too much.

Data will be relayed through the TDRS (tracking and data relay satellite) system to ground stations. To keep the shuttle in line of communication with the TDRS to permit as extensive coverage as possible, the shuttle will maneuver through a series of configurations: tail forward (STL); nose forward (STL), 90° roll tail forward and 90° roll nose forward.

SIR-C is planned for 1987. It will address the potential uses and applications of SAR with multiple frequencies and multiple polarizations. The payload will consist of an L- and C-band radar with parallel, cross and circular polarizations. Inclusion with this mission of the German MRSE X-band radar with dual polarization is being considered to expand the frequency scope of the payload [10].

SIR-D is tentatively scheduled for 1990. This mission will concentrate on the possibilities of electronic beam steering.

The German MRSE mission, with an X-band SAR, operating in the 9.6 GHz frequency and dual polarization, will make use of the Spacelab mission planned for this year.

The objectives of all of these missions are very clearly for research purposes.

THE EUROPEAN ERS-1 PROGRAMME

The European Space Agency (ESA) has deliberated for a number of years over the mission objectives of the first European earth observation satellite. Two options were available: a land applications satellite system (LASS) and a coastal and ocean monitoring satellite system (COMSS). Both options have been the subject of a number of preliminary system studies. It was finally decided in 1980 that priority would be given to ocean and ice monitoring. The final authorization for construction was given by the member countries in March 1982. The first earth observation satellite designed, constructed, launched and operated by ESA, under the designation ERS-1 (ESA), will therefore be devoted to the scientific understanding of coastal zone and global ocean processes and to the development and promotion of economic and commercial applications related to a better knowledge of ocean parameters and sea-state conditions [5].

The payload will consist of:

-C-band active microwave instrumentation (AMI), including a wind scatterometer and a synthetic aperture radar (SAR) to be employed in two different modes–as a wave scatterometer and a SAR–primarily for wind field and wave spectrum measurements and all-weather imaging.

-A radar altimeter (RA), primarily for the measurement of significant wave height and wind speed.

-Laser retroreflectors (for tracking).

The SAR (imaging mode) will have a spatial resolution of 100 m x 100 m or 30 m x 30 m, a swath width of 80 km (minimum) and an incidence angle of 23° (mid-track). The wind scatterometer will have a spatial resolution of 50 km² and measure wind speeds of 4 to 24 m per second with an accuracy of 10 percent and a wind direction accuracy of less than 20°. The swath width will be 400 km one side (25° to 55° incident angle) with two polarizations and two antenna beams. The radar altimeter (RA) will measure wave heights from 1 to 20 m altitude with an accuracy of less than 10 cm.

The ERS-1 orbit will be sun-synchronous, circular with a 98.5° inclination, allowing a three day repeat cycle. Orbit altitude will be 777 km. Minor changes of a few kilometres in the orbit altitude will permit major increases in the ground coverage–at the expense of repetitivity. The flexibility in tuning the orbit provides the possibility, for example, of achieving full coverage of Canada and Europe with the all-weather imaging instrument within a few days, or global coverage of the Atlantic Ocean with the wind and wave instruments.

The mission is planned to be launched in December 1987, for a life cycle of two years. It is considered as both an experimental and pre-operational project. A second satellite with an identical payload is planned to follow after the two year life of ERS-1 to give the system an operational status. A radar satellite for land applications is in the planning stage.

OTHER FUTURE RADAR SATELLITES

The National Space Development Agency of Japan (NASDA) is planning a marine observation satellite (MOS-1) with a microwave scatterometer and several radiometers but without an imaging radar. A later oceanographic satellite (MOS-2) will probably include a synthet-

230

ic aperture radar.

Canada is greatly interested in radar surveying, especially with respect to sea ice monitoring and ship routing in their northern waters. Land applications of spaceborne radar are also receiving their full attention. A Radarsat is planned for the 1990s carrying a C-band radar with electric beam steering, which will permit variation of the incidence angles between 20° and 45°. Much emphasis is put on near real-time processing, aiming at image availability a few hours after overflight.

TABLE 1 Characteristics of Spaceborne imaging radars

	Seasat	SIR-A	SIR-B	SIR-C	ERS-1 (ESA)
Launch date	1978	1981	1984	1987	1987
Altitude (km)	795	259	225	200-300	777
Inclination (°)	108	38	57	–	98.5
Radar characteristics:					
Wavelength (m)	0.235	0.235	0.235	0.235/0.057	0.057
Frequency (GHz)	1.275	1.275	1.275	1.275/5.28	5.3
Polarization	HH	HH	HH	HH,HV,VV, VH,RCP,LCP	HH
Incidence angle (°)	23±3	50±3	15-60	15-60	23±3
Swatch width (km)	100	50	35-50	35-120	80 (min)
Spatial resolution (m):					
Azimuth	25	40	25	15	100 or 30
Range	25	40	17-58	30	100 or 30
Signal correlation (O = optical, D = digital)	O+D	O	O+D	D	D
Data recording (G = ground real time, S = onboard)	G only	G+S	G via TDRSS + S	G via TDRSS + S	–

CONCLUSIONS

With microwave research picking up very quickly and airborne as well as spaceborne SAR experiments going on in North America and Europe, knowledge in the field of microwave scattering interaction with natural surfaces will improve considerably in the coming years. The shuttle experiments will provide SAR imagery of test sites all over the world to study backscatter characteristics over many diverse areas under different climatic conditions. This will make the remote sensing science community ready for the semi-operational earth observational radar satellites which we may expect around the end of this decade.

REFERENCES

1 Allan, T D (ed), 1983. Satellite Microwave Remote Sensing. Ellis Howard Ltd, 526 pp.

2 Allen, T D, 1983. A review of Seasat. In: Satellite Microwave Remote Sensing, T D Allen (ed). Ellis Howard Ltd, pp 11–44.

3 Apel, J R, 1983. A survey of some recent scientific results from the Seasat altimeter. In: Satellite Microwave Remote Sensing, T D Allan (ed). Ellis Howard, Ltd, pp 321–326.

4 Breed, C and G Schaber, 1983. Subsurface geology of the western desert in Egypt and Sudan revealed by shuttle imaging radar (SIR-A). Proc Spaceborne Imaging Radar Symposium, JPL Publ 83-11, Pasadena, Calif, USA.

5 European Space Agency (ESA), 1981. Programme proposal for the definition, design, development and exploitation of the ESA remote sensing satellite (ERS system). Remote Sensing Programme Board, Doc ESA/PB-RS (81)12, Paris.

6 Ford, J P, J B Cimino and C Elachi, 1983. Space Shuttle Columbia views the World with Imaging Radar: the SIR-A Experiment. JPL Publ 82-95, Pasadena, Calif, USA.

7 Hempenius, S A, B S Marwaha, A Murialdo and Wang Ren-Xiang, 1983. Characteristics of the second generation earth observation satellites. ITC Journal 1983-1, pp 21–33.

8 Jet Propulsion Laboratory (JPL), 1983. Proceedings of the Spaceborne Imaging Radar Symposium, January 1983. JPL Publ 83-11, Pasadena, Calif, USA.

9 Jet Propulsion Laboratory (JPL), 1982. The SIR-B Science Plan. Imaging Radar Science Working Group, JPL Publ 82-78, Pasadena, Calif, USA.

10 Jet Propulsion Laboratory (JPL), 1983. Shuttle Imaging Radar-C (SIR-C). Executive Summary, JPL Publ 83-47, Pasadena, Calif, USA.

11 Marsh, J G, R E Cheney, T V Martin, J J McCarthy and A C Brenner, 1982. Mean sea surface computations based upon satellite altimeter data. Proc 3rd Int Symp on Use of Artificial Satellites for Geodesy and Geodynamics, Ermioni, Greece.

12 Pravdo, S, B Huneycutt, B Holt and D Held, 1983. Seasat Synthetic Aperture Radar Data User's Manual. JPL Publ 82-90, Pasadena, Calif, USA.

RÉSUMÉ

L'évolution des images radar à partir de l'espace est décrite, du premier Seasat (1978) jusqu'au SIR-A actuel. SIR-A (1981) avec un angle de 43° d'incidence, fournit une bien meilleure configuration que le Seasat pour le levé terrestre. Des études comparatives montrent que l'image SIR-A est plus apte à montrer les détails morphologiques du terrain, aussi bien que les formes des linéaments et des failles. L'influence de l'inégalité de la surface du sol sur la dispersion des ondes radar retour rend possible la différentiation du terrain. Sa capacité de pénétration de la couverture de sable meuble dans les zones hyperarides a fourni en outre les plus intéressants résultats dans le programme SIR-A. SIR-B, SIR-C et les programmes de différentes autres agences spatiales sont également décrits.

RESUMEN

Se describe la evolución de las imágenes de radar originadas en el espacio, desde el primer satelite Seasat (1978) al actual SIR-A. El SIR-A (1981) con un ángulo de incidencia de 43°, aporta una configuración mejor que el Seasat para ejecutar surveys de terreno. Estudios comparativos muestran que la imágen del SIR-A esta mejor capacitada para mostrar los caracteres morfológicos del terreno, al igual que los patrones de rasgos y de fallas. La influencia en los reflejos del radar ("backscatter") de las irregularidades de la superficie hace posible la diferenciación de terrenos. Su capacidad de penetración en cubiertas de arena suelta en áreas hiperáridas, demuestra ser su ventaja más importante en el programa SIR-A. Se discuten también los programas SIR-B, SIR-C y de otras agencias similares.

231

Reprinted by permission of the Geological Society of America from H. C. MacDonald, A. J. Lewis, and R. S. Wing, *Geological Society of America Bulletin*, v. 82, no. 1 (1971), p. 345-357.

H. C. MACDONALD   *Department of Geology, Georgia Southern College, Statesboro, Georgia 30458*

A. J. LEWIS   *Department of Geography and Anthropology, Louisiana State University, Baton Rouge, Louisiana 70803*

R. S. WING   *Princeton Geological Group, Continental Oil Company, Princeton, New Jersey 08540*

Mapping and Landform Analysis of Coastal Regions with Radar

ABSTRACT

Although aerial photographs still constitute an important source of information for the coastal geomorphologist, data acquisition is severely limited in cloud-shrouded coastal environments. Sequential radar mapping of remote coastlines, however, will provide a practical and rapid method of updating maps and monitoring coastal processes. Contrasts in signal return from the land-water boundary produce a striking interface on radar imagery which is advantageous for the delineation and mapping of coastal features. Along the Pacific and Caribbean coasts of eastern Panama and northwestern Colombia, marked discrepancies exist between available maps and the actual configuration as revealed by radar.

In addition to the revision of coastal maps, radar imagery also provides the geomorphologist with most of the information interpretable from aerial photographs. Mud flats, beach ridges, natural levees, backswamp basins, beach features, and zones of breaking surf are examples of the observable features. Mangrove swamps, which abound on tropical shores throughout the world, provide a readily recognizable vegetation type which can be unmistakably interpreted from the imagery. Although offshore sediment transport cannot be directly interpreted, inferred trends can be obtained through sequential radar coverage. The Atrato River Delta of northwestern Colombia is cited as an example.

Radar imagery interpretation has indicated marked landform differences between the Pacific and Caribbean coasts of eastern Panama and northwestern Colombia. The partially submerged mountainous terrain of the southeastern coast is contrasted with the mangrove swamps, beaches, and barrier reefs of east-central Panama.

INTRODUCTION

During the past two decades, aerial photographs have been used increasingly for coastal geomorphic studies. Generally, more detail of the coastal configuration and near-shore features are observable on the photographs than can be recorded in field studies. Sequential aerial photography is an excellent tool for monitoring shoreline changes; however, in some regions of the world, photographic acquisition is extremely limited by daylight or weather conditions.

The use of radar imagery for identifying landforms, interpreting processes, and delimiting physiographic regions is a relatively new tool for the geomorphologist. Radar as a tool for the coastal geomorphologist has several interesting and advantageous characteristics, including the ability to scan a broad band of terrain (40 mi wide) with a single pass, and presenting the imaged area on a continuous strip of film. Continuous strip presentation of a 40-mi band of terrain of several hundreds of miles is of special importance to the coastal geomorphologist who is interested in a long, but relatively narrow, area.

Radar imaging systems have a near all-weather, 24-hr imaging capability, and this advantage is of special importance for coastal studies of areas masked by darkness as in polar regions, or by clouds in tropical environments. Coastal changes, which are greatest during high winds, waves and tides are best monitored during the period of maximum erosion and not after the storm. The ability of certain radar wavelengths to penetrate all but thick precipitating clouds may enable monitoring near maximum erosional conditions and, therefore, should result in a better understanding of the cause and process of coastal erosion.

Except for a few cursory studies, the potential of radar as a tool for coastal geomorphology has not been documented. It was the purpose of this study to define both the capabilities and limitations of radar for coastal geomorphic studies; specifically to determine the coastal landforms that are identifiable.

RADAR PARAMETERS[1]

For those readers not familiar with radar imagery it is important to note, that although the imagery is recorded on photographic film, we have recorded relative differences in signal return from the microwave portion of the electromagnetic spectrum. For example, Figure 1 shows an area being imaged by a typical SLAR system. The transmitter generates short bursts or pulses of radio frequency (RF) energy. These pulses are propagated into space by means of a directional antenna (A), and radiate from the antenna as a block of energy (B) at the velocity of light (3×10^8m/

sec). If a terrain feature capable of intercepting RF energy is irradiated at point (a), a fraction of the transmitted energy will be reradiated back to the antenna (A). An object at (b) will also reradiate energy back at a later time when it is illuminated by the pulse packet of RF energy. The same is true for features at (c) and (d). This portion of energy returned to the antenna from the terrain features (a), (b), (c), and (d) is converted to a video signal by the receiver. The signal return from these features is displaced from the origin as a function of the range from the antenna to the target, whereas the amplitude of each return is a function of the scattering properties of the terrain.

The antenna (A) is repositioned laterally at the velocity of the aircraft (V_a). Each transmitted RF pulse (B) returns signals from the targets within the beamwidth. These target returns are converted to a time/amplitude video signal (C) which is displayed on a cathode-ray tube (CRT) and then recorded as a single line (E) on photographic film (F). Returns from subsequently transmitted pulses are displayed on the CRT at the same position (D) as the previous scan lines. By moving the photographic film past the CRT display line

---

[1] For a more detailed discussion of radar systems operation, the reader is referred to MacDonald (1969, p. 6–19).

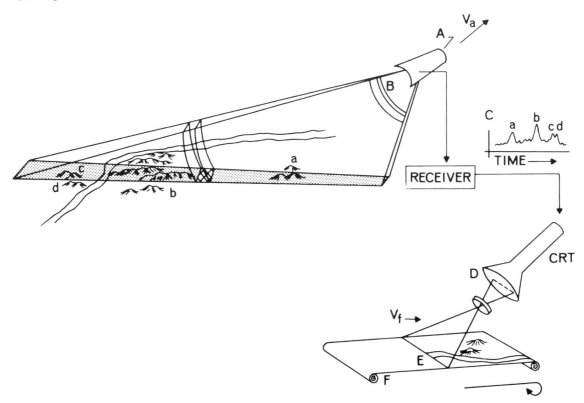

Figure 1. Operational schematic, typical side-looking radar system.

at a velocity (V_f) proportional to the velocity of the aircraft (V_a), an image of the terrain is recorded on the film (F) as a continuous strip.

SENSOR-TERRAIN INTERACTION

Depending on the relative surface roughness[2] of the terrain, the signal return from vegetation, rock, and soil is generally higher than the signal return from water. Thus contrasts in signal return from the land-water boundary produce a striking interface on the radar imagery which is advantageous for the delineation and mapping of coastal features. For example, in Figure 2 along the Texas Gulf Coast, the coastal waters are shown in black (low or no radar return) and the terrain in contrasting gray tones (relatively higher return). It is significant that the radar imagery shown in Figure 2 was taken with an obsolete, low-resolution, side-looking radar system, and even though we are using a relatively crude-resolution system, many coastal features are easily identifiable. For example, the distinctive pattern of a chenier plain can be

---

[2] Surface roughness is not an absolute roughness, but the relative roughness expressed in wavelength units. For a surface roughness much less than the wavelength ($\lambda/10$), the surface appears "smooth" to the imaging radar; for a surface roughness on the order of a wavelength or more, the surface appears "rough."

located south of Sabine Lake. The distinctive asymmetrical shape provides information of past and present shoreline processes. Some of the more obvious features of Figure 2 include: (1) beach ridges or growing lines of a chenier plain, (2) double jetty to contain the river in a narrower channel (to increase its erosional power and maintain a deep channel), (3) old meander scar, and (4) offshore drilling platforms.

Figures 3 and 4 are more representative of the quality of radar imagery that can be expected with the present-day commercial·imaging radar systems. In the vicinity of Seaside, Oregon, several coastal features have been illustrated on Figure 3. The bright offshore feature just south of South Jetty of the Columbia River designated as area (1), is a line of surface convergence probably caused by a combination of wave refraction and longshore current. An additional surface pattern has been defined in area (2). The surf zone is visible in area (3). The relatively bright return region (high signal) shown at area (4) represents vegetated foredunes. Beach ridges landward of the foredune can be delineated in area (5). These beach ridges are essentially parallel to the present shoreline and provide evidence of a prograding shoreline.

Figure 4 of San Diego, California, illustrates the ability of radar to detect moving sea-

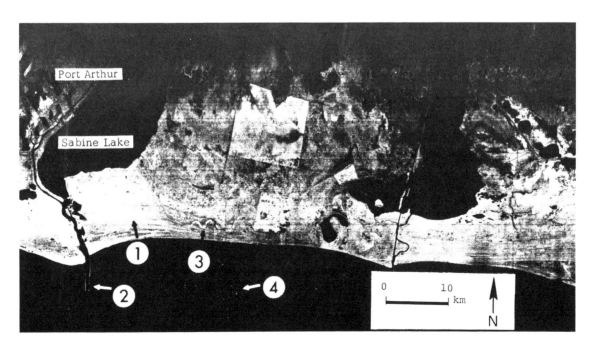

Figure 2.   Radar imagery of the Gulf Coast area near Port Arthur, Texas. Photograph from CRES, University of Kansas.

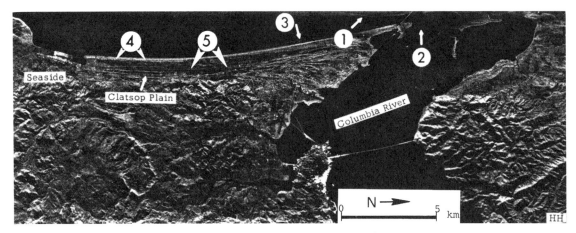

Figure 3. Radar imagery of the Seaside, Oregon area. Photograph courtesy of CRES, University of Kansas.

going vessels and their wakes (1), and kelp beds (2). The separation of sea state is also indicated where the open sea (3), was sufficiently rough to produce enough backscatter to be recorded, whereas in the more sheltered area (4), the water surface was smooth enough that the emitted radar signal was reflected away from the receiver thus lost and a no-return area recorded.

RADAR IMAGERY PRACTICALITY

Obviously, the three previous examples of SLAR imagery were obtained from geographic areas where aerial photography fully satisfies the reconnaissance needs of the coastal geomorphologist, and where radar provides only supplementary information. Consider now the role of an imaging radar in a coastal environment where aerial photography is difficult or impossible to obtain. For example, in east-central and southeastern Panama (Fig. 5),

attempts to obtain complete aerial photographic coverage of this area have been notably unsuccessful for several decades because parts of this region are perpetually covered by clouds. Particularly in Darien Province, the area northwestward from the Panama-Colombia border to north of San Miguel Bay (Fig. 5), gathering topographic data has defeated photographers for several decades.

To exploit radar's capability to penetrate clouds, in 1967 the U.S. Army Topographic Command, Engineer Topographic Laboratories successfully collected radar topographic map data and imagery of southeastern Panama and northwestern Colombia. The data were gathered during six separate days of flight operation, and a portion of the imagery was used to construct a radar mosaic (Fig. 6) encompassing approximately 17,000 sq km. Although not without error, the radar mosaic provides a much more accurate configuration

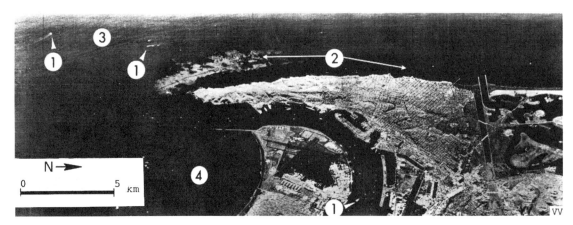

Figure 4. Radar imagery of San Diego, California. Photograph courtesy of CRES, University of Kansas.

of the Darien region than has been previously available and reveals numerous errors incorporated in earlier maps of the area. Figure 7 represents a reduction of the best available topographic map made prior to radar coverage in the area. Some of the more obvious discrepancies between the mapped features and the actual configuration as delineated on the radar include: (1) the meander pattern of the Rio Tuiria (upper center of the map), (2) the position of the Isla Mangle which lies upstream from where it was originally mapped, and (3) the incorrect orientation of the low hills east of the Rio Tuiria.

Because this study is concerned with coastal features, the selection of two areas along the Caribbean are of particular interest. There are several obvious discrepancies revealed on Figures 8A and 8B when the mapped coastline and the radar imagery are contrasted.

These illustrations provide evidence that sequential mapping by radar, and ultimately the production of orthomaps from imagery would result in a practical and rapid method of updating coastal maps. Figure 5 can be used for location of Figures 8A and 8B.

RADAR IMAGERY INTERPRETATION

Tidal flats usually exhibit striking patterns on air photos. On radar imagery the pattern is also unique because of its offshore location, herringbone texture-drainage, high tonal contrasts with the adjacent water, and marked textural contrast with the land. The tidal flats on the Pacific coast of east-central Panama (Fig. 9) are generally nonorganic, fluviomarine accumulations occurring in shoal areas, and like most coastal areas where such features are found they are protected from

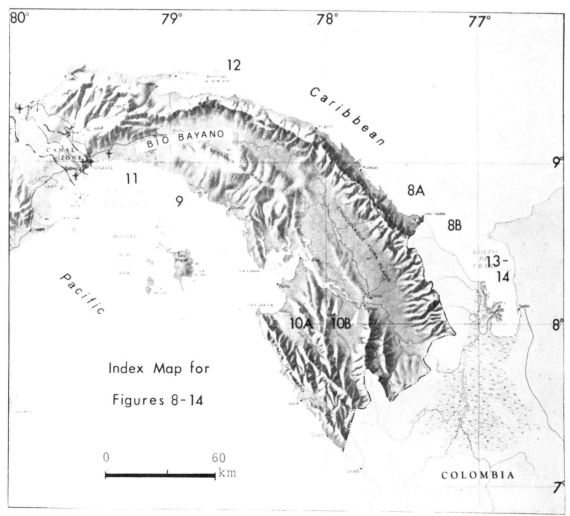

Figure 5. Physiographic sketch map, eastern Panama and northwestern Colombia. Photograph from CRES, University of Kansas.

Figure 6.   Radar mosaic, Darien Province, Panama and northwestern Colombia.

strong wind and current action. Tidal flats, such as those outlined by arrows on Figure 9, are believed to be the first stage in the formation of mangrove swamps and exhibit a slightly undulating surface devoid of vegetation. In this particular location, extensive mangrove swamps can be delineated shoreward of the tidal flats.

Mangrove swamps, which abound on tropical shores throughout the world, are a readily recognizable vegetation type unmistakably identifiable on radar imagery. On Figure 10A white arrows define the boundary between mangrove vegetation to the north and jungle vegetation to the south. Proximity to coastal waters and distinctive tone and texture pattern aid in delineation of the mangrove swamp limits.

The unique shape of estuarine meanders[3] are easily defined from radar imagery, and Figures 10A and 10B provide examples. The stippled areas on the silhouettes (traced from the imagery) of the major rivers outline the zones of estuarine meanders. These diagrams can be contrasted with the actual configuration on the radar imagery. Differing from either river or tidal meanders by their position in the fluvial-marine scheme of the hydrologic system, and by their relationship to a specific family of curves, the recognition of the estuarine meander form allows insight into the

[3] The identification of estuarine meanders is based primarily on planimetric shape. They were first described by Ahnert (1960) as a succession of oblong pools connected by narrow channels at the bends.

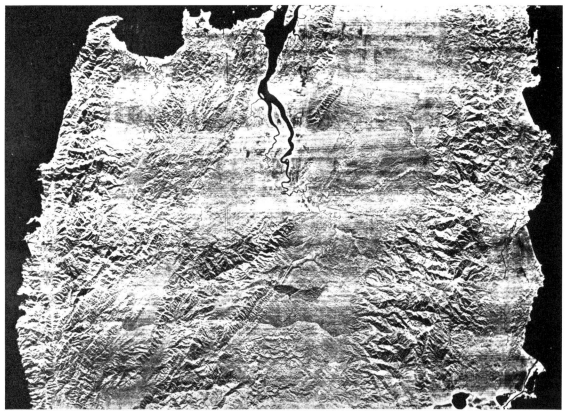

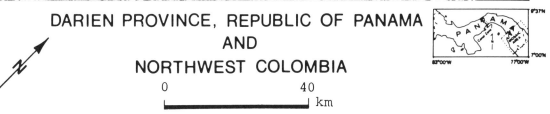

DARIEN PROVINCE, REPUBLIC OF PANAMA
AND
NORTHWEST COLOMBIA

0   40

km

Figure 7. Topographic map, Special Map No. 2, sheet III, Army Map Service, USAR   CARIB, 1:500,000, 1967.

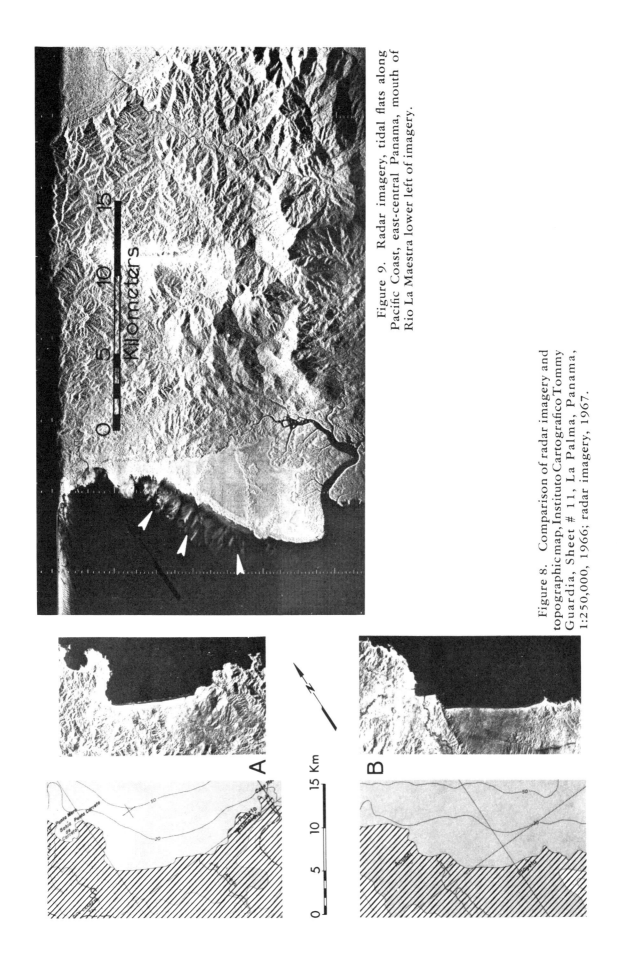

Figure 9. Radar imagery, tidal flats along Pacific Coast, east-central Panama, mouth of Rio La Maestra lower left of imagery.

Figure 8. Comparison of radar imagery and topographic map, Instituto Cartografico Tommy Guardia, Sheet # 11, La Palma, Panama, 1:250,000, 1966; radar imagery, 1967.

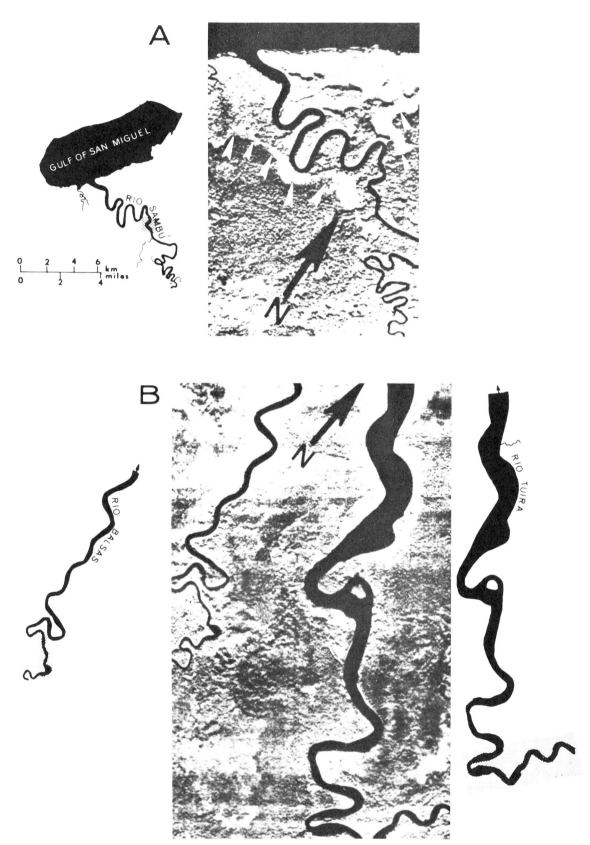

Figure 10. Radar imagery and silhouette diagrams, zones of estuarine meanders within stippled areas. A. White arrows define the boundary between mangrove vegetation to the north and jungle vegetation to the south.

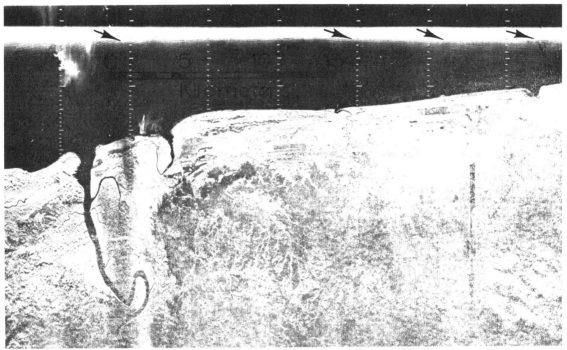

Figure 11. Radar imagery, beach ridges along Pacific coast, east-central Panama, mouth of Rio Bayano left center of imagery.

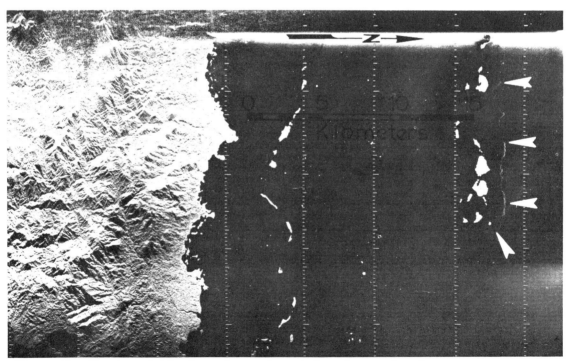

Figure 12. Coral sand islands on leeward slope of barrier reef, surf zone (arrows) outline windward part of reef.

type of coast, channel flow, availability of sediment, and increased knowledge of the balance between marine and fluvial processes. The estuarine meanders illustrated in Figures 10A and 10B had not been previously reported until acquisition of the radar imagery (Lewis and MacDonald, 1970).

Beach ridges provide evidence (Fig. 11) of progradation along the Pacific coast between the Rio Bayano Estuary and the Panama Canal (*see* Fig. 5 for location). Back swamp drainage is well defined north of the beach ridges; to the east the outline of a mangrove swamp is equally apparent. A wave refraction pattern can be delineated (black arrows) along the upper margin of the imagery.

On the Atlantic coast are the San Blas Islands (Fig. 12) which consist of coral sand that has collected on the leeward slopes of barrier reefs. The outline of the windward part of the reef is delineated by the surf zone (arrows, Fig. 12). Landward from the San Blas Islands, coral flats fringe the coast, but this cannot be determined from imagery interpretation; however, the occurrence of these dense clusters of dead corals at the present sea level was noted in a study by Tuan (1960, p. 24).

The delta of the Atrato River is located on the southwestern side of the Gulf of Uraba in northwestern Colombia (Fig. 5). The delta forms a coastal lowland belt bounded on three sides by mountains, thus the Atrato River receives large quantities of water from the mountain runoff. Vann (1959) has revealed that the Atrato delta exhibits physical features which permit recognition of recent changes by analyzing landform and vegetation. Landforms in the delta consist of mud flats, natural levees, backswamp basins, beach features, and round lakes.

Figure 13 represents a summary of Vann's investigation of the Atrato delta. Figure 14 provides a comparison between the 1954 coastal conditions constructed by Vann in the field and from aerial photographs, and the coastal configuration as recorded by radar imagery some 13 yrs later. Since 1954, the northern spit (area 1, Fig. 14) has been either eliminated by wave action or has been breached, then later connected from the south by a spitlike tombolo (2). Encroaching vegetation from the natural levee (3) in the main channel reflects a reduction in stream flow. The largest lake in the region (4) is now being reduced in area because of sedimentation, and

Figure 13. Atrato Delta coastline conditions, 1954.

appears to reflect the initial stages in the formation of an inland swamp. Contrasting Vann's study with that of the radar imagery suggests that Vann's projection of coastal retreat was correct.

COASTAL LANDFORM ANALYSIS

Interpretation of single strip radar imagery (Figs. 8, 9, 10, 11, 12, and 14) and the radar mosaic (Fig. 6) of the eastern half of Panama provides evidence of marked landform differences between the Pacific and Caribbean coasts. However, the present shorelines on both sides of the eastern Panamanian Isthmus are marked by over-all characteristics of submergence. The Pacific side of southeastern Panama and northwestern Colombia (Fig. 6), is dominated by a coastline of partially submerged mountainous terrain which extends from Colombia northwestward to San Miguel Bay. Although the Pacific coast of Darien Province is in an early youth stage of submergence, the Caribbean coast represents a more advanced stage of youthful submergence. For example, along the Darien mountain range (Fig. 5) partially submerged mountainous terrain is interrupted by straight stretches of sandy beaches (the bayhead beaches in Figs. 8A and 8B). To the southeast is the delta complex of the Atrato River.

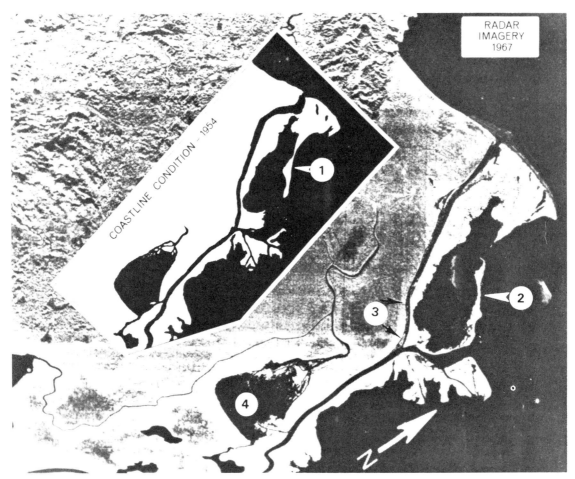

Figure 14. Comparison between radar imagery (1967) and coastline conditions constructed by Vann in the field in 1954.

The rise in sea level at the end of the Wisconsin glacial substage provided sufficient submergence to cover most of the narrow coastal plain sediments and to bring both the basement complex and sedimentary strata in contact with the sea. Along the submergent Pacific and Caribbean coasts of eastern Panama, the sedimentary strata of Eocene age or younger (MacDonald, 1969; Wing, 1970) provide a less rugged coastline than does the basement complex. Contrast, for example, the coastal terrain in Figure 8A and 8B, where the mountainous coastline (Punta Carreto and Puerto Obaldia of Fig. 8A) represents outcropping of the basement complex, but the sediments of the beach areas are juxtaposed with relatively less resistant strata of Eocene age.

The Bay of San Miguel (Fig. 5, 8°15′N, 78°20′W) is a compound estuary. Formerly it was the main trunk of a large river whose tributaries included all of the major rivers now supplying fluvial sediments to the estuary (Wing, 1970, p. 110). The shoreline of San Miguel Bay is submergent, but many bayhead beaches and especially bay deltas attest to a rapid fill-in process. This region contains the most extensive development of mangrove swamps in all of eastern Panama, and the entrapment of fluviomarine muds contributes considerably to the rapid sedimentation now taking place.

The coastlines of east-central Panama are more complex than those of Darien Province. The Caribbean shoreline from Punta Ailigandi westward to the Canal Zone (Fig. 5), appears to be both contraposed and compound. Here there is no continuous coastal plain, but Holocene sedimentation is much more prevalent than to the southeast. This fringe of Holocene sediments lends support to Terry's (1956) supposition of slight emergence of the Caribbean coast in post-Pleistocene time, but this subsequent emergence was relative to the submergence of the Pacific littoral. Nevertheless, numerous coral islands (Fig. 12) along the Caribbean coast of east-central Panama are not yet tied to land or each other by spits, bars, or tombolos. The net impression is that of a shoreline of contraposed submergence.

Along the Serrania De Maje (8°50'N, 78° 30'W) from San Miguel Bay to the mouth of the Rio Bayano (9°00'N, 79°05'W) a submerged coastline is interrupted by mangrove swamps and mudflats. Bay deltas building out from the mouths of nearly all rivers are filling in estuaries and forming extensive tidal flats. Along this coast, a shallow bottom extends far offshore protecting the deltas from wave erosion. This coastline appears to qualify for a neutral classification though net submergence is characteristic of the recent past.

Immediately east of the Canal Zone and extending to the mouth of the Rio Bayano a coastal plain has developed ranging in width from 4 to 25 km. The beach ridges previously illustrated in Figure 11 provide evidence of progradation, the causes of which are attributable to one or all three of the following possibilities (Wing, 1970, p. 46): (1) Acidic terrain drained by the Bayano tributaries has supplied an abundance of quartz sand debris to a locally regressive coastal segment; (2) slight progressive uplift (on the coastal side of a major fault) has resulted in regression and progressive abandonment of beach bars; and (3) a combination of high tides with anomalous wind and wave conditions provides unique sorting and sediment distribution action.

CONCLUSIONS

The spatial resolution capability of radar is less than that of aerial photogtaphy, and the coastal detail interpretable from radar imagery does not equal that obtainable from aerial photographs. For example, wave height and beach features such as dune shape, size, and orientation cannot always be observed; however, refinements in operational and system parameters such as gain-setting and resolution may improve certain aspects of data retrieval. Water depth information provided by color photography, and near-shore bottom detail interpretable from panchromatic aerial photography are not available with radar imagery.

Radar does, however, provide a rather precise basis for measuring the net changes in dimensions or shape by sequential imaging. A qualitative estimate of the amount of sediment deposited in the form of shoals, tidal deltas, and spits can be determined. Also, estuarine changes, delta formation, nearshore processes, and sea state can be observed. Thus in cloud shrouded environments where photography cannot be obtained, radar will provide an important source of information for the coastal geomorphologist and for updating coastal maps.

ACKNOWLEDGMENTS

Facilities for this study were provided by the Remote Sensing Laboratory, Center for Research in Engineering Science (CRES), University of Kansas. This study was sponsored by the Department of the Army, Engineer Topographic Laboratories under THEMIS Contract DAAK02–68-C-0089 and NASA Contract NAS9–10261.

REFERENCES CITED

Ahnert, F. 1960, Estuarine meanders in the Chesapeake Bay area: Geogr. Rev., Vol. 50, No. 3, p. 390–401, 1960.

Lewis, A. J., and MacDonald, H. C. Significance of estuarine meanders identified from radar imagery of eastern Panama and northwestern Colombia: Mod. Geol., Vol. 1, No. 3, p. 187–196, 1970.

MacDonald, H. C. Geologic evaluation of radar imagery from Darien Province, Panama: Mod. Geol., Vol. 1, p. 1–63, 1969.

Terry, R. A. A geological reconnaissance of Panama: Calif. Acad. Sci., Occas. Pap. 23, 91 p., 1956.

Tuan, Y. Coastal landforms of central Panama: Dept. Geogr. Univ. California, Berkeley, Contr. ONR–222(11) NR 388 067, AD No. 239 568, 30 p., 1960.

Vann, J. H. Landform-vegetation relationships in the Atrato Delta: Ann. Ass. Amer. Geogr., Vol. 49, No. 4, p. 345–360, 1959.

Wing, R. S. Structural analysis from radar imagery, eastern Panamanian Isthmus. Mod. Geol., 180 p., 1970, in press.

MANUSCRIPT RECEIVED BY THE SOCIETY JUNE 16, 1970
REVISED MANUSCRIPT RECEIVED SEPTEMBER 9, 1970

5 September 1980, Volume 209, Number 4461

Reprinted by permission of the American Association for the Advancement of Science from *Science*, v. 209, no. 4461 (1980), p. 1073-1082. Copyright 1980 by the AAAS.

Spaceborne Imaging Radar: Geologic and Oceanographic Applications

Charles Elachi

In June 1978, the Seasat satellite was put into orbit around the earth with a payload of active microwave sensors consisting of an altimeter, a scatterometer, and an imaging synthetic aperture radar (SAR). The objective of the mission was a proof-of-concept demonstration of the capability to monitor the ocean sur-

from the sun. The radar energy also penetrates cloud cover; consequently, the sensor operation is not constrained by weather conditions. The illumination angle and illumination direction can be controlled and selected (4), whereas in optical systems these parameters are constrained by the sun's location. The

Summary. Synoptic, large-area radar images of the earth's land and ocean surface, obtained from the Seasat orbiting spacecraft, show the potential for geologic mapping and for monitoring of ocean surface patterns. Structural and topographic features such as lineaments, anticlines, folds and domes, drainage patterns, stratification, and roughness units can be mapped. Ocean surface waves, internal waves, current boundaries, and large-scale eddies have been observed in numerous images taken by the Seasat imaging radar. This article gives an illustrated overview of these applications.

face and near-surface features such as surface waves, internal waves, currents, eddies, surface wind, surface topography, and ice cover (1). The imaging radar, which was operated in the synthetic aperture mode (2, 3), provided, for the first time, synoptic radar images of the earth's surface (both ocean and land areas) obtained from an orbiting platform. The resolution of these images is about 25 meters. The success of this complex sensor was a major technological advance, and it opened up a new dimension in our capability to observe, monitor, and study the earth's surface.

The SAR imaging sensor has some unique characteristics. It is an active system; that is, it uses its own energy to illuminate the surface, and it generates an image from the backscatter echoes. Thus it is not dependent on illumination

SAR imaging sensor thus is not limited by any environmental factor, is an all-time, all-weather sensor, and allows a great deal of flexibility in imaging geometry.

The SAR also provides images of the surface in a different region of the electromagnetic spectrum than infrared, visible, and ultraviolet sensors. Most SAR's operate in the spectral region from about 1-centimeter to 1-meter wavelength. The Seasat SAR operated at a 23-cm wavelength. In this spectral region, the backscatter energy is dependent primarily on the surface physical properties (slope and roughness) and on electric properties (complex dielectric constant, which is dependent on surface rock type, moisture, vegetation cover, and near-surface inhomogeneities).

Exploration companies, government agencies, geological surveys, and indus-

trial organizations use airborne radar images for large-scale mapping, particularly in equatorial regions with extensive cloud cover. Almost all of Brazil, Nigeria, Venezuela, Panama, Togo, and several other countries in the tropical regions, as well as many areas in the United States, have been mapped with radar, some for the first time (5).

Most of the radar imaging of the ocean surface in the past has been experimental in nature (6-10). Before the advent of Seasat, airborne SAR sensors had been used to image surface waves, internal waves, currents, weather fronts, polar ice (11), and ocean vessels. However, the usefulness of these observations was severely hindered by the limited temporal and spatial extent that could be achieved with an airborne platform. The Seasat SAR demonstrated that numerous ocean surface features can be observed and potentially monitored from a space platform that can provide synoptic coverage of large areas and fast repetitive observations of a particular region.

The Seasat SAR provided, for the first time, synoptic radar images of large areas with a homogeneous illumination geometry. Each Seasat SAR data path covers an area 100 kilometers wide and about 4000 km long with a resolution of 25 m. The incidence angle at the surface was about $20° \pm 3°$ from the vertical. The orbit inclination was 108°, which allowed two illumination directions (one during the ascending pass and one during the descending pass) for specific regions.

The Seasat SAR was designed to observe the ocean surface where there is little topography. However, it also provided useful images of large land regions in the continental United States, Canada, Alaska, Central America, and Western Europe. The land images are most useful in regions with low to moderate relief. In rugged mountainous regions, the foldover effect (12) gives exaggerated distortion, which complicates the interpretation of the image.

The purpose of this article is to give an overview of the characteristics and po-

The author heads the Radar Remote Sensing Group in the Earth and Space Sciences Division at the Jet Propulsion Laboratory, California Institute of Technology, Pasadena 91103.

tential applications of spaceborne imaging radar, particularly in geology and oceanography. The synthetic aperture technique and the unique requirements of spaceborne SAR sensors are reviewed first. Then I discuss the different applications of spaceborne radars in geologic mapping, using illustrations from the Seasat SAR images. The oceanographic applications are discussed next. The analysis of the Seasat SAR data is still at an early stage, and most of the discussions presented here are preliminary. A number of radar images are included, with only a brief discussion, to give the reader a broad idea of the different features that can be observed. All of the figures (except Fig. 1) were obtained with the Seasat SAR.

The SAR Concept

In the synthetic aperture technique, the Doppler information in the returned echo is used simultaneously with the time delay information to generate a high-resolution image of the surface being illuminated by the radar. The radar usually "looks" to one side of the moving platform (to eliminate right-left ambiguities) and perpendicular to its line of motion. It transmits a short pulse of coherent electromagnetic energy toward the surface. Points equidistant from the radar are located on successive concentric spheres. The intersection of these spheres with the surface gives a series of concentric circles centered at the nadir point (Fig. 1). The backscatter echoes from objects along a certain circle will have a well-defined time delay.

Points distributed on coaxial cones, with the flight line as the axis and the radar as the apex, provide identical Doppler shifts of the returned echo. The intersection of these cones with the surface gives a family of hyperbolas (Fig. 1). Objects on a specific hyperbola will provide equi-Doppler returns. Thus, if the time delay and Doppler information in the returned echoes are processed simultaneously, the surface can be divided into a coordinate system of concentric circles and coaxial hyperbolas (Fig. 1), and each point on the surface can be uniquely identified by a specific time delay and specific Doppler. The brightness that is assigned to a specific pixel (picture resolution element) in the radar image is proportional to the echo energy contained in the time delay bin and Doppler bin which corresponds to the equivalent point on the surface being imaged. The

resolution capability of the imaging system is thus dependent on the measurement accuracy of the differential time delay and differential Doppler (or phase) between two neighboring points on the surface.

In actuality, the situation is somewhat more complicated. The radar transmits a pulsed signal which is necessary to obtain the time delay information. To obtain the Doppler information unambiguously, the echoes from many pulses are required to meet the Nyquist (12) sampling criterion. Thus, as the moving platform passes over a certain region, the recorded series of echoes contains a complete Doppler history and range-change history for each point on the surface that is being illuminated. These complete histories are then processed to identify uniquely each point on the surface and to generate the image (2, 12). This is why a very large number of operations is required to generate one pixel in the image; such is not the case in optical sensors. A simplified comparison is that the radar sensor generates the equivalent of a hologram of the surface, and further processing is required to obtain the image. This processing can be done either optically or digitally.

In the case of spaceborne sensors, there are additional effects (3) that are not encountered with airborne sensors. (i) The rotation of the earth relative to the spacecraft adds a Doppler shift that must be accounted for during processing. This Doppler shift varies as a function of latitude and inclination of the orbit. (ii) The orbit eccentricity causes an altitude rate of change which translates into a Doppler shift that must be eliminated. (iii) The ionospheric granularities introduce

phase scintillations which induce errors in the Doppler measurements. (iv) The far distance to the surface requires that many pulses be transmitted before the echo from the first one is received, and attention must be given to the timing of the transmitted and received echoes.

The synthetic aperture imaging technique has one unique characteristic. The resolution capability is dependent on the measurement accuracy, in the range dimension, of the differential time delay between two different points, and in the azimuth dimension of the Doppler shift from a target. Neither of these measurements is related to the absolute distance from the radar to the surface. Thus, the resolution of an imaging SAR is independent of the altitude of the platform (2, 3). Spaceborne and airborne SAR's with similar characteristics will have the same resolution capability. The main difference is that spaceborne sensors require more transmitted power to be able to obtain the necessary echo signal-to-noise ratio. The size of the antenna aperture is usually determined by the width of the swath being imaged and the observing geometry, not by the resolution.

Application in Geologic Mapping

The brightness in the radar image is a representation of the surface backscatter cross section, which is a function of the surface slope, surface roughness at the scale of the observing wavelength, and surface complex dielectric constant. The geologic interpretation of the radar image is based on two general types of information, (i) geometric patterns and shapes and (ii) image tone and texture. Examples of the former are lineaments, joints, folds, domes, drainage pattern density, fracture patterns, and the spatial relationships between these features. These patterns, forms, and shapes are interpreted in a way similar to that used with regular photography (13). The radar technique has the advantage that the angle of illumination and direction of illumination are selectable (4). This is not the case with regular photography where the geometry is fixed by the position of the sun and the time of the year, and some illumination directions are not available at any time of the year. Image tone and texture are primarily a function of the surface roughness and sub-resolution small-scale topography, the surface complex dielectric constant, and surface variations on the scale of few resolution elements. The dielectric property is most useful in the study of vege-

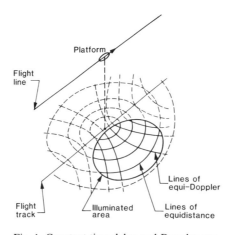

Fig. 1. Constant time delay and Doppler contour lines, which form the radar imaging coordinate system. Each point on the surface can be uniquely identified if energy in the appropriate time delay bin and Doppler shift bin is filtered out of the received echoes.

tated and moist surfaces. The tonal and textural data in the radar image provide new information that is not available with optical or infrared photography; their interpretation requires an understanding of the interaction of electromagnetic waves, in the microwave region, with natural surfaces.

The sensitivity of the amplitude of the radar echo to changes in the surface topography is very high in comparison to the optical and infrared albedo. A change in the surface slope of a few degrees can easily change the amplitude of the radar echo by a factor of 2 or more, particularly at small incidence angles (up to 30° from the vertical). At larger incidence angles (30° to 70°), the backscattered energy is proportional to the roughness power spectrum, which can easily change by a factor of 10 or more between two neighboring geologic units (14). In comparison, optical albedo rarely changes by more than a factor of 10 (from about 0.06 for basalts to 0.6 for the brightest salts). Thus, the radar sensor is most useful for the study of patterns and features that are expressed in changes of slope or roughness. This is why airborne

radar sensors have been used mostly for geomorphologic and structural mapping (4). Specific examples from the Seasat data will be discussed later in this section.

The surface roughness and dielectric constant are also useful indicators of changes in the surface rock type. Different rock types will erode differently in a similar environment. They also have different dielectric constants. The presence of moisture (because water has a very high dielectric constant) or vegetation could also help in the separation of lithologic units. However, it is not feasible at the present time to use the radar data to identify the surface rock type. In this application, the radar sensor is most useful in discriminating between areas with different rock types, in providing complementary information in conjunction with optical and infrared sensors, and in planning field investigations.

Lineaments, faults, fractures, and contacts. These features are usually expressed on the surface as sharp changes in the surface topography, morphology, or cover (for example, slope change, alignment of hills or valley sections,

small-scale roughness and texture change, alignment of stream segments, or vegetation cover change). All of these features have a strong effect on the radar wave scattering and are observed on the image as a tonal or textural change. Figure 2 shows the Seasat SAR image of the western Mojave, California, where the Garlock and San Andreas faults intersect. Each fault is clearly delineated as a long linear tonal change resulting from the change in the local topography. More subtle lineaments are observed in Fig. 3, which shows the southern Appalachians around Knoxville, Tennessee. This area is almost completely vegetated, and the lineaments represent the alignment of short valley segments. The localized tonal change is predominantly a result of surface slope changes, which, in some cases, may be as much as 30°. Detailed analyses by Ford (15) show that numerous lineaments are observed on the radar image that have not been detected on the Landsat images. Ford, and earlier investigators, have noted that, where a lineament is expressed by a topographic change, it is least well observed when the illumination direction is parallel to

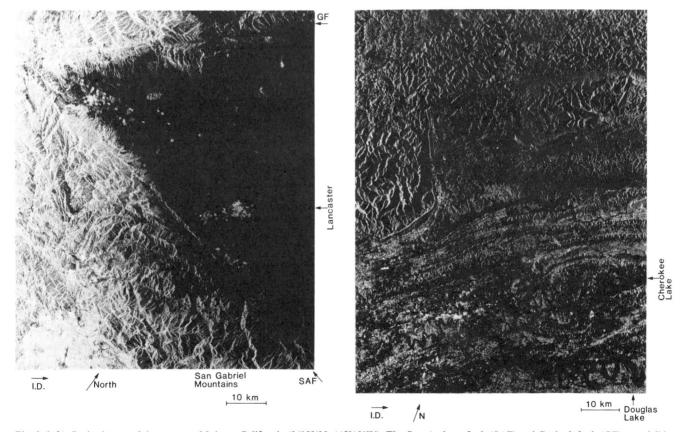

Fig. 2 (left). Radar image of the western Mojave, California (34°50′N, 118°10′W). The San Andreas fault (*SAF*) and Garlock fault (*GF*) are visible as abrupt linear tonal changes near the boundary between the bright mountainous region and the dark desert region. Lancaster is the very bright area in the center. The illumination direction is shown by the arrow labeled *I.D.*   Fig. 3 (right). Radar image of the valley and ridge province in northwest Tennessee and the Cumberland plateau in southeastern Kentucky (36°45′N, 83°W). The Pine Mountain overthrust is sharply demarcated across the upper part of the image. The Jacksboro fault at left center marks the southwestern margin of the Pine Mountain thrust block. Tennessee Valley Authority reservoirs are black on the image. Knoxville is the bright region at the bottom left.

the lineament trace. The availability of two different illumination directions from the Seasat SAR substantially reduced this problem (15).

Topographic structural features. Domes, cinder cones, anticlines, synclines, and folds are observed with the radar sensor because of the high sensitivity of the radar return to the slope change. Figure 4a is an image of the Obayos region (northeastern Mexico) where Mesozoic sedimentary rocks have been folded into plunging anticlinal structures. A large breached anticline

and two smaller, doubly plunging anticlines are clearly observed as a result of tonal changes in a recognizable pattern associated with this type of feature. The tonal differences are mainly the result of slope differences which arise from the erosion of dipping beds. Figure 4b shows a folded terrain near Harrisburg, Pennsylvania, which is also observed mainly as a result of topographic expressions from the erosion of different beds.

Sand dune fields. This is another type of geologic feature expressed in terms of local topographic variations. In this

case, the surface is homogeneous and very smooth at the scale of the radar wavelength. The scattering occurs primarily in the specular mode; that is, strong echoes are returned from dune facets that are normal to the incident wave vector. Thus, the radar images mainly the dune with facets which are appropriately oriented relative to the radar illumination. It should be expected then that variations in the image patterns are observed for different illumination directions.

Figure 5 shows two images of the Algodones dunes in southeastern California. The dune patterns are observed mainly as changes in the density of bright specular points. The large barchan dunes in the central region and longitudinal dunes to the west can be identified. The dark areas in the dune field correspond to the interdune flats consisting of relatively smooth gravel pavement. Appreciable variations are observed for the two different illumination directions. The barchan dunes are best identified in Fig. 5b, where the illumination is perpendicular to the crest line. Their visibility is enhanced by the foreshortening and foldover effects (16). The longitudinal dunes are observed well on both images; however, different directional components are emphasized in each image.

Canyons. Canyons such as the Grand Canyon in Arizona represent the extreme in topographic discontinuity. In this case, most of the tonal variation in the image is due to shadowing (17). Figure 6 shows the image of the southern rim of the Grand Canyon near Grand Canyon Village. The major rock formations (Kaibab limestone–Coconino sandstone, Redwall limestone, and Tapeats sandstone) are observed as dark bands in the canyon wall because of the shadow from their vertical profile. Indeed, the imaging geometry of the radar sensor is more favorable than the imaging geometry of optical sensors for observing strata exposed in vertical cliffs (17). Thinner strata are observed in Fig. 7 of the Berufjordur region in eastern Iceland. In this case, strata as thin as 50 m could be detected. Because of the foldover and foreshortening effects (16), the steep sides of the mountains which are oriented toward the radar appear as a very bright region in the image. The resulting distortion makes it impossible to observe details on those sides.

Volcanic lava flows. These are observed mainly because of their small-scale topographic (roughness) characteristics. Lava flows, particularly relatively recent ones, usually have much brighter

I.D.   N→   Oballos Pass   Sierra de Muzquiz

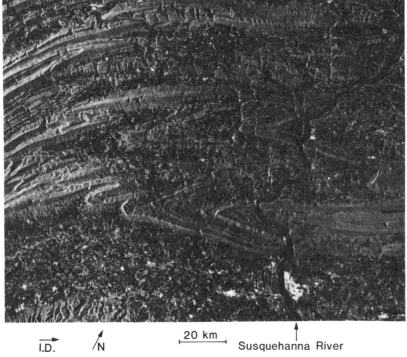

I.D.   N   20 km   Susquehanna River

Fig. 4. Examples of topographic structural features. (a) Three plunging anticlines in the Obayos region, Mexico (27°30′N, 101°30′W). (b) Folded terrain in the Harrisburg region of Pennsylvania (40°20′N, 77°W). Harrisburg is seen as a very bright area in the lower part of the image, situated on the bank of the Susquehanna River, which has a dark tone.

tone than their surroundings (see Fig. 6), because their extremely rough surfaces strongly backscatter the radar signal. The tone, form, and surrounding patterns allow us to identify the bright region in the center of Fig. 6 as a lava flow. Almost all the cinder cones and craters in the surrounding regions are also visible on the radar image, as well as the fronts of some older flows. A detailed analysis of the radar imagery of the SP lava flow has been reported by Schaber *et al.* (*18*).

Drainage patterns. The classification and interpretation of the different drainage patterns observed in a radar image are identical to what is done in photo interpretation (*13*). Drainage patterns are used to infer variations in the surface lithology (*13*). In bedrock areas these patterns depend, for the most part, on the lithologic character of the underlying rocks, the attitude of these rock bodies, and the arrangement and spacing of the planes of structural and lithologic weakness encountered by the runoff. Anomalous or abrupt changes in the drainage patterns in the region under observation are particularly important. In the case of the radar sensor, the drainage patterns are observed because of (i) variation in local slope at the edge of a river segment, (ii) variation in vegetation cover, (iii) difference in tone due to the different backscatter from the water surface, and (iv) strong scattering from the boulders and pebbles in the dry channel bottom in arid regions.

Figure 8 shows two examples of Seasat SAR images of drainage patterns in two different environments. Figure 8a shows a region in Pennsylvania, near Lock Haven, where vegetation cover is extensive and the drainage channels are observed primarily as a result of topographic expressions. The drainage pattern density is a good reflection of the lithology of the region. Dense drainage patterns to the north are in the region of the Catskill Formation (red to brown shale and sandstones), which is a relatively soft rock. The drainage pattern in the southern part is less dense. This is a dissected plateau of the Mississippian Pocono Group (conglomerate and sandstone with some shale), which is a relatively resistant rock. The boundary between the two rock types corresponds very closely to the boundary between the two different drainage pattern densities. Figure 8b shows an arid region west of Tucson, Arizona, near Silver Bell, where the drainage channels are observed mostly in terms of riparian vegetation along those channels. Variation in the drainage density is also associated

with variation in the coarseness of the alluvium in that region.

Lithologic mapping. Lithologic mapping cannot be based exclusively on the tonal change in the radar image. The most that can be done is to separate geologic units which have a different surface roughness (which is dependent on the rock type), different susceptibility to moisture content, or different type of vegetation cover. These differences are expressed as a tonal or textural change.

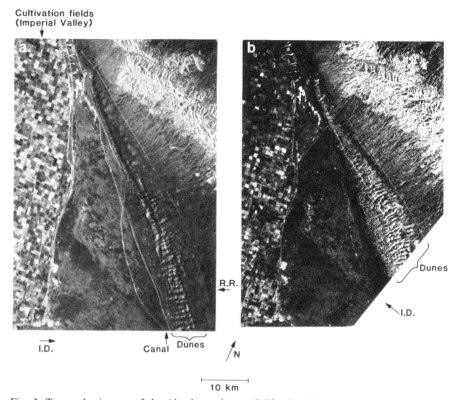

Fig. 5. Two radar images of the Algodones dunes, California (33°N, 115°W) taken from two different directions. The cultivated fields of the Imperial Valley are imaged as a checkerboard pattern on the left. The railroad (*R.R.*) (east of the dunes) and the water canal (west of the dunes) are imaged as bright linear features.

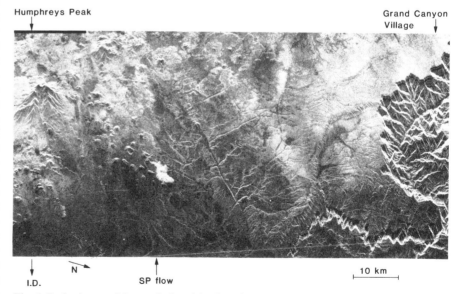

Fig. 6. Radar image of the south rim of the Grand Canyon (on the right part of the image) near Grand Canyon Village, and of the San Francisco volcanic field and Humphreys Peak (on the left part of the image). The location of the center of the image is 35°45′N, 111°30′W. The dark bands in the canyon wall correspond to the different strata which have a vertical slope and are therefore in shadow. The bright feature in the center of the image is the SP lava flow with its source, the SP cinder cone. Almost all the cinder cones in the area are identifiable on the image. The linear feature in the center of the image is the Mesa Butte graben (part of the Mesa Butte fault system).

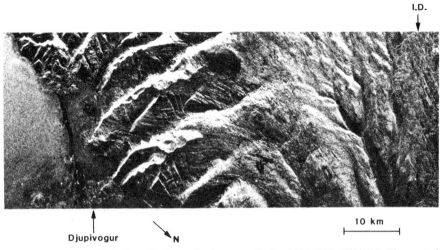

Fig. 7. Radar image of the Berufjordur region in eastern Iceland (64°40′N, 14°20′W). The post of Djupivogur is at the lower left. Numerous strata on the north sides of the fjords are observed as bands of different tones. Strata as thin as 50 m can be observed. The mountain slopes facing the radar (facing south) are drastically distorted because of the foldover effect.

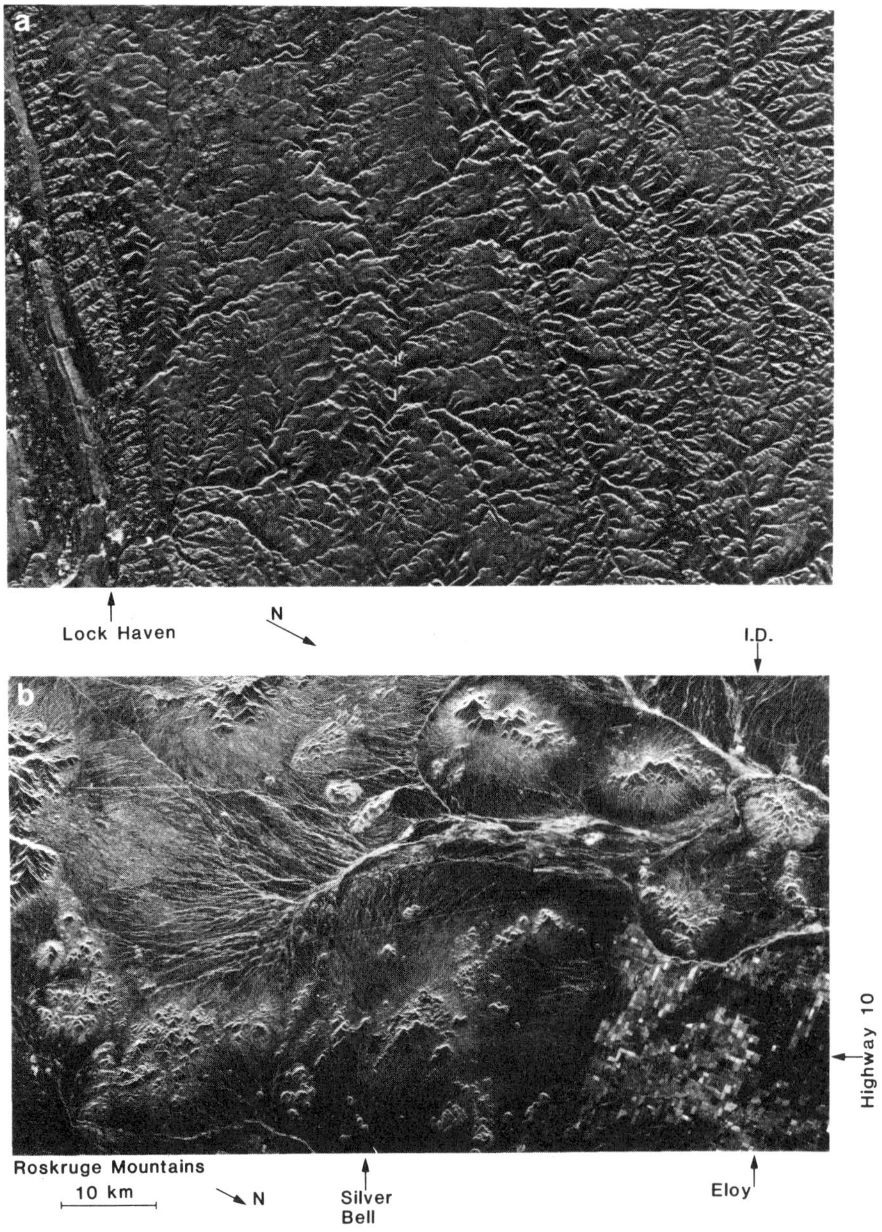

Specific determination of the rock type in each one of the different units would require additional information from multispectral infrared and visible sensors and ultimately from spot field measurements. Figure 9 shows an image of Death Valley, California, where the surface lithology is expressed in variations of the surface roughness. This is a situation that is particularly favorable for radar observation. Almost all of the geologic units that have been mapped as a result of field work (*14, 19, 20*) can be discerned on the radar image as tonal changes. In Fig. 3 also, variations in texture correlate well with variation in the rock type (*15*).

The radar images discussed in this section constitute a small sample of the types of geologic features that were observed with the Seasat SAR. Work is just beginning to assess the additional information that could be derived from this new type of data. The Seasat SAR had an observation geometry which was not necessarily the most favorable for geologic mapping. Future experiments, especially those in which the Space Shuttle is used as an orbiting platform, should allow us to assess the effects of the different radar sensor characteristics (that is, look angle, polarization, and frequency) on the information content of the images.

Oceanographic Applications

In oceanographic applications, the imaging radar sensor has a unique and essential characteristic: the capability of obtaining high-resolution surface images independent of the cloud cover and at any time of the day or night. It is essential because of the dynamic nature of almost all the features on the ocean surface. In actuality, the desired objective of oceanographic monitoring satellites is to be able to monitor, on a global basis, the ocean surface every few days.

The radar sensor provides an image

Fig. 8. Two examples of drainage patterns observed on the Seasat SAR images. (a) The Lock Haven region in Pennsylvania (41°N, 78°W). Notice the change in the drainage pattern density in the Catskill Formation on the right part of the image relative to the Mississippian Pocono Group in the center part. The drainage channels are observed mainly because of their topographic expression. (b) The Silver Bell region, Arizona (32°30′N, 111°40′W). Radial, centripetal, and annular patterns can be observed. The drainage channels are observed mainly because of the strong scattering from the vegetation along the channels.

that is representative of the surface backscatter characteristics. In the case of the ocean, the backscatter is completely controlled by the small-scale surface topography, the short gravity and capillary waves which scatter the radar energy by the Bragg scattering mechanism (21), and the local tilt of the surface, which is due to the presence of large waves and swells. Thus, the SAR is capable of imaging surface and near-surface phenomena that affect the surface roughness directly or indirectly. These phenomena include surface waves, internal waves, currents, weather fronts, wind or oil slicks, and eddies. Changes in surface temperature can be detected only if they affect the surface roughness. In this section, I briefly discuss examples of ocean features that have been observed with Seasat SAR. Verification of the observations on the radar image requires that surface observations be planned in advance and conducted simultaneously with the collection of radar data. Thus it is more difficult to verify ocean features than land features, which were observed on the Seasat SAR images..

Observations of ocean waves with airborne SAR were first reported by Brown *et al.* (6) and Larson *et al.* (7) in 1976. Since then, numerous other ocean features have been observed with airborne SAR sensors (6-9). However, aircraft observations are limited in aerial and temporal coverage, and the phenomena under observation are highly dynamic and variable. For the first time, the Seasat SAR provided a synoptic view of large ocean and ice-covered polar areas. In some cases, it also provided repetitive observations of the same region every 3 days.

Surface waves are visible on the radar image as a periodic regular change in the image tone (Fig. 10). The spatially periodic change in the surface-coherent backscatter cross section is a result of three surface effects that are modulated by the presence of a propagating surface wave or swell: (i) local slope, (ii) the intensity and bunching of the small gravity and capillary waves (22), and (iii) the wave orbital velocity, which affects the phase of the returned echo. The relative importance of these three effects is not yet well understood (10).

Figure 10 shows a Seasat image of ocean surface waves acquired over the northeastern Atlantic. In less than 5 minutes, the Seasat SAR provided an image of the Atlantic region between Scotland and Iceland 2000 km long and 100 km wide. In effect, it provided an almost instantaneous snapshot of the wave pattern (wavelength and wave direction) in that region.

Internal waves. These waves are observed as a result of their surface manifestations and their effect on the surface roughness. The rather large currents associated with these waves modify the capillary-ultragravity surface wave spectrum overlying the oscillations. The exact mechanisms by which the modifications take place are still the subject of discussion, but at least two hypotheses have been advanced (23). According to the first hypothesis, the high velocity of surface water arising from the internal wave amplitude can sweep surface oils and materials together to form a smooth strip near regions of surface water convergence. The second mechanism predicts that capillary and ultragravity wave energy is concentrated in the convergence zone by surface current stress, which then becomes a region of enhanced roughness rather than a smooth area as with the first hypothesis. When such smooth and rough regions are illuminated away from normal incidence and then viewed at nonspecular angles, the smooth region would appear darker and the rough one brighter than the normal sea surface. This geometry is the same for both imaging radar and multispectral (including optical) sensors.

Internal waves are usually observed on the radar image as a wave packet that consists of a series of convex strips, with the spatial periodicity becoming shorter toward the center of curvature (Fig. 11). The length of the crest may range up to many tens of kilometers. The leading

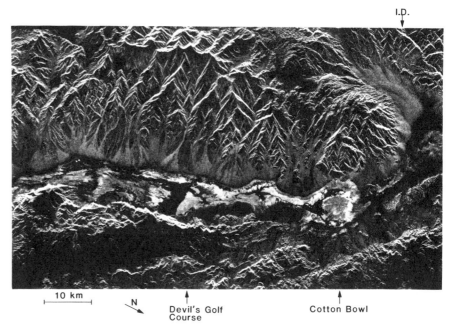

Fig. 9. Radar image of Death Valley, California (36°20′N, 116°50′W). The Cotton Bowl (silty rock salt) is the gray circular feature on the right. The dark continuous curvilinear band across the image corresponds to the edge of the alluvial fans at their intersection with the valley floor. The rough massive rock salt in the Devil's Golf Course shows a relatively bright tone.

Fig. 10. Radar images of ocean surface waves in the northeastern Atlantic (60°N, 6°W, 19 August 1978). The swells have a wavelength of 400 m and are visible as a periodic tonal change. Other large-scale curvilinear features are also visible, some of which could be associated with internal waves. The ocean bottom depth in this area is between 100 and 500 fathoms.

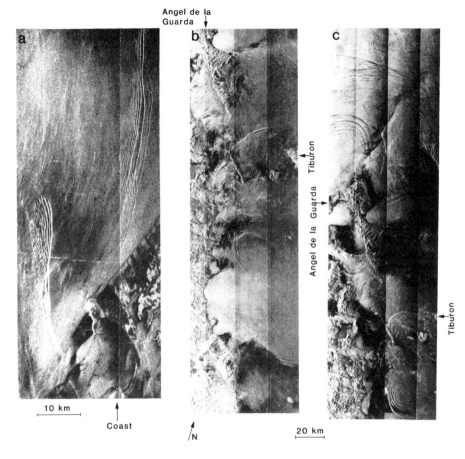

Angel de la Guarda

a

10 km

Coast

b

Tiburon

Angel de la Guarda

/N

20 km

c

Angel de la Guarda

Tiburon

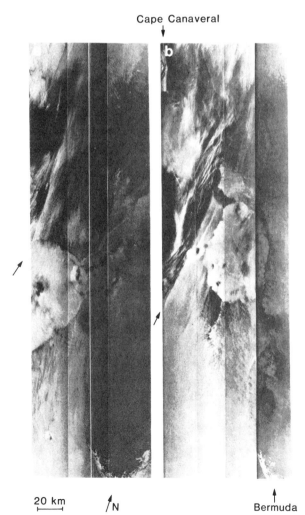

Cape Canaveral

a

b

20 km /N

Bermuda

Fig. 11 (above). Radar images of groups of internal waves. (a) The western coast of Baja California (7 July 1978) near Isla Magdalena (27°50'N, 115°W). (b) The Gulf of California off the coast of Baja California (14 September 1978) (upper part of image). The narrow long island in the center is San Lorenzo. (c) The Gulf of California (17 September 1978 at 17:20 G.M.T.) (29°N, 113°W). Fig. 12 (left). Two radar images of the region between Grand Bahama Island (lower right corner) and the Florida coast. The images were taken on (a) 8 July 1978 and (b) 25 July 1978. The linear tonal change boundary, which is diagonal across the upper half of the images (indicated by an arrow) corresponds to the expected location of the western edge of the Gulf Stream.

wavelengths are on the order of 1 to 2 km and decrease monotonically toward the rear. They usually occur in groups or packets, and they have been observed in numerous places along the western and eastern coasts of North America. These waves are visible apparently because the surface stress is active in the front of a given cycle (leading to a rough surface and strong backscatter), which shows up as a bright tone, and is trailed by a slick formation, which exhibits a dark tone. Similar observations have been conducted with aircraft SAR (6-8) and optical sensors (24). On some Seasat single swaths, more internal waves could be observed than the total number observed during dozens of aircraft flights over a period of 5 years. This illustrates the new insight that resulted from the Seasat SAR experiment on the extent and rate of occurrence of certain dynamic ocean phenomena.

Figure 11a shows a group of internal waves along the west coast of Baja California, near Boca de la Soledad and Isla Magdalena. They appear as a succession of bright narrow convex strips separated by dark regions. The bright strips are about 100 m wide, and the wavelengths range from about 1 km at the front to about 100 m at the back of the wave. The fronts are usually a well-defined, long bright strip, whereas the backs sometimes consist of short, broken strips.

Figure 11, b and c, shows the surface patterns observed in the Gulf of California near the islands of Angel de la Guarda, Tiburon, and San Lorenzo. The two images were taken 3 days apart. In Fig. 11b, some internal waves are visible, particularly the two next to San Lorenzo. Their characteristics are similar to the ones discussed above. Figure 11c shows a very large number of what seem to be internal waves with a wide variety of linear scales. Some of the waves just northwest of Angel de la Guarda had wavelengths up to 5 km, had highly convex fronts, and were more than 40 km long. The wave pattern was characterized by dark narrow strips, about 500 m wide, separated by wide, relatively brighter regions. Some of these waves overlap, leading to interference patterns. The waves northeast of Angel de la Guarda and south of San Lorenzo have characteristics similar to the ones observed in Fig. 11, a and b.

The Gulf Stream. The northwestern Atlantic region was imaged numerous times to observe the surface features that are associated with the Gulf Stream. Figure 12 shows two radar images of the region between Grand Bahama Island and the coast of Florida (near Cape Canav-

1080

eral) that were taken on 8 and 25 July 1978. No sea truth was obtained simultaneously with the radar observations. The linear boundary observed on both images corresponds closely to the expected location of the western edge of the Gulf Stream. The boundary in the image corresponds to an abrupt change in the image tone which corresponds to a change in the backscatter cross section. Numerous streaks in the same general direction as the boundary are also observed on the image, giving the impression of a flow. Possible explanations for the change in the backscatter return, and therefore image tone, include the following: (i) the water motion and temperature change lead to an abrupt discontinuity in the surface roughness which results in a change of the backscatter; or (ii) the abrupt change of the water motion leads to a differential Doppler shift in the returned echo which tends to create a bunching effect and a localized change in the image brightness at the boundary.

In Fig. 12b, a group of internal waves can be seen just off the coast of Cape Canaveral. In the same area, two vessels are visible as very bright point objects.

Polar sea ice. Mapping and large-scale dynamics measurement of polar sea ice is another major application of SAR that is being investigated. Of particular interest is the capability of an orbiting SAR to monitor on a global and repetitive basis the dynamics, structure, and extent of the polar sea ice cover. Figure 13 shows two Seasat images taken 3 days apart over a region just north of Banks Island in Canada. Image tone variations and geometric shapes and forms enable one to identify floes and ridges in the floating ice and the open water. Many leads froze or closed during the 3-day period. Even though the ice patterns changed appreciably over that period, many individual ice floes could be identified and their motion determined. An ice movement of up to 15 km per day, on the average, has been measured on these images.

Conclusions

The Seasat SAR has provided large-scale radar images of land and ocean surfaces for the first time. A preliminary analysis of the data indicates that spaceborne imaging radar will improve our capability to assess earth resources and monitor the ocean surface. It is expected that the radar sensor will add new types of information that will complement the geologic information presently being collected by optical and infrared sensors and by other conventional mapping techniques. The radar sensor is unique because it is capable of monitoring the ocean surface without environmental limitations. Much more work is needed to understand the geophysical information in the radar signature of different surfaces and to determine the optimum sensor characteristics (observation geometry, spectral coverage, polarization) for specific applications. The Space Shuttle is expected to be an appropriate platform for conducting such research, and experiments are being planned for the early 1980's that will use Shuttle imaging radar.

As with optical sensor data, the identification of different geologic units (and possibly ocean surface features) is more successful when multispectral observations are made. The L-band (wavelength about 25 cm) radar sensors, such as the one on Seasat, are most sensitive to roughness in the range from 2.5 to 25 cm. With the addition of an X-band (wave-

Banks Island

25 km

Fig. 13. Two radar images of a region of the Beaufort Sea, west of Banks Island in Canada (72°30′N, 126°W), acquired 3 days apart in late September 1978. The northern shore of Banks Island is to the left with open water along it. The rest of the image covers a region of floating ice. Some individual ice floes can be identified on the two images; they have moved as much as 45 km in 3 days, and the ridges and lead patterns have changed as a result of the ice motion. The striping is an artifact due to mosaicking.

length about 3 cm), for example, the range of roughness sensitivity can probably be expanded to 0.3 to 3 cm, thus improving the discrimination capability (*18, 20, 25*). The development of a multispectral spaceborne radar system is expected to take place in the next few years.

Orbital radar will also allow us to develop a data base for the interpretation of images to be obtained from planetary missions. For Venus, and possibly Titan, radar will be the only means of mapping the planet's surface through the continuous and complete cloud cover.

Because of the complex nature of SAR's on orbital platforms, the successful development of the Seasat SAR was a key technical advancement. Major technological developments are still needed before multispectral orbiting SAR sensors can become operational, particularly in the area of digital, real-time processing. Because the radar sensor basically provides a Doppler time-delay history of each point target, thousands of computational operations are required to generate a single image element. This processing requirement, combined with the desire to have large swath mapping with high resolution, requires extremely fast processing hardware which is just at the limit of present-day technology.

References and Notes

1. G. H. Born, J. A. Dunne, D. B. Lame, *Science* **204**, 1405 (1979).
2. For a detailed review of the SAR principle, see the following: W. M. Brown and L. J. Porcello, *IRE Trans. Mil. Electron.* **MIL-6**, 111 (1969); L. J. Cutrona, in *Radar Handbook*, M. I. Skolnik, Ed. (McGraw-Hill, New York, 1970), p. 23.1; R. O. Harger, *Synthetic Aperture Radar Systems, Theory and Design* (Academic Press, New York, 1970); H. Jensen, L. C. Graham, L. J. Porcello, E. N. Leith, *Sci. Am.* **237**, 84 (October 1977).
3. K. Tomiyasu, *Proc. IEEE* **66**, 563 (1978).
4. In the case of the Seasat SAR, the illumination angle was fixed at 20°. Two illumination directions were possible, one during the ascending orbit and one during the descending orbit. The angle between these two directions varied from 0° to 144°, depending on the latitude of the area being imaged.
5. H. C. MacDonald, *Mod. Geol.* **1**, 1 (1969); R. S. Wing, *ibid.* **1**, 173 (1970); *ibid.* **2**, 1 (1971); *ibid.*, p. 75; _____ and L. Dellwig, *Geol. Soc. Am. Bull.* **81**, 293 (1970).
6. W. E. Brown, C. Elachi, T. W. Thompson, *J. Geophys. Res.* **81**, 2657 (1976).
7. T. R. Larson, L. I. Moskowitz, J. W. Wright, *IEEE Trans. Antennas Propag.* **AP-24**, 393 (1976).
8. C. Elachi, *J. Geophys. Res.* **81**, 2655 (1976); *Boundary Layer Meteorol.* **13**, 165 (1978); _____, T. W. Thompson, D. King, *Science* **198**, 609 (1977).
9. C. Elachi and J. Apel, *Geophys. Res. Lett.* **3**, 647 (1976).
10. C. Elachi and W. E. Brown, *IEEE Trans. Antennas Propag.* **AP-25**, 84 (1977); W. R. Alpers and C. L. Ruffenach, *ibid.* **AP-27**, 685 (1979).
11. For more information, see the following: W. J. Campbell *et al.*, *Boundary Layer Meteorol.* **13**, 309 (1978); F. Leberl, M. L. Bryan, C. Elachi, T. Farr, W. Campbell, *J. Geophys. Res.* **84**, 1827 (1979).
12. R. K. Moore, in *Manual of Remote Sensing*, R. G. Reeves, Ed. (American Society of Photogrammetry, Falls Church, Va., 1975).
13. V. C. Miller, *Photogeology* (McGraw-Hill, New York, 1961); R. G. Ray, *U.S. Geol. Surv. Prof. Pap.* 373 (1972); T. E. Avery, *Interpretation of Aerial Photographs* (Burgess, Minneapolis, Minn., ed. 3, 1977); F. F. Sabins, *Remote Sensing, Principles and Interpretation* (Freeman, San Francisco, 1978).
14. G. G. Schaber, G. L. Berlin, W. E. Brown, *Geol. Soc. Am. Bull.* **87**, 29 (1976).
15. J. Ford, *Am. Assoc. Pet. Geol. Bull.*, in press. For lineament mapping with Seasat SAR, see also F. F. Sabins, R. Blom, and C. Elachi [*ibid.* **64**, 619 (1980)].
16. Foreshortening refers to the fact that areas with a slope toward the radar sensor will be spatially shortened as compared to areas of the same size with a slope away from the radar sensor. Foldover refers to the situation when two or more points on the surface are at exactly the same range distance from the radar. In this case the images of these points are superimposed. This situation usually occurs when imaging rugged terrain at a small look angle (*12*).
17. C. Elachi and T. Farr, *Remote Sensing Environ.* **9**, 171 (1980).
18. G. G. Schaber, C. Elachi, T. Farr, *ibid.*, p. 149.
19. Based on a geologic map by C. B. Hunt and D. R. Mabey, *U. S. Geol. Surv. Prof. Pap. 494-A* (1966).
20. M. Daily, C. Elachi, T. Farr, G. Schaber, *Geophys. Res. Lett.* **5**, 899 (1978); *J. Photogramm. Remote Sensing* **45**, 1109 (1979).
21. M. W. Long, *Radar Reflectivity of Land and Sea* (Lexington, Lexington, Mass., 1975); G. R. Valenzuela, *Boundary Layer Meteorol.* **13**, 61 (1978).
22. J. W. Wright, *Boundary Layer Meteorol.* **13**, 87 (1978).
23. G. Ewing, *J. Mar. Res.* **9**, 161 (1950); A. E. Garrett and B. A. Hughes, *J. Fluid Mech.* **52**, 179 (1972).
24. J. R. Apel, H. M. Bryne, J. R. Proni, R. L. Charnell, *J. Geophys. Res.* **80**, 865 (1975).
25. L. F. Dellwig and R. K. Moore, *ibid.* **71**, 3597 (1966); L. Dellwig, *Mod. Geol.* **1**, 65 (1969).
26. This article represents the results of one phase of research carried out at the Jet Propulsion Laboratory, California Institute of Technology, under contract NAS7-100, sponsored by the Non-Renewable Resources Office, Office of Space and Terrestrial Applications, NASA. I thank M. Abrams, R. Blom, M. L. Bryan, M. Daily, T. Dixon, J. Ford, R. S. Saunders, and H. Stewart from Jet Propulsion Laboratory for their helpful discussions.

Reprinted by permission of the Environmental
Research Institute of Michigan from *Proceedings of
the Fifth Thematic Conference on Remote Sensing for
Exploration Geology*, 1986, v. 1, p. 15-34.

APPLICATION CF SYNTHETIC APERTURE RADAR (SAR)

TO SOUTHERN PAPUA NEW GUINEA FOLD BELT

EXPLORATION*

James M. Ellis and Frank D. Pruett

Chevron Overseas Petroleum Inc.
6001 Bollinger Canyon Road
San Ramon, California 94583-0946, U.S.A.

ABSTRACT

Niugini Gulf Oil Pty. Ltd., as operator for two exploration licenses within the southern Papuan Basin fold and thrust belt, has successfully used synthetic aperture radar (SAR) to map surface structure, stratigraphy, and to help in planning a hydrocarbon exploration program.

Rugged karst topography developed on massive Tertiary limestone causes the region to be exceedingly dangerous, if not impossible, to traverse on the ground. The area is seldom cloud free, is covered with tropical rain forest, and geologic field studies are limited. The region is ideally suited to geologic analysis using remote sensing technology. Landsat images and vertical aerial photographs complement SAR but provide subdued structural information because of the jungle cover and minimal shadowing (due to high sun angles). SAR provided our explorationists with an excellent data base because (1) structure is enhanced with low illumination, (2) resolution is 6 x 12 m, (3) digital reprocessing is possible, and (4) clouds are penetrated by the SAR.

Stereoscopic analysis of SAR provided essential geologic information that was used to guide ongoing field work, modeling of subsurface structure, and selecting of well locations. Surface data, including SAR, are used in place of seismic technology as the primary exploration tool in this area because surface conditions limit acquisition of acceptable seismic data at reasonable cost.

SAR imagery revealed significant mass wasting that led to re-evaluation of previously acquired field data. Lithologies were recognized on the radar imagery by textural and tonal changes in spite of the continuous canopy of jungle. The characteristic radar signature of karst topography enabled some limestone-capped, fold and thrust structures to be interpreted beneath thin veneer of volcanics. Reprocessing and contrast stretching of the digital radar imagery allowed additional geologic information to be extracted from the survey in oversaturated (bright) or flared zones.

1.0 INTRODUCTION

Synthetic aperture radar (SAR) imagery was acquired over an area covering Petroleum Prospecting Licenses PPL-17 and PPL-18 during February 1985. These data were acquired in order to help Niugini Gulf Oil Pty. Ltd. and Joint Venture Participants to: (1) map surface geologic structure and stratigraphy, and (2) provide data to assist in planning and conducting the exploration program for hydrocarbons. More than 26,000 km² were surveyed with airborne radar (Figure 1). Individual flight strips side-lapped by ~60%, permitting stereoscopic analysis of the terrain. The imagery was

---

*Presented at the Fifth Thematic Conference: "Remote Sensing for Exploration Geology," Reno, Nevada, September 29-October 2, 1986.

15

recorded digitally and then plotted at a scale of 1:250,000. A mosaic at the same scale was also constructed.

Interpretation of the radar data was facilitated by previously acquired and ongoing field geology programs within the two permit areas and was complemented by Landsat MSS imagery, aerial photographs, topographic maps, and published geologic maps.

The purpose of this paper is to demonstrate how interpretation of SAR and stereoscopic SAR markedly improved our understanding of surface geology in the southern fold and thrust belt of Papua New Guinea. This paper discusses some of the strengths and weaknesses of other data sets (Landsat, aerial photographs, existing maps, field work) in this isolated, tropical mountainous environment. The limitations of airborne radar imagery (and ways to minimize these limitations) will also be discussed.

The region is largely uninhabited, and is covered by a dense rain forest that approaches 30-40 m (±) in height (Figure 2). Massive Darai Limestone (Tertiary) crops out extensively across the exploration area. The limestone and high rainfall have produced a rugged karst topography that is exceedingly dangerous, if not impossible, to traverse on the ground. The rocks are folded and thrust into large anticlines. Several exceed 25 Km in length (Figure 3). The relief of some of these anticlines approaches 1500 m. The area is seldom cloud free and geologic field studies are limited in extent. Obviously the region is ideally suited for geologic analysis using remote sensing technology.

Sabins (1983) demonstrated the usefulness of space shuttle SAR for interpreting the geology of nearby Indonesia, a country also characterized by persistent cloud cover. Airborne, real-aperture radar (SLAR) was successfully used to compile geologic reconaissance maps for petroleum exploration in cloud-covered Eastern Panama and Northwestern Colombia (Wing and MacDonald, 1973) and in Irian Jaya, Indonesia by Chevron (1972-75).

Exploration evaluation, including drilling, is underway within these licenses in search of liquid hydrocarbon reserves which may be trapped in economic quantities in some of the large structures. Hydrocarbon traps within most exploration provinces are usually defined with seismic technology prior to costly drilling; however, deep karst weathering of surface and near-surface limestones within these licenses limit acquisition of interpretable seismic data at reasonable cost. Typical subsurface information that is required prior to drilling (depth and configuration of reservoir, thickness of stratigraphic units, fault attitudes and throw) is primarily interpreted from imagery, surface observations and sparse well data.

It has been shown that surface-geologic data (dip and strike of bedding, structural plunge, stratigraphic thickness, etc.) can be used to geometrically model subsurface structure in many fold and thrust belts (Dahlstrom, 1969; Hobson, 1986). The better the surface data, the better the subsurface interpretation. We have found stereoscopic SAR to be of significant value in our surface mapping and a tool that has improved subsurface geologic interpretation. In addition, the capacity of modern, airborne SAR imagery to be clearly and accurately enlarged to a working scale of 1:50,000 facilitates: (1) accurate location of field measurements, (2) evaluating potential well locations, (3) checking prior mapping, and (4) planning future field programs to solve critical geologic questions.

2.0 GEOLOGIC DATA AVAILABLE PRIOR TO SAR

2.1 Geologic Maps

Three geologic maps with explanatory notes that cover most of the exploration area have been published at a scale of 1:250,000 by the Australian

16

Bureau of Mineral Resources and the Papua New Guinea Department of Lands, Surveys and Mines. These maps are the Blucher Range (or OK Tedi), Wabag, and Kutubu sheets (Sheets SB/55-7, -8, and -12, International Index, respectively). The three sheets are compilations of arduous field work and careful photogeology carried out since the 1940's (see Figure 3 for sample). Limited traverses across the rugged anticlines provide dip, strike, stratigraphic contacts, and plunge information. These data, along with more recent field surveys, helped provide constraints on the interpretation from stereoscopic radar strips.

Estimating dip from radar imagery is difficult because no real parallax exists. Vertical exaggeration varies across each radar flight strip due to changing depression angle of the microwave beam (from about 8° to 27° below the horizontal; see Figure 4). Stereoscopic viewing is possible because of small displacements in terrain features and differences in shadow length on adjacent flight strips (R. H. Gelnett and F. F. Sabins, 1986, pers. comm.). A 30° slope on the ground will appear much different to the viewer depending on its position on the radar flight strip. In addition, inherent problems with radar arise because of "layover" (steep slopes leaning into the direction of radar acquisition). Layover causes geometrical inconsistencies (Sabins, 1986). Experience interpreting stereoscopic radar images is required to minimize error in estimating structural dip and plunge.

Geologic mapping indicates surface strata in the area to be predominantly massive Tertiary limestones (the Darai Formation), younger Tertiary silici-clastics, and late Pliocene-Pleistocene volcanics. Typically the anticlines have been formed as the structurally competent Darai Limestone was moved southward, above decollemont surfaces, and thrust over itself or the younger Orabadi (predominantly shale) Formation. This limestone is relatively more resistant to erosion and forms the surface outcrop and topographic relief over most of the anticlinal structures in the licenses. The major structural trend and topographic grain is northwest-southeast.

Volcanic cones began to form during late Pliocene time. Most are steep-sided, conical, strato-volcanoes with deeply dissected slopes and central craters; they are thought to be dormant or extinct (Brown and Robinson, 1982). Published geologic maps indicate a considerable portion of the study area is covered with volcanic rocks. Structural information associated with the folded and thrusted Tertiary rocks is largely lacking where volcanic debris has been mapped. However, as will be shown, the SAR data provide information in some areas which enabled some of the buried stratigraphy and structure to be interpreted beneath the volcanics.

2.2 Topographic Sheets

Excellent topographic sheets were generated from stereoscopic aerial photographs by the Royal Australian Survey Corps (scale 1:100,000; 40 m contour interval). Inherent distortions in the radar imagery due to layover in the high relief areas were minimized by transferring the SAR interpretation onto these topographic sheets. Subtle topographic contour inflections, trends, and slopes frequently have geologic significance which can be observed employing stereoscopic analysis of the SAR flight strips (see Figure 3). Changes in slope and elevation due to lithology or structure; linear trends revealed by fractures, karst trends, streams, and ridges; and oversteepened curvilinear features such as landslide scarps that were seen on the radar image could easily be located within the contour pattern of the 1:100,000 topographic sheets.

2.3 Landsat Imagery

Landsat MSS imagery is available with <20% cloud cover over the southern foldbelt of Papua New Guinea. Chevron Oil Field Research Company digitally processed two scenes to enhance color contrast and emphasize linear features.

17

As of this date, no TM imagery has been acquired over the southwestern Pacific because no data relay satellite (TDRS) is available for the region.

Standard color IR composites (BGR = 457) of the MSS data are almost completely red because of the extensive forest cover. Nevertheless, there is more information in the interactively processed, color IR images than in standard product images. Agricultural patterns and plots of land that have been slashed and burned dominate the lower flanks of the volcanoes north of the exploration licenses. Recent landslides on the limestone-capped anticlines are clearly revealed as white patches in an otherwise red landform. Vegetation changes with altitude across major anticlines can be detected, and in the lowlands tonal variations that may relate to drainage/topography anomalies and subsurface structure have been observed.

However, Landsat (and vertical aerial photographs) have subdued structural information because of the near-continuous jungle cover and minimal shadowing (the sun is 45 to 50° above the horizon on our Landsat MSS images at 9:30 AM local time).

2.4 Literature

Well illustrated publications for the southern fold belt of Papua New Guinea include the previously mentioned geologic maps with explanatory notes (Davies and Norvick, 1974; Davies, 1983; Brown and Robinson, 1982), Dow's (1977) geological synthesis, and Jenkins' (1974) structural analysis.

Radar technology and geologic interpretation of radar imagery are explained by Sabins (1983, 1986), Wing and MacDonald (1973), and Jensen and others (1977). The latter reference pertains to synthetic aperture radar data that are recorded as a hologram on film. For this survey the SAR data were recorded digitally.

3.0 SAR SURVEY DESIGN

Topography rises steeply toward the north, structure trends northwest-southeast, and thrusting is predominantly directed toward the southwest. Flights paths were east-west with the microwave beam directed northward to minimize shadowing and enhance detection of south-facing bedding and thrust fault traces. Also, by directing the beam at an acute angle to the structural grain, excessive signal returns (signature flare), caused by the many slopes and cliffs that face southwest, were reduced.

Although this survey design provided for the most economical and complete data acquisition, viewing the imagery with north up (shadows directed toward the top margin of the image) typically results in topographic inversion. Most of the radar imagery in this paper is arranged conventionally with north up. The reader may need to rotate this imagery 180° so that north and the shadows are projected downward in order to see the topography in proper perspective (ridges as highs and valleys as lows).

The airplane flew at an altitude of 9 km (30,000 ft) above ground level. Recording of the microwave returns was time-delayed an average of 18 km ground distance, resulting in a near range depression angle of 27° (Figure 4). Data were recorded from 18 km to 63 km for an image swath of 45 km. The depression angle of the radar beam in the far range was 8°. The stereoscopic overlap allowed shadowing to be relatively consistent near joint lines on the mosaic; it was constructed with an average depression angle of 17° (R. H. Gelnett, 1985, pers. comm). A non-stereo survey would have resulted in a poorer mosaic with "joint lines" between flight strips having long shadows of the far range juxtaposed next to short shadows of the near range.

The angle of incidence that the microwaves made with the terrain was

18

low, accentuating (through shadowing) subtle topographic relief, fracture patterns, and shallow-dipping beds. The drawback of such a low beam was an increase in the amount of area concealed in shadow in regions of high relief. This detrimental shadowing was minimized by analyzing the near range portion of the flight strips (Figure 5). Digital reprocessing of near-range radar imagery cannot improve those areas that are in shadow because there are no data in shadowed areas.

4.0   COMPARISON OF RADAR WITH COLOR PHOTOGRAPHS

Photographs taken during a helicopter field trip in November 1985 allow a comparison between oblique color photographs and corresponding radar images (Figure 6). Visible light and the microwave radar beam (0.4-0.7 μm and 3 cm long wavelengths, respectively) are both reflected from the near-continuous jungle canopy. Radar does not penetrate vegetation because the structure and water of leaves and limbs reflect the microwave energy (see Sabins, 1986).

Figure 6 shows a color photo contrasted with a close-up of radar imagery that covers the same geographical location; however, scale and viewing angle are different. The color photo demonstrates that the jungle canopy conforms with the relief of the underlying terrain; however, the top of the canopy is smoother than the ground surface. The radar-imaging process "smooths" the terrain even more because it averages the returned energy into synthetic 6 x 12 m ground-size cells. These cells have the effect of a low-pass filter. They minimize high frequency relief/reflectance signals. The resultant enhancement of longer wavelength features appears to reveal fundamental geologic features better than the color photos.

5.0   MINIMIZING RADAR SHADOW

The radar mosaic, which utilized the mid-range portion of each flight strip (17° depression angle), has shadows on the north side of major topographic highs that obscure terrain. However, in the near-range portion of the strips much more terrain is illuminated by the steeper depression angle (Figure 5). Within areas on adjacent flight strips that have unequal shadowing (such as the stereopair of Figure 7), it was found that during stereoscopic viewing the "illusion" of topographic relief was maintained in areas where one strip had the terrain illuminated (near-range portion) and the other strip had total shadow (mid- to far-range portion). This sense of topographic relief in areas that were illuminated on only one strip greatly enhanced the interpretation. The foregoing is important for exploration purposes because topography in such a young fold and thrust belt largely reflects geological structure (topographic highs often relate to surface structural culminations or anticlines; topographic lows tend to be zones of weaknesses, faulting, or synclines).

In the near range of individual flight strips, the side of a topographic high that faces away from the radar beam and the area that is cast in shadow cannot provide any information to the interpreter. Flying an entire survey with two look directions (opposite or orthogonal) can nearly eliminate this problem. However, a single, reverse-look flight strip across terrain heavily shadowed in the regional survey would markedly improve the data content across an area 45 km wide. A recent SAR survey planned jointly by Chevron and the contractors utilized a single, reverse-look swath to markedly reduce lost information due to shadowing at a nominal cost.

A Landsat image and a corresponding radar flight strip can be viewed as a stereopair to integrate their different data content, provided both data sets are at the same scale (Figure 7). However, because this exploration area is south of the equator, shadows can be cast toward the west-southwest on Landsat. These shadows oppose the northward-directed radar shadows and create eye strain during stereoscopic viewing.

19

395

6.0 STRUCTURAL ANALYSIS WITH SAR

In the subject area the radar image, especially when viewed stereoscopically, is superior to Landsat and aerial photographs for interpreting geologic structure. SAR clearly reveals more geologic information than shown on the published geologic maps (Figures 7-9). With SAR, strikes and dips, anticlines and synclines, and structural plunge can be mapped across terrain where little or no field work has been done.

SAR imagery was used successfully to identify major structural zones throughout the area using standard photogeologic mapping techniques and the stereoscopic radar flight strips. In one area, stratigraphic and structural continuity was interpreted for over 100 km along strike. These interpretations greatly simplify the modeling of subsurface structure associated with hydrocarbon entrapment.

The Amdi, Emuk, and Kaban Ranges are obscured in cloud cover on Landsat and lack detail on the published geologic map (Figure 7). However, the SAR imagery reveals a major fracture pattern that cuts across structure, provides a sense of displacement on many faults, and shows at least two anticlines not mapped on the published geologic sheets. We outlined these previously unmapped structures on the topographic sheets of the area to generate a detailed, surface-structure map using SAR stereoscopic imagery.

Volcanic cones imaged with SAR expand our knowledge of volcanic activity. The degree of dissection (moderate in Figure 8) gives a clue as to the time that volcanic activity ceased. The relationship of the volcanic strata with underlying structure gives a sense of timing for the development of anticlines and synclines in the area; this may be determined from the stereoscopic SAR. The volcanic strata may be (1) in stratigraphic continuity (folded) with the folded structures underneath (suggesting volcanism accompanied or predated the folding), (2) undisturbed or draped over pre-existing folds (suggesting the cones post-date the folding), or (3) markedly disrupted by faults and stratigraphic unconformities (suggesting the cones pre-date the folding).

The age of the volcanic cones and the relationship of cones with the underlying folds provide an estimate for the time of possible trap formation within anticlines.

Determining the extent of faulting (vertical and horizontal displacement, overall length) is of interest to our exploration programs. An example of correlating faults on geologic maps with radar imagery is shown in Figure 9. Two arrows on the radar image highlight a band of upturned, resistant strata that form a sharp topographic break. In the southeastern portion of this zone, a thrust fault is mapped on the geologic sheet. This fault is terminated as it trends northwest on the published map.

The radar image reveals the complex nature of this mapped fault and the surrounding terrain. The upturned strata appears to be topographically higher toward the northwest (left arrow on the radar image) where the Darai Limestone dipslope (Tr on Figure 9) grades smoothly into the plunging nose of the Juha anticline. Here there is no sharp topographic break at the Tr/Tma stratigraphic contact (see Figure 9) as there is farther east where the fault is mapped at the surface. The radar image can be interpreted with the mapped fault losing throw toward the northwest and becoming a blind thrust fault (not reaching the surface).

Two geologic sheets are joined along a north-south line in Figure 9. The SAR image helps extrapolate the greater detail of the eastern sheet onto the western sheet and enables smooth continuation of geology across the boundary.

20

7.0 STRUCTURES BENEATH THIN VOLCANIC FLOWS

The widespread Darai Limestone weathers to rugged karst topography that has a characteristic radar signature. The radar image of this limestone has an uneven, pitted appearance that is often dissected by long, linear grooves. Tones on the SAR images range from dark to bright on Darai-capped structures (see Figure 3) because of the irregular topography. Resistant limestone pinnacles over 40 m high, sinkholes with surface openings in excess of 100 m, and deeply weathered fractures dominate the surface of the Darai. Resistant strata within the Darai Limestone crop out with an irregular pattern due to the rugged surface morphology.

Pliocene to Pleistocene volcanics cover large portions of the fold and thrust belt. On the radar imagery, the surface morphology of some of the volcanic deposits can appear as karst terrain with well-developed sinkholes, pinnacles, and linear troughs. Here it is inferred that the Darai Limestone is buried beneath the volcanic cover. The radar is reflected from trees growing on the volcanics, generating a subdued but characteristic radar signature for the Darai Limestone (Figure 10).

The timing of the volcanic deposition and the formation of karst topography is poorly understood. The development of karst in the Darai Limestone may have been before or after the deposition of the volcanic debris (see Figure 10). Radar imagery is not conclusive; field work is required. Speleological expeditions along the northeastern margin of the PPL-17 and -18 exploration licenses indicate karsting was initiated in Late Pliocene to Early Pleistocene time (Francis and others, 1980, p. 111).

In addition, the timing of volcanism with folding and thrusting of the underlying Darai Limestone is not clear. As noted before, the volcanics may unconformably overlie the folds but in some areas are not thick enough to conceal them, or the volcanics may have been deposited as a "flat" blanket that subsequently was deformed by folding and thrusting (Figure 11).

Pliocene to Pleistocene volcanics are mapped (Brown and Robinson, 1982) across most of the area shown in Figure 11 (for regional see Figure 3). Two volcanic units dominate: TQvs (andesitic and basaltic agglomerates, tuff and lava...) and TQsl (volcaniclastic andesitic and basaltic breccia, reworked agglomerate, tuff, ...exotic blocks of limestone...). Agglomerates are chaotic assemblages of coarse, angular pyroclastic material - formed by volcanic explosion or aerial expulsion from a volcanic vent.

The geologic map (Figure 11) shows two faults, trending NW-SE, that converge toward the southeast, near "Landslide Mountain". The Darai Limestone (Tmd) caps an anticline (seen toward the northwest) that is mapped with an anticlinal axis symbol; this symbol terminates at the edge of the volcanics (TQsl). With the accompanying SAR image, this axis can be extrapolated southeast under the volcanics, and the plunge of the anticline can be inferred. Dip and strike information along the trend of this buried anticline also can be determined stereoscopically from SAR.

8.0 MASS WASTING

The high relief, heavy rainfall, pronounced undercutting of steep slopes by surface rivers, and extensive subsurface erosion caused by karst phenomena (solution and sapping) have resulted in widespread mass wasting in the fold and thrust belt. Giant landslide blocks and scarps can be confused with tectonic features or in-situ bedding, which leads to poorer structural interpretations. The radar images enabled us to recognize numerous mass-wasting zones and alerted us to re-evaluate published dips and strikes in the affected areas.

The detection of mass-wasting features on SAR helps on-going field

21

work by locating areas with fresh outcrops. Detection of faults, fractures, or stratigraphic unconformities within areas of mass wasting is effectively accomplished with stereoscopic SAR, unless the features are hidden by radar shadows. Satellite imagery, aerial photographs, and/or field work must be employed in shadowed areas.

One of the largest anticlines in the subject area (Mananda, Figure 3) has significant mass wasting along its flanks. The topographic slope along the northeastern nose is extreme (see simplified topographic map - 400 m contour interval, Figure 3). Vertical relief changes >1000 m over a horizontal distance of <4000 m. Here the Hegigio River is cutting a steep gorge. Rock slides are readily seen on the Landsat image (Figures 3 and 12). The bright white patch on the Landsat is a recent rockslide that exposes 1 km of fresh Darai Limestone. The limestone strata are steeply dipping and unvegetated. These rock slides are evident but partially hidden in shadow on the radar mosaic, and they are not mapped on the geologic sheet.

On the radar image, large WSW-ENE topographic breaks are recorded that cut the eastern nose of the anticline; these are very subtle on Landsat and unmapped on the geologic map (Figure 12). The downward extent of these breaks is unknown. They may affect reservoir continuity at depth, or they may be restricted to the Darai Limestone cap and represent giant blocks moving downslope toward the down-cutting Hegigio River (see Figure 3). The anticline steps down hundreds of meters toward the southeast across these blocks. The Hegigio River gorge, where it crosses and truncates the nose of the Mananda anticline, may reveal a zone of weakness at depth.

At other anticlines in the area, SAR clearly reveals large slump blocks caused by rivers undercutting and oversteepening the anticlines' flanks. These blocks have an arcuate headward scarp opening toward the downslope valley. Accurate mapping of these actively eroding landforms is a prerequisite for pipeline and logistical planning within the valleys of the area.

A north-south field traverse crossing the southern flank of Mananda is shown on the geologic map by dip and strike symbols (Figure 3). Examination of the radar image reveals an apparent topographic depression along this southern flank that is also marked on the topographic map by an increase in the horizontal distance between the 800 and 1200 m contour lines (Figure 3). Elevations decrease by 1000 m from the anticlinal axis to this synclinal axis, across a horizontal distance of only 6000 m. This steep relief has facilitated massive rock slides and slumps of the Darai Limestone caprock. Fissures and erosional scarps (generally parallel to topographic contour) and lobes of slumped material occur along the entire southern flank of Mananda. Many fractures cut across the axis of this anticline. These fractures can be followed from the stable crest southward where they are often aligned with an edge of a slump or slide scar.

Dips and strikes determined in the field on the southern flank of Mananda must be reconciled with the radar evidence of severe surface disruption (Figure 3). Interpretation of reservoir geometry and depth based on surface evidence may be in error if SAR is not used to detect mass wasting.

Down-dropped blocks of Darai Limestone (to 2 x 2 km in area) along the extended crests of anticlines have been detected with SAR (Figure 13). Such depressions may reflect (1) structurally controlled, downdropped grabens due to crestal extension during folding, (2) zones of thinner limestone created by increased subsurface erosion (water, solution, and sapping concentrated by associated fracture systems), (3) incipient rock slides caused by fluvial erosion at the base of the anticline's flanks, or (4) the surface expression of a deep fault and fracture system that has major exploration significance. Only field work and drilling can determine what geological process(es) controls the development of crestal depressions.

22

9.0 LITHOLOGIC DISCRIMINATION

Although the terrain is covered with a jungle canopy, many lithologies can be recognized by their characteristic radar signatures (see Sabins, 1983). The typical slope, relief, and resistivity to weathering of each lithologic category affects the morphology of the terrain. This typical morphology is reflected in the surface of the jungle canopy and is imaged with SAR.

The Darai Limestone, a resistant ridge former that is prone to intense karst weathering, is easily identified (see Figure 3). Non-resistant slope formers (Tertiary mudstones, shales and siltstones) are shown in Figure 6. Quaternary alluvium, a medium-gray, flat surface on SAR, is also easily identified in the bottoms of valleys (Qk on Figure 11).

However, localized field work associated with our exploration program differentiated the Tertiary sedimentary rocks into (1) stratified volcaniclastics, (2) siltstones, (3) a weak limestone unit, and (4) the more resistant Darai Limestone (Figure 14). These four units crop out along relatively flat to gently sloping terrain (Figure 14). The accompanying radar image has labeled arrows pointing to areas where the field work confirms the type of rock cropping out beneath the jungle canopy.

The stratified volcaniclastics (generally Tpw on the geologic map, Figure 14) are seen on the SAR and in the field as beds with characteristic resistant strata. This rock type crops out along the southern margin of the Juha anticline and also along the northern flank in a northwest-southeast trending band. On the geologic map, Tpw is only mapped along the southern perimeter of the anticline.

Darai Limestone has a more limited outcrop pattern on the radar imagery as compared with the map (Figure 14). The radar shows it cropping out within a gorge that cuts across the anticline, but there does not appear to be the typical radar signature of the Darai along the crest of the anticline as depicted on the map. Along the crest the radar records a grainy, coarse texture that is seen in the field as a siltstone. A weaker limestone (generally Tma on the map) is imaged by SAR with a lighter tone and slightly smoother texture than the siltstone. These subtle radar and field observations about lithology can be extrapolated to areas where no field data exist to improve surface stratigraphic maps and, therefore, subsurface structural interpretations.

The radar image in Figure 14 is from a mosaic composed of 1:50,000 enlargements of the near- and mid-range portions of two flight strips. The splice lines can be seen intersecting near the center of the SAR image in Figure 14. It can be seen that radar signatures for the same lithology vary across the east-west splice line. The northern and the southern strips have ~8° and ~17° depression angles along the east-west splice line, respectively. Differences in the intensity of radar returns and; therefore, the resulting image signatures (Sabins, 1983) may result from the different depression angles.

10.0 DIGITAL REPROCESSING OF SAR IMAGERY

Reprocessing of the digital data allowed additional geologic information to be extracted from the survey. The original photographic strips were contrast-balanced for the entire scene. However, cliffs that faced the radar antenna reflected (as expected) a relatively high amount of radar energy, which resulted in oversaturated (bright) or "flared" zones on the original strips (Figure 15A).

The radar digital data shown in Figure 15A and B were resampled during processing and 10 x 10 m pixels were generated from the original ~4.2 x11.4 m

23

pixels. The subscene in Figure 15B is a full resolution image of this resampled data as it was displayed on the image processing monitor. The number of pixels in this undecimated subscene is 262,000 (the monitor displays 512 x 512 pixels); therefore the area displayed is only ~5 x 5 km². This area is outlined in Figure 15A. Although this contrast stretching was accomplished on a relatively small area, full-resolution stretching was found to be more informative than working with decimated images that covered a larger ground area.

The oversaturated (bright) cliffs outlined on Figure 15A had reflectances that ranged from 100 to 210 on a scale of 0 (pure black) to 255 (pure white). The subscene (Figure 15B) was interactively manipulated and the example shown is one where reflectance values below 100 were saturated to pure black (0) and those from 100 to 210 were stretched over the full range of 0 to 255. Contrast stretching of the brightest digital values (at the expense of the rest of the scene) does reveal subtle topographic and structural information along south-facing cliffs. This new information could not be discerned on the original, balanced imagery.

11.0 REFERENCES

Brown, C. M. and G. P. Robinson (compilers), 1982, Kutubu, Papua New Guinea 1:250,000 Geological Series - Explanatory Notes, Sheet SB/54-12 International Index: Department of Minerals and Energy, Australia and Geological Survey of Papua New Guinea, 43 p., 1 map sheet.

Davies, H. L., (compiler), 1983, Wabag, Papua New Guinea 1:250,000 Geological Series - Explanatory Notes, Sheet SB/54-8 International Index: Department of Minerals and Energy, Australia and Geological Survey of Papua New Guinea, 84 p., 1 map sheet.

Davies, H. L. and M. Norvick (compilers), 1974, Blucher Range, Papua New Guinea 1:250,000 Geological Series - Explanatory Notes, Sheet SB/54-8 International Index: Department of Minerals and Energy, Australia and Geological Survey of Papua New Guinea, 29 p., 1 map sheet (OK Tedi is revised name of sheet).

Dahlstrom, C. D. A., 1969, Balanced cross sections: Canadian Journal of Earth Sciences, v. 4, no. 6, p. 743-757.

Dow, D. B., 1977, A geological synthesis of Papua New Guinea: Bureau of Mineral Resources, Australia, Bulletin 201, 41 p.

Francis, G., J. M. James, D. S. Gillieson, and N. R. Montgomery, 1980, Underground geomorphology: in J. M. James and H. J. Dyson, eds., Caves and Karst of the Muller Range - Exploration in Papua New Guinea: A. T. Sutton & Co. (ISBN 086758 042 9), Australia, p. 111.

Hobson, D. M., 1986, A thin-skinned model for the Papuan thrust belt and some implications for hydrocarbon exploration: Australasian Petroleum Exploration Association Ltd. Conference, Adelaide, April 7-9, 1986, v. 26, part 1, p. 214-224.

Jenkins, D. A. L., 1974, Detachment tectonics in western Papua New Guinea: Geological Society of America Bulletin, v. 85, p. 533-548.

Jensen, H., L. C. Graham, L. J. Porcello, and E. N. Leith, 1977, Side-looking airborne radar: Scientific American, v. 237, no. 4, p. 84-95.

Sabins, F. F., 1983, Geologic interpretation of space shuttle radar images of Indonesia: American Association of Petroleum Geologists Bulletin, v. 67, no. 11, p. 2076-2099.

24

Sabins, F. F., 1986, Remote sensing - principles and interpretation, second edition: W. H. Freeman & Co., New York, 449 p.

Smith, J. G., 1965, Orogenesis in Western Papua and New Guinea: Tectonophysics, v. 2, no. 1, p. 1-27.

Thornbury, W. D., 1969, Principles of Geomorphology, second edition: Wiley, New York, p. 303-344.

Wing, R. S. and H. C. MacDonald, 1973, Radar geology - petroleum exploration technique, Eastern Panama and Northwestern Colombia: American Association of Petroleum Geologists Bulletin, v. 57, no. 5, p. 825-840.

13.0 ACKNOWLEDGEMENTS

We are grateful to Chevron Overseas Petroleum Inc., Niugini Gulf Oil Pty. Ltd. (a Chevron-owned company and operator), and the Joint Venture Partners of the PPL-17 and PP1-18, Papua New Guinea exploration licenses:

    Merlin Petroleum Company
    Pioneer Concrete (Bougainville) Pty. Ltd.[1]
    BP Petroleum Development Australia Pty. Ltd.[2]
    Ampol Exploration Pty. Ltd.
    Australasian Petroleum Company Pty. Ltd.
        (composed of BHP Petroleum Pty. Ltd., Oil Search Ltd.,
        and 1 and 2 above)

for granting permission to publish this paper. We also express our appreciation to the Bureau of Mineral Resources, Australia and the Geological Survey of Papua New Guinea for approval to reproduce their excellent geological maps. The staff at Niugini Gulf Oil Pty. Ltd. (in particular, E. H. Gurney and H. E. James) were instrumental in providing geological information to us and arranging for a field visit to the southern fold and thrust belt of Papua New Guinea. We acknowledge proprietary data made available to the operator.

The Chevron Overseas Petroleum Inc. graphics and word processing personnel were extremely helpful in preparing this paper.

Mars Associates, Inc. constructed an excellent radar mosaic and completed the initial interpretation of the SAR data. R. H. Gelnett of Mars Associates, Inc. was especially helpful in organizing the survey and ensuring we understood the interpretative aspects of SAR imagery. Intera Technologies Ltd. acquired the outstanding imagery using an X-band, synthetic aperture STAR-1 digital radar system. Rob Inkster of Intera has assisted in improving our understanding of the technical aspects of SAR acquisition.

F. F. Sabins, W. S. Kowalik and T. F. Battey of the Remote Sensing Lab at Chevron Oil Field Research Company accomplished the digital manipulation of the SAR and Landsat data. F. F. Sabins also provided timely geological and technical advice, and reviewed our paper.

25

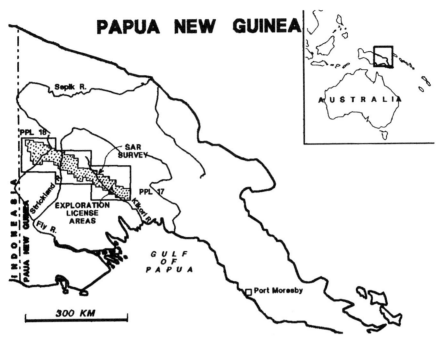

Figure 1. Location map showing Papua New Guinea, area of exploration licenses PPL-17 and -18, and area surveyed by SAR (~26,000km²).

Figure 2. Darai Limestone dips steeply on the flank of Kutubu anticline. Small huts can be seen along the shoreline for scale. Topographic relief from lake to crest of anticline is 500 m. Rugged karst topography has developed on the limestone, making seismic technology impractical.

26

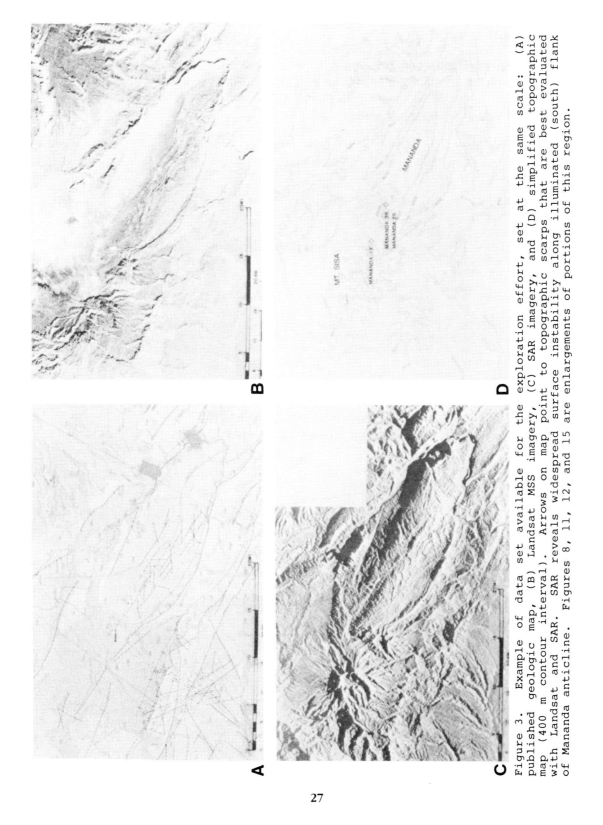

Figure 3. Example of data set available for the exploration effort, set at the same scale: (A) published geologic map, (B) Landsat MSS imagery, (C) SAR imagery, and (D) simplified topographic map (400 m contour interval). Arrows on map point to topographic scarps that are best evaluated with Landsat and SAR. SAR reveals widespread surface instability along illuminated (south) flank of Mananda anticline. Figures 8, 11, 12, and 15 are enlargements of portions of this region.

27

AIRBORNE RADAR SHADOWING EFFECT

LOW DEPRESSION ANGLES ENHANCE DETECTION OF SUBTLE RELIEF CHANGES (FRACTURES, STREAM CHANNELS, VEGETATION, FAULTS, DOMES, INTRUSIVE BODIES, FOLDS, JOINTS)

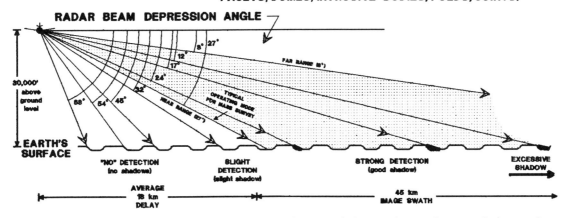

Figure 4. Profile of airborne radar beam with various depression angles and shadowing of subtle topographic relief. This survey's depression angles ranged from 27° in the near range to 8° in the far range; an average of 17° was used in the mosaic.

SIDE-LOOKING AIRBORNE RADAR WITH AVERAGE 17° DEPRESSION ANGLE SHOWING DECREASING SHADOWS DUE TO OVERLAP BETWEEN FLIGHT LINES.

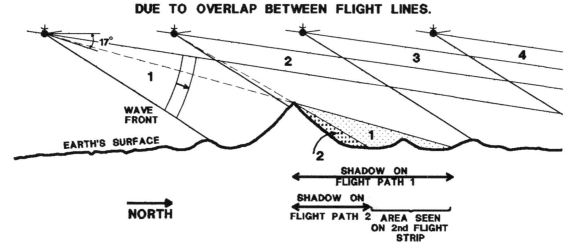

Figure 5. Profile of overlapping SAR flight lines resulting in stereoscopic flight strips and reduction in the area hidden by radar shadow. The detrimental effect of shadowing due to topography was minimized by mapping in the near-range portion of the stereo flight strips.

28

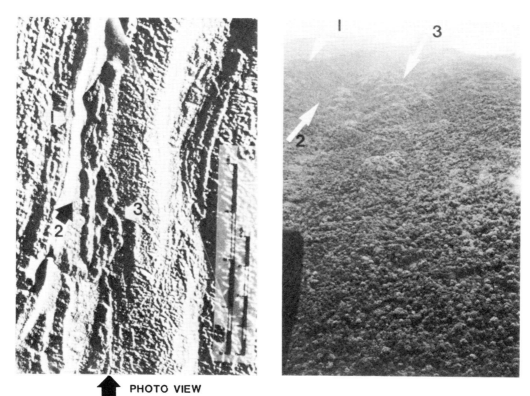

PHOTO VIEW

Figure 6. Comparison of radar image with color photo of a (1) thrust fault carrying Darai Limestone over shales, (2) stratigraphic contact between shales and resistant limestone, and (3) outcrop pattern of Darai strata folded into an anticline. Pitted and irregular radar signature of deeply weathered limestone contrasts with smooth, medium-gray signature of shales.

29

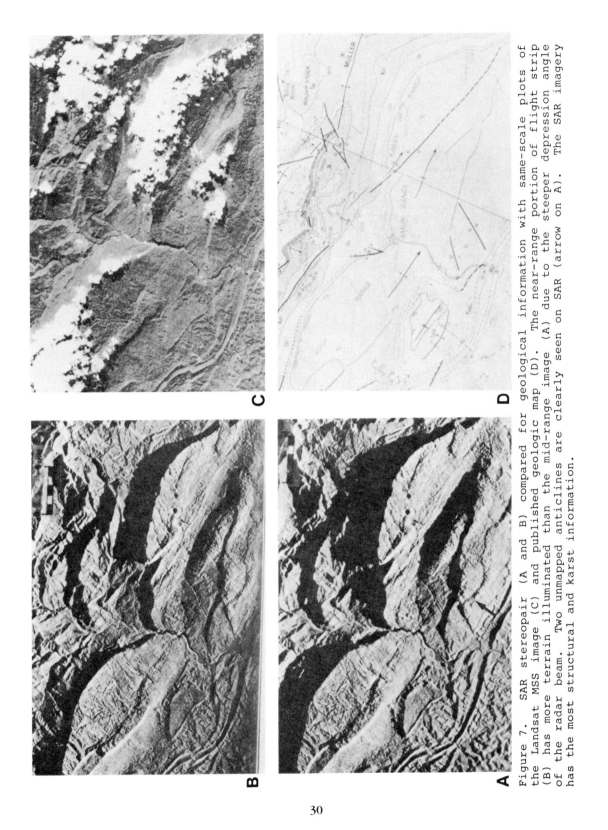

Figure 7. SAR stereopair (A and B) compared for geological information with same-scale plots of the Landsat MSS image (C) and published geologic map (D). The near-range portion of flight strip (B) has more terrain illuminated than the mid-range image (A) due to the steeper depression angle of the radar beam. Two unmapped anticlines are clearly seen on SAR (arrow on A). The SAR imagery has the most structural and karst information.

30

406

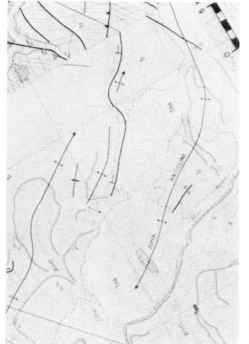

Figure 9. A band of resistant, upturned Darai Limestone (Tr) is highlighted with 2 arrows on the SAR image (NE is toward the upper margin of the same-scale image and map). A thrust fault is mapped only along the SE base of this strata. See text for discussion (Section 6.0).

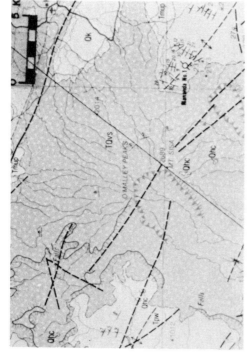

Figure 8. Suspected faults mapped on the geologic sheet as trending NW-SE through craters can be expanded upon with SAR. SAR shows valleys that trend NE-SW as deeply eroded into flanks of Mt. Sisa, suggesting a major zone of structural weakness.

31

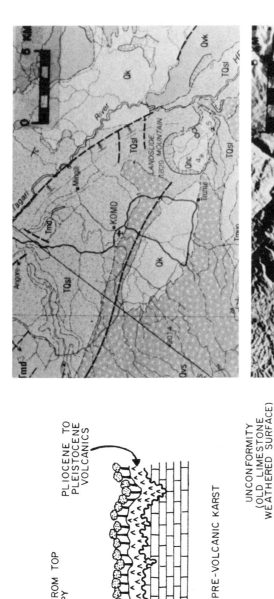

Figure 11. Folded Darai limestone (Tmd) can be inferred, and plunge determined, beneath Pleistocene volcanics (TQsl, TQvs). The geologic sheet properly shows volcanics at the surface; however, for hydrocarbon exploration the buried structure evident on SAR is of paramount importance.

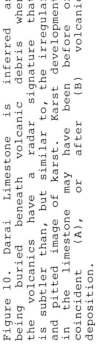

Figure 10. Darai Limestone is inferred as being buried beneath volcanic debris when the volcanics have a radar signature that is subtler than, but similar to, the irregular and pitted image of karst. Karst development in the limestone may have been before or coincident (A), or after (B) volcanic deposition.

32

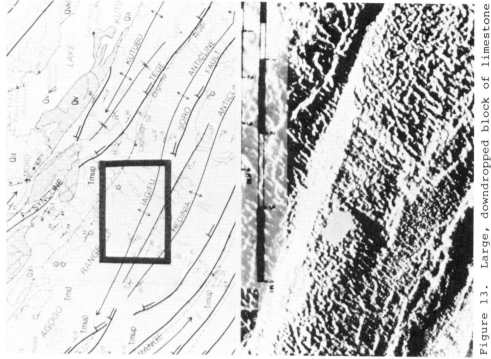

Figure 13. Large, downdropped block of limestone caprock seen along crest of an anticline on radar, but unmapped on smaller-scale map. The importance of these depressions to exploration depends on the geological process(es) controlling their formation (see Section 8.0).

Figure 12. Recent rockslides along the NE nose of Manada anticline are best seen on Landsat as white scars. Major WSW-ENE topographic breaks across nose of anticline obvious on radar but very subtle on Landsat. See text for discussion (Section 8.0).

33

Figure 15. Digital reprocessing (contrast stretching) of SAR subscene (B) obtained additional geologic/topographic information from oversaturated (bright) cliffs that faced the radar antenna. This contrast stretching degraded (as expected) the rest of the scene (see Section 10.0).

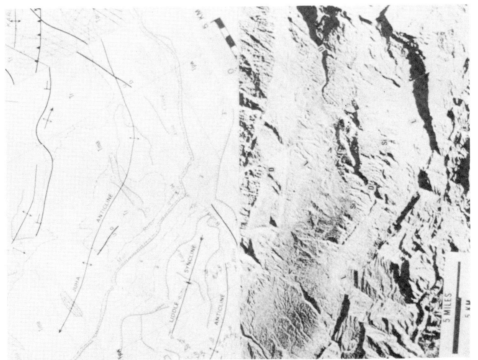

Figure 14. SAR image shows volcaniclastics (V), siltstones (Si), weak limestones (L), and tough Darai Limestone (D) at surface. NE is toward the upper margin of the same-scale image and map. Radar signatures for the same lithology vary across the E-W splice line (see Section 9.0).

34

410

The American Association of Petroleum Geologists Bulletin
V. 57, No. 5 (May 1973), P. 825–840, 10 Figs., 2 Tables

Radar Geology—Petroleum Exploration Technique, Eastern Panama and Northwestern Colombia[1]

RICHARD S. WING[2] and HAROLD C. MACDONALD[3]
Princeton, New Jersey 08540, and Fayetteville, Arkansas 72701

Abstract Petroleum exploration in eastern Panama and northwestern Colombia has gained impetus by recent side-looking-radar, geologic reconnaissance mapping. Radar-derived geologic information is now available for approximately 40,000 sq km where previous reconnaissance investigations have been extremely limited because of inaccessibility and almost perpetual cloud cover.

With radar imagery as the sole source of remote-sensing data, the distribution, continuity, and structural grain of key strata provide evidence that the eastern Panamanian Isthmus can be divided into three main physiographic-structural parts: two composite coastal mountain ranges separated by the taphrogenic Medial basin, which trends southeastward from the mouth of the Bayano River to the Atrato River valley of northwestern Colombia. Within the Medial basin, most of the clearly exposed surface structures are not particularly attractive petroleum prospects because prime reservoir strata have been stripped from their crests. However, several large geomorphic anomalies which have been mapped in the Medial basin may be reflections of subsurface structures having a complete stratigraphic section. Possibilities for gravity-type hydrocarbon accumulations in fractured organic shales, siltstones, and carbonate rocks are suggested within several synclinal elements along the axis of the Medial basin. The southwestward extension of the Medial basin trend, coincident with the western Gulf of Panama, may have potential as a future petroleum-producing province. A relatively thick marine stratigraphic section should be present here, with associated paralic and deltaic clastic rocks derived from acidic San Blas terrace since mid-Miocene time. The occurrence of active shell bars in the Bay of San Miguel and present reef trends on the northern Caribbean coast suggest possible offshore sites for geophysical surveying.

INTRODUCTION

Petroleum exploration techniques, which in previous years were highly dependent on exploitation of visible-spectrum sensors, have been augmented by a new family of other-wavelength remote sensors. No longer is the petroleum geologist restricted to the analysis of conventional aerial photography for reconnaissance studies, although cameras *per se* have by no means been displaced or abandoned. Aerial photography, although it provides valuable data for the compilation of highly accurate maps, is severely limited by light and weather conditions. Thus, many regions of the world, especially the tropics, are poorly mapped because adequate photographs cannot be obtained. Radar imaging systems provide the capability of obtaining terrain data independent of most weather conditions, and this information may be sufficient to provide a data base for geologic reconnaissance map construction.

Radar has been used extensively in aerospace, astronomy, meteorology, and military studies, but its use for oil and gas exploration has not received wide publicity. Most of the radar imagery used in past and present geologic investigations has been collected over areas in the U.S. where temperate climatic conditions prevail and where adequate geologic field data are generally available. The extension of our experience and knowledge to a climatic environment where geologic data are extremely limited has been made possible through the availability of extensive radar imagery covering approximately 40,000 sq km in eastern Panama and northwestern Colombia. This part of Central America serves adequately as an area typical of tropical, cloudy environments with dense vegetal cover, characteristic of most countries in equatorial latitudes where adequate geologic mapping is missing.

The primary objective of this study was to determine the utility of radar in the compilation of geologic reconnaissance maps for petroleum exploration. That radar is useful in certain terrain studies was known, but the degree of geologic reconnaissance information that radar will provide the petroleum geologist in the tropical environment had not been well documented. To determine this, a geologic reconnaissance map of eastern Panama and northwestern Colombia was constructed and evaluated. This region is characterized by areas where existing geologic maps have been compiled from such limited and scat-

[1] Manuscript received, September 12, 1972; accepted, November 15, 1972. Paper presented before the Association at the 57th Annual Meeting, Denver, Colorado, April 17, 1972.

[2] Advance Geology Group, Continental Oil Company.

[3] Department of Geology, University of Arkansas.

This study was conducted at the Center for Research in Engineering Science, University of Kansas. Supported by NASA contract No. NAS-9-7175, and USAETL contract No. DAAK 02-68-C-0089. Monitored by U.S. Army Engineer Topographic Laboratories Geographic Information Systems Branch, Geographic System Division, Ft. Belvoir, Virginia.

tered field data that they usually fail to delineate regional continuity of geologic formations and structure.

RADAR IMAGERY ACQUISITION

The radar imagery, covering all of Darien Province, Panama, part of northwestern Colombia, and much of east-central Panama, was obtained in 1967 and 1969 by Westinghouse for the U.S. Army Engineer Topographic Laboratory (USAETL) during less than 20 hours of actual imaging time. Prior to this, during the preceding 30-year period, there had been minimal success in obtaining conventional aerial photographic coverage because of almost perpetual cloud cover, especially over the mountains and foothills.

The 1967 and 1969 radar imagery is the product of a high resolution AN/APQ-97 K-band Side Looking Airborne Radar (SLAR) system. The 1967 imagery provided a slant-range display that progressively incorporates compression in the near range, i.e., a geometric distortion just the opposite of that inherent in conventional oblique photographs (MacDonald and Waite, 1971). In contrast to the 1967 data, the 1969 imagery consists of a ground-range display, which means that a hyperbolic sweep correction was applied to radar signals returned from the ground. Theoretically, this would result in a planimetric display of the terrain (i.e., no compression in the near range), if the terrain were perfectly flat. In areas having rugged topography, however, considerable radar imagery distortion can occur (MacDonald and Waite, 1971). Much of the ground-range 1969 imagery, of same-look and with up to 50 percent overlap, lent itself to stereoscopic viewing. Individual radar strips of approximately 18 km width have been joined to provide the uncontrolled radar mosaic in Figure 1.

The operation of side-looking radar can be summarized by examining Figure 2, where a microwave antenna (A) is repositioned laterally by the velocity of an aircraft (Va) carrying the SLAR system. Each radar-frequency pulse transmitted (B) returns signals from the targets within the beam width. These target returns are converted to a time/amplitude video signal (C), which is imaged as a single line (E) on photographic film (F). Returns from subsequently transmitted pulses are displayed on the cathode ray tube (CRT) at the same position (D) as the previous scan lines. By moving the photographic film past the CRT at a velocity (Vf) proportional to the velocity of the aircraft, an image of the terrain is recorded on the film (F) as a continuous strip of radar imagery.

HISTORY OF GEOLOGIC MAPPING—EAST-CENTRAL PANAMA AND NORTHWEST COLOMBIA

East-central Panama has been the least mapped, geologically, of any part of eastern Panama. Terry (1956) provided a geologic reconnaissance map which was lacking in all but the grossest structural aspects of this part of Panama, and included only speculative versions of even supposed major faults. This fact is mentioned, not to detract from his epic work, but only to illustrate the state of geologic knowledge prior to the recently available radar imagery, which has made far more detailed and more reliable reconnaissance geologic mapping possible. The climate is tropical, with vegetation canopy continuous and terrain marked by extremes of inaccessibility. Roads do not exist in most of the region; even the Pan American Highway has not been constructed through this part of Panama. Outcrops commonly are deeply weathered and poorly exposed. Only during the 3 or 4 months of the so-called dry season, beginning around late January, is field work feasible; however, the term "dry season" is misleading, because in some years up to 30 cm of precipitation have been recorded during this period.

Terry's (1956) geologic observations are based on field work and overflights in the Republic of Panama between the years 1920 and 1949. In 1966, a geologist with the Panama Canal Commission, published a report on the general geology of Bayano basin, based on field work through the period from 1955 to 1965 (Stewart, 1966). More recent ground reconnaissance studies have been completed in easternmost Darien Province, Panama, and northwestern Colombia by the Corps of Engineers (Interoceanic Canal Study Comm., 1968, 1969).

Prior to radar imaging, Darien Province was better mapped geologically than other provinces in eastern Panama, however, this previous mapping was quite general and in many places incorrect. For example, entire landforms were offshape or misoriented. Nevertheless, various gold mining operations and some petroleum exploration had resulted in a smattering of publications and mapping of gross elements as summarized on Terry's (1956) map. Shelton (1952) provided geologic data for the central part of the Tuira-Chucunaque subbasin. Some additional data have accrued from limited wildcat drilling (Johnson and Headington, 1971); however, the Interoceanic Canal Study Comm. (1968) study represented the most recent geologic data available for Darien prior to radar geologic reconnaissance mapping (MacDonald, 1969). Since completion of the 1969 radar mapping a new geologic map

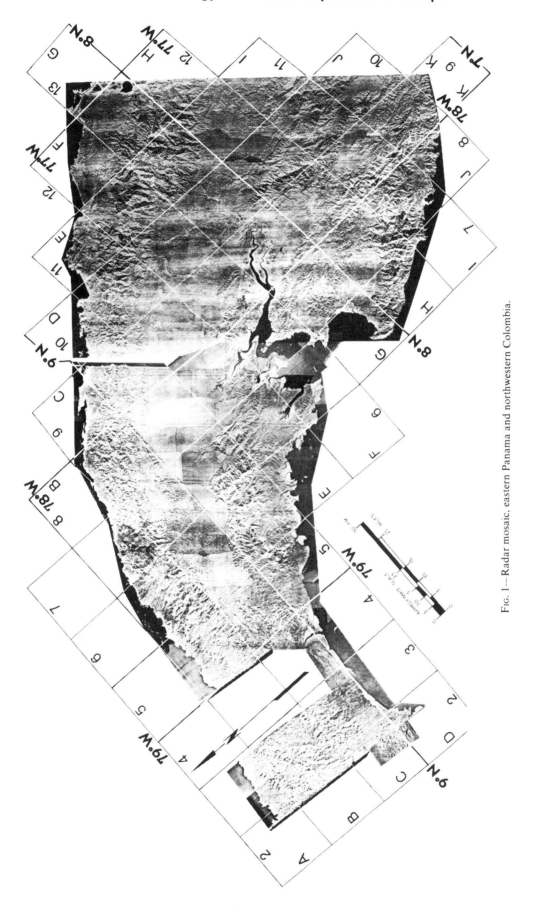

FIG. 1—Radar mosaic, eastern Panama and northwestern Colombia.

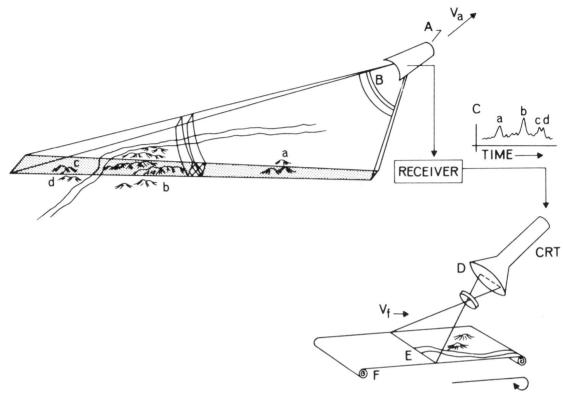

FIG. 2—Sketch diagram, typical side-looking airborne radar system.

has been issued which covers a small part of eastern Panama (Rió Pito, 1:40,000, Direccion General de Recursos Minerales, República de Panamá, 1972).

IMAGERY ANALYSIS

Radar imagery has been shown to be a most valuable supplementary tool in mapping surface-expressed geologic phenomena in many diverse physiographic and structural environments (Wing and Dellwig, 1970; Wing et al., 1970; Gillerman, 1970). In the tropical environment, where field work is at best difficult and sometimes impossible, radar may be the only adjunct to field surveys. Radar imagery is often superior to vertical air photos for display and detection of surface fractures—i.e., faults and megajoint zones (MacDonald et al., 1969). Although vertical air photos may reveal more clearly the smaller details of structural elements and patterns, radar imagery can show, as well or better, the true nature and extent of structural patterns—the proverbial forest as well as the trees. Radar imagery is also superior to conventional photography for detection of land-water boundaries, because smooth water is a specular reflector (for radar wavelengths) and consequently returns little of the incident signal to the aircraft antenna.

There are, of course, limitations. For example, most earth-science interpreters have found that radar imagery favorably and preferentially displays those topographically expressed linears most nearly parallel with the flight track of the aircraft (orthogonal to radar-look direction). Conversely, linears most nearly parallel with the look direction are minimized for lack of shadowing, and some to such an extent that a small percentage of linears are not detectable at all (MacDonald, 1969). For areas of extremely rugged terrain, where extensive radar shadowing is encountered, trade-offs between loss of geologic data and radar geometry must be taken into account during mission planning sessions (MacDonald and Waite, 1971).

There is essentially no penetration of vegetation by the SLAR imaging system when operating at K-band radar frequencies (MacDonald, 1969), and consequently, in the jungle environment, only the canopy surface of the dense foliage is imaged. Whether depicted on air photos or displayed on radar imagery, the canopy surface mirrors topography, reflecting outcrops which have been weathered and eroded differentially according to varying resistance of diverse rock types, structural attitudes, and fractures. Although conventional vertical air photos tend to show a welter of canopy irregularities, the inte-

Table 1. Summary of Key Stratigraphic Data, Eastern Panama

Age	Formation	Thickness	Description
			TUIRA-CHUCUNAQUE SUBBASIN
Early Pliocene-late Miocene	Chucunaque		Sandstone and shale containing foraminiferal fauna; occupies narrow inner trough of SD-VII
Middle Miocene	Pucro	610 m	Calcareous sandstone and thin limestone beds
	Gatun	1,070 m	Shale, underlain by conglomerate, sandstone, and thin limestone lenses
Early Miocene	Aquauqa	610-1,370 m	Bentonitic and bituminous shale, limestone, and sandstone
Late Oligocene	Arusa	460 m	Massive, brownish, tuffaceous marlstone
		310 m	White-weathering, tuffaceous, massive mudstone
Middle early Ologocene	Clarita	300 m	Tuffaceous carbonate rocks (mostly limestone) and sandy limestone
Late Eocene	Corcona	350-400 m	Shale, sandy, marine, calcareous; massive agglomerate, conglomerate, limestone (includes volcanic material)
Middle early Eocene			Chert (possibly equivalent to Quayaquil chert of Eduador), limited distribution (Pacific Hills area)
Pre-late Eocene	Basement		Andesite flows; crystalline rocks (somewhat silicic on Caribbean side)
			BAYANO SUBBASIN
Late Miocene		915-1,220 m	Soft, argillaceous siltstone, Foraminifera
Middle Miocene		460 m	Reef-type nearshore limestone beds, 15 km long; present on south side of SD-III syncline
Early Miocene			Sandstone and siltstone
Late Oligocene			Tuffs, conglomerates, calcareous and tuffaceous sandstone, and tuffaceous siltstone
Middle early Oligocene			Limestone (calcareous algae, Foraminifera)
Late Eocene			Sandy shale, agglomerate, conglomerate, tuffaceous and calcareius sandstone, tuffaceous siltstone
Pre-late Eocene	Basement		Andesite flows; crystalline rocks (silicic on Caribbean side; mafic on Pacific side)

gration inherent in the relatively coarse SLAR resolution cell has a smoothing effect which facilitates discrimination of large-scale structural patterns and major elements thereof. Contrasting types of vegetation grow selectively on the most favorable types of strata or on regolith derived from different types of underlying bedrock. Thus, some vegetal patterns and types can be differentiated on SLAR imagery and lithic inferences can be made (Lewis and MacDonald, in press).

Radar terrain return depends on: (1) surface roughness, (2) incidence angle of the slanted microwave beam, (3) polarization, (4) wavelength of the electromagnetic signal used to "illuminate" the landscape, and (5) the complex dielectric constant. As with aerial photography, interpretation of the radar display involves considerations of tone, texture, shape, and pattern. Tone is the intensity of signal backscatter recorded on film as shades of gray, from black to white. Texture, in simplest terms, is the degree of erosional dissection, and variously textured drainages can be used to infer lithologic type. For example, igneous outcrops generally have a massive appearance on radar imagery and are distinguished by "rugged and peaked divides." Shape alludes to telltale outlines of surface-expressed features such as those of alluvial fans, dikes, igneous plugs,

volcanos, and folds. Pattern is the arrangement of geologic, topographic, or vegetal features such that one may distinguish linears—e.g., fractures (faults and megajoints), bedding-trace grains, en échelon folds, etc. Thus, it is possible to extract considerable geologic data through careful interpretation of this imagery (Wing, 1971).

Probably the most important interpretive phase of geologic reconnaissance mapping with radar imagery is selecting rock units that have sufficient areal extent to be significant and are still distinctive enough on the imagery to be easily mapped. One is generally restricted by the scale of the imagery, but the primary limitation for the differentiation of rock units (either on aerial photographs or radar imagery) is topographic relief. Depending on the rock type involved, climatic conditions, thickness of the regolith and vegetal canopy, radar imagery may reveal abundant interpretive data, or it may show almost none. The geologic data interpreted from radar imagery for this study were derived from surrogates rather than through direct identification. Similarly, the time-stratigraphic assignments used for geologic map units must be inferred from corroborative data, and are obviously not interpretable from radar imagery. A summary of key stratigraphic data from the previously mentioned geologic reports is presented in Table 1.

Geologic reconnaissance mapping of the area shown in Figure 1 was presented in a study by Wing (1971) which incorporated prior radar mapping of Darién Province by MacDonald (1969).

PETROLEUM PROVINCES

Through examination of the radar mosaic (Fig. 1), eastern Panama and northwestern Colombia can be divided into three main physiographic-structural parts: two composite coastal ranges separated by a taphrogenic Medial basin (Wing, 1971), which extends southeastward the length of the eastern Isthmus, from the Gulf of Panama on the Pacific, to the Atrato River Valley of northwestern Colombia. A more detailed subdivision of major structural and physiographic elements has been derived from the interpretation of individual imagery strips (Wing, 1971). The resultant classification has been outlined in Figure 3 and indexed in Table 2.

Medial Basin

The Medial basin can be divided into three main segments. The transverse Canazas platform (SD-V and VI of Fig. 3) structurally separates the Bayano subbasin (west) from the Tuira-Chucunaque subbasin (east). The Bayano structural subbasin includes Medial basin subdivisions SD-I, II, III, and IV. The Tuira-Chucunaque structural subbasin includes Medial basin subdivisions SD-VII and VIII. The Bayano subbasin is a great half graben (*e.g.*, SD-III), whereas the Tuira-Chucunaque subbasin (*e.g.*, SD-VII) is an asymmetric synclinorium. The basin contains up to 4,600 m of strata, ranging in age from late Eocene to Holocene (Stewart, 1966). A great belt of en échelon folds evidences substantial past wrench movements between the Medial basin and the Pacific coastal ranges (Wing, 1971).

Bayano Subbasin—The oldest strata within the Bayano subbasin are of late Eocene age and the youngest, widely distributed, are of Miocene age (Terry, 1956). Pliocene-Pleistocene strata are reported in the axial parts of the subbasin, especially in the SD-IV syncline. Holocene sediments are notably present in the SD-I area.

Medial basin subdivision SD-I is a synclinal trough which plunges south-southwest and is continuous with a shallow trough submerged beneath the western Gulf of Panama. It is flanked by the Pearl Islands on the east and the Azuero Peninsula on the west. SD-I includes the mouth of the Bayano estuary and related Holocene beach-ridges, marshlands, and low-lying jungle (Fig. 4).

The unique wide band of beach ridges near the mouth of the Bayano estuary (Fig. 4) indirectly suggests the possible presence of quartz sands derived from silicic rocks of the San Blas Range, which may long have contributed debris for the periodic formation of reservoir strata in the western part of the Gulf of Panama. SD-I and this western part of the Gulf of Panama are actually a south-southwest-trending extension of the Medial basin, and this particular area may have considerable potential for hydrocarbon entrapments. A thick, prospective, marine, stratigraphic succession of paralic and deltaic clastic rocks, is anticipated here in water generally less than 50 m deep. The reservoirs are very likely to be found in quartz sandstone deposited in a high-energy environment. The incipient Bayano delta just south of the Upper Maje fault, in the S 1/2, S1/2, C-4 (Fig. 1), may be especially prospective. At this same site, deltas and offshore bars probably have been present sporadically from the Miocene to present time.

Medial basin subdivision SD-II is a projected northwestward subsurface extension of the Maje Range anticlinorium. The relatively thin stratigraphic section (610-915 m) makes this part of the Medial basin unattractive as a petroleum prospect. Medial basin subdivision SD-III is an east-trending synclinal segment of the Bayano subbasin. Stewart (1966) estimated that the Bayano subbasin deepens to approximately 2,135 m near the Maje River in SD-III. The radar imagery displays an inner closed structural low in the NE 1/4, NW 1/4, SE 1/4, C-5 (Fig. 1), expressed in outcropping Miocene strata that outline the structurally lowest part of the SD-III syncline.

The arcuate Igandi anticline (Fig. 5) trends approximately N60°W through the SW 1/4, NW 1/4, C-6, separating the SD-III and SD-IV synclines. The anticline is asymmetric and its steeply-dipping northeast flank probably is bounded by a reverse fault. It is less than 3 km north of a northward bulge (Pavo prominence) in the adjacent front of the Maje Range (Fig. 5), to which it is almost certainly genetically related.

Medial basin subdivision SD-IV is a northwest-trending synclinal segment of the Bayano subbasin, which extends from the Igandi anticline (east) to the west side of the Canazas platform (SD-V, Fig. 3). Stewart (1966) estimated that the Bayano subbasin stratigraphic section is 3,050 m deep in the vicinity of the Ipeti River in SD-IV.

Canazas structural platform—The Canazas platform is composed of Medial basin subdivisions SD-V and SD-VI. Along the northwest side of the Canazas platform is the distinctive outcrop of lower-middle Oligocene carbonate rocks (Fig. 6). The Canazas anticline (W 1/2, SE 1/4, SW 1/4, C-7; also "a" Fig. 6) trends north-northwest, subparallel with the Congo-Torti fault zone, to which it may be genetically related (Wing, 1971).

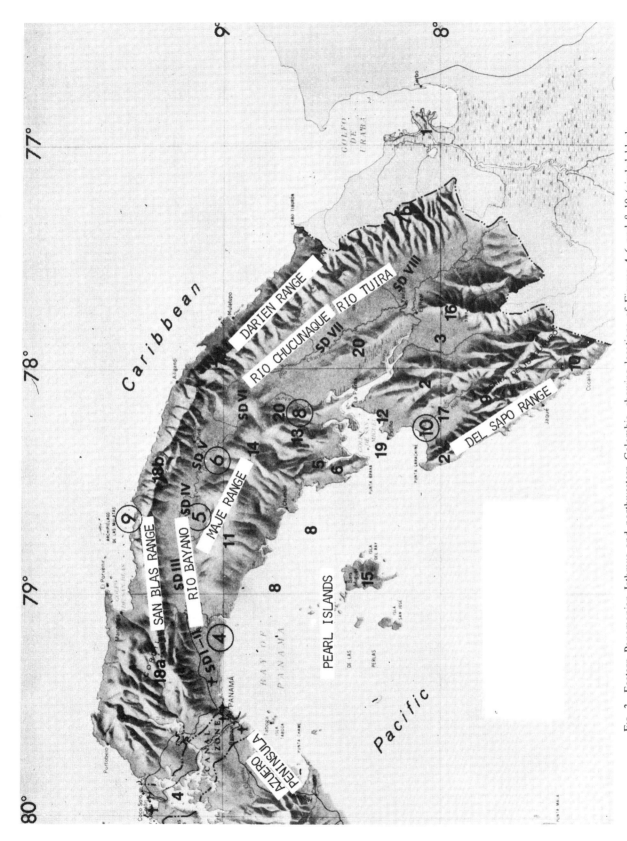

Fig. 3—Eastern Panamanian Isthmus and northwestern Colombia, showing locations of Figures 4-6 and 8-10 (circled black numbers). Major structural and physiographic elements are shown in black numbers and keyed to Table 2. Medial basin subdivisions are shown in Roman numerals.

Table 2. Major Structural and Physiographic Elements,
Eastern Panama and Northwestern Colombia

Nos. on Fig. 3		Locations on Fig. 1
1	Atrato delta	G-13
2	Bagre Range	G-8
3	Balsas embayment	G-9
-	Bayano subbasin (Medial basin)	C-5
4	Canal Zone	C-1
-	Canazas platform (Medial basin)	C-7
5	Congo basin	E-7
6	Congo Peninsula basement block	F-7
7	Darien Range	D-9
7a	Darien Range-Morti segment	C-9
7b	Darien Range-Limon segment	G-12
8	Gulf of Panama basin	E-5
9	Jungurudo Range	I-8
10	Jurado Range	J-9
11	Maje Range	D-6
12	Mogue embayment	F-8
13	Pacific Hills anticlinorium	E-7
-	Medial basin	C-5
14	Pan American Hills	D-7
15	Pearl Islands	F-5
16	Pirre (anticline) Range	H-10
17	Sambu basin	G-7
18	San Blas Range	B-4
18a	San Blas Range-Azucar segment	B-6
18b	San Blas-Diablo segment	B-7
19	San Miguel Bay	G-7
20	Sanson Hills fold belt	F-9
21	Sapo Range	H-7
-	Tuira-Chucunaque subbasin (Medial basin)	E-9

SD-I of Bayano subbasin	C-3
SD-II of Bayano subbasin	C-4
SD-III of Bayano subbasin	C-5
SD-IV of Bayano subbasin	C-6
SD-V of Canazas platform	C-7
SD-VI of Canazas platform	D-8
SD-VII of Tuira-Chucunaque subbasin	E-9
SD-VIII of Tuira-Chucunaque subbasin	G-10

A similar anticline, which culminates in a prominent dome ("b" of Fig. 6), also has been delineated on the radar imagery. Between these two prominent anticlines, a more subtle northeasterly-trending arcuate fold belt has been mapped (NE 1/4, SW 1/4, C-7, Fig. 1) which must be related to the northeast-trending shear zone, expressed in the basement core of the San Blas Range on the north. The Medial basin subdivision SD-VI is a transverse, northeast-trending topographic divide.

Tuira-Chucunaque subbasin—The Tuira-Chucunaque subbasin is composed of Medial basin subdivisions SD-VII and VIII. Paralleling the continental divide along the northern part of eastern Panama, moderately dipping strata eventually flatten out to form the broad structural and topographic depression containing the Tuira-Chucunaque basin. The boundary between SD-VII and SD-VIII, within the Tuira-Chucunaque basin, is the transverse Chico River fault (N 1/2, G-10) upthrown on the southeast. This fault is part of a trans-isthmian disturbance that includes the Pirre anticline (H-10) and the anomalously uplifted southeastern segment of the Darien Range, shown by a widened basement outcrop (Fig. 7, starting in the S 1/2, F-11). Stewart (1966) estimated the Tuira-Chucunaque subbasin

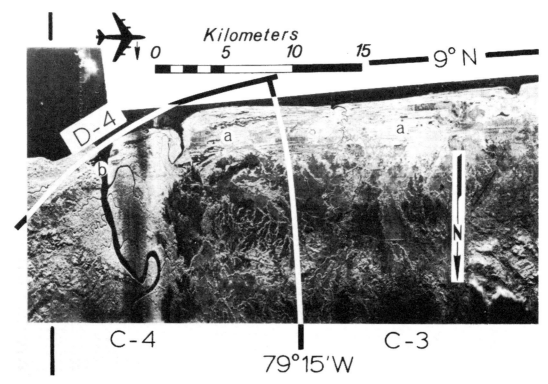

FIG. 4—Radar imagery of Gulf of Panama Coast, showing beach ridges (a) in proximity to mouth of Bayano Estuary (b).

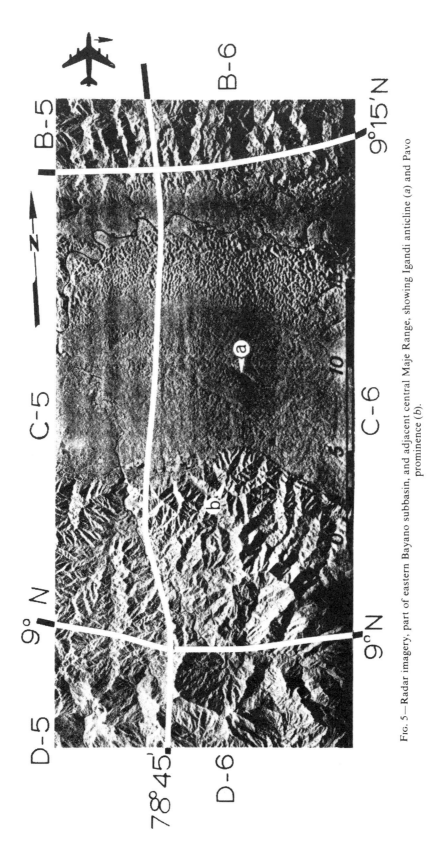

Fig. 5—Radar imagery, part of eastern Bayano subbasin, and adjacent central Maje Range, showing Igandi anticline (a) and Pavo prominence (b).

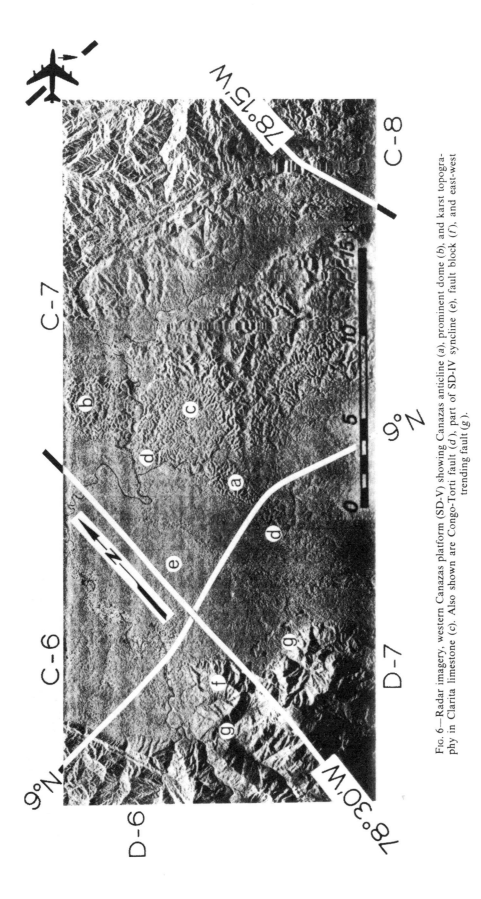

Fɪɢ. 6—Radar imagery, western Canazas platform (SD-V) showing Canazas anticline (a), prominent dome (b), and karst topography in Clarita limestone (c). Also shown are Congo-Torti fault (d), part of SD-IV syncline (e), fault block (f), and east-west trending fault (g).

to be approximately 4,575 m deep in the Darien area near the Membrillo River (center E-9), and this is probably the greatest depth to basement in the Medial basin.

Sanson Hills Fold Belt

The Sanson Hills fold belt is comprised of a series of genetically related en échelon anticlines which, individually, trend an average N21°W (Fig. 8). Most of these anticlines are asymmetric and bounded by reverse faults on their southwest sides. The youngest strata exposed in the anticlinal cores are probably of Oligocene or early Miocene age (Interoceanic Canal Study Comm., 1968).

Petroleum Prospects

There have been no oil or gas discoveries in Panama to date. Most of the drilling thus far has been in Darien Province, where significant shows were found in the Arusa and Gatun Formations (upper Oligocene and middle Miocene, respectively, Table 1), which include some organic shales. The Aquaqua Formation (lower Miocene) contains some possibly prospective sandstone reservoirs. However, reservoirs are sparse, inasmuch as most of the subsurface stratigraphic succession is quite tuffaceous, and most clastic material was derived from quartz-poor basic igneous rock types. The Clarita carbonate rocks (lower-middle Oligocene) have been considered to be potential pay zones by some, but there is little to substantiate their merit. They are exposed over broad areas on the north side of the Medial basin; also, there is a change of facies to shale south and west of the Tuira-Chucunaque subbasin. On the south side of the basin, the anticlines of the Sanson Hills fold belt generally are stripped at least to Oligocene strata on their crests, except for the Upper Sabana and northwestern Santa Fe closures (Fig. 8). This leaves but a very narrow part of the Medial basin which could be prospective, and it is not particularly well protected against flushing, especially from north flank exposures.

Thus, the outlook for eventual terrestrial Medial basin hydrocarbon production does not seem bright—at least not with respect to conventional entrapments. However, there is the possibility of gravity-type entrapments in fractured organic shales, siltstones, or carbonate rocks, within closed structural lows, such as the SD-III syncline. The SD-IV and SD-VII synclines could be similarly prospective, but depths probably would be greater. The radar imagery displays an inner closed structural low, in the NE 1/4, NW 1/4, SE 1/4, C-5 (Fig. 1), expressed in outcrops of Mio-

cene strata that outline the structurally lowest part of the SD-III syncline. It is of considerable potential importance as the possible locus of a Canon City embayment type hydrocarbon accumulation. What is envisaged is the possibility of gravity entrapment within the closed structural low in fractured Oligocene carbonate or siltstone strata, the source being overlying late Oligocene carbonaceous shales. Stewart (1966) described a major carbonate section that might be prospective for gravity hydrocarbon accumulations in fractures where it is present in the subsurface of the inner synclinal deeps of SD-III, or SD-IV, or both.

Five small anticlines were mapped by Shelton (1952) within the confines of the Tuira-Chucunaque basin; however, because of the lack of topographic expression, only two of them could be inferred from radar imagery interpretation. The anticlines mapped by Shelton are located as follows:

Yape anticline 8°8′N, 77°34′W—NW 1/4, SE 1/4, G-10
Quebrada Sucia dome 8°15′N, 77°35′W—SE 1/4, SE 1/4, F-10
Capete anticline 8°4′N, 77°34′W—NW 1/4, SE 1/4, G-10
Rancho Ahogado anticline 8°35′N, 77°52′W—NE 1/4, SW 1/4, E-9
Tuira anticline 8°08′N, 77°34′W—NW 1/4, SE 1/4, G-10

Geologic exploration for oil in the Tuira-Chucunaque basin was initiated by Sinclair Panama in 1925, when three shallow test holes were drilled to 332, 1,100, and 1,200 m. The deepest of these three tests was drilled on Yape anticline, NW 1/4, SE 1/4, G-10. A basal 110-ft oil-stained sandstone (Aquaqua Formation) was tested by Sinclair and recovered 1,500 bbl/day of salt water, with good shows of gas (Johnson and Headington, 1971).

Shelton (1952), in field work associated with petroleum exploration, provided geologic data for the central part of the Tuira-Chucunaque basin, and Terry (1956) reported on the regional geology of eastern Panama. Subsequent to the work of Shelton and Terry, the most recent and deepest drilling was completed (dry hole) by the Delhi-Taylor Corporation in 1959 at a total depth of 3,498 m. This test hole was on the Rancho Ahogado anticline centered at the NE 1/4, SW 1/4, E-9. Sufficient gas was found in the thin sandstones of this well to cause considerable drilling problems. Well-log and sample analysis provides evidence that the total stratigraphic section is still untested over this large structure (Johnson and Headington, 1971). A geomorphic anomaly of substantial size, embracing at least 400 sq km, has been mapped from the radar imagery, and is centered at 8°22′N, 77°41′W, F-10. This anomaly

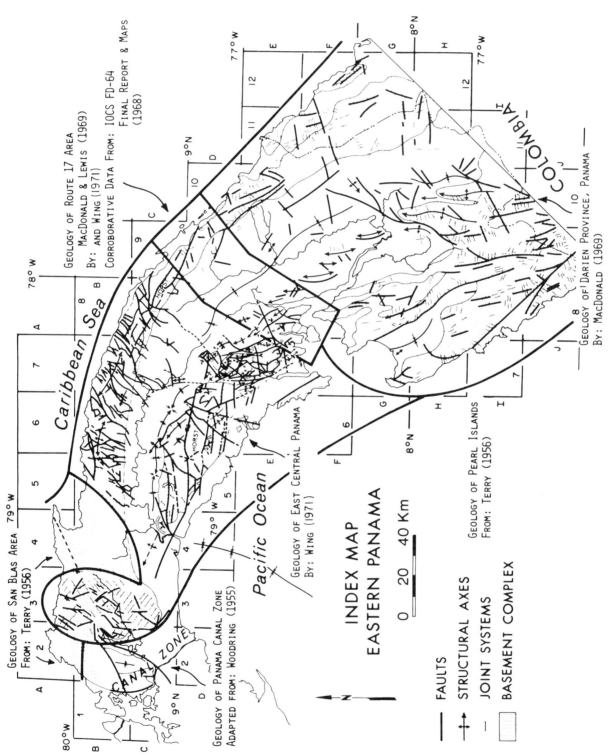

Fig. 7—Generalized geologic reconnaissance map, eastern Panama and northwestern Colombia.

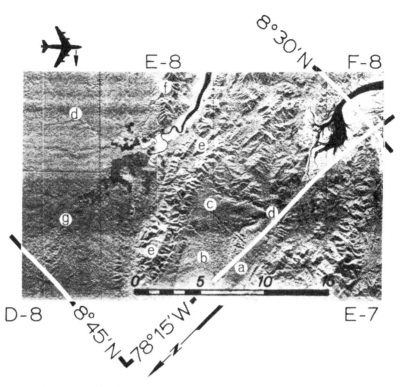

Fig. 8—Radar imagery, part of Sanson Hills fold belt showing Cucunati anticline (a) and syncline (b), Trans-Cucunati anticline (c), unnamed fault (d), Santa Fe anticline (e), Arreti anticline (f), and unnamed fold (g). Sabana Estuary can be seen in upper right.

appears to be reflecting a subtle, subsurface, structural high which is the result of the northerly plunge of Pirre anticline (SW 1/4, G-10; W 1/2, H-9). The outline of this anomaly is especially evident when one examines the drainage of the Rios Tuquesa and Tupisa on the radar mosaic (Fig. 1). Not only does this structure provide excessive size, but an almost complete stratigraphic section should be found by the drill.

In the Sambu basin (location SE 1/4, G-7; NW 1/4, H-8; Fig. 1) Gulf Oil Company drilled three shallow wells between 1922 and 1927. Two of these tests were in the vicinity of 8°04'N, 78°15'W (NE 1/4, SE 1/4, G-7) and the third well at approximately 8°03'N, 78°18'W (C, SE 1/4, G-7). Oil shows were present in the Gatun beds of Miocene age, and the anticlinal trends located on land by field mapping have been projected offshore in San Miguel Bay (Johnson and Headington, 1971).

The Gulf of Panama is generally less than 100 m deep. Terry (1956) illustrated the extensive area of the shallow Gulf, which is several thousands of meters higher than the adjacent bottom of the Pacific basin. He stated: "If the ocean were withdrawn to the 100 fathom (183 meters) isobath, the area of Panama would be increased by about one-third, most of the addition being on the Pacific side."

The west side of the Gulf of Panama is a trough, plunging gently south-southwest, a shelfward extension of Medial basin SD-I. That part of the Gulf of Panama which is west of the Pearl Islands is considered to be highly prospective for hydrocarbons. As previously mentioned, a wide band of beach bars near the mouth of the Bayano Estuary (N 1/4, D-4) is indirect evidence of quartz sands derived from silicic rocks of the San Blas Range, which may long have contributed to the periodic formation of reservoir strata in the western part of the Gulf of Panama. Beach ridges (trending east-west along area "a" of Fig. 4) provide evidence of progradation along the Pacific Coast between the Rio Bayano and the Panama Canal. Back-swamp drainage north of the beach ridges and mangrove swamps (area "b") are easily delineated on the radar imagery.

The Caribbean coast is flanked by offshore reef chains manifested by the archipelagos north of the San Blas Range (Fig. 9). In areas where Holocene reef chains are present, there probably have been middle and late Tertiary reef chains as well, though not necessarily in exactly the same places. Thus, geophysical exploration for buried reefs and carbonate bands would likely be successful offshore in the Caribbean, north of the San Blas Range, where the water is less than 100 m deep.

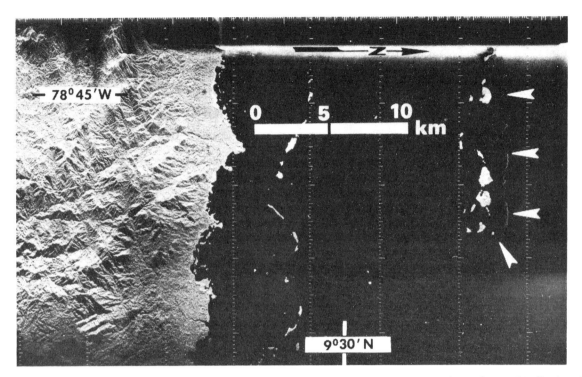

Fig. 9—Radar imagery of Caribbean coast north of San Blas Range, showing windward part of active reef zones (a). Coral sand islands are on leeward slope of barrier reef north of (b).

The shelf involved has a fairly irregular bottom, probably reflecting the intense folding in evidence on shore along the north side of the Caribbean coastal range; and some such irregularities may even mark prospective structures offshore where the stratigraphic cover has not been stripped away. Edgar (1968) cited considerable structural disturbance in deep water on the north side of the eastern Panamanian Isthmus.

One other phenomenon of note is the presence of shell bars in the Gulf of Panama (C, SE 1/4, G-7), especially near the mouths of some rivers (Fig. 10) and around the Pearl Islands (Wing, 1971). Transgressions and regressions of middle and late Tertiary time should have resulted in migrations of the shell bars, such that their distribution *in toto,* in the subsurface, may be in the form of long, linear, porous carbonate belts generally transverse to the land.

CONCLUSIONS

The term "reconnaissance" is applied to incomplete or generalized mapping which usually precedes more detailed or localized studies. Similarly, reconnaissance mapping can enlarge the scope of local studies by providing a general geologic picture of the surrounding region. From a practical standpoint, reconnaissance mapping

may be the only feasible method for geologic exploration because of the limitations of time, funds, adequate base maps, and accessibility.

The interpretation of radar imagery facilitates physiographic differentiation and geologic reconnaissance mapping on a regional scale. At the very minimum, a ready subdivision generally can be made between igneous and sedimentary rocks. Large-scale structural units can be synoptically studied, and the single-strip imagery format used in conjunction with a radar mosaic enables the petroleum geologist to become quickly familiar with the essential features of structural provinces. On a regional scale, gross lithologic and structural subdivisions can be interpreted at levels of detail which commonly exceed those available on existing small-scale geologic maps in many parts of the world.

Stereoscopic viewing, using same-look flight strips of 50 percent overlap, can provide a wealth of detail, including semiquantitative dip components and relative fault offsets. This method is currently the primary interpretive technique; some such stereo work was part of the subject Panama mapping effort.

On a more detailed scale, a relative stratigraphic sequence can be determined by using radar imagery, but only if the lithic units are

0   4   8 Kilometers

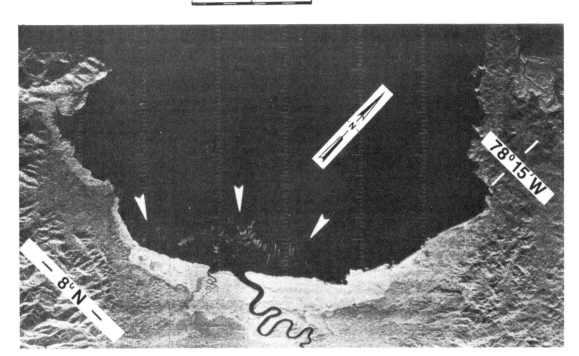

Fig. 10—Radar imagery near mouth of Rio Sambu in San Miguel Bay. Imagery taken at low tide, exposing shell bars growing perpendicular to shoreline (outlined by white arrows).

expressed in the terrain configuration and the structure is not too complicated. In this regard, stereo viewing of large-scale opaque prints is a great help. Where collateral field data are available and are used in conjunction with radar imagery interpretation, the problem becomes less complicated.

With the exception of those data provided by field investigation, the geologic information interpretable from the radar imagery of eastern Panama far exceeds those data previously available through conventional reconnaissance methods. Certainly, radar remote sensing offers the only practical technique for reconnaissance mapping in the wet tropics; however, even where conventional aerial photographic coverage can be obtained, radar imagery can be a valuable supplement because of its unique data content. Radar geologic reconnaissance—preferably with, but even without, air-photo support—can serve the petroleum geologist as an important exploration tool because of its substantial physiographic-geologic data content.

REFERENCES CITED

Edgar, N., 1968, Seismic refraction and reflection in the Caribbean Sea: Ph.D. dissert., Columbia Univ., 159 p.
Gillerman, E., 1970, Roselle lineament of southeast Missouri: Geol. Soc. America Bull., v. 81, p. 975-982.

Interoceanic Canal Study Commission, 1968, Geology final report, route 17, v. 1: IOCS Memo FD-64, Field Director, Office Interoceanic Canal Studies, Panama, 32 p.
———— 1969, Geology final report, route 25, v. 1: IOCS Memo FD-80, Field Director, Office Interoceanic Canal Studies, Panama, 110 p.
Johnson, M. S., and E. Headington, 1971, Panama—Exploration history and petroleum potential: Oil and Gas Jour., April 12, p. 96-100.
Lewis, A. J., and H. C. MacDonald, in press, Radar mapping of mangrove zones and shell reefs in southeastern Panama: Photogrammetria.
MacDonald, H. C., 1969, Geologic evaluation of radar imagery from Darien Province, Panama: Modern Geology, v. 1, p. 1-63.
———— and A. J. Lewis, 1969, Terrain analysis with radar—a preliminary study: Interim Tech. Progress Rept., Proj. THEMIS, Univ. Kansas, Lawrence, Kansas, p. F1-F12, October 1969.
———— and W. P. Waite, 1971, Optimum radar depression angles for geological analysis: Modern Geology, v. 2, p. 179-193.
———— J. N. Kirk, L. F. Dellwig, and A. J. Lewis, 1969, The influence of radar look-direction on the detection of selected geological features: 6th Symp. Remote Sensing of Environment Proc., v. 1, Univ. Michigan, Ann Arbor, Michigan, p. 637-650.
———— A. J. Lewis, and R. S. Wing, 1971, Mapping and landform analysis of coastal regions with radar: Geol. Soc. America Bull., v. 82, no. 2, p. 345-358.
Shelton, B. J., 1952, Geology and petroleum prospects of Darien, southeastern Panama: Master's thesis, Oregon State Univ., 61 p.
Stewart, R., 1966, The Bayano basin, a geological report: Panama Canal Commission Rept. (IOCS Memo PCC-r, File PCC-200.02), 17 p.

Terry, R. A., 1956, A geological reconnaissance of Panama: California Acad. Sci. Occasional Paper 23, 91 p.

Wing, R. S., 1971, Structural analysis from radar imagery of the eastern Panama Isthmus: Modern Geology, v. 2, p. 1-21, 75-127.

———— and L. F. Dellwig, 1970, Radar expression of Virginia Dale Precambrian ring-dike complex, Wyoming/Colorado: Geol. Soc. America Bull., v. 81, no. 1, p. 293-298.

———— W. K. Overbey, Jr., and L. F. Dellwig, 1970, Radar lineament analysis, Burning Springs area, West Virginia—an aid in the definition of Appalachian Plateau thrusts: Geol. Soc. America Bull., v. 81, no. 11, p. 3437-3444.

Woodring, W. P., 1955, Geology and paleontology of Canal Zone and adjoining parts of Panama: U.S. Geol. Survey Prof. Paper 306-A, 145 p.

Reprinted by permission from WORLD OIL, April
1973, p. 67-70.

Radar imagery identifies hidden jungle structures

Dr. R. S. Wing, Director of Advanced Geology, Continental Oil Co., Princeton, N.J., and **Harold C. MacDonald,** Associate Professor of Geology, University of Arkansas.

10-second summary

An aircraft radar survey in eastern Panama gave surface imagery of nearly inaccessible terrain that could not be photographed due to heavy cloud cover. Examples show how typical jungle terrain reflects data on potential hydrocarbon formations.

RADAR IMAGERY obtained from a high resolution Side Looking Airborne Radar (SLAR) system in less than 20 hours of recording time was successfully used for geologic reconnaissance mapping of 40,000 square miles in eastern Panama and northwestern Colombia. This area is typical of many cloud-covered tropical environments with dense vegetation in which adequate surveys cannot be made with

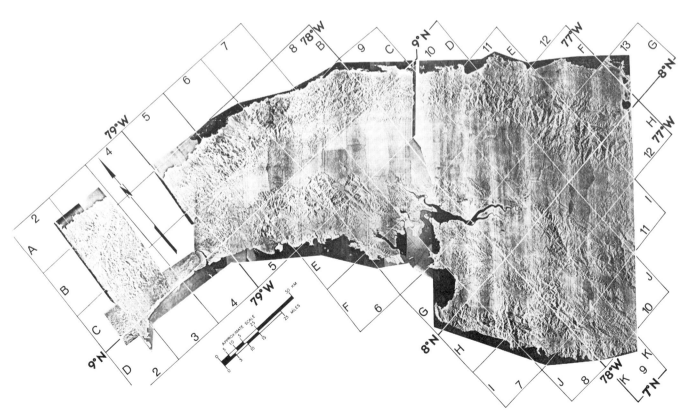

FIG. 1—Radar imagery mosaic of Panama's eastern Isthmus is made from individual strips approximately 18 km in width. Data were collected with the Side Looking Airborne Radar System, involving only 20 hours of actual recording time. Compared to aerial photography, imagery is not affected by the nearly constant cloud cover in this tropical area. Grid system here is used to define areas described in greater detail from interpretation of individual strips. See Fig. 3 for an example description of the southern half of area C-7.

$\longrightarrow$

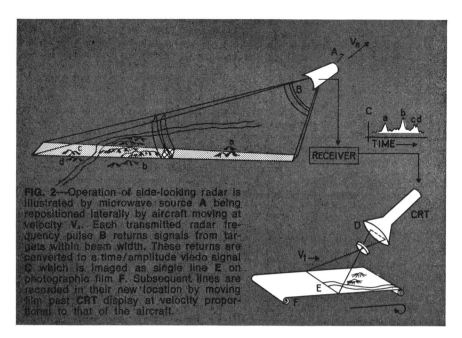

FIG. 2—Operation of side-looking radar is illustrated by microwave source A being repositioned laterally by aircraft moving at velocity V$_a$. Each transmitted radar frequency pulse B returns signals from targets within beam width. These returns are converted to a time/amplitude video signal C which is imaged as single line E on photographic film F. Subsequent lines are recorded in their new location by moving film past CRT display at velocity proportional to that of the aircraft.

aerial photography or other visible spectrum sensors. Existing maps compiled from limited and scattered field data had failed to delineate regional continuity of some promising geologic formations.

This article explains how imagery is acquired and describes how various features such as water, vegetation and outcrops appear in completed pictures. A radar mosaic of the eastern Isthmus of Panama is presented (Fig. 1) along with one detailed example of an area containing anticlines and outcroppings worthy of further exploration.

Reconnaissance is the term applied to incomplete or generalized mapping, normally preceding more detailed or localized studies. Reconnaissance mapping also can enlarge local studies, providing a general geologic picture of a region. Such mapping also may be the only feasible geological exploration method because of limitations in time, funds, adequate base maps and accessibility.

Radar imagery facilitates physiographic differentiation on a regional scale. At the minimum, quick subdivision of igneous and sedimentary rocks can generally be made. Large scale structures can be synoptically studied, and the single strip format used in conjunction with a radar mosaic helps familiarize geologists with features of structural provinces. Regionally, gross lithologic and structural subdivisions give interpretive details often exceeding those available on existing small scale geologic maps.

Stereoscopic viewing, using same-

look flight strips with 50% overlap give important detail, such as semiquantitative dip components and relative fault offsets. This is currently the primary interpretive technique. Some stereo work was part of the Panama effort.

On a more detailed scale, a relative stratigraphic sequence can be determined if lithic units are expressed in the terrain configuration and the structure is not too complicated. Here, stereo viewing of large-scale opaque prints is a great help. Where supplementary field data are available, the problem is simplified.

With exception of data provided by field investigation, geologic information interpreted from the radar imagery of eastern Panama far exceeds that previously available. Radar remote sensing offers the only practical technique for reconnaissance mapping in the wet tropics. However, even where aerial photographs can be obtained, radar imagery is a valuable supplement, with its own unique data content.

The radar imagery, covering all of Darien Province, Panama, part of northwestern Colombia, and much of east-central Panama, was obtained in 1967 and 1969 by Westinghouse for the U.S. Army Engineer Topographic Lab in less than 20 hours of actual imaging time. During the preceding 30-year period, there had been minimal success in obtaining aerial photographic coverage because of almost perpetual cloud cover, especially over mountains and foothills.

The imagery is the product of a high resolution AN/APQ-97 K-band Side Looking Airborne Radar (SLAR) system. The 1967 data provided a slant range display that incorporates near range compression, geometric distortion just the opposite of that found in conventional oblique photographs.[1] In contrast, 1969 imagery is a ground range display in which a hyperbolic sweep correction was applied to returning signals, theoretically resulting in planimetric terrain display, where terrain is flat. However, in areas with rugged topography, considerable distortion can occur.[1] Operating principles of side looking radar are summarized in Fig. 2.

Imagery analysis. While radar has been a valuable supplementary tool in many diverse environments,[2,3,4,5] in the tropics it may be the only aid to *field* surveys.

Radar imagery can be superior to vertical air photos for display and detection of surface fractures, i.e., faults and megajoint zones. Although photos may more clearly reveal smaller details of structural elements and patterns, radar often shows true nature and extent of structural patterns better. Radar is also superior to conventional photography for detecting land/water boundaries because smooth water returns little of the incident signal.

There are limitations. For example, interpreters know that radar preferentially displays those topographically expressed linears most nearly paralleling the aircraft flight track. Conversely, linears paralleling the *look* direction are minimized for lack of shadowing, some to such an extent that a small percentage are not detectable at all.[6] In extremely rugged terrain with extensive radar showing, trade-offs between loss of geologic data and radar geometry must be considered during mission planning.[1]

There is essentially no vegetation penetration by SLAR when operating at K-band frequencies.[6] In the jungle, only the canopy surface of dense foliage is imaged. Whether depicted on air photos or radar imagery, this canopy surface mirrors topography, reflecting outcrops which were weathered and eroded differentially according to varying rock resistances, structural attitudes and fractures. However, while air photos tend to show a welter of canopy irregularities, inherent integration in the relatively coarse SLAR resolution cell has a smoothing effect which aids definition of large-scale structures or their major elements.

Contrasting types of vegetation grow on sediments derived from different bedrocks. Thus, some vegetal patterns can be differentiated on SLAR and lithic inference made.

Radar terrain return depends on:

1. Surface roughness
2. Incidence angle of the slanted microwave beam
3. Polarization
4. Wave length of the electromagnetic signal used, and
5. The complex dielectric constant.

As with aerial photography, interpretation of the radar display involves considerations of tone, texture, shape and pattern.

Tone is the intensity of signal backscatter recorded on film as shades of gray, from black to white.

Texture, in simplest terms, is the degree of erosional dissection. Variously textured drainages can be used to infer lithologic type. For example, igneous outcrops generally have a massive appearance distinguished by "rugged and peaked divides."

Shape is telltale outlines of surface expressed features such as those for alluvial fans, dikes, igneous plugs, volcanoes and folds.

Pattern is the arrangement of geologic, topographic or vegetal features such that linears may be distinguished. For example: faults and megajoints, bedding trace grains, en echelon folds, etc.

Thus it is possible to extract considerable geologic data through careful imagery interpretation. [7]

The most important interpretive phase of reconnaissance mapping with radar is selecting rock units that have sufficient areal extent to be significant and are distinctive enough to be mapped. The primary limitation for differentiating rock units, either on photographs or imagery, is topographic relief. Depending on rock type, climatic conditions, regolith thickness and vegetal canopy, imagery may reveal abundant data, or almost none.

Summary of key stratigraphic data, eastern Panama

Tuira-Chucunaque subbasin		
Early Pliocene Late Miocene	Chucunaque	Sandstones and shales containing foraminiferal fauna; occupies narrow inner trough of plain D-8 to F-10.
Mid. Miocene	Pucro	610 m calcareous sandstone with thin ls. beds.
	Gatun	1070 m sh., underlain by conglom., ss., and small ls. lenses.
Early Miocene	Aquaqua 610-1,370 m	Bentonitic and bitum. shales, limestones, and sandstones.
Late Oligocene	Arusa	460 m massive brownish tuffaceous marlstone
	770 + m	310 m wh. weathering tuffaceous massive mudstone
Mid. Early Oligocene	Clarita 300± m	Tuffaceous carbonates, mostly ls. and sandy limes.
Late Eocene	Corcona	Sandy shale, marine, calcareous.
	350-400 m	Massive agglom., conglom., ls. incl. volcanic matter, etc.
Mid. Early Eocene		Chert, equiv. to Guayaquil chert of Ecuador, limited distribution in Pacific Hills area.
Pre-late Eocene	basement	Andesite flows
		Crystallines, somewhat acidic on Caribbean side.

Bayano subbasin		
Late Miocene		915-1220 m of soft arq. ststs. foraminifera.
Mid. Miocene		460 m reef-type near-shore ls. beds. 15 km long; on south side of syncline in C-5.
Early Miocene		Ss. and sltst., sharks teeth and fish vert. and shells.
Late Oligocene		Tuffs, congl., calc. and tuff. ss., and tuff. sltst.
Mid. Early Oligocene		Limestones, calc. algae and forams.
Late Eocene		Sandy shale
		Agglom., conglom., tuff. and calc. ss., tuff. sltst.
Pre-late Eocene	basement	Andesite flows.
		Crystallines, silicic on Caribbean side; mafic on Pacific side.

Geologic data interpreted from radar imagery for this study were derived from surrogates rather than through direct identification. Similarly, time-stratigraphic assignments used for geologic map units must be inferred from corroborative data, and are obviously not interpretable from radar imagery. A summary of key stratigraphic data from previously mentioned reports is presented in the accompanying table. Geological reconnaissance mapping of the area shown in Fig. 1 was presented in a study which incorporated prior radar mapping of Darien Province.[6]

History of geological mapping. East-central Panama has been the least mapped area, geologically, in eastern Panama. Terry's 1956 map[8] has only the most obvious structural aspects and includes only speculative versions of major faults. This illustrates the state of geologic knowledge prior to the recently available radar imagery.

The climate is tropical, vegetation is continuous and terrain is extremely inaccessible. Even the Pan American Highway has not been constructed through this part of Panama. Out-crops are often deeply weathered and poorly exposed. Only during the three or four-month "dry season," beginning in late January, is field work feasible. In some years up to 12 inches rain has fallen during this period.

Terry's observations are based on field work and overflights from 1920 to 1949. In 1966, a Panama Canal Commission geologist published a report based on field work from 1955 to 1965.[9] More recent ground reconnaissance studies have been completed in easternmost Darien Province, and northwestern Colombia.[10,11]

Before radar, Darien Province was the best mapped area in eastern Panama. This mapping was quite general and often incorrect. Some entire landforms were off-shape or misoriented. Still, gold mining operations and petroleum exploration resulted in some recognition of gross elements as summarized by Terry. Shelton provided geologic data for the north central part of the Tuira-Chucunaque subbasin.[12] Additional data has resulted from limited wildcat drilling.[13] The IOCS study[11] represented the most recent geologic data available for Darien prior to MacDonald's 1969 radar geologic-reconnaissance.[6]

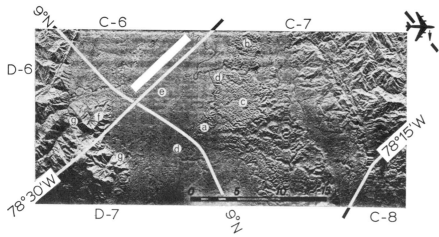

FIG. 3—Detailed analysis of one radar imagery strip illustrates important features of the Canazas structural platform. Note how superimposed grid lines locate this area on the large map in Fig. 1. Two prominent structures are the Canazas anticline "a" and a prominent dome "b." Other features shown are Karst topography in Clarita limestone "c," Congo-Torti fault "d," synclinal trend "e," fault block at "f" and east-west trending fault at "g." This example is typical of the high quality imagery available.

Petroleum provinces. Utilizing radar imagery as the sole source of remote sensing data, the distribution, continuity and structural grain of key strata provide evidence that the eastern Isthmus of Panama can be divided into three main physiographic-structural parts, the two coastal mountain ranges and the separating taphrogenic Medial basin which trends southeastward from the mouth of the Bayano River to the Atrato Rivery valley of northwestern Colombia, Fig 1.

Within the Medial basin, most clearly exposed surface structures are not attractive petroleum prospects because prime reservoir strata have been stripped from their crests. However, several large geomorphic anomalies which have been mapped may be reflections of sub-surface structures having a complete stratigraphic section.

Gravity type hydrocarbon accumulations in fractured organic shales, siltstones and carbonates are possible in several closed synclinal elements along the Medial basin axis. And, the southwestward extension of the Medial basin trend, coincident with the western Gulf of Panama, may have petroleum potential, considering likelihood of a thick prospective, marine, stratigraphic succession of paralic and deltaic clastics derived from silicic San Blas terrain since mid-Miocene time.

The occurrence of active shell bars in the Bay of San Miguel and present day reef trends on the northern Caribbean coast suggest possible *offshore* sites for geophysical surveying.

Detailed subdivisions of major structural and physiographic elements have been derived from interpretation of individual imagery strips. These individual areas are described in the literature.[7,14] One example from the Canazas structural platform area is discussed here to illustrate the application and technique of radar imagery interpretation.

Distinctive outcropping of lower middle Oligocene carbonates are found along the northwest side of the Canazas platform, "c" in Fig. 3. The Canazas anticline, shown as "a" in Fig. 3 and located in SE/4, SW/4 of C-7 in Fig. 1, trends north-northwest and subparallel to the Congo-Torti fault zones, to which it may be genetically related.[7]

A similar anticline, which culminates in a prominent dome, Fig. 3 "b," has also been delineated on the imagery. Between these two prominent anticlines, a more subtle northeasterly trending arcuate fold belt has been mapped (NE/4, SW/4 of C-7, Fig. 1) which must be related to the northeast trending shear zone, expressed in the basement core of the San Blas Range to the north.

About the authors

DR. RICHARD S. WING *received the B.S. (1949) and M.S. (1950) in geology from the University of Wisconsin. He served as an exploration geologist with Texaco from 1950 to 1967, at which* time he left the position of district geologist at Casper, Wyo., to undertake Ph.D. training at the University of Kansas. There, he was a research assistant at the Remote Sensing Lab where he developed interpretive techniques for structural analysis with side-looking radars. He completed the Ph.D. in 1970, and joined Conoco where he is presently director of the Advanced Geology Group at Princeton, N.J. Dr. Wing is a member of AAPG, GSA, Rocky Mountain Association of Geologists and North Dakota Geological Society.

DR. HAROLD C. MAC-DONALD *received the B.A. degree in geology from the State University in New York in 1960, and the M.S. and Ph.D. (1969) degrees in geology from the University of Kansas. Before college, he served six years with* the U.S. Air Force as navigator and SHORAN test project officer. From 1962 to 1965, he was employed by Sinclair Oil and Gas Co., Denver, Col., as an exploration geologist. While completing the Ph.D., he was a research assistant at the Remote Sensing Lab, University of Kansas. From 1969-1970, he was a research associate at the university where he evaluated the geoscience potential of side-scanning radar systems. He is now an associate professor, Department of Geology, University of Arkansas. Dr. MacDonald is a member of AAPG, Sigma Xi, Sigma Gamma Epsilon, GSA and the American Society of Photogrammetry.

[6] MacDonald, H. C., "1969, Geologic evaluation of radar imagery from Darien Province, Panama," *Modern Geology*, Vol. 1, pp. 1-63.
[7] Wing, R. S., "1971, Structural analysis from radar imagery of the Eastern Panama Isthmus," *Modern Geology*, Vol. 2, pp. 1-21 and 5-127.
[8] Terry, R. A., "1956, A geological reconnaissance of Panama," California Acad. Sci. Occasional Paper 23, p. 91.
[9] Stewart, R., "1966, The Bayano Basin, a geological report," Panama Canal Commission Report (IOCS Memorandum PCC-r, File PCC-200.02) p. 17.
[10] Interoceanic Canal Study Commission, 1968, Geology final report, route 17, Vol. 1, IOCS Memor FD-64, Field Director, Office Interoceanic Canal Studies, Panama, 32 p.
[11] Interoceanic Canal Study Commission, 1969, Geology final report, route 25, Vol. 1, IOCS Memo FD-80, Field Director, Office Interoceanic Canal Studies, Panama, 110 p.
[12] Shelton, B. J., 1952, "Geology and petroleum prospects of Darien, Southeastern Panama," Unpublished Masters Thesis, Oregon State University, 61 p.
[13] Johnson, M. S., and Headington, E. 1971, "Panama: Exploration history and petroleum potential," *Oil and Gas Journal*, April 12, 1971, pp. 96-100.
[14] Wing, R. S., and MacDonald, H. C., "Radar geology—a petroleum exploration technique, eastern Panama and northwestern Colombia," Paper presented at AAPG 57th Annual Meeting, Denver, Col., April 17-19, 1972.

LITERATURE CITED
[1] MacDonald, H. C., and Waite, W. P., "Optimum radar depression angles for geological analysis," *Modern Geology*, Vol. 2, pp. 179-193.
[2] Wing, R. S., and Dellwig, L. F., "1970, Radar expression of Virginia Dale Precambrian ring—dike complex, Wyoming/Colorado," *Geol. Soc. America Bull.*, Vol. 81, No. 1, pp. 293-298.
[3] Wing, R. S., Overbey, W. K., Jr., and Dellwig, L. F., "1970 Radar lineament analysis, Burning Springs area, West Virginia—and aid in the definition of Appalachian Plateau thrusts," *Geol. Soc. America Bull.*, Vol. 81, No. 11, pp. 3437-3444.
[4] Gillerman, E., "1970 Roselle lineament of southeast Missouri," *Geol. Soc. America Bull.*, Vol. 81, No. 3, pp. 975-982.
[5] MacDonald, H. C., Lewis, A. J., and Wing, R. S., "1971 Mapping and landform analysis of coastal regions with radar," *Geol. Soc. America Bull.*, Vol. 82, No. 2, pp. 345-358.

CASE HISTORIES

Reprinted by permission of the Environmental
Research Institute of Michigan from *Proceedings of
the Seventh Thematic Conference on Remote Sensing
for Exploration Geology*, 1989, v. 1, p. 299-313.

LINEAMENT AND GEOMORPHIC ANALYSIS OF REMOTE SENSING DATA
AS AN AID TO HYDROCARBON EXPLORATION, SIRT BASIN, LIBYA*

Al Fasatwi, Y.A.
Petroleum Research Centre (P.R.C), P.O.Box 6431
Tripoli, Libya

van Dijk, P.M.
International Institute for Aerospace Survey and Earth Sciences
(ITC), P.O.Box 6, 7500 AA Enschede, The Netherlands

ABSTRACT

The Sirt basin, in semi-arid north-central Libya, provides an
ideal case study to evaluate different remote sensing data
sources for hydrocarbon exploration.

Remote sensing images, including Landsat TM, SIR-A and aerial
photos, were compared with subsurface data including approximately 200 exploratory wells, a Bouger gravity map and
some seismic cross sections. The comparison covers the area of
Landsat TM frame WRS 185/040.

A strong relationship exists between basement faults and
lineaments detected in remote sensing images. Nine oil fields
of a total of twenty two in the area are located in areas
dominated by lineaments.

Geomorphic analysis indicated that seven of the oil fields
coincide with geomorphic anomalies. On the basis of their
characteristics another five anomalies were detected.

This study suggests that prospective areas could be delineated
by using the surface characteristics of known oil fields,
together with the available subsurface information, especially
gravity data.

----

* Presented at the Seventh Thematic Conference on Remote Sensing for
Exploration Geology, Calgary, Alberta, Canada, October 2-6, 1989.

1. INTRODUCTION

The study area (Fig. 1), located in the Sirt basin, covers parts of the Gattar ridge, the Beda platform and the Zallah through. It is delimited by the Landsat TM frame WRS 185/040 (Fig. 2). The Sirt basin is considered as the youngest and smallest of the sedimentary basins of Libya.

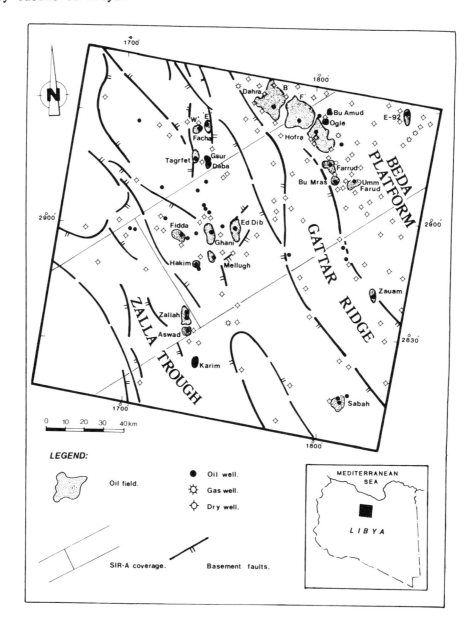

Figure 1: Index map of the study area showing major basement faults (after N.O.C., 1983) and oil fields.

Most of the known oil fields are situated in the southern part of the basin, and twenty two of them are included in this study. The Sirt basin has the greatest reserves of all African basins. Ultimate reserves are estimated at 37,600 MMBOE which represents 29% of Africa's total of fifteen giant oil fields (Chattelier and Slevin, 1988).

The Precambrian basement in the basin is overlain by an approximately 5000 m thick sedimentary sequence of mainly upper Cretaceous to Miocene age (Fig. 3). The hydrocarbon source rock is dominantly upper Cretaceous in age.

Oil reservoirs are located in traps associated with a NW trending horst and graben system that is related to the Post-early Cretaceous subsidence of the basin. Present oil production is mainly from five different stratigraphic horizons of Paleocene and Eocene age (Fig. 3). Figure 3 shows that the reservoir horizons of the producing oil fields are progressively deeper when going from the SW to the NE of the area, i.e. from the margin towards the centre of the Sirt basin (cf. Fig. 1)

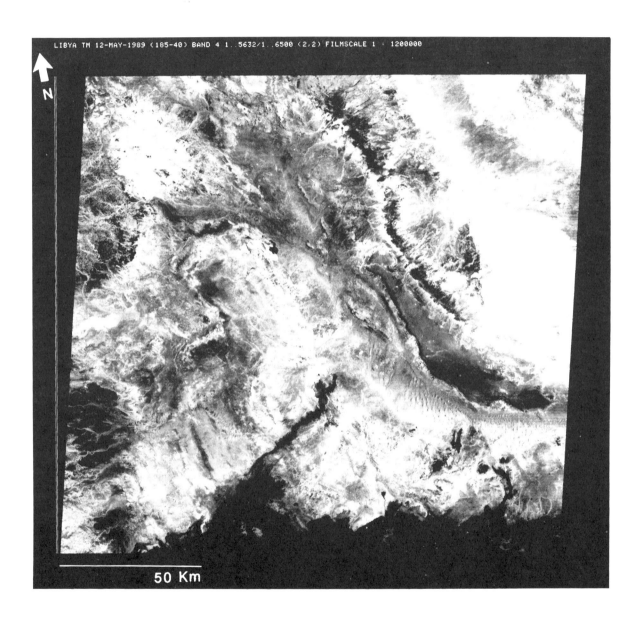

Figure 2: Landsat TM image of the area (band 4).

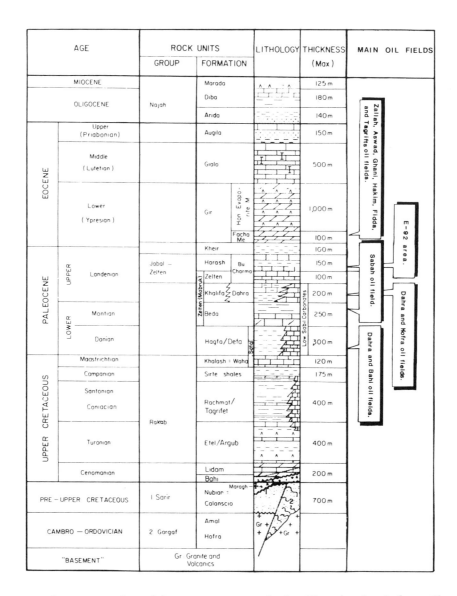

Figure 3: Composite stratigraphic sequence of the Sirt basin (after Chatellier and Slevin, 1988).

2. LINEAMENT ANALYSIS

Lineament analysis and tectonic model

A major (NW-SE) and a minor (NE-SW) lineament trend can be distinguished from the length frequency rose diagrams (Fig. 4) of Landsat TM (Fig. 2) SIR-A (Fig. 5, shows SW part). The main trend ranges from 300° to 335° in a single maximum of the TM-scene. It is represented by four maxima in the SIR-A image (290° to 300°; 300° to 310°; 310° to 330° and 330° to 340°), of which the 310° to 330° is the strongest. The peak occurs in both cases (TM and SIR-A) at 320°. Compared with the NW-SE trend, the NE-SW trend, peaking around 035°, is weakly represented in TM (Fig. 4a) and not detected in the area covered by SIR-A (Fig. 4b).

It should be noted that the scanning direction of SIR-A (335°) is close to parallel to the observed lineaments. Nevertheless, the 330-340° maximum of SIR-A (Fig. 4b) is reasonably well developed.

Comparison of these trends with the tectonic history of the basin indicates that the NW-SE trend is related to the late Cretaceous to early Tertiary formation of the Sirt basin by NE-SW directed extension. The resulting horst and graben structure continued to be active and influenced the deposition of the later Tertiary units.

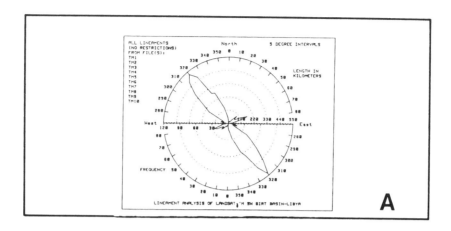

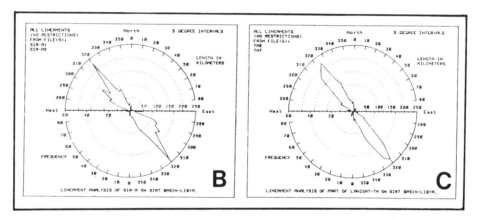

Figure 4: Lineament rose diagrams of the entire TM scene (A), the SIR-A image (B) and the equivalent part of TM (C).

The NE-SW lineaments are parallel to the so-called Hercynian trend. These faults may be reactivated as cross faults during the Sirt basin formation.

Some lineaments that are detected on SIR-A are not clearly visible on the Landsat-TM image (nor on the ground). These lineaments are apparent mainly in areas covered by surficial deposits and they are mostly parallel to the major NW-SE trend. For example, a number of lineaments detected on SIR-A in the Zallah oil field area, with outcrops of sandstone, claystone and conglomerate of Miocene age and partly covered by Quaternary sand, are poorly detected on Landsat TM. The SIR-A lineaments in this area correspond better to the number of faults known from subsurface mapping (Fig. 5 and 6).

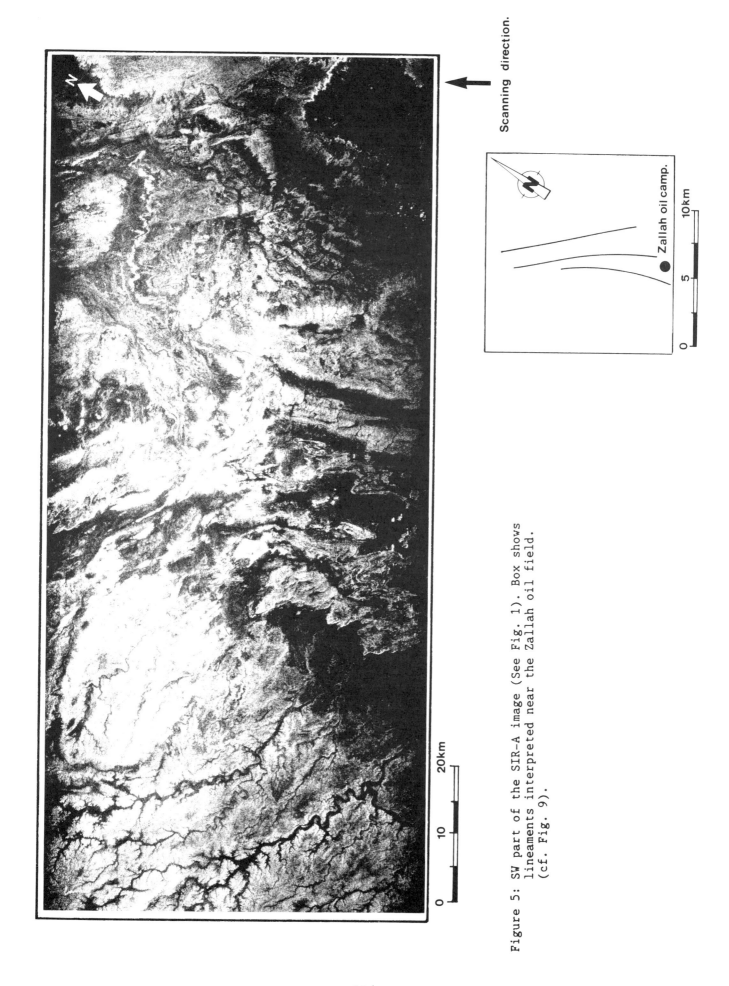

Scanning direction.

Zallah oil camp.

0 5 10km

0 10 20km

Figure 5: SW part of the SIR-A image (See Fig. 1). Box shows lineaments interpreted near the Zallah oil field. (cf. Fig. 9).

304

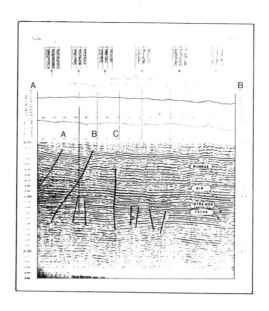

Figure 6: Seismic section through Zallah oil field (see Fig. 9 for location) showing the faults detected in SIR-A (A, B and C) and Landsat TM (A and B only)

Basement faults and their surface expression

Comparison of the lineament pattern of the whole Landsat-TM frame (Fig. 7) with a structure contour map of the basement (Goudarzi, 1978) and a Bouger gravity map (Fig. 8), shows a strong relationship between basement faults and surface lineaments. This is noteworthy since the faults are transferred through an approximately 1000 m. thick evaporite-bearing formation (e.g. Gir formation, Fig. 3). Gradually decreasing fault movements influenced deposition throughout the infill episode of the basin until the Oligocene. One example is the Gattar fault, which shows a monoclinal flexure associated with numerous small lineaments at surface (Alfasatwi, 1989).

Relation of lineaments to oil field locations

Migration and accumulation of hydrocarbons occurred in different locations coinciding with the detected lineaments. A group of nine oil fields (Ghani, Mellugh, E-92, Ed Dib, Zauam, Facha "W", Facha "E", Farrud and Umm Farrud oil fields; Fig. 7) coincides with the NW-SE trending lineaments, in particularly along the eastern edge of the Zallah through. This zone can be recognized as an approximately 7 km wide strip with numerous lineaments (Fig. 7). These oil reservoirs are located in fault traps or combined fault/stratigraphical traps of this fault zone. This zone can therefore be defined as a prospective area, and could also lead to investigation of similar, less well-known, structures.

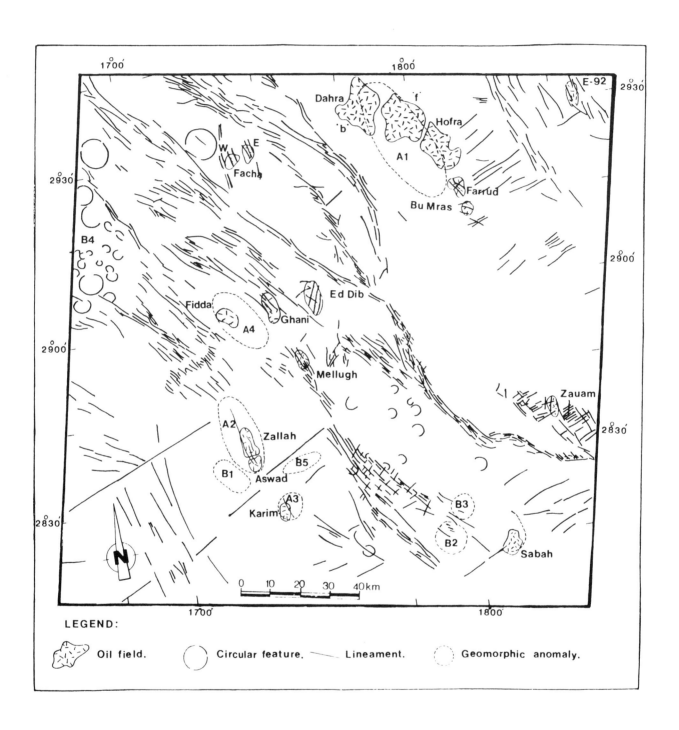

Figure 7: Lineaments and geomorphic anomalies interpreted from Landsat TM
(Fig. 2)

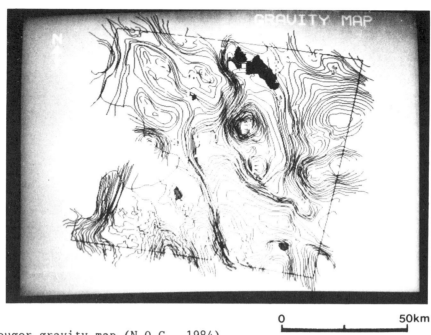

Figure 8: Bouger gravity map (N.O.C., 1984)

0   50km

3. GEOMORPHIC INTERPRETATION

Geomorphic anomalies, as described here, are circular or oval-shaped surface areas with contrasting tonal/textural characteristics and specific drainage patterns (mostly radial centrifugal). The term "circular features" is used in a wider sense, and applies to any (semi-) circular surface. A geomorphic anomaly may consist of several circular features (e.g. B4 on Fig. 7).

Several geomorphic anomalies were detected in the area of investigation (Fig. 7). Seven of the oil fields coincide with geomorphic anomalies (Hofra, Dahra B, Dahra F, Fidda, Karim, Aswad and Zallah oil fields). Most of the anomalies are elongated in the NW-SE direction, parallel to the Sirt basin trend. Only one anomaly has a north-east trend (Fig. 7, anomaly B5; Table 1). This anomaly is most probably related to cross faults oriented in the NE-SW Hercynian direction. The northwest trending geomorphic anomalies are parallel to the orientation of the basement faults, suggesting that these anomalies are located over the crests of the basement highs.

An example of the northwest trend is the Zallah oil field (Figs 2, 7 and 9). This field produces from the lower Eocene Facha member of the Gir formation (Figs 3 and 10). The average depth of the reservoir is 1940 m below surface (Garoussi, 1988). The drainage is dendritic with an oval radial centrifugal pattern. Seismic sections indicate that the field is located on a gentle anticline (Figures 6, 9 and 11). The flanks of the anomaly are dipping in the range of 5° (at surface). Correlation of Landsat TM interpretation and the structure contour map on top of the Facha member indicates that the Zallah oil field is located on the SE part of the anomaly. This suggests that the field could be developed towards the NW part of the anomaly. Furthermore, an adjoining smaller anomaly (Fig. 7, B1; Table 1) may be the surface expression of a subsidiary anticline.

307

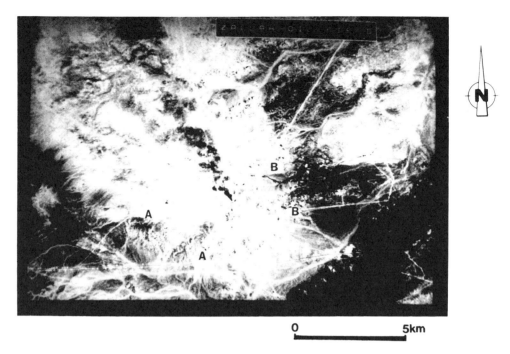

Figure 9: Landsat TM subscene (band 4) over the Zallah oil field.

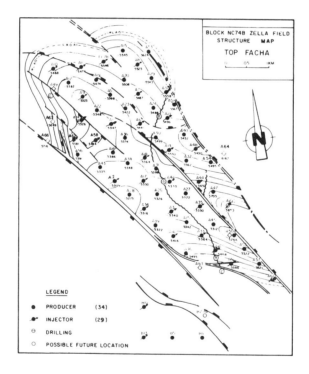

Figure 10: Structure contour map on top of the producing horizon (Facha member) of the Zallah oil field, showing faults truncating the domal structure of the field (cf. Fig. 6).

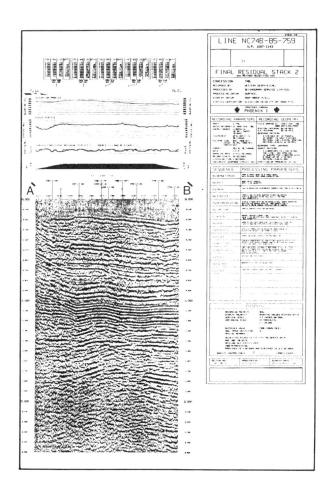

Figure 11: Seismic cross section through Zallah oil field, showing faults and gentle domal structure of the field (for location see Fig. 9).

A second example of the NW-SE trending anomalies is Karim oil field (Fig. 12), which is relatively small in size. Karim oil field is located south-east of Zallah oil field, the reservoir is the Facha member of the Gir formation, the average depth of the reservoir is 1880 m (below surface). The drainage pattern is dendritic with a semicircular deflection of the main stream and a radial centrifugal pattern of tributaries. It is interpreted to be a subsurface structural high. As shown in the structural contour map on top of the Facha member (Fig. 13; Oxey, 1978), and the seismic section (Fig. 14), the field is producing from a domal structure. This gentle anticline is truncated by the Karim fault. A trace of this fault can be clearly observed on the Landsat-TM image (Fig. 12). This fault forms an extension to the Zallah oil field faults.

As a last example a large geomorphic anomaly is evident on the Landsat-TM image over the Hofra and the Dahra oil fields, the largest in the area. This anomaly is characterized by semi-radial dendritic drainage patterns; it is elongated in the northwest direction and measures approximately 47.5 x 20 km. The location of these three fields in a single large geomorphic anomaly suggests that they are producing from the same domal structure. The multiple reservoir Dahra/Hofra oil fields form the only geomorphic anomaly in the northeast part of the area. The lack of other geomorphic anomalies there suggests that optimal benefit from remote sensing data for smaller oil fields can be obtained in areas of limited reservoir depth (cf. Figs 3 and 7) around the edge of the Sirt basin.

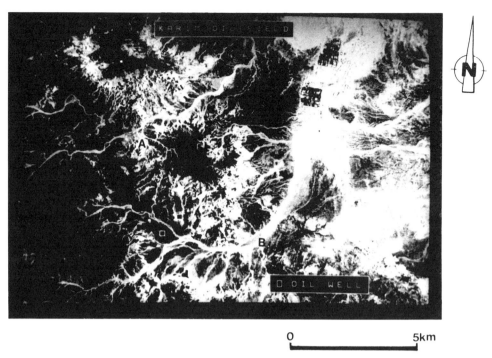

0   5km

Figure 12: Landsat TM subscene (band 4) over the Karim oil field.

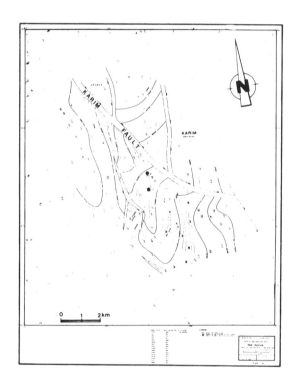

Figure 13: Karim oil field. Structure contour map on top of the producing horizon (Facha member).

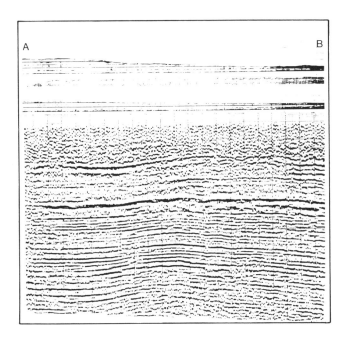

Figure 14: Karim oil field. Seismic section (for location, see Fig. 12).

The geomorphic anomalies over the oil fields were used in a next stage to assess other anomalies that are not producing. Five geomorphic anomalies were delineated (Fig. 7). Comparison of producing geomorphic anomalies and those which are not producing suggests that the reservoir targets for the non-producing anomalies are similar (Table 1). Therefore, these could be regarded as new prospects. A detailed structural study of an area in the SW part of the Landsat TM frame by Eabadi (1989) mentions phenomena that are of interest to the modeling of oil traps.

The oil fields which are not related to lineaments or geomorphic anomalies are Sabah, Bu Amud, Ogle, Umm Farrud, Gsur, and Diba. These fields are not evident because of their small size and Quaternary cover, especially in the NE corner of the Landsat TM frame where most of the small oil fields are located. Sabah oil field can be regarded to possess a geomoprhic anomaly because of the semi-circular deflection of a lava flow of the Al-Haruj volcanic province.

4. CONCLUSION

In this study, different remote sensing data were examined for hydrocarbon exploration in the Sirt basin, Libya. A lineament analysis was made using Landsat TM and SIR-A images to study the surface expression of basement faults and their relationship to known oil fields. The result indicate:
(1) that a strong relationship exists between basement faults and surface lineaments, despite an approximately 1000 m thick evaporite-bearing formation, and
(2) that nine of a total of twenty-two oil fields in the area are located near major lineaments.

Geomorphic interpretation of the Landsat TM image indicated that another seven of the oil fields coincide with geomorphic anomalies (other than lineaments). Furthermore, five new areas of geomorphic anomalies were delineated.

Further analysis combined with geophysical data, subsurface data and surface geology will help to determine prospective areas for more detailed study.

ACKNOWLEDGEMENTS

This presentation is based on MSc. research work, performed by Y.A. Alfasatwi at ITC, The Netherlands. The authors wish to express their appreciation to the Petroleum Research Centre (P.R.C.) Tripoli, Libya, for financial and logistic support and permission to publish this aricle. Ann Stewart is thanked for reviewing the manuscript. Various ITC-staffmembers assisted in the preparation.

REFERENCES

Alfasatwi, Y.A., 1989, Unpublished MSc. Thesis, ITC The Netherlands, in prep.

Chatellier, J.Y., and Slevin, 1988, Review of African Petroleum and Gas Deposit, Journal of African Earth Science, 7(3), pp. 561-578.

Eabadi, A.M., 1989, Unpublished MSc. Thesis, ITC, The Netherlands, in prep.

Garroussi, K.A., 1988, Eocene spheroidal dolomite from the Western Sirte basin, Libya. Sedimentology, 35, pp. 577-585.

Goudarzi, G.H. and Smith, J.P., 1978, Preliminary structure-contour map of the Libyan Arab Republic and adjacent areas; 1:2.000,000. U.S. Geol. Surv. Misc. Geol. Invest, Map I-350-C.

National Oil Corporation (N.O.C.), 1983, Tectonic elements map of Libya.

National Oil Corporation (N.O.C.), 1984, Bouger gravity map of Sirt basin, Libya.

Oxey, 1978, Reservoir data Summary of Hakim and Karim oil fields, Internal Report.

| ANOMALY | SURFACE EXPRESSION | | | | | EROSION SURFACE | | RESERVOIR (proposed for B anomalies) | |
| | SHAPE | TONE | DRAINAGE | | AGE | ELEVATION (above sea level) | DIP OF LAYERS | AGE | DEPTH (below surface) |
			PATTERN	DEFLECTION					
A1 DAHRA & HOFRA	NW/SW 48 x 20km	dark	dendritic semi-radial	---	Eocene	350 - 450 m	---	multiple reservoir from Paleocene to Eocene	
A2 ZALLAH & ASWAD	NW/SE 20 x 5km	light	dendritic semi-radial	---	Miocene & Oligocene	250 - 300 m	5 - 10°	Eocene Facha member, Gir Fm	± 1950 m
A3 KARIM	Circular 7km Ø	light to medium	radial	circular	Miocene & Oligocene	250 - 300 m	5 - 10°	Eocene Facha member, Gir Fm	± 1900 m
A4 FIDDA	NW/SE 20 x 10km	medium	dendritic semi-radial	---	upper Miocene-Pleistocene	250 - 300 m	---	Eocene Facha member, Gir Fm	± 1500 m
B1	NW/SE 9 x 4km	light to medium	radial	against lava flow	Miocene	250 - 300 m	7°	Eocene Facha member, Gir Fm	?
B2	Circular 6km Ø	light to medium	dendritic	southern part	Eocene	300 - 350 m	5 - 10°	lower Paleocene Beda Fm	?
B3	Circular 5km Ø	light to medium	dendritic	southern part	Eocene	250 - 300 m	---	lower Paleocene Beda Fm	?
B4	Several circular feature from 2 to 10 km Ø	dark	dendritic compressed meanders	circular	Eocene	350 - 450 m	---	?	?
B5	NE/SW 6 x 4km	light to medium	dendritic structurally controlled	---	Oligocene	250 - 300 m	5°	Eocene Facha member, Gir Fm	?

Table 1. Surface and subsurface characteristics of geomorphic anomalies from Landsat TM. A 1-4 are producing oil fields; B 1-5 are prospective areas from this study.

The American Association of Petroleum Geologists Bulletin
V. 66, No. 9 (SEPTEMBER 1982), P. 1348-1354, 9 Figs.

Applications of Landsat Imagery to Problems of Petroleum Exploration in Qaidam Basin, China[1]

G. BRYAN BAILEY[2] and PATRICK D. ANDERSON[3]

ABSTRACT

Tertiary and Quaternary nonmarine, petroleum-bearing sedimentary rocks in the Qaidam basin of remote western China have been extensively deformed by compressive forces. These forces created many folds which are current targets of Chinese exploration programs. Manual techniques of image analysis and interpretation were applied to computer-enhanced Landsat images of the western part of the Qaidam basin in an effort to evaluate the contributions of Landsat imagery in defining the geologic conditions of the basin and to determine its usefulness as an exploration tool in the region. Most success was realized in defining the structural geologic setting of the region.

Image-derived interpretations of folds, strike-slip faults, thrust faults, normal or reverse faults, and fractures compared very favorably, in terms of locations and numbers mapped, with Chinese data compiled from years of extensive field mapping. The image studies resulted in the identification of at least one subsurface fold that had not been detected by field mapping. The results of this study have direct exploration significance. Many potential hydrocarbon trapping structures were precisely located and information was obtained that may have significant implications with respect to fluid migration or attempts to locate offset reservoirs and buried folds. In addition, the orientations of major structural trends defined from Landsat imagery correlate well with those predicted for the area based on global tectonic theory. These correlations suggest that similar orientations exist in the eastern half of the basin where folded rocks are mostly obscured by unconsolidated surface sediments and where limited exploration has occurred.

INTRODUCTION

In November 1978, scientists from the Earth Resources Observation Systems (EROS) Data Center and the Scientific Research Institute for Petroleum Exploration and Development, Ministry of Petroleum Industry of the People's Republic of China began an informal joint study to demonstrate and evaluate the application of Landsat imagery to exploration for oil and gas in the Qaidam basin, located in Qinghai Province of western China.[4] Although cooperative studies of this region are not yet complete and have, in fact, been expanded in scope under a protocol of scientific and technological exchange, this paper reviews work done by EROS Data Center scientists through November 1981, discusses techniques of analysis and interpretation employed, and presents initial results.

GENERAL GEOLOGIC AND PHYSIOGRAPHIC SETTING OF QAIDAM BASIN

The Qaidam basin is a large intermontane sedimentary and structural basin located in the remote western part of China (Fig. 1). The basin covers an area of approximately 38,600 mi² (100,000 km²) and is enclosed in a triangular shape by three major mountain ranges. The basinal surface has an average elevation of about 9,800 ft (3,000 m). Annual precipitation averages less than 1.38 in. (35 mm), and the average relative humidity is less than 3%. Temperature variation is extreme and the winter season is long. Wind velocities are typically high, particularly during the months of March through May, resulting in a landscape dominated by eolian erosional and depositional landforms. The entire region is generally devoid of vegetation because of the harsh climatic conditions.

The Qaidam basin has been the site of sediment deposition since at least Late Jurassic time when uplift of the surrounding mountain ranges and coincident basin subsidence began. Subsequent to that time, as much as 39,400 ft (12,000 m) of lacustrine and other associated nonmarine sediments were deposited. Depositional and subsequent thermochemical conditions were periodically favorable for the formation of hydrocarbons. Compressional deformation of the basin sediments, which occurred mostly during the late Tertiary and Quaternary, resulted in the formation of many large folds that were potentially favorable for the entrapment of hydrocarbons.

OBJECTIVES

Chinese scientists have been studying the geology of the Qaidam basin and exploring for petroleum in the region since the mid-50s. To date, they have mapped the surface geology of the entire basin in detail, and they have conducted gravity, aeromagnetic, and seismic surveys throughout much of the basin. As a result, many oil and

[1]Manuscript received, November 30, 1981; accepted, April 16, 1982.

Official transliterations are used for the spelling of Chinese proper names in the text of this paper. Traditional transliterations of Qaidam, such as Tsaidam and Chaidamu, are common in previous literature. English, rather than Chinese, is used for words such as basin, mountain, etc, throughout this paper.

[2]U.S. Geological Survey, Sioux Falls, South Dakota.

[3]Technicolor Graphic Services, Inc., EROS Data Center. Current affiliation, Love Oil Company, Inc., Denver, Colorado. Work done under U.S. Geological Survey Contract No. 14-08-0001-16439.

The writers thank Maurice J. Terman, Thomas D. Fouch, Charles M. Trautwein, Raymond A. Byrnes, and Robert C. Davis for their many helpful suggestions during the preparation of this manuscript. We also express our sincere appreciation to our Chinese colleagues from the Scientific Research Institute for Petroleum Exploration and Development for their cordial hospitality and for the assistance they gave us during our stay in China.

[4]See footnote 1 for transliteration policy.

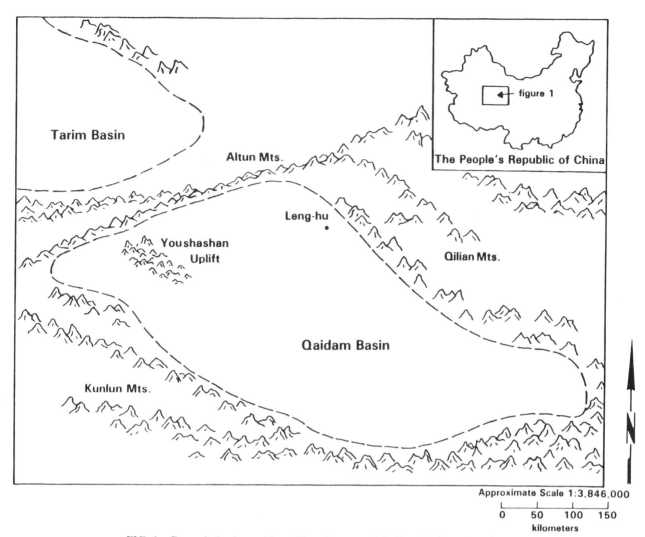

FIG. 1—General physiographic and location map of Qaidam basin region, China.

gas shows have been located, and several fields have been discovered by drilling. Yet, much exploration remains to be done, and the Chinese want to determine the usefulness of Landsat data for petroleum exploration. In the Qaidam basin, they are particularly interested in determining the capabilities of Landsat data to provide some indication of subsurface geologic structures that may have acted as hydrocarbon traps.

EROS Data Center geologists initially had only general knowledge of the geologic setting of the Qaidam basin region from brief discussions with visiting Chinese geologists in late 1978. Thus, the cooperative studies provided an opportunity to apply and evaluate certain techniques of Landsat image analysis and interpretation in an effort to gain a better understanding of the geologic framework and petroleum exploration potential of a relatively unknown region of the world. The main objective of this study was to develop a comprehensive geologic interpretation of the basin by focusing on structural, lithologic, and geodynamic relations that are expressed on or can be inferred from Landsat imagery of the area. Another objective was to develop a predictive geologic model which would be useful in exploration in this and other geologically similar regions of the world.

PROCEDURE

The initial step in this study was to construct a 32-scene, 1:1,000,000 scale mosaic of Landsat band 7 images covering the general region of the Qaidam basin (Fig. 2). This mosaic was used primarily to interpret regional structural relationships, particularly as they may have related to the development of the basin within the regional geologic framework. The mosaic was also used to select a six-scene area covering the western part of the basin for which detailed image analysis and interpretation were conducted (Fig. 2).

Digital data for the six-scene intensive study area were computer processed to produce enhanced false-color images. Processing included destriping, other radiometric corrections, and contrast enhancement using interactively determined, scene-dependent stretch limits. Manual analysis and interpretation techniques were applied to the study of these enhanced images, both as individual scenes and as a six-scene mosaic (Fig. 3). Surface drainage channels, landforms, natural cover patterns, lineaments, curvilinear features, and tonal anomalies were delineated on the images in an effort to extract all information potentially significant to a geologic interpretation of the area.

FIG. 5—Photograph from western part of Qaidam basin showing typical exposure of folded, nonmarine, clastic sedimentary rocks. Photograph illustrates the low topographic relief typical of the folded sedimentary rocks.

INTERPRETIVE RESULTS

Geologic interpretations of the Landsat images were greatly facilitated by the fact that the Qaidam basin has little or no vegetative cover and rocks are typically well exposed. Soils, windblown sands, and alluvial deposits do cover the rocks in parts of the basin, but these materials were not a severe hindrance to the geologic interpretations of the western part of the basin.

The aspect of the interior basin geology that is most readily interpreted from the imagery is the folding in the sedimentary rocks. The basinal sediments have been folded into a large number of generally northwest-southeast trending, commonly doubly plunging anticlines and synclines. On the images, these folds are typically defined by a series of semielliptical features which on the ground represent the traces of continuous sedimentary beds or of their associated weathered surfaces. The axes of the folds identified on the images are plotted on Figure 4. Most of the folds identified were determined to be anticlines on the basis of associated landform characteristics and of the apparent dips of the exposed stratigraphic units. Stereoscopic analysis of image sidelap areas proved to be extremely valuable in establishing some of these interpretations.

Figure 5, a photograph taken in the west-central part of the basin, illustrates a typical exposure of the sandstone, siltstone, and shale units that comprise the folds which are

so readily visible on the imagery. In contrast, Figure 6 is a photograph taken in the northwest part of the basin (point A, Fig. 4) looking southeast at a drill rig that is exploring a subsurface anticlinal structure. The anticline has no surface outcrop expression and was discovered by reflection seismic surveys. Nevertheless, during our study of the Landsat imagery we independently identified the area as being underlain by a folded structure on the basis of some subtle, semisymmetrical tonal variations that have a pattern similar to the patterns displayed by the exposed folded structures. Our field observations revealed that the area is expressed as a very subtle topographic high. We suggest that the observed semisymmetrical tonal variations result from precipitation-induced leaching, outward and downward migration, and subsequent redeposition of evaporite salts contained in the soil cover. Thus, the observed image patterns are indirectly related, because of the subtle topographic expression, to the anticlinal structure buried beneath the soil cover. Similar patterns, possibly indicating the presence of additional subsurface structures, were delineated elsewhere in the basin, but the significance of these patterns has not yet been confirmed by other data.

Lineament analysis of the Landsat imagery also provided important information for geodynamic and structural interpretation of the western Qaidam basin region. Figure 7 shows the distribution of the interpreted faults and fractures occurring within the study area. Lineaments

FIG. 6—Photograph taken near point A on Figure 4, looking east at operating drilling rig (in far distance, center) that is exploring subsurface anticlinal structure. Note total lack of exposed bedrock; surficial material is encrusted, saline soil.

observed on the images were interpreted and classified as belonging to one of four general groups: strike-slip faults, thrust faults, normal or reverse faults, and fractures. Lineaments were not interpreted as faults unless apparent offset along them could be observed or inferred.

Strike-slip faults.—The most readily apparent lineament observed on the imagery is part of the east-northeast–trending lineament system that forms the northwest boundary of the Qaidam basin and clearly offsets the Altun Mountains in a left-lateral manner. In the Qaidam basin region, we estimated offset along this major strike-slip fault zone, known as the Altun zone, to be a few tens of kilometers.

Lineament analysis revealed a generally north-south–trending zone in the west-central part of the basin (Fig. 7) where there exists a distinctly higher density of observed lineaments than was found for the interior of the basin in general. This zone closely corresponds to what appears to be a general, but rather subtle, right-lateral offset of axial trends of the major folds within the basin. We interpreted this zone as being a right-lateral strike-slip fault zone, although it is certainly much less clearly defined than is the left-lateral Altun zone.

Numerous shorter lineaments in the basin that were interpreted as faults are displayed on Figure 7 with directional arrows indicating a strike-slip movement along the fault. These notations should be viewed as indicating only *apparent* horizontal offsets, because it is generally not possible to determine with certainty whether the offsets

actually resulted from a strike-slip motion or whether they are manifestations of a vertical offset, or a combination of both.

Thrust faults.—Figure 7 also shows the locations of lineaments and curvilineaments that have been interpreted, primarily on the basis of image studies, to be traces of thrust faults. The most prominent of these is a northwest-trending lineament or curvilineament in the southwest part of the basin near the Youshashan uplift. It seems to define the leading edge of a major thrust plate which appears to have moved from the southwest to the northeast. The circular thrust-fault symbols southeast of the Youshashan uplift on Figure 7 mark the locations of what we have interpreted to be windows in the upper plate of this thrust sheet. The thrust fault that bounds the southwest edge of the Youshashan uplift was interpreted from our image studies as being a high-angle fault, but information provided to us by our Chinese colleagues indicates that the fault plane dips beneath the Youshashan uplift at a relatively shallow angle. Such a relationship makes our interpretation related to the previously discussed thrust fault somewhat difficult, but not impossible, to explain and points to the need for further field-verification of certain image interpretations.

Normal or reverse faults and fractures.—Lineaments interpreted as traces of high-angle normal or reverse faults, as well as those believed to represent major fractures, have also been plotted on Figure 7. As noted earlier, some faults shown on Figure 7 as displaying apparent horizontal offset may, in fact, be normal or reverse faults. Those faults shown as normal or reverse faults were so interpreted because the primary indication of offset was a difference in topographic expression associated with the fault. If no apparent vertical or horizontal offset could be observed or inferred along a lineament, it was interpreted to be a fracture. High-angle faults and fractures occurring within the basin appear to display northwest and northeast preferred directions of strike. However, statistical analyses of fault and fracture orientations have not been performed.

Attempts to interpret precise lithologic characteristics on the basis of image analysis were generally unsuccessful. However, lithologic variations could be established on the basis of differences in color, drainage characteristics, and apparent resistance to erosion. These variations were most useful in defining the structural and major stratigraphic relationships within the basin. Numerous small color anomalies were delineated during image analysis, but their potential significance with respect to specific lithologic associations or to the possible occurrence of hydrocarbons has not yet been evaluated. Digital processing techniques will be applied later in the project to more fully address the problems of lithologic discrimination and identification.

SIGNIFICANCE OF INTERPRETIVE RESULTS

Geologic data derived from studies of Landsat imagery of the western Qaidam basin are significant in several respects. They resulted in an accurate interpretation, based on comparison with Chinese geologic maps as well as our own field observations, of the structural geologic setting for an area greater than 19,300 mi^2 (50,000 km^2). The image interpretations were accomplished during a period of just a few months, compared to the many years required to derive similar interpretations from field-acquired data. Many major, exposed folded structures were precisely located, thus identifying potential targets for detailed exploration work. Of even greater significance to exploration efforts specific to the Qaidam basin was the demonstration of the capability of Landsat to detect subsurface folded structures that were not identified in detailed surface geologic mapping.

Interpretations of faulting within the basin have less direct significance for petroleum exploration than do interpretations related to folding, yet several important observations can be made. The recognition of an apparent major thrust sheet is significant from the standpoint that there exists a potential for discovering favorable structural traps in the lower plate rocks even though there may be no indication of their presence on the surface. The existence of the proposed thrust sheet should at least be considered in interpretations of seismic and other geophysical data collected for the area.

The recognition of high-angle faults which offset certain of the major folds may be potentially significant both in locating offset hydrocarbon reservoirs and because some of these faults could themselves be acting as hydrocarbon traps. The major significance of interpreted fractures lies in their possible effects on the migration of hydrocarbons into or out of structural or stratigraphic

traps, as well as in their possible effects on increased reservoir permeability.

The structural geologic interpretations derived from these studies have been used to develop a predictive structural model for application in the eastern part of the Qaidam basin. The eastern part of the basin appears to have a greater amount of unconsolidated surficial cover than does the western part of the basin. This cover obscures surface indications of folded structures which nevertheless may be present there. The model, which is based on global tectonic theory, may be useful in guiding exploration efforts in the eastern part of the basin.

Global tectonic theory suggests that the principal direction of horizontal compressive stress in the Qaidam basin region, which resulted with the collision of the Indian and Eurasian plates, was in a northeast direction as shown by the short, thick arrows on Figure 8. If a pure shear model of deformation is assumed, we would expect strike-slip faults, thrust faults, and major folds to develop in the general orientations depicted on this figure. Comparison of these predicted orientations with those actually observed in the western Qaidam basin (Fig. 9) reveals a close agreement between the two. It is not unrealistic to assume that rocks in the eastern part of the basin also deformed in response to the same general conditions and orientation of stress. Thus, we can predict with reasonable confidence that the same general types and orientations of deformation, and most importantly the folded structures, that exist in the western Qaidam basin should also be found in the eastern part of the basin even though they may be obscured there by surficial cover (Fig. 8).

SUMMARY AND CONCLUSIONS

This paper has presented initial results of studies being carried out by the U. S. Geological Survey and the People's Republic of China to define and evaluate the application of various remote sensing techniques to petroleum exploration. Analysis and interpretation of enhanced Landsat images of the western Qaidam basin resulted in an accurate definition of the structural geologic setting of that area and have provided information which will be useful in further exploration of the basin. The work completed to date has clearly demonstrated the value of Landsat imagery as a regional exploration tool, particularly for areas where little geologic information exists. Further work on this project will attempt to refine the geologic interpretations of this and other regions of western China by applying additional techniques of digital data processing and by integrating various remotely sensed and ground-based data.

SELECTED REFERENCES

Balley, A. W., et al, 1980, Notes on the geology of Tibet and adjacent areas--report of the American Plate Tectonics Delegation to the People's Republic of China: U. S. Geological Survey Open-File Report 80-501, 100 p.

Masters, C. D., et al, 1980, A perspective on Chinese petroleum geology: U. S. Geological Survey Open-File Report 80-609, 21 p.

Stone, D. S. , 1975, A dynamic analysis of subsurface structure in northwestern Colorado: Rocky Mountain Association of Geologists, Symposium Proceedings, p. 33-40.

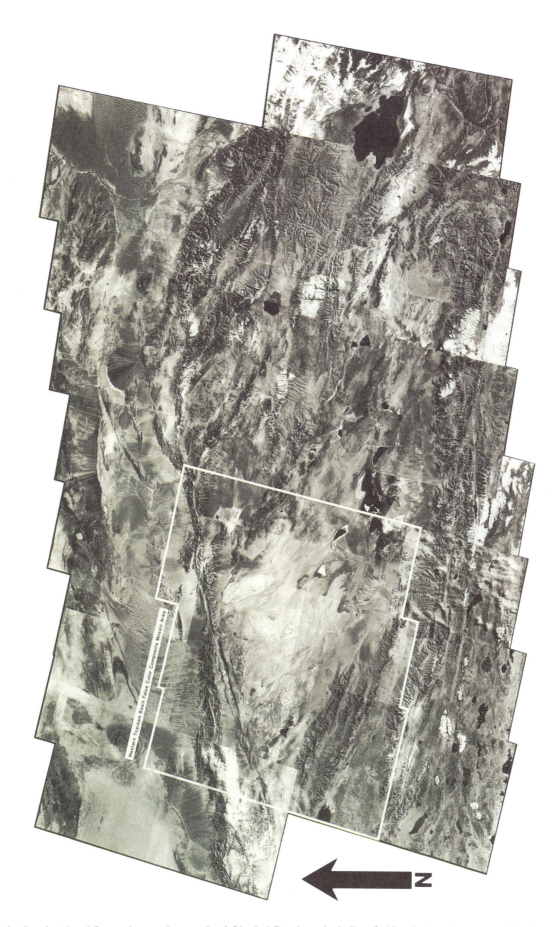

FIG. 2—Landsat band 7 mosaic covering much of Qinghai Province, including Qaidam basin. Area covered by six-scene mosaic made from digitally enhanced false-color images (Fig. 3) is outlined.

453

FIG. 3—Landsat mosaic of western part of Qaidam basin made from digitally enhanced false-color images.

454

FIG. 4—Landsat mosaic image of Western Qaidam basin showing locations of fold axes interpreted from image studies. Standard notations are used to designate type of fold and direction of plunge.

455

FIG. 7—Landsat mosaic image map of western Qaidam basin showing locations of faults and fractures interpreted from image studies. Standard notations are used to designate fault type or associated relative movement. Fractures are designated by non-annotated solid lines.

456

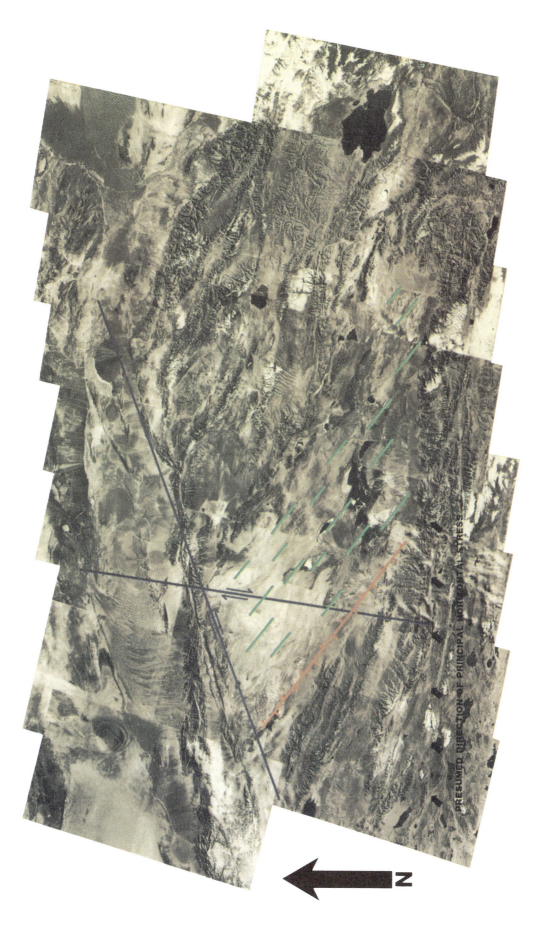

FIG. 8—Qaidam basin region, showing presumed direction of horizontal stress associated with collision of Eurasian and Indian plates and predicted orientations of resulting strike-slip faults (blue), thrust faults (yellow), and fold axes (green), based on a pure shear model of deformation.

457

FIG. 9—Landsat mosaic image map of Qinghai Province showing existing orientations of strike-slip faults, thrust faults, and fold axes interpreted from image studies. These orientations can be compared with predicted orientations shown in Figure 8.

458

Reprinted by permission of the American Society for
Photogrammetry and Remote Sensing from *Technical
Papers of the American Society of Photogrammetry*,
1981, p. 259-262.

LANDSAT IN THE SEARCH FOR APPALACHIAN HYDROCARBONS

H. W. Blodget
Eastern Regional Remote Sensing
Applications Center (ERRSAC)
NASA/Goddard Space Flight Center
Greenbelt, MD 20771

BIOGRAPHICAL SKETCH

Herbert W. Blodget is Projects Group Leader for the Eastern Regional Remote Sensing Applications Center at the NASA Goddard Space Flight Center. He received his training in geology, receiving a B.S. from Rutgers University and an M.S. and Ph.D. from the George Washington University. His prior experience includes employment as Exploration Geologist with the Arabian American Oil Company and Texaco, Inc.

ABSTRACT

An investigation to assess the application of enhanced Landsat imagery for hydrocarbon exploration in Appalachia was carried out jointly by ERRSAC, the Appalachian Regional Commission, and geologists from seven Appalachian states. The primary objective was to determine the utility of lineaments identified on Landsat imagery as an exploration tool in the search for petroleum within three Appalachian test sites.

The study was conducted in two stages. The initial stage was designed to identify the optimum Landsat imagery enhancement technique for displaying lineaments. The second stage was an analysis of the Landsat lineament data and correlation of the results with oil and gas field information for each of three test sites. Good correlations were found for several areas. Successful techniques can be incorporated into a broader exploration model.

INTRODUCTION

A study to determine the applicability of enhanced Landsat imagery for oil and gas exploration in Appalachia was conducted jointly by ERRSAC, the Appalachian Regional Commission, and geologists from seven Appalachian states. The immediate objective of this investigation was to determine if lineaments identified on Landsat imagery could be used as a tool in petroleum exploration, within three Appalachian test sites (Figure 1).

259

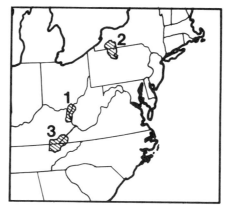

Figure 1:   Appalachian Lineament Study Test Sites

Lineaments, as defined here, are linear or sub-linear traces seen on Landsat imagery. They frequently reflect subsurface geological structures, the most common of which are joints and faults. Such faults are significant in some types of hydrocarbon accumulation, both for improving reservoir characteristics and in the formation of certain classes of strucutral traps (Levorsen, 1967, p. 267, and Kostura and Ravenscroft, 1977).

METHODOLOGY-RESULTS

The investigation was accomplished in two stages; the first was designed to identify the specific Landsat imagery enhancement best suited for displaying lineaments; the second consisted of plotting lineaments and making an analysis of the lineament trends in respect to both local geological structure and regional tectonics. This included correlation of the data with oil and gas field production information.

In the first phase, each of the state teams of geologists was provided with 12 images of its test site. The set included four differently enhanced images for each of three seasons--spring, summer, and fall (Wescott and Smith, 1979). The results of the image analysis were compiled by Smith and Miller (1980). Even though there were wide differences in the training and background of the interpreters, and no collaboration among teams, there was almost unanimous agreement in the conclusions. In particular, it was indicated that winter or "leaves-off" imagery provided the best seasonal data, and that a simple digital linear stretch provided optimum tonal contrast for all three test sites. In areas of highly dissected topography, as was characteristic in the two southern sites, edge enhancement further improved lineament display, and the Laplacian (3 x 3 filter) algorithm was generally preferred. In the flatter terrain of the New York-Pennsylvania test site, however, edge enhancement tended to emphasize field boundary lines and other cultural features in detriment to geology-related lineaments. Printing on

260

Cibachrome film consistently produced a tonal and textural
sharpness not present on paper prints. "Standard" enhanced
Landsat imagery of the optimum types is now available from
several commercial vendors.

During the second phase of the study, lineaments were
plotted on Landsat overlays at 1:250,000 scale for each of
the test sites. These data were correlated with published
geologic maps, and checked in the field where significant
new information appeared to be indicated. In New York,
Ohio, and Virginia, some Landsat lineaments coincided with
the reservoir limits of a number of known producing oil and
gas fields--a correlation which can be significant in
exploring for hydrocarbons.

In Allegany County, New York (Figure 2), for example,
Landsat lineaments appear to truncate the hydrocarbon production on at least one margin of 13 oil or gas fields.

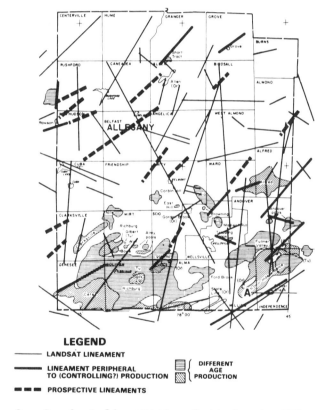

LEGEND

——— LANDSAT LINEAMENT

———— LINEAMENT PERIPHERAL
TO (CONTROLLING?) PRODUCTION

▬ ▬ ▬ PROSPECTIVE LINEAMENTS

DIFFERENT
AGE
PRODUCTION

Figure 2: Landsat lineaments plotted onto Oil and Gas Map
of Allegany County, New York (modified after Van Tyne and
Peterson, 1979; Van Tyne and Wickerham report summarized in
Blodget, 1979).

261

These lineaments have been broadened in Figure 2 for emphasis. Nine of these significant lineaments trend in a northeasterly direction, suggesting a genetic relationship among these lineaments, regional geologic structure and the associated reservoirs. In the same county, 13 additional similarly-trending lineaments have been identified which are not associated with known petroleum accumulation. These lineaments are shown as broad dashes. Because of the apparent significance of this lineament directional trend, the undrilled areas around these 13 lineaments are considered highly attractive petroleum prospects. These areas should be examined in detail using traditional geophysical exploration techniques in order to obtain a detailed definition of the subsurface geology.

CONCLUSIONS

The close relationship between lineaments identified on Landsat imagery and known oil and gas reservoirs in three diverse Appalachian test sites strongly suggests that lineaments can be related to the structural control of oil and gas reservoirs throughout Appalachia. These relationships should be exploited in developing all reconaissance exploration programs. Geologic interpretation of Landsat imagery is essential to developing the most complete understanding of regional geology and must be used as one of the guides for the application of more costly, definitive geophysical techniques.

REFERENCES

Blodget, H. W. (ed.), 1979, Summary of Proceedings, Appalachian Lineament Analysis Workshop, NASA, Goddard Space Flight Center, October 29-30, 41 p.

Kostura, J. R. and J. H. Ravenscroft, 1977, Fracture-Controlled Production: AAPG Reprint Series No. 21, The American Association of Petroleum Geologists, Tulsa, Oklahoma, 221 p.

Levorsen, A. I., 1967, Geology of Petroleum, second ed.: W. H. Freeman & Company, San Francisco, California, 724 p.

Smith, A. F. and J. E. Miller, 1980, Evaluation of Computer Enhanced Landsat Imagery for Lineament Analysis and Other Geological Applications: General Electric Company Phase 1 report for NASA contract NAS 5-25364.

Van Tyne, A. M. and J. C. Peterson, 1979, Oil and Gas Fields in New York--as of July 1978: Morgantown Energy Technology Center METC/EGSP series 102, scale 1:250,000.

Wescott, T. F. and Smith, A. F., 1979, Digital Image Processing and Enhancement of Landsat Data for Lineament Mapping and Natural Gas Exploration in Appalachian Region: in Barlow, H. (ed.), Proceedings Third Eastern Gas Shales Symposium, p. 39-49, Morgantown, West Virginia, October 1-3.

262

The American Association of Petroleum Geologists Bulletin
V. 73, No. 9 (September 1989), P. 1053-1064, 6 Figs.

Subsurface Exploration via Satellite: Structure Visible in Seasat Images of North Sea, Atlantic Continental Margin, and Australia[1]

R. C. BOSTROM[2]

ABSTRACT

Satellite observations of the sea-surface height are sensitive to crustal structure, overcoming the inability of previous remote-sensing techniques to penetrate the Earth's surface. Seasat gravity images have been compared with the structure of offshore sedimentary basins. In northwestern Europe, Seasat images display primary structural features (e.g., the Anglo-Dutch and smaller basins, down to the scale of the Manx-Furness basin), and numerous secondary structural features (e.g., the Halibut horst). Additionally, because of their synoptic viewpoint, Seasat images reveal features hard to distinguish. I cite as an example a basement ridge extending from Scotland to Norway, in position to have controlled marine circulation during deposition of the principal source rock, which is equivalent to the Kimmeridge Clay. In North America and South America, Seasat images have been used to examine the relation of the Atlantic fracture zones to the structure of the continental margin of the United States and Brazil. In Australia, known basins of the northwest continental margin and the Bass Strait have been compared with Seasat projections of shallow structure.

In comparison with reconnaissance seismic, magnetics, and ship-borne gravity surveys, Seasat images are valuable in discovering the presence of thick sediment accumulations; Seasat images are potentially more valuable than seismic and magnetics in the presence of basement masking by intrasectional volcanics or massive carbonates. In known basins, Seasat data deliver a uniform, synoptic view, permitting the identification of large structural features; in this form, these images give the structural geologist a tool like that given the stratigrapher by regional seismic sections. In remote unexplored regions, Seasat images provide a means of optimizing the layout of reflection seismic.

The gravity data from Seasat offer the first of increasingly accurate structural images to be expected from altimetric satellites.

INTRODUCTION AND OVERVIEW

I have compared the known structure of some offshore basins with their gravity images. The images were prepared from gravity data extracted by William Haxby and colleagues (Haxby et al, 1983; Haxby, 1985) using the radar on the Seasat orbiter.

Seasat images have been compared with the geometry of explored and partially explored basins. Those basins chosen are in the North Sea, the United States and Brazil Atlantic continental margins, and offshore Australia. Because the comparative data was obtained from drilling and reflection seismic, yet gravity is not depth-restricted, the "known" structure of explored basins is more shallow than that shown by gravity. An outline of the technical basis of the Seasat observations is provided in Appendix 1.

In the figures, shades of blue represent low gravity values, shades of green, red, and orange represent successively increasing gravity values. Thus, shades of blue mark thick sediments and their associated porosities, and shades of buff and orange indicate basement highs. For reasons explained in Appendix 1 (e.g., on Bouguer maps it is not possible to distinguish troughlike basins from mountain roots), I have constructed free-air images. I have indicated those images for which filters have been used, for example to pick up tectonic strike.

For each basin, two questions should be asked: (1) if the area were unexplored (e.g., devoid of uniform magnetic coverage), would the Seasat data have helped in its exploration, and (2) for basins with geophysical and drill data available, does Seasat contribute new tectonic information or leads for ongoing exploration.

NORTH SEA

North Sea regional structure (Figure 1a) is the product of Paleozoic convergent orogenesis followed by late Paleozoic extensional stress (Kent, 1975; Ziegler, 1980; Glennie, 1984; Brown, 1986). Gross crustal extension and oceanic crust emplacement (Roberts, 1974), which helped form the site of the North Atlantic Ocean, developed to the west during the Mesozoic and continues

[1]Manuscript received, October 15, 1987; accepted, May 1, 1989.
[2]University of Washington, AJ-20, Seattle, Washington 98195.
The writer thanks W. F. Haxby, Lamont-Doherty Geological Observatory, Columbia University, for providing a shaded-relief print, prepared from Seasat altimeter data, of the gravity field of the world ocean. Thanks go to R. Ludwin, S. D. Malone, and E. J. Mulligan, all of the University of Washington, who greatly assisted in data transfer and image screening. The writer thanks the AAPG Foundation for help in publication of this paper. I also gratefully acknowledge receipt of Atlantic-seaboard data sent by K. D. Klitgord, U.S. Geological Survey, Woods Hole, Massachusetts 02543. Figures 1a; 3b, c; 4b; and 5b, d were prepared from National Oceanic and Atmospheric Administration/ National Geophysical Center, Denver, Colorado, tape 00978/H/001 (1987), using image processing system PCIPS donated by IBM.

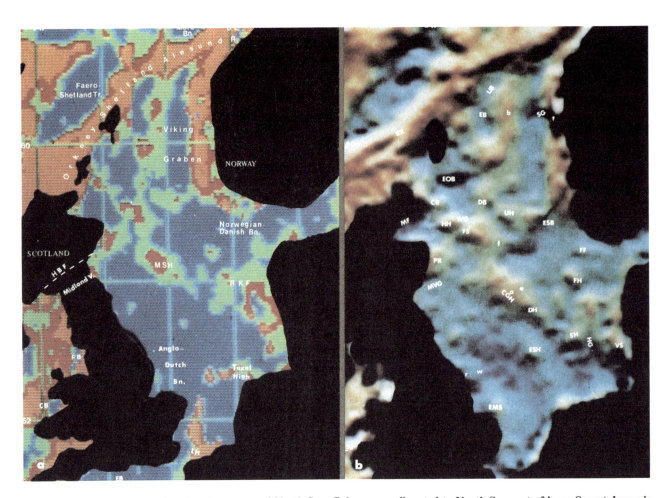

Figure 1—(a) Seasat image of regional structure of North Sea. Colors were allocated to North Sea part of large Seasat dynamic range. Blue tones are associated with basinal areas, buff tones show highs that overlie basement ridges. CB = Cardigan Bay basin, EB = English Channel basins, FB = Furness basin, LH = London-Brabant structural high, MSH = mid-North Sea high, RKF = Ringkobing-Fyn structural high, HBF = Highland boundary fault. (b) Seasat image of secondary and intrabasinal structure. Shaded image supplied by W. F. Haxby, Columbia University, New York, New York, smoothed (low-pass filtered). CGH = Central graben high, CR = East Orkney high, DB = Dutch Bank basin, DH = Dogger Bank high, EB = East Shetland basin, EOB = East Orkney basin, EMS = East Midland Shelf, ESB = Egersund basin, ESH = Elbow Spit high, FF = Fjerritslev fault, FH = Fjerritslev high, FS = Buchan horst–Glen horst, HH = Halibut horst, HG = Horn graben, MF = Moray Firth basin, MVG = Midland Valley graben, PR = Peterhead Ridge, RR = Rona Ridge, SH = Schill ground high, SG = Sogne graben, UB = Unst basin, UH = Utsira high, VS = West German basin high, WG = Witch Ground graben. Some productive localities: a = Auk, by = Brent fields, e = Ekofisk, f = Forties fields, r = Rough, t = Troll, w = West Sole.

through the Holocene. Epicontinental basins marked by gravity lows (blue in Figure 1) have developed in the crustal floor of the North Sea. These basins are separated by buoyant, thick crustal welts (green and buff, Figure 1) over which post-Paleozoic sedimentary deposits are relatively thin. The position of major grabens, such as the low-gravity (blue, Figure 1) Viking graben, is on a basement ridge, appearing as a background gravity high. The northwest-southeast central North Sea graben is the site of a string of gravity highs marking the underlying crustal ridge.

To evaluate the role of Seasat data at an early stage of exploration, Figure 2a, b shows the regional structure of the North Sea as seen at two exploration periods. In 1968 (Figure 2a), exploration necessarily proceeded without awareness of the Viking graben and structures dominating the post-Paleozoic sedimentary history of the north-

ern North Sea. At the time, aeromagnetic surveys were available as a basinal mapping tool. The penetration range of reflection seismic still is limited to the upper sedimentary column, and in many localities only to the Upper Cretaceous and younger rocks. Comparing Figures 1a and 2a, one can see that had Seasat been available, a preview of the structural framework of the North Sea would have been possible.

Figures 1a and 2b indicate that the North Sea sea floor is divisible into two provinces marked by a line running from the Midland Valley of Scotland to southern Norway (Egesund). The string of structural highs extending northeast from central Scotland indicates that the Midland Valley, bounded on the northwest by the Highland Boundary fault, is the on-land portion of crustal structure extending across the North Sea.

Seasat reveals major features, such as the crustal ridge

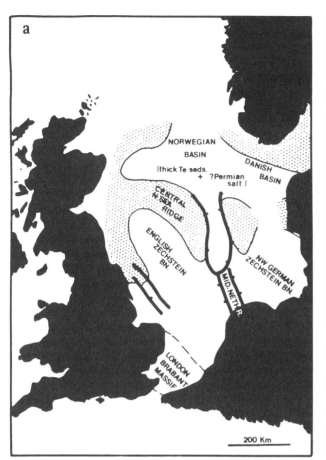

Figure 2—(a) North Sea structural elements as viewed at end of first stage of exploration (after Brennand and van Hoorn, 1986; Donovan, 1968). Shaded area shows pre-Mesozoic highs with thin Mesozoic cover. (b) Principal structural elements based on reflection-seismic exploration and drilling (after Brown 1986; Brennand and van Hoorn, 1986; Ziegler 1982); compare with Figure 1a. Not all workers agree that Great Glen fault continues from Scotland into Shetland Islands. In northern region, seismic encounters difficulties in mapping below basal Cretaceous. Generally, seismic has outlined only shallow sedimentary structures rather than underlying structures controlling them.

extending from Orkney to Norway near Alesund (Figure 1a), whose existence may not have been recognized in earlier North Sea exploration. The Orkney-Norway crustal ridge most nearly accords with structure postulated by Price and Rattey (1984; see also Smythe et al, 1983). The structure displayed on Figure 1a constitutes the natural division between the North Sea basins and basins such as the More basin farther north. In the Jurassic, North Sea access to the ocean was via the area of the ridge. The ridge may have controlled water circulation during deposition of the Kimmeridge clay equivalent, "the most important petroleum source rock known in the North Sea" (Dore et al, 1985). Debate is ongoing (Proctor, 1980; Brown, 1986) as to whether Middle Jurassic sedimentation was the product of southward progradation from a northerly landmass, or the converse.

Seasat images created so as to emphasize small features in shaded relief (Figure 1b) display a wealth of secondary intrabasinal structures, such as the Halibut horst. By allocating color to a selected range of values, mild structures, such as the East Midland Shelf, may be brought into relief. I believe that the Seasat images provide a more

useful view of the regional gravity than contoured regional gravity maps (see, for example, the excellent maps prepared by Hospers et al, 1985).

As might be expected, hydrocarbon accumulations tend to develop on the flanks of structural highs next to their generative basins. The western flank of the Viking graben consists of a zone of en echelon faults rather than normal faults parallel to the graben (Johnson and Dingwall, 1980). Gravity data depict the deep-seated structure on top of which sedimentary faulting has developed rather than the individual faults. In migration studies, the sea level equipotential surface, from which the Seasat dataset has been prepared (Appendix 1), parallels the interfaces of intrasectional fluids prior to disturbance.

The Seasat images show that unexplored structures in the North Sea are numerous. Despite Seasat delineation of structure, exploration in frontier areas like the Unst basin and the offshore continuation of the Midland valley (Brennand and van Hoorn, 1986) still is handicapped by limited seismic penetration. I am unaware of the nature of many features, for instance, the intense shallow-

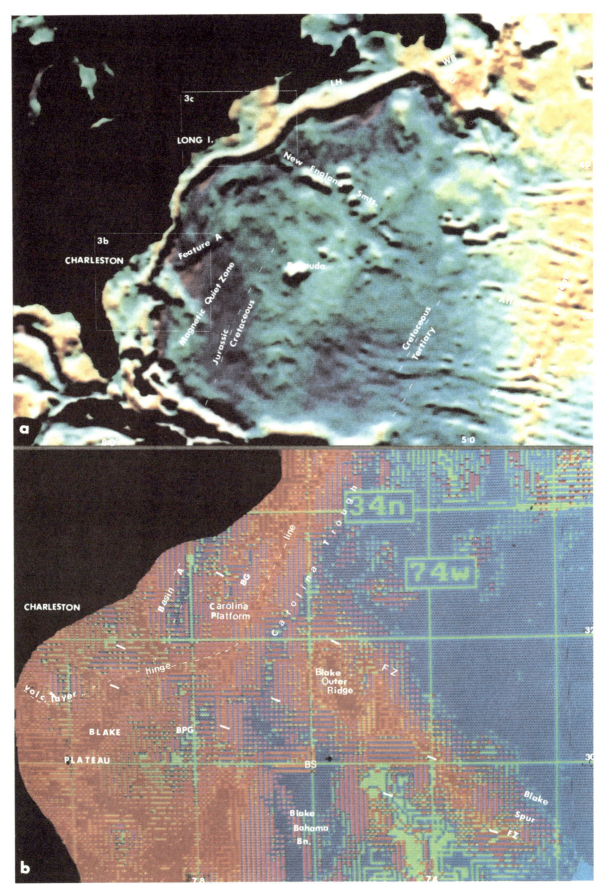

(Continued on next page)

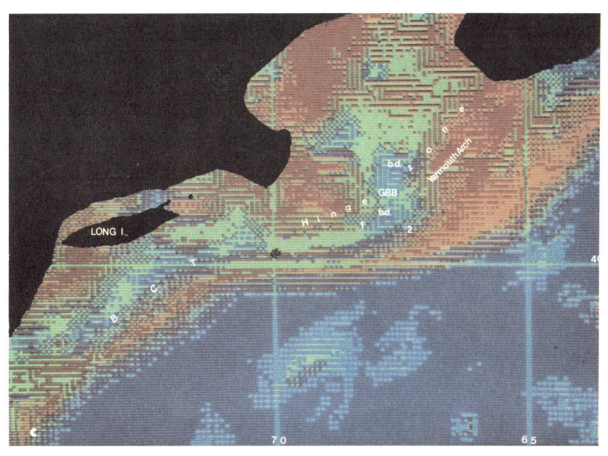

Figure 3—(a) Seasat image of western North Atlantic Ocean displaying fracture zones extending from Mid-Atlantic Ridge and fault (marked "Feature A") into north-bounding deep-basement region (magenta). Identification of crustal age after Klitgord and Schouten, 1986; Vogt and Einwich, 1979. AU = Avalon uplift, LH = LaHave platform, WB = Whale basin, F2 = fault zone. Regional background image courtesy of W. F. Haxby, Columbia University, New York, New York. (b) Seasat image of Blake Plateau–Carolina platform where intersected by Blake spur fracture zone (FZ) using mild edge filter and allocating spectrum to selected portion of Seasat value range. BG = Brunswick Graben, BPG = Blake Plateau basin, BS = Blake Spur. Area underlain by intrasectional volcanic layer (Dillon et al, 1983) offshore Charleston is shown by white double dots. White dashes show tectonic hinge line. Brunswick magnetic anomaly (Klitgord et al, 1988) parallels hinge line on landward side. (c) Seasat image of Gulf of Maine, allocating spectrum to Seasat values representing only continental shelf. BCT = Baltimore Canyon Trough, GBB = Georges Bank basin. Within Georges Bank basin, "b.d." is basinal deep with respect to Yarmouth sag, west of Yarmouth Arch; Georges Bank main basin (Klitgord et al, 1988) is centered on part of this basinal deep. 1 and 2 mark COST G-1 and COST G-2 well sites, respectively.

seated structure striking southwest and bounding the Egersund basin on the northwest. On the magnetic map (Hospers and Rathore, 1984), the Seasat feature separates an area of short-wavelength shallow magnetic features from the deep basement area to the southeast.

UNITED STATES ATLANTIC CONTINENTAL MARGIN

The eastern margin of the North American continent was formed by extension that began with graben formation at the end of the Paleozoic. Other workers have shown (Klitgord and Behrendt, 1979; Grow et al, 1979) that the joining of oceanic fracture zones (Figure 3a) and the continental margin has played an important role in the formation of shelf basins.

The Seasat regional image for this area correlates with the basement-depth map constructed by Tucholke et al (1982), who used 200,000 km of seismic reflection pro-

files. As might be expected, gravity is low (blue, Figure 3a) where sediments are thickest. A tightening of Tucholke et al's (1982) contours that extend northeast-southwest through lat. 33°N, long. 73°W is revealed on the Seasat image to be a linear feature, apparently marking a major fault.

Southwest of this region, a more detailed Seasat image (Figure 3b) delineates the intersection of the Blake spur fracture zone (FZ) and the continental margin. The area has been studied using refraction and common-depth-point (CDP) reflection seismic lines (Dillon et al, 1983), marine (surface) gravity surveys (Grow et al, 1979), and high-sensitivity airborne magnetics (Behrendt and Klitgord, 1980). Basinal mapping (basement depth determination) is handicapped by the existence of intrasectional volcanics (Figures 3b) (Dillon et al, 1983). These represent high-susceptibility high-velocity matter masking the underlying structure in respect to the use of seismic and magnetics. In respect to basement mapping, deep high-

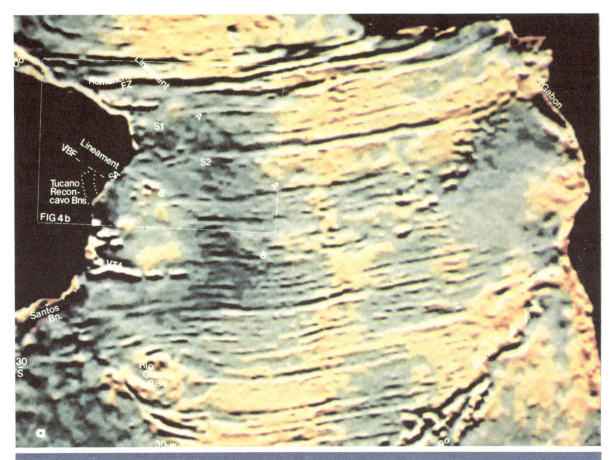

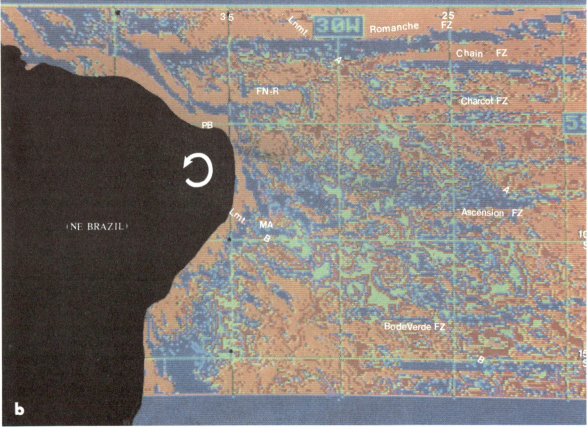

Figure 4—(a) Fracture zones, lineaments, and marginal basins in South Atlantic Ocean. Regional image supplied by W. F. Haxby. (See Cande et al, 1988; and Ojeda, 1982.) Szatmari et al (1985) believed that southern continuation of Sergipe petroleum province now is to be found in west Africa Gabon basin. VBF = Vasa-Barris fault, VTA = Vitoria-Trinidade alignment. Lineaments AA and BB show cross-cutting linear features. S1 and S2 mark circular features that resemble surface expression of mantle plumes. Box outlines area of (b). (b) Seasat image of northeastern Brazil continental margin and adjacent ocean floor. Seasat image using filter sensitive to structures aligned northwest-southeast. Orthogonal filter fails to reveal structures, e.g., conjugate faults, aligned northeast-southwest. FN-R = Fernando de Noronha Ridge, MA = Maceio alignment, PB = Potiguar basin. Lnmt. AA and Lmt. BB = lineaments bounding region of shear (Figure 4a). White arrow indicates rotation direction of northeastern Brazil at time of separation from Africa (Szatmari et al, 1985).

velocity sedimentary strata cannot be distinguished from basement.

The detailed Seasat image correlates with the marine gravity data but appears to have higher resolution. The existence of the small Brunswick graben (Klitgord et al, 1988) is suggested by the Seasat data. These data also correlate with the total-field magnetics. Both Seasat and magnetics delineate both the tectonic hinge line landward of the Carolina Trough and the zone of low values marking the Brunswick magnetic anomaly. Seasat is more valuable than both seismic and magnetics in mapping the low marked "Basin A" in Figure 3b. Seasat also illustrates the advantage of gravity in this tectonic setting because intervening formations are transparent to gravity. The unresolvable structure below seismic basement at Basin A (Figure 3b) may represent granitic plutons (Klitgord et al, 1983) or thick sedimentary rocks, and invites further investigation.

Georges Bank, including Georges Bank basin, Gulf of Maine (Figure 3c), has been surveyed using CDP seismic, accurate magnetic coverage, marine (surface) gravity, and deep-test drilling (Klitgord et al, 1982). Reflection seismic cannot map basement. Basement form has been delineated through the use of spot, magnetically determined depth estimates to extrapolate overlying structure mapped by seismic.

Marine gravity (Klitgord et al, 1982) indicates a Georges Bank basinal deep displaced about 60 km west of the location of the deep as mapped by magnetics (Klitgord et al, 1982, their Figure 7b) and seismic, and confirmed by deep-test wells (Klitgord et al, 1982; see also Mattick et al, 1981). Marine gravity shows positive values over the basinal deep. The detailed Seasat image places the basinal deep (blue, Figure 3c) in the same locality as magnetics, seismic, and test wells. Had this been an unexplored region, Seasat would have revealed the existence of Georges Bank basin, the northwest-adjoining Yarmouth arch, and other structures not yet explored.

BRAZIL ATLANTIC CONTINENTAL MARGIN

The South Atlantic (Figure 4a) began forming during the Cretaceous (Valanginian), approximately 100 m.y. later than the rifting that initiated the development of the North Atlantic. Strongly expressed fracture zones extend from the Mid-Atlantic Ridge to the Brazil continental margin.

Important production has been located where a defunct rift arm strikes inland to form the Reconcavo basin (Szatmari et al, 1985; see also Burke, 1976). The

major northwest-striking Vasa-Barris strike-slip fault separates cratonic basement from the Sergipe Trough. The regional image (Figure 4a) displays cross-cutting linear features (lineament AA and BB). The lineaments are believed to be real rather than artifacts associated with an orbital ground track direction because they define provinces having distinctive structural texture. Circular features marked S1 and S2 (Figure 4a) resemble the surface expression of mantle plumes elsewhere (Bostrom, 1988).

A detailed image using a filter sensitive to structures aligned northwest-southeast (Figure 4b) suggests that the entire region between the lineaments is affected by shearing or antecedent fracture zones. Within the region, fracture zones, such as the Charcot FZ, are so disjointed as to lose their identity. The basins located along the northeastern shore of Brazil are aligned transverse to the shore rather than parallel to it as elsewhere. The productive Potiguar basin (Kumar, 1981) is bounded on the north by the continental extension of the Fernando de Noronha Ridge. The Potiguar and kindred basins are bounded on the east and west by reactivated Precambrian faults.

Northeastern Brazil tectonics have been interpreted as the result of north-south compression (Szatmari et al, 1985) producing conjugate northeast- and northwest-aligned faulting. Experimental use of a Seasat filter sensitive to structures aligned northeast-southwest indicates that offshore, these structures are not significant. Regional shearing of the ocean floor accords with counterclockwise rotation of northeastern Brazil relative to Africa at the time of separation, but not with the reported sense of the Vasa-Barris fault. The structure of the oceanic region adjoining northeastern Brazil, therefore, suggests that the primary motion has been simple shearing; furthermore, that without first resolving the structural history of the ocean floor, the continental structure cannot accurately be determined.

OFFSHORE AUSTRALIA

A regional image of offshore southeastern Australia (Figure 5a) displays fracture zones already mapped on the basis of sea-floor magnetics and displays more clearly than magnetics oblique features such as those entering the continental margin near Otway basin (Figure 5a). When compared with other gravity maps of the ocean surrounding Australia (e.g., see Circum-Pacific Council for Energy and Minerals Resources Geodynamic maps of the Circum-Pacific Region, Doutch et al, 1985), Seasat images more clearly delineate large-scale structures.

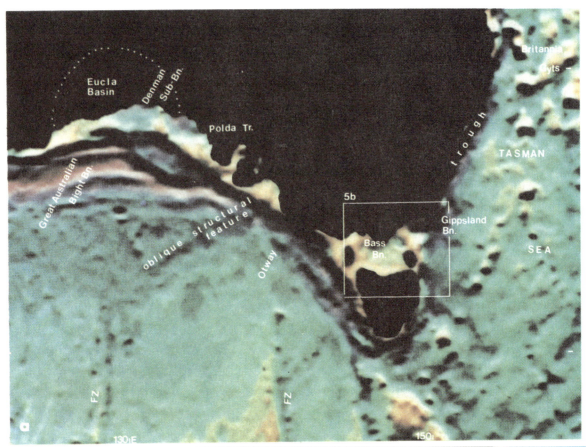

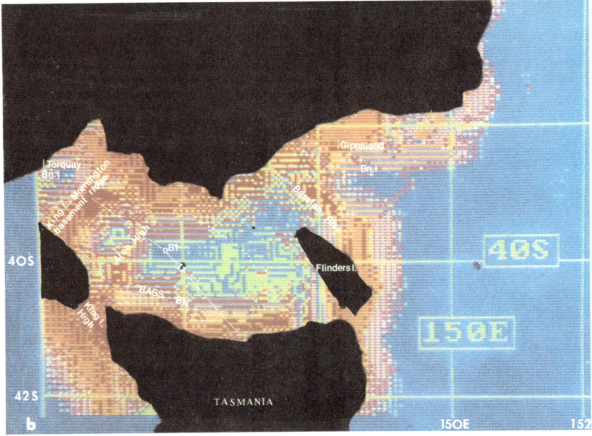

(Continued on next page)

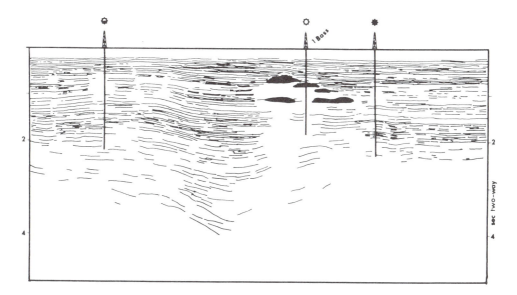

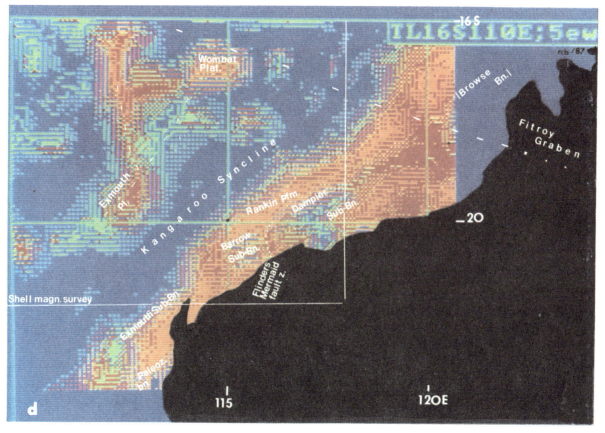

Figure 5—(a) Seasat image of regional sea-floor structures, southeast Indian Ocean and Tasman Sea adjacent to southeastern part of Australian continent. In Tasman Sea, Seasat image tends to confirm orientation of synthetic fracture zones derived from Weissel and Hayes' (1977) analysis of this region. Tasman Sea is sharply separated from tectonic province of north-south fracture zones formed at Indo-Australian spreading axis by line extending southeast from Tasmania. Box shows area of (b). (b) High-resolution image of Bass basin (dotted line) and surrounding area. Image formed by use of high-pass filter and allocating spectrum to local value range. White line shows seismic line 11 (Williamson et al, 1987), B1 = test well 1 Bass. (c) Line drawing of portion of seismic line 11 (Williamson et al, 1987) showing local, intrasectional volcanics (black) encountered in test well 1 Bass. Volcanics are too thin to contribute to regional gravity signal, but prevent transmission of seismic pulse and, if susceptible, produce false basement-depth values using magnetics. Well symbols: empty circle = dry, abandoned well; half-filled circle = well with show of oil and gas, abandoned well; filled circle = oil and gas discovery well. (d) Continental margin of northwestern Australia. Seasat image made using mild edge filter and allocating color spectrum to shallow-water area. White box shows area of marine magnetic survey (Willcox, 1981). Thick dashes show locus of alignment apparent on regional images of northwestern Australia interpreted as master fracture zone (see text), which passes between Wombat plateau and Joey Rise (not on image) and northbounds Exmouth plateau.

The Great Australian Bight basin (mauve, Figure 5a) is known to contain more than 10 km of sedimentary deposits (Talwani et al, 1979). A lengthy, intense gravity low (marked "trough," Figure 5a), hard to establish on the basis of Bouguer maps, parallels the New South Wales shoreline. The implied structural trough is undesirably close to the northeast-southwest direction of a Seasat orbital track. Nevertheless, irregularities in the direction and form of the imaged feature suggest that it is real and terminates between the mainland and the Britannia Seamount group. Shelf-crossing seismic lines in this region (Colwell and Coffin, 1987) are overwritten by sea-bottom multiples, and do not map basement.

The Bass basin, detailed in Figure 5b, is the site of production and ongoing exploration. Seismic line 11 (Williamson et al, 1987) (Figure 5c) which crosses the basinal deep, illustrates the limitations when attempting to map basement depth (basinal delineation) for the first time using magnetics or seismic vs. satellite gravity observations. Test well 1 Bass (Figure 5c) penetrated intrasectional volcanics, destroying reflections below this point in the section. Where widespread, as on the Carolina platform (Figure 3b), thin volcanics invalidate seismic and magnetics, but are transparent to the gravity signal. Although outlining the Bass basin as shown in Figure 5b, the Seasat image also suggests an extension of low-density crust to the northeast (toward the Bassian Rise). Surface geophysics (P.E. Williamson, 1989, personal communication) has found granitic basement in this area.

Seasat imagery of the northwestern Australian continental margin (Figure 5d) during early exploration would have delineated shelf basins more directly than marine magnetics (Willcox, 1981). Regional Seasat images display a master fracture zone separating the western Australia tectonic province from a northern province (Willcox, 1981; Boote and Kirk, 1989) offset by fracture zones having a more westerly orientation (Figure 5d). The master fracture zone appears to cross the continental shelf and is colinear with the onshore Fitzroy graben. A region of thick sediment accumulation is present at the junction of the transform and a northeastern extension of the Barrow and Dampier shelf basins.

the magnetic field at orbital altitude, with no possibility of providing equivalent resolution at surface.

The reason Seasat seems to localize offshore basins more accurately than marine gravity (e.g., Figure 3c) is not fully understood. It may be significant, however, that marine gravity surveys viewing the boat-level (geoid) rather than the equilibrium ellipsoid as datum introduce an elevation error. Offshore Massachusetts, the geoid is some 30 m below the ellipsoid and strongly warped. This placement is equivalent to an error of 92 gravity units. Elsewhere, e.g., in the Indian Ocean and at the western Pacific margin, the elevation difference and the warping are much more severe. If unaccounted for, this factor is capable of erroneously displacing mapped gravity anomalies by tens of kilometers. Another factor is that in calculating specialized anomalies, such as the isostatic anomaly, assumptions are made as to the way the worker believes the Earth should behave under load (Simpson et al, 1986). In contrast, interactive screening of the Seasat image takes place strictly in relation to the local gravity field, avoiding assumptions.

In remote-area reconnaissance, Seasat images delineate well the structural framework of offshore sediments. If low-cost Seasat images had been made available during early exploration, they would have provided a preview of the regional structure (Figures 1a, 2a), permitting optimal layout of reflection seismic. Seasat images prevent some ambiguities and false leads inherent in seismic/magnetic basinal exploration (Figures 3b; 5b, c).

Worldwide, Seasat imagery provides a vivid impression of the few subbasins that have provided large quantities of hydrocarbons, as at Ekofisk, compared with the large number of unexplored subbasins now becoming detectable through use of Seasat images. Regions in which gravity images have the potential to delineate structure are exemplified by offshore west Africa, the Campbell plateau, and the intra-arc parts of Sundaland.

In partly explored basins, the synoptic view provided by Seasat images can reveal large-scale structures not hitherto evident (e.g., Orkney-Norway ridge, Figure 1a) and innumerable local structures below the penetration depth of seismic (Figure 1b).

DISCUSSION AND CONCLUSIONS

As recently as 1984, remote-sensing techniques were restricted to observations of the Earth's surface (Lang and Nadeau, 1984). Through ingenious use of the sea surface as a gravimeter, the Seasat dataset prepared by W. F. Haxby (Haxby et al, 1983; Haxby, 1985; Appendix 1) depicts mass anomalies at depths limited only by the filter employed in interpretation. However, the deep penetration of the gravity signal has a disadvantage in that it can define only maximum depth of the anomaly. Forthcoming altimetric satellites (Appendix 1) will improve data accuracy and resolution, but cannot alter the penetration characteristic. In contrast to Seasat's measurement at surface level, Magsat (LaBrecque et al, 1985) measures

APPENDIX 1
Satellite Measurement of Gravity

Several workers have used satellite data to prepare bathymetry and gravity maps (Dixon and Park, 1983; Haxby et al, 1983; Rapp, 1983; Sandwell, 1984). Earlier, satellite orbital elements could define only mass anomalies of very large extent (approximately >1,500 km). Anticipating this limitation, these workers used a radar on the Seasat orbiter (Figure 6) to measure the distance to the sea surface. Allowing for the effect of currents and tides, sea level can define the geoidal surface and describe a constant value of the geopotential $U(\phi, \lambda, \gamma)$, in which U is the dimension energy, and ϕ, λ, γ are south, east, and vertical, respectively; $-\text{grad } U$ constitutes gravity. To recover $-\text{grad}U$ from observations of the geometry of U_{const}, Haxby et al (1983) employed the Laplace relation,

$$(\partial^2 U/\partial \phi^2) + (\partial^2 U/\partial \lambda^2) + (\partial^2 U/\partial \gamma^2) = 0,$$

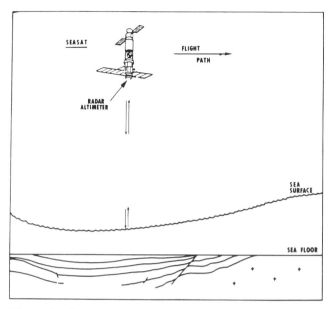

Figure 6—Operational schematic, Seasat altimetric radar. After allowing for effect of currents and tides, sea surface represents equipotential surface (geoid) of Earth's gravity field. Seasat's altimeter measured distance from orbit to sea surface using 3-nsec, 135-GHz pulses. Reception was gated to block side returns. Sea surface is drawn up over mass concentrations such as shallow basement and depressed over thick low-density sediments. Water layer acts as gravimeter read from orbit. Figures 1b, 3a, 4a, and 5a display values in 15 colors (violet = −40 mgal, blues = −45 to 0 mgal, yellows = 0 to 20 mgal, buffs through orange-red = >20 mgal, each having many shades representing relief illuminated from the north. In detailed images (Figures 1b; 2b, c; 3b; 4b,d) spectrum was allocated to restricted portion of dynamic range of data and filtered operators employed as noted.

obtaining a grid of gravity values, at 0.1° intervals, over both the world's oceans and epicontinental seas.

Limitations and Assumptions

Seasat measured sea level with 5 cm precision and 65 cm accuracy along track. The spatial resolution of sea-surface topography is estimated to average 30 km, decreasing as latitude increases and with the intensity of gravity anomalies. Haxby et al (1983) estimated that the accuracy is sufficient to resolve features with amplitudes of 10 and 5 mgals of extent 50 and 100 km, respectively. I found validation of the method is best provided by the coincidence of image features with those mapped on the basis of seismic and drill data. Thus, in the North Sea, features suspected originally to be noise correlated with structure mapped in petroleum exploration. Seasat images can represent real structures in unexplored areas, but limitations must be kept in mind.

An important limitation is presented by orbital coverage. The geologic masses detected by Seasat themselves ensured that its orbit became sufficiently disturbed so as to cause uneven trajectory spacing; thus, some trajectories are more closely spaced than others. Data interpolation was aided by the linearity of many tectonic features, but at the same time, those tectonic features nearly aligned with an orbital track must be viewed with caution. I cite as an example the nearshore trough paralleling the southeastern shore of Australia (Figure 5a).

Bouguer maps may be plotted from Seasat data using known bathymetry, but for tectonic analysis, unreduced Seasat data, akin to free-air gravity, are preferable to the use of Bouguer maps. The preparation of a Bouguer map fictitiously annihilates the expression of real topographic features, including mountain ranges and the continental

shelf. Such features control the structure of a region, which, without them, would profoundly change. Thus, on Bouguer maps, the expression of mountain roots (crustal thickening) cannot be distinguished from that of an important basin. In practice, I found that the attraction of the continental shelf was prominent in Seasat data, but was insufficient to mask the existence of other structures.

Finally, since Seasat was launched (1978), other geodetic satellites have been developed. Undoubtedly, these will show that Seasat data are in the developmental stage comparable to that reached by reflection seismic before World War II. Newer satellites already functional are Geosat (Sandwell and McAdoo, 1988) and ERS-1 (Anderson et al, 1988), providing data at latitudes at which Seasat returns were obscured by ice. Another system, GRM (Geopotential Research Mission) works by employing twin satellites to obtain continental gravity data (Douglas et al, 1980), but may not be launched due to budget problems. However, continental data will be obtained through the use of a planned orbital gravity gradiometer.

REFERENCES CITED

Anderson, A. J., G. Marquart, and H.-G. Schernek, 1988, Arctic geodynamics: a satellite altimeter experiment for the European Space Agency remote-sensing satellite: EOS, v. 69, p. 873-881.
Behrendt, J. C., and K. D. Klitgord, 1980, High-sensitivity aeromagnetic survey of the U.S. Atlantic continental margin: Geophysics, v. 45, p. 1813-1846.
Boote, D. R., and R. B. Kirk, 1989, Depositional wedge cycles on evolving plate margin, western and northwestern Australia: AAPG Bulletin, v. 73, p. 216-243.
Bostrom. R. C., 1988, Mediterranean tectonics: Balearics-Gibraltar-Madeira convergence zone and the Cape Verde plumes (abs.): EOS, v. 69, p. 1155.
Brennand, T. P., and B. van Hoorn, 1986, Historical review of North Sea exploration, in K. W. Glennie, ed., Introduction to the petroleum geology of the North Sea (2d ed): Boston, Blackwell Scientific Publications, p. 1-24.
Brown, S., 1986, Jurassic, in K. W. Glennie, ed., Introduction to the petroleum geology of the North Sea (2d ed): Boston, Blackwell Scientific Publications, p. 133-159.
Burke, K., 1976, Development of grabens associated with the initial rupture of the Atlantic Ocean: Tectonophysics, v. 36, p. 93-111.
Cande, S. C., J. L. LaBrecque, and W. F. Haxby, 1988, Plate kinematics of the South Atlantic; Chron C34 to present: Journal of Geophysical Research, v. 93, no. B11, 13479-13492.
Colwell, J. B., and M. F. Coffin, 1987, RIG seismic research cruise 13: structure and stratigraphy of the northeastern Gippsland basin and southern New South Wales margin: Australian Department of Primary Industries and Energy Report 283, p. 1-56.
Dillon, W. P., K. D. Klitgord, and C. K. Paull, 1983, Mesozoic development and structure of the continental margin off South Carolina: USGS Professional Paper 1313, p. N1-N16.
Dixon, T. H., and M. E. Parke, 1983, Bathymetry estimates in the southern oceans from Seasat altimetry: Nature, v. 304, p. 407-409.
Donovan, D. T., ed., 1968, Geology of shelf seas: Edinburgh, Scotland, Oliver and Boyd, 160 p.
Doré, A. G., J. Vollset, and G. P. Hamar, 1985, Correlation of the offshore sequences referred to the Kimmeridge clay formation, in B. M. Thomas, A. G. Doré, S. S. Eggen, P. C. Home, and R. M. Larsen, eds., Petroleum geochemistry in exploration of the Norwegian Shelf: London, Graham and Trotman, p. 27-38.
Douglas, B. C., C. C. Goad, and F. F. Morrison, 1980, Determination of the geopotential from satellite-to-satellite tracking: Journal of Geophysical Research, v. 85, p. 5471-5480.
Doutch, H. F., R. H. Rapp, M. L. Zoback, D. Denham, T. Simkin, L. Siebert, G. W. Moore, F. J. Mauk, A. C. Tarr, D. R. Soller, and R. D. Brown, 1985, Geodynamic map of the Circum-Pacific region: southwest quadrant, in G. W. Moore, ed., Geodynamic maps of the Circum-Pacific region: Tulsa, Oklahoma, Circum-Pacific Council for Energy and Mineral Resources, scale 1:10,000,000, 1 sheet.
Glennie, K. W., 1984, Structural framework and pre-Permian history of the North Sea area, in K. W. Glennie, ed., Introduction to the petroleum geology of the North Sea: Boston, Blackwell Scientific Publications, p. 17-39.
Grow, J. A., C. O. Bowin, and D. R. Hutchinson, 1979, The gravity field of the U.S. continental margin: Tectonophysics, v. 59, p. 27-52.

Haxby, W. F., 1985, Gravity field of the world's oceans: U.S. Office of Naval Research, one sheet, scale 1:40,000,000 at equator.

—— G. D. Garner, J. L. LaBrecque, and J. K. Weissel, 1983, Digital images of combined oceanic and continental data sets and their use in tectonic data studies: EOS, v. 64, p. 95-1004.

Hospers, J., and J. S. Rathore, 1984, Interpretation of aeromagnetic data from the Norwegian sector of the North Sea: Geophysical Prospecting, v. 32, p. 929-942.

—— E. G. Finnstrom, and J. S. Rathore, 1985, A regional gravity study of the northern North Sea (56N-62N): Geophysical Prospecting, v. 33, p. 533-566.

Johnson, R. J. and R. G. Dingwall, 1980, The Caledonides: their influence on the stratigraphy of the north-west European continental shelf, in L. V. Illing and G. D. Hobson, Petroleum geology of the continental shelf of north-west Europe, London Heyden and Son, p. 85-97.

Kent, P. E., 1975, The tectonic development of Great Britain and the surrounding seas, in A. W. Woodland, ed., Petroleum and the continental shelf of north-west Europe I: New York, John Wiley, p. 3-28.

Klitgord, K. D., and J. C. Behrendt, 1979, Basin structure of the U.S. Atlantic margin, in J. S. Watkins, L. Montadert, and P. W. Dickerson, eds., Geological and geophysical investigations of continental margins: AAPG Memoir 29, p. 85-112.

—— and H. Schouten, 1986, Plate kinematics of the central Atlantic, in P. R. Vogt and B. E. Tucholke, eds., The geology of North America, the western North Atlantic region: Boulder, Colorado, GSA, p. 351-377.

—— W. P. Dillon, and P. Popenoe, 1983, Mesozoic tectonics of the southeast U.S. coastal plain and continental margin: USGS Professional Paper 1313, p. P1-P45.

—— D. R. Hutchinson, and H. Schouten, 1988, U.S. Atlantic margin; structural and tectonic framework, in R. E. Sheridan and J. A. Grow, eds., The geology of North America, the Atlantic continental margin: Boulder, Colorado, GSA, p. 19-55.

—— J. S. Schlee, and K. Hinz, 1982, Basement structure, sedimentation and tectonic history of the Georges Bank basin, in P. A. Scholle and C. E. Wenkam, eds., Geological studies of the COST Nos. G-1 and G-2 wells, U.S. North Atlantic outer continental shelf: USGS Circular 861, p. 160-193.

Kumar, N., 1981, Geological history of north and northeastern Brazilian margin, in J. W. Kerr, ed., Geology of the North Atlantic borderlands: Canadian Society of Petroleum Geologists Memoir 7, p. 527-542.

LaBrecque, J. L., S. C. Cande, and R. D. Jarrard, 1985, Intermediate-wavelength magnetic anomaly field of the North Pacific and possible source distributions: Journal of Geophysical Research, v. 90, no. B3, p. 2549-2564.

Lang, H. R., and P. H. Nadeau, eds., 1984, Petroleum commodity report, in M. J. Abrams, J. E. Conel, H. R. Lang, and H. N. Paley, eds., Joint NASA/Geosat test case project final report: Tulsa, Oklahoma, AAPG, pt. 2, v. 2, p. 10-1 to 10-28.

Mattick, R. E., J. S. Schlee, and K. C. Bayer, 1981, The geology and hydrocarbon potential of the Georges Bank–Baltimore Canyon area, in J. W. Kerr, ed., Geology of the North Atlantic borderlands: Canadian Society of Petroleum Geologists Memoir 7, p. 461-486.

National Oceanic and Atmospheric Administration/National Geophysical Data Center, 1987, Marine gravity tape 00978/H/001.

Ojeda, H. A. O., 1982, Structural framework, stratigraphy, and evolution of Brazilian marginal basins: AAPG Bulletin, v. 66, p. 732-749.

Price, I., and R. P. Rattey, 1984, Cretaceous tectonics off mid-Norway: Journal of the Geological Society of London, v. 141, p. 985-992.

Proctor, C. V., 1980, Distribution of Middle Jurassic facies in the East Shetlands basin and their control of reservoir capability, in Sedimentation of the North Sea reservoir rocks, Geilo: Norsk Petroleumsforening, v. 15, p. 1-22.

Rapp, R. H., 1983, The determination of geoid undulations and gravity anomalies from Seasat altimeter data: Journal of Geophysical Research, v. 88, p. 1552-1562.

Roberts, D. G., 1974, Structural development of the British Isles, the continental margin, and the Rockall Plateau, in C. A. Burke and C. L. Drake, eds., The geology of continental margins: New York, Springer-Verlag, p. 343-359.

Sandwell, D. T., 1984, A detailed view of the South Pacific geoid from satellite altimetry: Journal of Geophysical Research, v. 89, p. 1089-1104.

—— and D. C. McAdoo, 1988, Marine gravity of the southern ocean and Antarctic margin from Geosat: Journal of Geophysical Research, v. 93, no. B9, p. 10389-10396.

Simpson, R. W., R. C. Jachens, R. J. Blakely, and R. W. Saltus, 1986, A new isostatic residual gravity map of the conterminous United States with a discussion on the significance of isostatic residual anomalies: Journal of Geophysical Research, v. 91, p. 8348-8372.

Smythe, D. K., J. A. Chalmers, A. G. Skuce, A. Dobinson, and A. S. Mould, 1983, Early opening of the Atlantic, I: Geophysical Journal of the Royal Astronomical Society, v. 72, p. 373-398.

Szatmari, P., E. Miliani, M. Lana, J. Conceicao, and A. Lobo, 1985, How Atlantic rifting affects Brazilian oil reserves distribution: Oil and Gas Journal, v. 82 (January 14), p. 107-113.

Talwani, M., J. Mutter, R. Houtz, and M. Konig, 1979, The crustal structure and evolution of the area underlying the magnetic quiet zone on the margin south of Australia, in J. S. Watkins, L. Montadert, and P. W. Dickerson, eds., Geological and geophysical investigations of continental margins: AAPG Memoir 29, p. 151-175.

Tucholke, B. E., R. E. Houtz and W. J. Ludwig, 1982, Sediment thickness and depth to basement in western North Atlantic Ocean basin: AAPG Bulletin, v. 66, p. 1384-1395.

Vogt, P. R., and A. M. Eimwich, 1979, Magnetic anomalies and sea-floor spreading in the western North Atlantic, and a revised calibration of the Keathley (M) geomagnetic reversal chronology: Initial Reports Deep Sea Drilling Project, v. 43, p. 857-876.

Weissel, J. K., and D. E. Hayes, 1977, Evolution of the Tasman Sea reappraised: Earth and Planetary Science Letters, v. 36, p. 77-84.

Willcox, J. B., 1981, Petroleum prospectivity of Australian marginal plateaus, in M. T. Halbouty, ed., Energy resources of the Pacific region: AAPG Studies in Geology 12, p. 245-272.

Williamson, P. E., C. J. Pigram, J. B. Colwell, A. S. Scherl, K. L. Lockwood, and J. C. Branson, 1987, Review of stratigraphy, structure, and hydrocarbon potential of Bass basin, Australia: AAPG Bulletin, v. 71, p. 71, 253-280.

Ziegler, P. A., 1980, Evolution of sedimentary basins in north-west Europe, in L. V. Illing and G. D. Hobson, eds., Petroleum geology of the continental shelf of north-west Europe: London, Heyden and Son, p. 3-42.

—— 1982, Geological atlas of western and central Europe: Amsterdam, Shell/Elsevier, 130 p.

THE COKEVILLE NORMAL FAULT - AN EXAMPLE OF INTEGRATION
OF LANDSAT AND PHOTOGEOLOGICAL DATA*

P. Don Erickson
TGA DIVISION
Petroleum Information Corporation
Littleton, Colorado

J. Frank Conrad
Veezay Geoservice Inc.
Englewood, Colorado

ABSTRACT

Integration of detailed geological information
with Landsat data is necessary in order to make an
effective interpretation from space images. Although
the present sophistication in computer enhancement
significantly increases the value of the Landsat images for a geologic investigation, this regional Landsat perspective is still gained at the expense of detail, and its proper utilization requires as much
geologic ground control as possible. The recognition
of the Cokeville normal fault is a result of the synergistic use of detailed geologic mapping and Landsat
image interpretation.

INTRODUCTION

The Idaho-Wyoming thrust belt extends from the Uinta Mountains north to
the Snake River Plain (Figure 1). Nearby features include the vast Basin and
Range Province to the west, the Green River Basin and Rock Springs uplift to
the east, and the Wind River Mountains to the north. These major geologic and
physiographic features, as well as the outline of the Landsat image used for
this project, are shown on Figure 1. This paper expands upon and refines an
aspect of thrust belt structure first reported by Conrad, Erickson, and Trollinger (1977) in a paper based on a mapping project by TGA Division of Petroleum Information Corporation.

AERIAL PHOTOGRAPHY AND LANDSAT IMAGERY

Color aerial photographs taken with a Zeiss camera with an 8 1/2" focal
length lens at a scale of 1:24,000 were used for the detailed evaluation of
this area.

The Idaho-Wyoming thrust belt lends itself to photogeologic-geomorphic
interpretation using color photography because many of the rock units weather
to vivid, distinctive colors. Examples include the Cretaceous Gannett and
Preuss red beds, the reddish-brown Wasatch Formation, the light-gray Fowkes
Formation, the pale-green and light-gray Green River Formation, the reddish-
purple Nugget Sandstone, and the yellow-gray beds of the Twin Creek Limestone.

The identification number of the Landsat image used is 1068-17355, taken
29 September, 1972 (Figure 2). The scene was chosen by TGA and digitally enhanced by GeoImage Division of Aero Service in Houston, Texas. Although

---

*Presented at the International Symposium on Remote Sensing of Environment,
Third Thematic Conference, Remote Sensing for Exploration Geology, Colorado
Springs, Colorado, April 16-19, 1984.

425

several ratios were used in the study, the majority of structural information was obtained from a false-color composite print, a band 7 black and white print, and a filtered band 7 black and white print, all at a scale of 1:250,000.

The computer enhancement significantly increased the value of the Landsat image for geologic investigation because of the increase in resolution and brightness and the virtual elimination of the characteristic Landsat "striping" or "scanlines".

INTERPRETATIONS AND RESULTS

The generalized geologic maps in Figures 3 and 4 cover the study area. Note the extent of cover by Eocene and younger rocks, which masks details of structure in Cretaceous and older, oil-bearing formations.

Major thrust faults, in order of youngest to oldest, that is, from east to west, include the Hogsback-LaBarge-Darby thrust complex, the Absaroka thrust, the Tunp thrust, and the Crawford thrust. Major normal faults include the Evanston fault (Figure 4) and the Rock Creek fault (Figure 3). Also note in Figure 3 the unnamed normal fault on the west flank of the Sublette Range and a similar unnamed normal fault on the west flank of the Crawford Mountains.

It should be noted that normal faults of the thrust belt are younger than the Eocene Wasatch and Green River Formations, and in many cases offset those beds. In contrast, the thrust faults are all older than the Eocene formations and are locally concealed by the Eocene cover. The detailed photogeologic analysis, using color aerial photographs, revealed evidence indicating that structural elements mapped in Eocene beds may provide useful clues to deformational trends in the older rocks.

The Evanston, Wyoming, Area

The Evanston area of southwest Wyoming is an example intended to illustrate the expressions of known geologic features as they appear on both aerial photographs and on the Landsat image. Note the relationship of the Tunp thrust and the Evanston normal fault in Figure 4. This sub-parallel relationship is most important and will be discussed in conjunction with the more detailed geologic map of the Evanston area in Figure 5 (area A of Figure 4) and cross-section A-A' in Figure 6. The geologic map of the Evanston, Wyoming area (Figure 5) shows the westerly-dipping Evanston normal fault and Tunp thrust. Discoveries in the area include the Yellow Creek/Evanston field and Painter Reservoir field (the shape and size of these fields are diagrammatic). The sub-parallel relationship of the Evanston normal fault and the Tunp thrust is remarkable because first movement along the Evanston fault occurred some 25 to 40 million years after cessation of movement along the Tunp thrust. The relationship between major thrust faults, such as the Absaroka and Tunp and the much younger, westerly-dipping normal faults such as the Evanston, were pointed out in an excellent paper by Royse, Warner, and Reese (1975), in which they devised several empirical rules for interpreting thrust belt structural form. The rules include a discussion of the geometry of normal faults. Evidence indicates many of the normal faults of the thrust belt are listric, that is, they flatten with depth and "sole" into an older, underlying thrust plane or bedding plane. (The word "listric" is derived from the Greek word "listron" meaning shovel.) Portions of major thrust planes have two episodes of movement (Figure 6), an early compressive motion or thrusting, as along the Absaroka thrust, and a later period of extensional faulting as seen on the Evanston normal fault. The net throw is up-to-the-west with repetition demonstrated by the Nugget Sandstone, but the most recent movement is down-to-the-west about 8000 feet (2500 meters) with offset at the base of the Eocene Wasatch Formation. The characteristics of shallow-listric normal faults are:

426

first, they merge with, but do not offset the related thrusts; second, in plan-view they are aligned roughly subparallel with the major thrust with which they are associated; third, some of these shallow-listric normal faults appear to be related to a step or ramp in an underlying, older thrust; fourth, often the beds on the down-thrown side of the shallow-listric faults dip into the fault as a result of rotation, and thus form an anticline in the upper plate of the thrust (similar to folds associated with growth faults in the Gulf Coast). Therefore, a normal fault at the surface may indicate the presence of a ramp in the underlying thrust and an anticlinal fold in the upper plate. The Yellow Creek discovery is in such an anticline apparently formed in post-Wasatch, late Tertiary time. Evidence for this is seen in cross-section A-A' east of the Hatch No. 1 well, where the dip of Eocene Wasatch beds parallels that of Cretaceous and older units below. Since the Wasatch beds were rotated in late Tertiary time as part of movement of the Evanston normal fault, the anticline seen in Jurassic and Cretaceous beds must be, in part, a late Tertiary feature.

The portion of the Evanston area outlined in Figure 7 is the location of the aerial photograph and geologic interpretation illustrated in Figure 8. Note the expression of the Tunp thrust where Jurassic and Cretaceous units of the Gannett, Stump, and Preuss Formations have been thrust over the Paleocene Evanston Formation. Note also the Evanston normal fault, where Eocene beds have been down-faulted to the west about 10,000 feet (3000 meters), and are in fault contact with Jurassic and Cretaceous rocks to the east. The pronounced northeast dip of the Eocene beds into the normal fault is the result of rotation of the entire block during normal faulting. On the Landsat sub-scene of Figure 7, the northeast-trending portion of the Evanston and Tunp faults are well-expressed on both sides of the Bear River Valley (Lineament A-A'). However, the northwestern extension of the Evanston normal fault seen in Figure 8 is less well-expressed. What appears to be the fault is in reality the distinctive contact (B, Figure 7) between the Wasatch and Fowkes Formations, both of Eocene age. The Evanston normal fault is marked by a faint color contrast a short distance to the northeast. This emphasizes the importance of integrating surface detail obtained from high-quality color aerial photographs into the interpretation of Landsat images.

The Sublette Range - Crawford Mountains Area

The next example is a feature that can be best seen on Landsat imagery. This is an unnamed normal fault that is known at several localities west of the surface trace of the Crawford thrust. Portions of this normal fault in the Sublette Range and in the Crawford Mountains were mapped during the detailed photogeologic evaluation. Figures 9 through 13 show the surface expression of this normal fault on both aerial photographs in the Sublette Range and Crawford Mountains, and on Landsat images for the areas in between.

That part of the Sublette Range outlined in Figure 9 is the location of the aerial photograph and geologic interpretation of Figure 10. Note the distinctive, light-banded units of the Twin Creek Formation and the massive, poorly-bedded sandstones of the Nugget Sandstone. Also note the distinctive scarp crossing the central part of the photo from north to south. This scarp is interpreted as the erosional remnant of the footwall of the normal fault under discussion. This westerly-dipping normal fault lies west of, and is subparallel with, the Crawford thrust (the position of the Crawford thrust is inferred in the subsurface two to three miles east of the area of the aerial photograph). The fault is interpreted to be structurally related to the Crawford thrust and to represent the northern end of a much longer feature that has been locally recognized farther south in the Crawford Mountains.

427

Figure 11 is a Landsat subscene showing the location of the aerial photograph and geologic interpretation of a portion of the Crawford Mountains seen in Figure 12. The western scarp of the Crawford Mountains is interpreted as an erosional remnant, marking the footwall of this normal fault. The scarp displays about 1400 feet (430 meters) of relief from the mountain peaks to the flood plain of Bear River. The geologic interpretation of the aerial photograph shows steeply-dipping Mississippian and Devonian beds exposed in the Crawford Mountains. The bottom of Figure 12 is as far south as the normal fault could be mapped on conventional color photographs.

The geologic interpretation of the Landsat image indicates a much greater extent for these normal faults. On the Landsat band 7 subscene of Figure 13, the normal fault can be traced southward along the distinctive topographic escarpment on the west flank of the Sublette Range as far south as the flood plain of Bear River, immediately northwest of Cokeville, Wyoming. The geomorphology of the flood plain adjacent to the Sublette Range and detailed geologic mapping of the area northwest of Cokeville suggest that displacement along this particular portion of the normal fault system dies out. Faulting in the vicinity of Cokeville is complicated. Immediately east of Cokeville is a zone of displacement transfer along the Crawford thrust. Here, as displacement decreases along the northern section of the Crawford thrust, displacement increases along a parallel splay to the east. The associated normal fault east of Cokeville can be traced southward along the valley wall to a point where the southern portion of the Crawford thrust and the normal fault appear to join. Geomorphological evidence suggests that there has been normal movement on the Crawford thrust plane south of this point for at least 5 to 6 miles (8 to 10 kilometers). It should be noted that at no place does the normal fault cut the Crawford thrust.

As normal displacement dies out along this section of the fault, it increases along a parallel fault to the west. This section of the fault zone can be traced on the Landsat image southward on the basis of drainage, color, and tonal features within the Bear River flood plain until it connects with the mapped portion of the normal fault on the western flank of the Crawford Mountains. South of the Crawford Mountains, the normal fault can be traced primarily on the bases of drainage and topography for an additional 25 miles (40 kilometers).

CONCLUSIONS

Landsat images provide the regional perspective to expand upon detailed photogeologic evidence between two localities, the Sublette Range and Crawford Mountains, and suggests the interpretation of a normal fault system of more than 82 miles (133 kilometers) in length. We propose the name Cokeville fault for this feature as it passes just west of the town of Cokeville, Wyoming.

The significance of the Cokeville fault is that this westerly-dipping, normal fault is subparallel to the older Crawford thrust. Therefore, the normal fault may be listric and may merge with the Crawford thrust in the subsurface. It then might be suspected that a ramp or step may occur in the footwall of the Crawford thrust. Additionally, several anticlines mapped in the Tertiary rocks to the west might have been formed as a result of rotation of beds along this fault, and might also reflect prospective structures in the older rocks as in the Evanston area.

In summary, the synergistic use of both Landsat image interpretation and detailed photogeological mapping has made possible the recognition of a potentially important structural feature which could not have been delineated in its entirety using either tool alone.

428

478

ACKNOWLEDGEMENTS

The authors wish to thank Stephanie B. Urban and James R. Muhm for their constructive criticisms and TGA Division of Petroleum Information Corporation for permission to print this paper.

REFERENCES

Conrad, J. F., Erickson, P. D., and Trollinger, W. V., 1977, Landsat, a useful tool in the Idaho-Wyoming thrust belt, in Application of satellite data to petroleum and mineral exploration, Pecora III Conf., Abs. with Program: Am. Assoc. Petrol. Geol., U. S. Geol. Survey, and National Aeronautics and Space Administration; Sioux Falls, South Dakota.

Royse, F., Jr., Warner, M. A., and Reese, D. L., 1975, Thrust belt structural geometry and related stratigraphic problems, Wyoming-Idaho-northern Utah; in Bolyard, D. W. (ed.), Deep drilling frontiers of the central Rocky Mountains: Rocky Mountain Assoc. Geologists Symposium, p. 41-54.

429

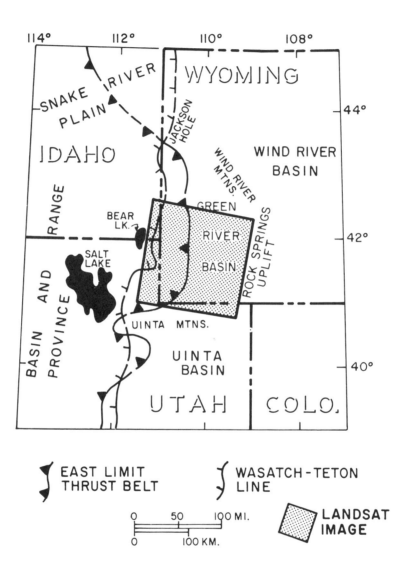

Figure 1. Location map showing major geologic and physiographic features. Stippled area indicates coverage of Landsat image used.

430

480

Figure 2. Landsat band 7 image used in study (location shown
            on Figure 1). Areas A through D are locations of
            Figures 7, 9, 11, and 13.

431

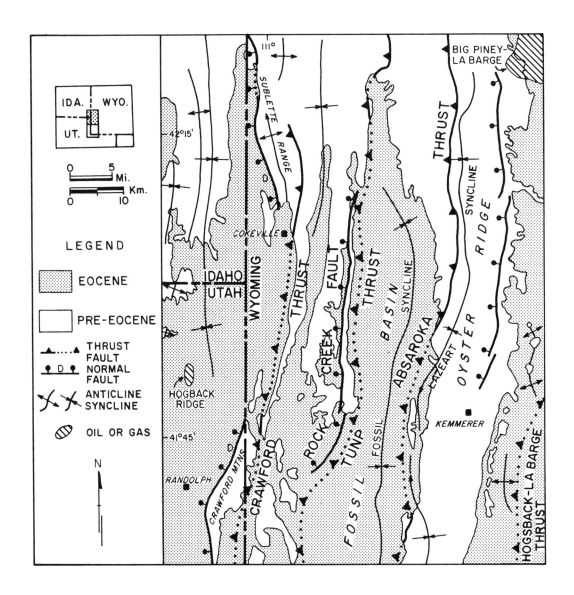

Figure 3. Generalized geologic map of northern half of study area.

432

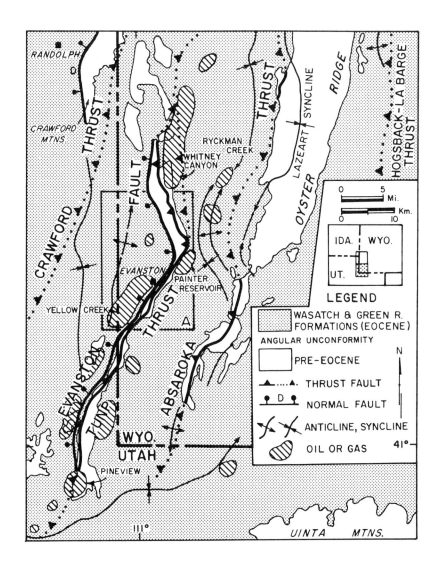

Figure 4. Generalized geologic map of southern half of study
area. Area A is the location of Figure 5.

433

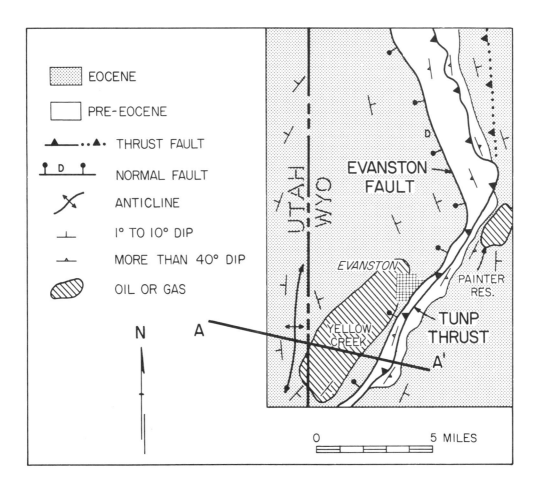

Figure 5. Generalized geologic map of the Evanston, Wyoming, area (area A of Figure 4).

434

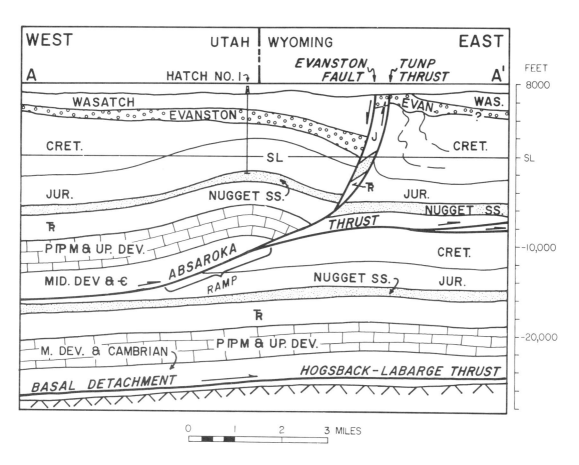

Figure 6.   Cross Section A-A' seen in plan view in Figure 5.   (after Royse, Warner, and Reese, 1975).

435

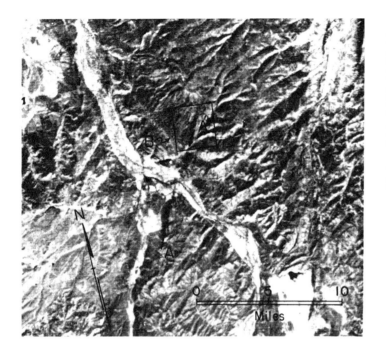

Figure 7. Landsat band 7 subscene of the Evanston, Wyoming, area (area A of Figure 2). Square represents location of aerial photograph seen in Figure 8. Linear feature A-A' reflects the position of a portion of the Evanston and Tunp faults. The color contrast at B is the contact between the Wasatch and Fowkes Formations.

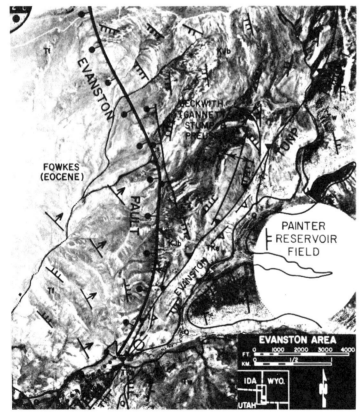

Figure 8. Aerial photograph and geologic interpretation of the area outlined in Figure 7. The size and shape of Painter Reservoir field is diagrammatic.

436

Figure 9. Landsat band 7 subscene of the Sublette Range area,
Wyoming (area B of Figure 2).
Square indicates location of aerial photograph seen in Figure 10.

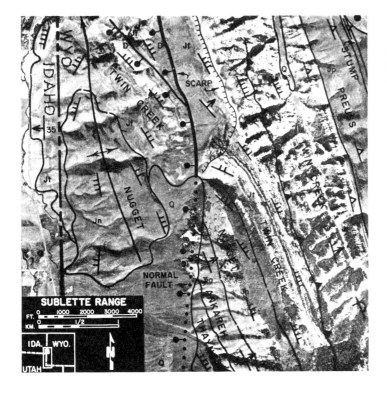

Figure 10. Aerial photo-
graph and geologic inter-
pretation of the area outlined in Figure 9.

437

Figure 11.   Landsat band 7 subscene of the Crawford Mountains
              area, Utah (area C of Figure 12).   Square indicates
              location of aerial photograph seen in Figure 12.

438

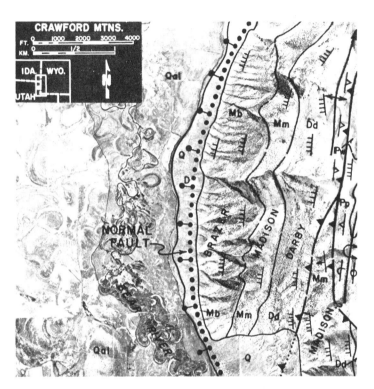

Figure 12. Aerial photograph and geologic interpretation of the area outlined in
Figure 11.

439

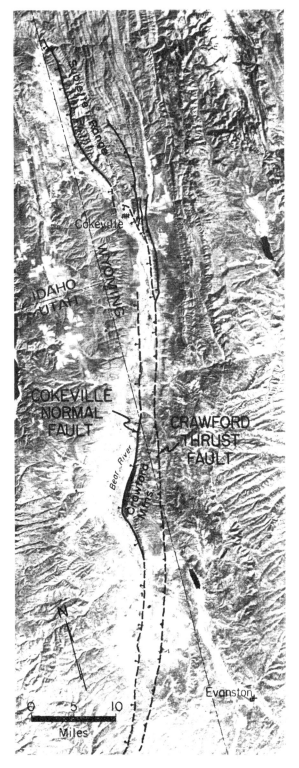

Figure 13. Extent of Cokeville normal fault system as mapped using aerial photographs and Landsat images (area D of Figure 2).

440

490

The American Association of Petroleum Geologists Bulletin
V. 71, No. 4 (April 1987), P. 389-402, 11 Figs., 1 Table

Multispectral Remote Sensing as Stratigraphic and Structural Tool, Wind River Basin and Big Horn Basin Areas, Wyoming[1]

HAROLD R. LANG, STEVEN L. ADAMS,
JAMES E. CONEL, BARBARA A. McGUFFIE,
EARNEST D. PAYLOR, and RICHARD E. WALKER[2]

ABSTRACT

Stratigraphic and structural analyses of the Wind River basin and Big Horn basin areas of central Wyoming are in progress. One result has been the development of an approach to stratigraphic and structural analysis that uses photogeologic and spectral interpretation of multispectral image data to characterize the attitude, thickness, and lithology of strata.

New multispectral systems that have only been available since 1982 are used with topographic data to map upper Paleozoic and Mesozoic strata exposed on the southern margin of the Bighorn Mountains. Landsat-acquired thematic mapper (TM) data together with topographic data are used to map lithologic contacts, measure dip and strike, and develop a stratigraphic column that is correlated with conventional surface and subsurface sections. Aircraft-acquired airborne imaging spectrometer (AIS) and thermal infrared multispectral scanner (TIMS) data add mineralogic information to the TM column, including the stratigraphic distribution of quartz, calcite, dolomite, smectite, and gypsum.

Results illustrate an approach that has general applicability in other geologic investigations that could benefit from remotely acquired information about areal variations in attitude, sequence, thickness, and lithology of strata exposed at the earth's surface. Application of our methods elsewhere is limited primarily by availability of multispectral and topographic data, and quality of bedrock exposures.

---

[1]Manuscript received, April 21, 1986; accepted, December 2, 1986.
[2]Jet Propulsion Laboratory, California Institute of Technology, 4800 Oak Grove Drive, Pasadena, California 91109.

This paper presents the results of one phase of research carried out at the Jet Propulsion Laboratory, California Institute of Technology, under contract with the National Aeronautics and Space Administration. We thank Cindy Inouye and Mary Jane Bartholomew for providing x-ray diffraction and spectral analyses of field samples. Gordon Hoover insured that "undependable" field equipment operated dependably. Jerry Solomon was responsible for development of computer programs used to extract spectra from imaging spectrometer data. Reviews of the manuscript by Michael Abrams, Ralph Baker, James Helwig, James Miller, William Moran, Frank Palluconi, and Alfredo Prelat are appreciated.

Reference herein to any specific commercial product, process, or service by trade name, trademark, manufacturer, or otherwise does not constitute or imply endorsement by the United States government or Jet Propulsion Laboratory, California Institute of Technology.

INTRODUCTION

Purpose

We are studying the formation and evolution of the Wind River basin and Big Horn basin areas, Wyoming (Figure 1), in the National Aeronautics and Space Administration (NASA)-funded Multispectral Analysis of Sedimentary Basins Project (Lang, 1985). One result of this investigation has been the development of new remote-sensing tools for structural mapping and stratigraphic analysis. We use multispectral-image data acquired from aircraft and satellite, combined with topographic data, to determine remotely the attitude, sequence, thickness, and lithology of strata exposed at the earth's surface.

Results are contributing to our understanding of the tectonic and depositional history of the Wind River basin and Big Horn basin areas; but more importantly, the results illustrate an approach that has general geologic applicability. Multispectral-image data can augment conventional surface and subsurface information to better constrain quantitative models of basin evolution that employ information about areal variations in the attitude, sequence, thickness, and lithology of strata.

This paper describes the development and application of a new multispectral remote-sensing approach to stratigraphic and structural mapping, as well as recent advances in remote-sensing instrumentation that make our work possible. Geologic examples from our Wyoming investigation are used to illustrate our methods. Correlation of remotely acquired stratigraphic information with conventional surface and borehole data is demonstrated. Technical descriptions of image-processing facilities used in the study are provided in appendices.

Recent Advances in Multispectral Remote Sensing

Remote sensing "refers to all the arts and techniques of measurement and interpretation of phenomena from afar . . . specifically . . . the use of electromagnetic radiation to obtain data about the nature of the surface of the earth" (Goetz and Rowan, 1981, p. 781). Physical principles and methods of remote sensing are summarized by Goetz and Rowan (1981) and Watson (1985), and described in detail in Colwell (1983).

More than 50 years of experience with aerial photography has established remote sensing as a geologic mapping

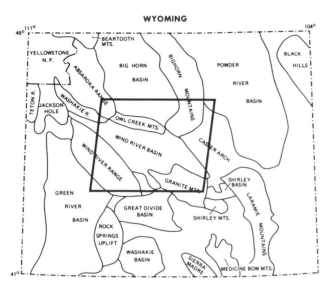

Figure 1—Map of Wyoming showing outline of Wind River basin and Big Horn basin study areas. Major features are after Love (1970).

tool. Stereoscopic aerial photographs are a well understood and routinely used source of structural and stratigraphic information (Miller and Miller, 1961; Miall, 1984, p. 213-214).

Digital, multispectral remote-sensing data first became available on a global basis during 1972 with the launch of the first Landsat (formerly ERTS) satellite. This and the four subsequent Landsats carried the multispectral scanner (MSS) (Table 1), an instrument that provides grid-sampled geophysical measurements. Solar radiation reflected from the earth's surface is measured in four spectral bands that cover the 0.5-1.1 µm (visible to near-infrared) region of the electromagnetic spectrum, extending outside the range of human vision or photographic film. Radiance data are digitally recorded in 6 bits, providing a 64 gray-level (digital number or DN) record for each spectral band at every 57 m by 79 m ground resolution spot (picture element or pixel). A single MSS image provides a synoptic picture of the earth's surface approximately 180 km (110 mi) on a side, an area of coverage equivalent to about 500 conventional aerial photographs.

An important result of the availability of digital, multispectral Landsat data was the development of systems for computer-image enhancement and display. These image-processing systems provide pictures from digital scanner data that are suitable for photogeologic interpretation and that accentuate spectral contrasts extending outside the spectral range of human vision or photographic film. One such system, the multimission image processing laboratory (MIPL), developed at the Jet Propulsion Laboratory (JPL), was used in our Wind River basin and Big Horn basin investigation and is described in appendix A. Other systems are described by Bracken (1983).

Regions inaccessible to field or conventional aerial-photograph observation may be mapped photogeologically using pictures produced from MSS data (Williams and Carter, 1976). Available published geologic investigations using MSS data focused on reconnaissance structural studies (e.g., Bailey and Anderson, 1982), lineament analysis (e.g., Halbouty, 1976), and detection of so-called tonal anomalies (e.g., Halbouty, 1980). Three published examples of the use of MSS data for structural mapping and stratigraphic analysis (Bailey et al, 1982; Viljoen et al, 1983; Leith and Alvarez, 1985) show that the coarse spatial resolution limits the applicability of MSS pictures to stratigraphic and structural problems that can be addressed by photogeologic mapping at scales of approximately 1:250,000 or smaller.

During the 1970s, laboratory and field studies established the potential use of multispectral-image data for direct

Table 1. Selected Multispectral Remote-Sensing Systems

Sensor (Date Available)	Platform	Altitude (km)	Swath Width (km)	Wavelength (µm)	Pixel Size (m)
MSS[a] (1972)	Landsat 1-5	900*/700**	185	0.50-0.60 0.60-0.71 0.69-0.80 0.80-1.10	80
TM[b] (1982)	Landsat 4-5	700	185	0.45-0.52 0.52-0.60 0.63-0.69 0.76-0.90 1.55-1.75 2.00-2.36 10.4-12.5	30 (at 0.45-2.36 µm) 120 (at 10.4-12.5 µm)
AIS[c] (1982)	aircraft	5†	0.290†	1.2-2.4 (128 bands)	9†
TIMS[d] (1982)	aircraft	10†	4†	8.1-8.5 8.5-8.9 8.9-9.3 9.5-10.1 10.2-10.9 11.2-11.7	25†

*Landsat 1-3.　†Typical.　[b]Thematic mapper.　[d]Thermal infrared multispectral scanner.
**Landsat 4-5.　[a]Multispectral scanner.　[c]Airborne imaging spectrometer.

determination of the lithology of strata exposed at the earth's surface. Measurement of variations in reflected and/or emitted energy with wavelength (i.e., spectral data) provides characteristic "fingerprints" for many materials. Hunt (1977, 1980) showed that many rock-forming minerals have diagnostic spectral reflectance features in the 0.4-2.5 μm (visible to short-wavelength infrared) region of the electromagnetic spectrum. In particular, the 2.0-2.4 μm interval was shown to contain spectral absorption features associated with important sedimentary rock-forming minerals, including kaolinite, smectite, gypsum, calcite, and dolomite. These diagnostic spectral features potentially could be detected using multispectral scanners with higher spectral resolution (i.e., more and narrower bands) than provided by the MSS.

Hunt and Salisbury (1975) and Hunt (1980) reported on laboratory measurements of spectral emissivity of rocks and minerals for the 8-12 μm (mid-infrared) region. Their results demonstrate that sedimentary rock-forming minerals, including quartz, kaolinite, gypsum, smectite, hematite, calcite, and dolomite, exhibit diagnostic, mid-infrared spectral features.

Thus, by 1980, less than 10 years after multispectral satellite data first became available, it was apparent that sensors with better spatial and spectral resolution than the MSS and increased spectral coverage, including the 2.0-2.4 μm and 8-12 μm wavelength region, could improve the geologic use of multispectral data by making remote lithologic determination possible. The availability during 1982 of data from three new instruments, the Landsat thematic mapper (TM), airborne imaging spectrometer (AIS), and thermal infrared multispectral scanner (TIMS) (Table 1), accomplished these improvements.

The TM on Landsats 4 and 5 provides increased spectral, spatial, and geometric performance compared to the MSS. Detailed description of the system is provided by Salomonson et al (1980) and Barker (1985).

The TM has six spectral bands in the visible and short-wavelength infrared and one band in the mid-infrared region of the electromagnetic spectrum. Images are in the same 180 km by 180 km synoptic format provided by the MSS, but have improved cartographic fidelity. The 8-bit (256-DN) digital system of the TM provides 30 m by 30 m pixels in visible and short wavelength infrared bands for five times better spatial resolution than the MSS (a TM pixel covers 900 m^2 compared to a 4,503 m^2 MSS pixel).

The AIS, flown on NASA Ames Research Center's C-130 aircraft, provides 128 contiguous 9.5 nm-wide bands of digital spectral data in the 1.2-2.4 μm wavelength. Detailed description of the system is provided by Vane and Goetz (1985). The narrow, 32-pixel groundtrack, typically only 300 m wide, limits the use of AIS pictures for photogeologic interpretation. Spectral resolution is comparable to laboratory spectrometers, so that spectra containing potentially diagnostic mineral reflectance features are provided for every 9 m by 9 m pixel.

The TIMS, flown on NASA National Space Technology Center's (NSTL) Learjet or NASA-Ames Research Center's C-130 aircraft, provides digital, 6-channel multispectral data in the 8-12 μm wavelength. Detailed description of the system is provided by Palluconi and Meeks (1985). The TIMS is one of the highest spectral-resolution multispectral

scanners available for mapping diagnostic spectral emission features in the thermal infrared portion of the electromagnetic spectrum. The 5-km groundtrack and 25 m by 25 m pixels of typical TIMS data are adequate for making pictures that are suitable for photogeologic interpretation.

Approach

We used a six-step procedure to remotely map strata exposed in the Wind River basin and Big Horn basin areas.

1. Acquisition of digital TM data covering the area of interest.

2. Reconnaissance photogeologic interpretation of 1:250,000-scale, TM color-composite images to identify regional stratigraphic markers and to locate relatively undeformed reference localities for detailed stratigraphic interpretation.

3. Detailed photogeologic interpretation of 1:24,000-scale, TM color-composite pictures covering reference localities, to determine resistance to erosion, attitude, sequence, thickness, and spectral characteristics of photogeologically defined stratigraphic units. This step requires topographic data. Results are used to construct a photogeologic map and stratigraphic column.

4. Acquisition of high spectral-resolution and spatial-resolution, aircraft multispectral data on flightlines that traverse TM-defined stratigraphic units that were identified in step 3. Spectral and photogeologic analyses of these data refine the stratigraphic column by adding mineralogic and lithologic information that cannot be obtained from TM data alone.

5. Correlation of stratigraphic columns constructed in steps 3 and 4 with stratigraphic columns defined elsewhere in the TM scene, and with conventional surface and borehole sections. Correlation with conventional sections provides a means for assigning geologic ages and formation names to the informal stratigraphic units that were defined using multispectral data. The photogeologic map and stratigraphic column from step 3 may be labeled with formation names obtained from the correlation to create a conventional geologic map.

6. Field checking of results.

In the following section, we illustrate this procedure with results from the Wind River basin and Big Horn basin study.

WIND RIVER BASIN AND BIG HORN BASIN RESULTS

Geologic Setting

Most of Wyoming is in the foreland region of the Cordilleran orogene. The foreland is flanked on the west by the Cordilleran geosyncline and on the east by the stable North American craton (Grose, 1972). Keefer (1965a, b) and Love (1960, 1970) described the stratigraphic and structural evolution of the region that includes the Wind River basin and Big Horn basin areas (Figure 1). These reports showed that the Wind River and Big Horn basins are the structural and topographic expression of Late Cretaceous–Eocene and subsequent deformation of the Cordilleran foreland. The most detailed, published geologic map coverage of the entire study area is a 1:500,000-scale Wyoming compilation by Love and Christiansen (1985).

Uplifts, cored by Precambrian granitic and metasedimentary rocks and Tertiary volcanic and shallow-intrusive rocks, separate Wyoming basins. Structurally deformed, Cambrian to Holocene (excluding Silurian) strata crop out around the uplift margins of the Wind River and Big Horn basins. Approximately 10 km of marine and nonmarine strata, predominantly sandstone, shale, limestone, dolostone, and bedded gypsum, record Phanerozoic sedimentation. Tertiary to Holocene, nonmarine strata blanket basin interiors. The complex history of Laramide deformation, sedimentation, and erosion is expressed in the study area by over 6 km of relief measured between basin interiors and uplift margins on the top of the Permian strata.

1:250,000 Photogeologic Interpretation of TM Data

We selected the most cloud- and snow-free TM data available, a November 21, 1982, Landsat 4 scene (Figure 2), that covered the 32,000 km² area of interest in central Wyoming. We reported elsewhere (Conel et al, 1985) on the quality and radiometric characteristics of this scene. Two attributes of this particular scene that negatively affect spectral interpretation, low (21°) sun angle, and snow and cloud cover at higher elevations, did not significantly impair our photogeologic interpretation. In fact, the negative effect of low sun angle shadows on spectral interpretability was offset by enhanced structural interpretability resulting from shadows highlighting subtle topographic features.

Reconnaissance photogeologic interpretation of a false-color infrared composite (CIR) of this scene at 1:250,000 scale revealed many areas of well-exposed strata around the margins of the snow-covered Owl Creek, Bighorn, Wind River, and Granite Mountains (Figure 2). The CIR image format that combines TM bands 2 (0.45-0.52 μm, printed in blue), 3 (0.63-0.69 μm, green), and 4 (0.76-0.90 μm, red) was used because many geologists are familiar with this standard format of Landsat multispectral data. The CIR image portrays vegetation in red and red beds in yellow-gold. We used the image as a photogeologic base to map a red-bed unit cropping out on the margins of the Big Horn and Wind River basins. This unit served as a marker bed for regional structural interpretation and stratigraphic correlation.

Structural interpretation of the image demonstrated that an area of well-exposed and relatively undeformed strata that includes the red-bed marker exists at the northern end of the Casper arch, on the southern margin of the Bighorn Mountains uplift. This area was selected as a stratigraphic reference locality for 1:24,000-scale photogeologic and spectral analyses.

1:24,000 TM Photogeologic Interpretation

Figure 3 is a TM CIR image of the 190 km² stratigraphic reference locality outlined in Figure 2. The picture was enlarged photographically and printed at 1:24,000 scale for analysis. The enlarged picture geometrically matched part of the area covered by the Grave Spring, First Water Draw, Roughlock Hill, Deadman Butte, Three Buttes, and Flat Top Hill USGS 7½-min topographic quadrangles in T38-40N, R84-86W.

Initial examination of the picture confirmed that it covered an area of relatively undeformed and generally well-exposed strata dipping southeast. Detailed photogeologic interpretation identified 35 stratigraphic horizons. These were traced on a transparent mylar overlay and numbered in ascending stratigraphic order. A simplified version of this interpretation is illustrated in Figure 4. (Figure 3 and subsequent figures depicting the stratigraphic reference locality are printed at the same scale as Figure 4. All location references are keyed to marginal letter and number scales shown in Figure 4.)

Geomorphology expressed by topographic shadowing is a major influence in this photogeologic interpretation. For example, horizon 60 is defined by a shadow on the northwest side of a prominent northeast-trending escarpment. Units are also recognized because of spectral characteristics that are portrayed by image color. For example, red beds (displayed in yellow-gold on Figure 3) are generally confined to the interval between horizon 40 and 60. Gray rocks appear gray on Figure 3. The interval between horizon 100 and 110 is characterized by dark-gray to black rocks, and horizon 150 is the base of a light-gray to white stratum, in the upper part of darker, medium-gray strata between horizon 110 and 170. Strata supporting vigorous vegetation, such as those between horizon 90 and 100 in the vicinity of C7, are portrayed in red, reddish brown, or pink.

Dips and strikes shown on Figure 4 were determined using three-point problem solutions (Compton, 1962, p. 31; Ragan, 1973, p. 15-22) constrained by: (1) the trace of stratigraphic horizons obtained from the 1:24,000-scale photogeologic interpretation; and (2) elevations obtained from 7½-min topographic maps. Except for the east-plunging anticline near F5 that was first discovered in the field, all folds were identified by dip and strike determinations and all faults by offset image horizons in the photogeologic interpretation. The F5 anticline was recognized during a field examination of the fault on its northern limb.

Construction of TM Stratigraphic Column

The image and photogeologic map were used with topographic information provided by 7½-min quadrangles to construct a TM stratigraphic column (Figure 5, left side of center panel labeled "TM"). A cross section (not shown) along a line of section from D1 to J10 was the principal source of information. The D1 to J10 line of section may be considered a reference section in the conventional stratigraphic sense. The resulting column depicts the resistance to erosion, stratigraphic thickness, sequence, and spectral

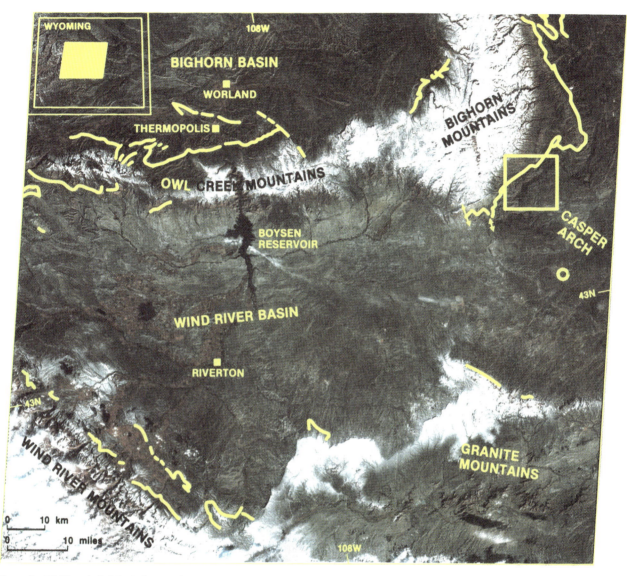

Figure 2—TM bands 2 (blue), 3 (green), and 4 (red) false-color infrared (CIR) composite image of study area in central Wyoming. This image, enlarged to 1:250,000 scale, was used as base for regional photogeologic interpretation. Yellow square outlines northern Casper arch stratigraphic reference locality. Locations of reported composite surface section (arrow-tipped lines southwest of box) and of reported borehole section (circle south of box) are shown. Irregular yellow lines highlight interpreted distribution of red-bed stratigraphic marker.

(CIR image color) characteristics of strata in the reference section. Total measured stratigraphic thickness is approximately 3,400 ft (1,040 m).

Assuming no topography, and given the 30-m spatial resolution of the TM image and the 5° dip typical of strata in the reference section, the nominal stratigraphic resolution of the image is approximately 3 m. However, the actual accuracy of thickness measurement for an individual unit is dependent upon the quality of bedrock exposure, topography, illumination geometry, and precision of photogeologic horizon location, limited by the spectral contrast between adjacent units in the sequence. For this example, we estimate accuracies of ± 20-50 ft (6-15 m).

The resistance of units as portrayed in Figure 5 is relative, and is based upon their geomorphic expression on (1) the CIR image, and (2) the corresponding topographic maps.

No direct determination of lithology is possible using the TM image alone; however, inferences can be made using geomorphic expression and spectral characteristics. On the basis of known weathering characteristics of sedimentary rock in arid regions, resistant strata are inferred to be either sandstones or carbonates; nonresistant units, either shales, siltstones, or bedded gypsum. Strata that appear red on the CIR image are vegetated and therefore have better soil development and/or higher permeability and water saturation than nonvegetated units.

If the stratigraphic sequence is not overturned and thrusting is absent (assumptions that are supported by the regional structural interpretation), relative ages of strata can be determined directly from the TM spectral stratigraphic column. Assignment of conventional formation names or geologic ages to stratigraphic units defined by

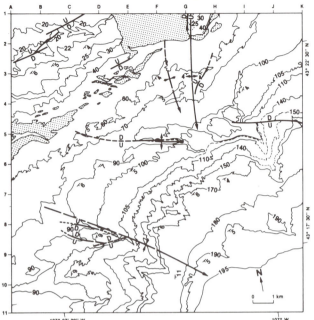

Figure 3—TM bands 2 (blue), 3 (green), and 4 (red) false-color infrared (CIR) composite image of northern Casper arch stratigraphic reference locality (see Figure 2 for location). This image, enlarged to 1:24,000 scale, was used with USGS 7½-min topographic maps as base for detailed photogeologic interpretation.

Figure 4—Photogeologic map from interpretation of Figure 3 covering northern Casper arch stratigraphic reference locality. The traces of 17 selected photogeologic horizons are numbered in ascending stratigraphic order. Dips and strikes of strata were determined using three-point problem solutions constrained by trace of photogeologic horizons and topographic data. Traces of structural axes and faults were mapped based upon dips and strikes and offsets of photogeologic horizons. Terrace deposits are depicted by stippled pattern; alluvium is not shown.

bounding photogeologic horizons, however, is not possible with TM data alone. Correlation of the TM stratigraphic column with conventional surface or subsurface columns may resolve these nomenclature problems.

The left panel of Figure 5 illustrates a reported conventional stratigraphic section (Woodward, 1957), measured by the Jacob-staff method, approximately 10 km southwest of the TM type section (Figure 2). Unique and complementary aspects of the TM and conventional surface columns are apparent. The two can be correlated, thus providing formation names and chronostratigraphic assignments for TM stratigraphic units. For example, Lower Cretaceous Cloverly-Mowry strata of the surface section are equivalent to the thicker, horizons 90-to-150 interval in the TM column. Lower Mowry–equivalent strata of the TM column (the interval between horizon 107 and 110) are more similar spectrally to Thermopolis-equivalent strata (100-105) than they are to upper Mowry–equivalent strata (110-150). A thin white stratum above horizon 150 is correlated with the Clay Spur Bentonite, a volcanic ashfall that marks the Lower Cretaceous–Upper Cretaceous boundary throughout much of Wyoming (McGookey, 1972).

The right panel of Figure 5 illustrates the reported section penetrated by a borehole, approximately 20 km south of the TM reference section (Figure 2). The TM stratigraphic column is also correlated with the cuttings and electric logs for this subsurface section.

After preparation of the Figure 4 photogeologic map and construction of the Figure 5 TM stratigraphic column, we visited the northern Casper arch reference locality to check results. Dip and strike measurements were made and the traces of photogeologic horizons were examined in the field.

Field dip and strike measurements were consistent with results from three-point problem solutions obtained from TM and topographic data. Strike determinations agreed within 5°, dips within 2°. Strike differences reflect irregularities on bedding surfaces measured in the field. Field dip determinations were consistently 1°–2° less than TM determinations. The discrepancy was caused by the TM interpreter's selection of points for three-point problem solution that yielded the maximum possible dip. Once identified, the problem was corrected easily in subsequent TM interpretation by selection of points for three-point problem solutions that yielded the minimum possible dip. Dips and strikes shown on the final Figure 4 interpretation reflect this correction.

Field examination demonstrated that all photogeologic horizons that were traced on the TM picture correspond to mappable lithologic contacts. Figure 6 is a field photograph with two such contacts identified. Correlations depicted on Figure 5 were also confirmed in the field. For example, horizon 150 does indeed correspond to a prominent bentonite bed (Clay Spur Bentonite); horizon 170 corresponds to the base of the first sandstone above the Clay Spur Bentonite (basal sandstone of the Frontier Formation).

Spectral Reflectance Analysis of AIS Data

On October 30, 1983, AIS data (Table 1) were acquired on a flightline that coincided with the line of section used to

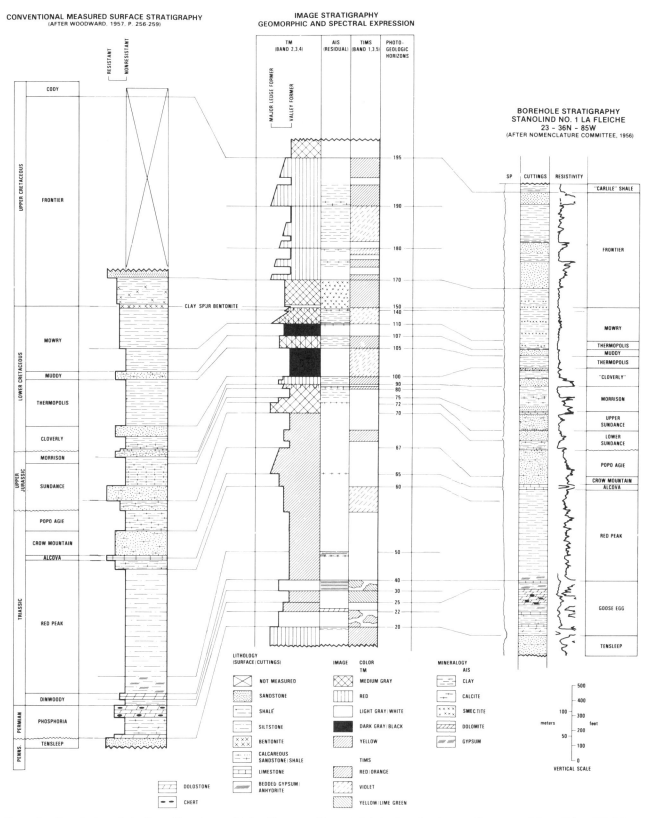

Figure 5—Correlation of stratigraphic column obtained from multispectral data alone (center panel), measured from D1 to J10 (Figure 4), with conventional surface section (left panel) and subsurface section (right panel). Center panel summarizes lithostratigraphic information obtained from three multispectral instruments. TM and topographic data provide information regarding thickness and resistance to erosion of strata. TM image colors identify red beds (yellow), vegetation (red), and white, medium gray, and black strata. AIS data identify stratigraphic distribution of lithologically diagnostic minerals. TIMS image colors identify silica-rich strata (red/orange), clay (violet), and gypsum (yellow/lime green). Photogeologic horizons provide stratigraphic markers that were sequentially numbered by photointerpreter.

Figure 6—Field photograph showing trace of two TM photogeologic horizons near F9 (Figure 4). Field observations confirmed accuracy of bedding attitudes and surface traces of photogeologic horizons, faults, and fold axes as mapped in photogeologic interpretation of 1:24,000 TM image.

construct the TM stratigraphic column (from D1 to J10, Figure 4). These spectral data were acquired to determine the stratigraphic distribution of clay, gypsum, and carbonate minerals. Previously reported laboratory spectra (Hunt, 1977, 1980) demonstrate that diagnostic spectral features for these minerals exist in the 2.1-2.4 μm wavelength interval. Our analysis therefore focused on the last 32 channels of AIS data covering 2.1-2.4 μm. Two methods were used to analyze AIS data covering this spectral interval: (1) interactive sampling of image spectra using spectral analysis manager (SPAM) software; and (2) along-track evaluation of spectral variability using a residual image. SPAM and the procedure used to produce the residual AIS image are described in appendices B and C, respectively. Results provide information about the stratigraphic distribution of lithologically diagnostic minerals.

SPAM was used to retrieve interactively from AIS data the spectra of surface materials along the northern Casper arch reference section. Based upon preliminary correlation of TM units with reported conventional stratigraphic units, lithostratigraphically diagnostic minerals that might be expected in the section and that exhibit diagnostic spectral features in the 2.1-2.4 μm wavelength interval include:

1. Gypsum, with an absorption band near 2.2 μm, in gypsum beds of the basal Red Peak, Dinwoody, and equivalent Goose Egg lithostratigraphic units;

2. Calcite, with an absorption band near 2.33 μm, in limestone beds of the Goose Egg and Alcova, and calcite-cemented sandstone and thin limestone beds in the super-Alcova lithostratigraphic units;

3. Dolomite, with an absorption band near 2.30 μm, in dolostone beds of the Dinwoody and Phosphoria, and equivalent Goose Egg lithostratigraphic units;

4. "Clay," including (a) kaolinite, with a double absorption band near 2.20 μm in clastic and carbonate beds of the entire section, and (b) smectite, with a single absorption band near 2.20 μm in bentonites and other volcaniclastic beds of the super-Cloverly stratigraphic interval.

Spectra of all four types were identified in SPAM analysis. No spectra exhibited the 2.2 μm kaolinite doublet. Spectra exhibiting a poorly defined 2.2 μm absorption band were given a generic "clay" assignment; spectra with a strong enough 2.2 μm band to preclude the existence of a kaolinite doublet were given a smectite assignment.

The absence of any detectable absorption band in the 2.1-2.4 μm interval also may be lithologically diagnostic. Quartz and feldspar exhibit no major absorption bands in this wavelength interval. Strata characterized by flat spectra (left blank on the Figure 5 AIS column) may therefore represent outcrops or residual soils derived from quartz sandstones and siltstones, or arkosic sandstones and siltstones.

Field samples were collected to confirm SPAM results. These were analyzed in the laboratory using a Beckman UV5240 spectrophotometer and a Phillips x-ray diffractometer. Results confirmed SPAM spectral and mineralogic interpretations. Figure 7 compares AIS and laboratory spectra of four selected samples.

Residual processing of AIS data was also used to evaluate spectral variability in the northern Casper arch reference locality. Marsh and McKeon (1983) reported the utility of the residual display for spectral interpretation of high-resolution airborne spectroradiometer data. Residual images are produced by subtracting the average spectrum for an entire flightline from each spectrum along the flightline. Differences (residuals) are then color coded so that negative values, associated with potentially diagnostic spectral absorption features, are distinguishable.

Figure 8 includes an AIS residual image. Spatial variations are depicted along the length of the strip and spectral residuals across its width. As labeled on Figure 8, the left edge of the image represents residual values at 2.0 μm and the right edge, 2.33 μm. The trace of the flightline is represented by the left edge of the strip. Negative residual values (absorption features) are coded in blue and green colors.

The residual picture facilitates assessment of mineralogic and lithologic variations in the reference section. For example, the interval between horizon 30 and 40, near E3 on the residual strip, exhibits a deep-absorption feature at 2.2 μm that can be attributed to gypsum; and the interval between horizon 150 and 170, near H7, exhibits a shallower and narrower feature at 2.2 μm that can be attributed to smectite.

This mineralogic information helps refine correlations with the conventional surface and subsurface columns. For example, horizon 65 is confidently correlated with the Triassic Alcova Formation, because of the presence of the diagnostic calcite absorption band associated with this resistant limestone marker bed.

Mineralogic information from SPAM and residual spectral analyses was incorporated into the stratigraphic column. Results are summarized in the column labeled "AIS" in the center panel of Figure 5.

Spectral Emission Analysis of TIMS Data

On July 18, 1983, TIMS data (Table 1) were acquired on a flightline that coincided with the line of section used to construct the TM stratigraphic column. This line was the same

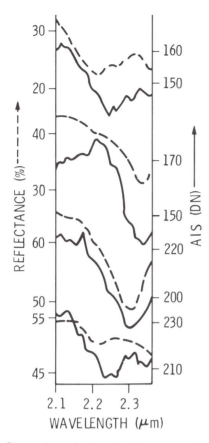

Figure 7—Comparison of selected AIS spectra (solid) to laboratory spectra (dashed). In descending order, samples are: bedded gypsum from Goose Egg Formation (sample collected near E3, Figure 4); marly limestone from Alcova Formation (F3); dolostone from Goose Egg Formation (D2); and bentonite from Clay Spur Bentonite (H7). These spectra illustrate physical basis for spectral interpretation of AIS residual images such as Figure 8. Spectral absorption bands due to gypsum and smectite are apparent near 2.2 μm, and absorption bands due to calcite and dolomite near 2.33 μm and 2.30 μm, respectively. (Apparent shift in position of absorption features between laboratory and AIS spectra for gypsum and bentonite examples reflect shift in AIS wavelength calibration.) DN = digital number.

Figure 8—Residual display of AIS data superimposed on Figure 3 TM image. Left side of strip locates center of AIS ground track. Residual values are determined by subtracting average spectrum for entire flightline from average spectrum for each 32 sample-wide (10 m $\times$ 320 m) line of AIS data. Spectral absorption features (negative residual values) are portrayed in green and blue. Residual strip can be used to determine stratigraphic distribution of calcite, dolomite, smectite, and gypsum on basis of diagnostic spectral reflectance features. For example, interval exhibiting absorption band at 2.3 μm near northern end of strip corresponds to dolomite, associated with dolostone bed in Goose Egg Formation. Compare to Figures 3 and 4.

as the AIS flightline (from D1 to J10, Figure 4). The TIMS data were digitally registered to the Figure 3 TM image using MIPL software.

Photogeologic and spectral interpretation focused on a band 1 (blue), band 3 (green), band 5 (red) color decorrelation contrast stretch version of the registered data (Figure 9). The procedure used to create the picture is described in appendix D; an understanding of the physical meaning of colors displayed on the picture requires additional background information.

The color decorrelation contrast stretch (dstretch) is an image-processing algorithm that was developed by Soha and Schwartz (1978) to enhance spectral differences in multispectral data that show statistically high interband correlations. TIMS thermal data are dominated by the temperature of the material sampled. Uncorrelated interband differences, caused by spectral emissivity differences among the materials sampled, unfortunately, are muted by temperature. Kahle and Rowan (1980) and Gillespie et al (1984) demonstrated the utility of dstretch enhancement for making pictures from multiband thermal data that are suitable for photogeologic and spectral interpretation. They reported that dstretch pictures are especially useful for mapping variations in silica content. Based upon laboratory spectral emission data reported by Hunt (1980) and Hunt and Salisbury (1975), silica-rich minerals should exhibit higher emissivity in TIMS channel 5 (10.2-10.9 μm, displayed in red in Figure 9) relative to channels 1 and 3 (8.1-8.5 and 8.9-9.3 μm, blue and green, respectively). "Redness" of the Figure 9 picture therefore should correlate with silica content. Quartz-rich rocks should appear red; clay-rich rocks, violet; and silica-poor rocks such as limestone and dolostone, shades of blue and green. Based upon an examination of reported laboratory spectra, we predicted that bedded gypsum should be displayed in greenish-yellow shades in Figure 9.

These simple color assignments are complicated by the presence of vegetation. Vegetation is nearly spectrally flat in the wavelength region sampled by the TIMS; therefore, vegetation appears either dark or bright (depending on temperature), and tends to dilute color differences associated with spectral emissivity variations of the soil and bedrock in partially vegetated areas.

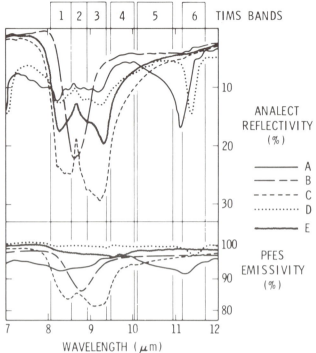

Figure 9—TIMS band 1 (blue), 3 (green), and 5 (red) decorrelation stretch image that is digitally registered to Figure 3 TM picture. Silica-rich strata appear red, clay-rich strata appear violet, and gypsum-rich strata appear yellow–lime green in this image. Compare to Figures 3 and 4.

Figure 10—Comparison of laboratory thermal-reflectivity and field thermal-emissivity spectra for five rock surfaces: A, dolostone from Goose Egg Formation (sample collected near D2, Figure 4); B, bedded gypsum from Goose Egg Formation (D3); C, orthoquartzite boulder from Quaternary terrace gravel (E2); D, marly limestone from Alcova Formation (B6); and E, bentonite from Clay Spur Bentonite (F9). These spectra confirm spectral interpretation of Figure 9 TIMS image. For example, lower emissivity of quartz (spectrum C) in TIMS bands 1 and 3 (blue and green, respectively, in TIMS image) relative to band 5 (red), should result in quartz-rich areas appearing red in image. PFES = JPL portable field emission spectrometer.

Based upon these ideas, the Figure 9 TIMS picture was interpreted spectrally at 1:24,000 scale. Diagnostic spectral information (image color) was added to the stratigraphic column (TIMS column, Figure 5). Stratigraphic variations in TIMS color provide lithologic information: strata displayed in red are silica-rich quartz sandstones, siliceous shales, or chert beds; those displayed in violet, clay-rich shales or clay siltstones; and those displayed in yellow–lime green, gypsum beds. Strata exhibiting no diagnostic color on the TIMS picture are given no color designation on Figure 5.

To check the spectral interpretation, a field-laboratory spectral study was conducted. Emissivity spectra were measured at selected field sites using the JPL portable field emission spectrometer (PFES) (instrument described by Hoover and Kahle, 1986). Measurements were also made on the same natural surfaces in the laboratory with an Analect model 6200, Fourier transform spectrometer. X-ray diffraction analysis was used to determine mineralogy of spectral samples.

Examples of field and laboratory results for four selected samples are provided in Figure 10. These spectral and mineralogic results confirm the lithologic interpretation of the TIMS image.

Utility of Digital Elevation Data

We used photogeologic interpretation of 1:24,000-scale TM images, drawn on mylar overlays, and topographic maps to manually prepare the photogeologic map (Figure 4) and TM stratigraphic column (Figure 5). The availability of

USGS topographic data in digital form as digital elevation models (DEM) (Elassal and Caruso, 1983) makes possible the interactive geologic interpretation of compatible digital image and topographic data at a computer image-display terminal. To test the utility of such an approach, we developed the interactive geologic interpretation system (IGIS) (appendix E). Figure 11 illustrates results of an interactive session using IGIS. Photogeologic interpretation and automatic dip and strike determination were accomplished using IGIS in that part of the northern Casper arch reference locality covered by the First Water Draw and Three Buttes DEMs.

IGIS results can be compared to the results obtained by manual procedures as shown in Figure 4. Attitudes determined by IGIS (shown in red) are equivalent to those obtained in manual analysis of the 1:24,000 TM picture and topographic maps, and to those obtained in the field. Differences are attributed primarily to differences in location of points used for three-point problem solutions. Two photogeologic horizons, shown in blue, are also traced partially in the IGIS example: number 30, near B4; and 140, near G7. Also shown on the IGIS example are a program-generated north arrow and bar scale, and the axial trace of an anticline near G3.

Figure 11—Results from 15-min session using IGIS software for interactive photogeologic interpretation of coregistered TM and digital elevation models (DEM) data. Compare to Figures 3 and 4. This example illustrates that geologists, with aid of IGIS software, can rapidly trace photogeologic contacts (shown in blue) and fold axes (shown in purple), and benefit from program's automatic dip-and-strike, scale, and north-arrow capabilities (shown in red).

DISCUSSION

As depicted in Figure 5, stratigraphic columns constructed using multispectral and topographic data alone can be used in conjunction with conventional surface and subsurface data for constructing correlation charts, stratigraphic cross sections, and panel diagrams. Thickness data obtained remotely can augment conventional control for preparing isopach maps; lithologic information can augment conventional control for preparing lithofacies maps. In areas less well studied than the Wind River and Big Horn basins, multispectral data can be a principal source of lithostratigraphic information.

Our procedure for combined analysis of multispectral and topographic data provides a powerful tool for acquisition of structural information, including the dip and strike of beds. Figure 4 depicts structures not shown on published maps available for our Wind River basin and Big Horn basin areas investigation. These structures show that although the northern Casper arch reference locality is relatively undeformed compared to other areas on the margins of the Wind River basin and Big Horn basin, the area does contain subtle structures that potentially could provide important information for tectonic analyses.

For our Wyoming investigation, we spent considerable time in the field checking structural and stratigraphic results. Dip and strike measurements were made and samples were collected for laboratory mineralogic and spectral analyses. These studies confirmed results that were obtained remotely using multispectral and topographic data alone. We are therefore confident that for future geologic investigations, field and laboratory resources can be allocated primarily for detailed analysis of key stratigraphic and structural problems rather than geologically unproductive reconfirmation of our basic approach. Based upon our experience, we estimate that in an operational mode, work summarized in Figures 2, 4, and 5 would require one man-month of office work and five man-days of fieldwork.

Based upon the successful application of IGIS software for interactive geologic analysis of digital elevation and multispectral data, we are adding capabilities to generate: (1) geologic cross sections constrained by image, field, topographic, borehole, and geophysical data; (2) rotatable 3-D block diagrams using IGIS-created cross sections plus image and topographic data; and (3) stratigraphic columns using information from IGIS-generated cross sections. These additional capabilities should make IGIS a powerful tool for interactive lithologic and structural mapping using digital image, topographic, and other geological and geophysical data.

Major factors that limit the applicability of our methods elsewhere are related to (1) the nature of the terrain, and (2) the availability and cost of multispectral data and image-processing facilities.

Photogeologic interpretation of multispectral images, using well-established principles of geomorphology, can be successfully conducted, even in totally vegetated terrain. This is not the case, however, for spectral interpretation. The multispectral images used for the Wind River basin and Big Horn basin examples depict surface spectral properties only. Strata must be exposed for direct spectral measurements. In terrain with total vegetation or transported soil and alluvial cover, interpretation of bedrock lithology is not possible using multispectral data alone. Spectral analysis in such terrain therefore requires models that relate spectral properties of the plant canopy or transported soil and alluvial blanket to lithology of subjacent strata. Such models presently do not exist. Their development represents an important topic for future research.

The three types of multispectral and DEM data used for the Wind River basin and Big Horn basin examples do not exist for all areas of the earth's surface. TM data now exist for most of North America, Europe, and north Africa. Global coverage by the TM should be accomplished by 1989. The Earth Observation Satellite Company (EOSAT) recently became the commercial operator of the Landsat system and distributor of Landsat data, including TM data. EOSAT has promised rapid and nondiscriminatory availability of TM data.

Aerial photographs could have been used instead of TM data as a base for the photogeologic interpretations (Figures 2, 4). However, approximately 500 color aerial photographs costing about $4,500 would have been required to cover the same area as the digital TM data (Figure 2) that cost approximately $3,300. From a data-cost standpoint, TM data are less expensive than color aerial photography. Cartographic fidelity, wider spectral coverage, quantitative radiometry in seven spectral bands, and the wide choice of image formats provided by image processing make digital TM data a more flexible base for photogeology than color aerial photographs. The significantly better spatial resolution of color aerial photographs (1 m) compared to TM data (30 m) is not considered significant for the purposes of our investigation.

Although experimental TIMS and AIS data theoretically can be acquired on a contractual basis from NASA, the priority assigned to NASA-funded research severely limits acquisition and distribution for non-NASA research or application projects. As the geologic usefulness of these experimental systems is demonstrated, however, it is expected that equivalent commercial systems will proliferate. For example, Geophysical Environmental Research, Inc., (GER) now operates commercially an airborne spectroradiometer with spectral resolution in the 1.5-2.5 μm interval that is spectrally equivalent to AIS data (Collins et al, 1981; Marsh and McKeon, 1983). A similar situation exists regarding DEM data. Many commercial firms are today producing DEM-equivalent digital topographic data from topographic maps or aircraft-acquired or satellite-acquired stereophotographs in areas where USGS DEM data do not exist.

By 1995, the NASA earth observation system (EOS) satellites will be a new source of multispectral data (Goetz et al, 1985). EOS will provide global multispectral coverage with image data having similar bandpasses in the spectral regions used for the Wind River basin and Big Horn basin examples.

Acquisition of TM images for conventional photogeologic interpretation requires no image-processing. TM photographic products (both prints and negatives) that are equivalent to the TM pictures used in the Wind River basin and Big Horn basin examples are available as standard products from EOSAT at a cost of $160 to $800 for a full TM image. The use of topographic maps similarly requires no image-processing capability. Other versions of TM data, requiring more sophisticated processing but probably more useful for both spectral and photogeologic interpretation, have been described elsewhere (Abrams et al, 1985; Conel et al, 1985; Paylor et al, 1985). Such digital manipulation of TM data, and use of AIS, TIMS, and DEM data require image-processing facilities that are not presently available to most geologists. However, the recent proliferation of commercial, governmental, and academic processing facilities, including microcomputer-based systems, should soon change this situation.

CONCLUSIONS

1. Stratigraphic and structural analyses of the Wind River and Big Horn basins and Casper arch areas of Wyoming are in progress using satellite (TM), and aircraft high-spectral resolution (AIS) and thermal (TIMS) multispectral data.

2. A remote-sensing approach for determining the lithology and structure of layered sedimentary rock sequences exposed at the earth's surface has been developed. The method is photogeologic in character but, in addition, uses differences in spectral reflectance and emittance properties of rocks as represented in the multispectral data to discriminate strata and map clay species, gypsum, and carbonate minerals.

3. The high geometric fidelity of Landsat TM images when coupled with topographic maps or digital elevation data provide for determination of strike and dip of strata from the TM images themselves. Attitudes determined by these methods are within a few degrees of those determined by field measurement with Brunton compass.

4. Using what is termed here a residual AIS image (an enhanced display of spectral reflectance variations), the stratigraphic distribution of clay species, gypsum, calcite, and dolomite were determined by the position of characteristic absorption bands.

5. Using what is termed here a dstretch TIMS image (an enhanced display of thermal infrared spectral variations), the stratigraphic distribution of clay, quartz, and gypsum were represented by emissivity contrasts.

6. A stratigraphic column based on multispectral and topographic data alone was developed for upper Paleozoic and Mesozoic strata exposed in the Casper arch area. Lithologic units identified in the image data were subsequently correlated with lithostratigraphic units defined by conventional surface and subsurface stratigraphic measurements outside the area of image analysis.

7. Lithologic identifications and structural interpretations agreed with subsequent field observations, and augment conventional stratigraphic and structural information reported in the literature.

8. Our results show that it proves feasible to measure remotely areal variations in attitude, thickness, and lithology of strata, thereby aiding development of quantitative models of the structural and stratigraphic evolution of a sedimentary basin.

APPENDICES:
IMAGE PROCESSING FACILITY AND SOFTWARE

Appendix A. Multimission Image Processing Laboratory

The Multimission Image Processing Laboratory (MIPL) is JPL's image-processing facility. This Digital Equipment Company computer-based installation consists of two VAX 11-780 and one VAX 8600 computers with a total of 28 megabytes of core memory. Ten gigabytes of shareable disk storage are available under the VMS operating system. A Floating Point Systems array processor is available on one of the 11-780 computers. Video image communication and retrieval/image-based information system (VICAR/IBIS) software provides over 300 image-processing programs written in Fortran-77, C, and MACRO that run under an executive user interface, the transportable applications executive (TAE) (COSMIC, 1985, p. 175). By June 1987, software described in appendices B-E should be available through NASA's Computer Software Management and Information Center (COSMIC, 112 Barrow Hall, University of Georgia, Athens, Georgia 30602) as part of the newest VICAR/TAE VMS version 4.

MIPL images are displayed on Gould DeAnza, Ramtek, or Adage frame buffers. Three image memory planes, connected to 3 lookup tables (red, green, and blue), are available for color or black-and-white processing. A fourth image memory plane is used for graphics display. Interactive input/output devices are a trackball for the DeAnza device and a pen and tablet for the Adage device.

Photographic output is provided by film recorders, including: (1) a McDonald Detweiler MDA Color Fire 240 that provides color pictures on large-format, 9 × 9-in. film; (2) three small-format (4 × 16-in.), black-and-white DICOMED D46J devices; (3) a color DICOMED D47J CRT system; and (4) an on-line Dunn Instruments model 631 color camera for quick-look, 8¹⁄₂ × 11-in. film positives. (The two DICOMED devices have been modified considerably, thus the "J" suffix.)

Appendix B. Spectral Analysis Manager

The spectral analysis manager (SPAM) is a package of more than 30 functions that enable geologists to interactively analyze digital-imaging spectrometer data provided by the AIS. SPAM is under development by the JPL Image Analysis Systems (IAS) Group. The current version is described in Mazer et al (in press) and Jet Propulsion Laboratory (1985). The VICAR version of SPAM and two other versions that operate as stand-alones under VAX/VMS or UNIX may be obtained free-of-charge from the IAS Group. All three versions of SPAM will soon be available through COSMIC.

Present SPAM capabilities for interactive analysis of AIS data include: (1) visual interaction and statistical analysis; (2) display of spectra for individual pixels or average spectra for user-designated areas; (3) interaction with a library containing laboratory spectra of more than 200 minerals; and (4) use of fast algorithms for identification and mapping of geologic materials through spectral-signature matching and image classification.

Appendix C. AIS Residual-Image Processing

Four VICAR programs used to produce AIS residual images are GRAD-REM, SIZE, F2, and LOOKUP. The procedure starts with 32-channel, GPOS-4 (nominal 2.1-2.4 μm), radiometrically corrected AIS data in VICAR MSS format. The six steps used to create residual pictures for geologic interpretation are as follows.

1. GRADREM removes systematic and nonlinear artificial brightness gradients along spatial scan lines of wide-angle aircraft scanner data. For residual processing of AIS data, GRADREM removes the systematic, wavelength-dependent decrease in DN that results from falloff of solar radiation from channel 1 to channel 32 that is caused by a decrease in solar radiation with wavelength and by a strong atmospheric absorption band at 2.74 μm (Goetz and Rowan, 1981).

First, a region of the AIS image that is assumed to be spectrally flat, only slightly varying with wavelength, or to contain no spectral features of geologic interest is selected. Image lines in this region are averaged across the spectrum by GRADREM to produce a "standard" DN at each sample. Each wavelength pixel in the image is divided by its appropriate standard DN. DN values for the entire image are then subjected to a linear transformation for convenient display. The result of this step is an MSS-format image with scene-wide spectral features removed.

2. SIZE expands or reduces an image and/or changes its aspect ratio by decreasing the number of pixels in the sample or line direction by averaging or interpolation. In this step, SIZE, using the interpolation option, reduces the image from step 1 by 32 pixels in the sample direction. This produces an image in which each line contains 32 samples that represent the average DN for each AIS spectral band.

3. SIZE, also using the interpolation option, reduces the image from step 1 by 1,024 pixels (32 bands × 32 pixels per line) in the sample direction. This produces an image in which each line contains one sample that represents the average DN for all 32 AIS bands.

4. SIZE expands the image produced in step 3 by repeating DN values 32 times in the sample direction. This produces an image that is the same size as the image produced in step 2.

5. F2 performs mathematical operations on images. In this step, it subtracts the image produced in step 2 from the image produced in step 4. An offset is added to the result to remove negative values. The resulting image is the AIS "residual."

6. LOOKUP provides color tables for coloring single-channel (black-and-white) images by density grouping DNs. In this step, it produces a pseudocolor "residual image" from the step 5 output. A color table is developed that portrays negative residuals (samples of relative spectral absorp-tion) in dark or "cool" colors, and positive residuals (samples of relative reflectance peaks) in bright or "warm" colors.

Appendix D. TIMS Dstretch Image Processing

Five VICAR programs are used in five steps to produce dstretch TIMS images.

1. VTIMSLOG separates each TIMS file on the computer compatible data tapes provided by NSTL into two new files. One file contains the image data (scene DNs). The other contains calibration data, including a line-by-line accounting of the temperature of two calibration black bodies and their DNs as measured in each of the six TIMS bands.

2. TIMSCAL uses the two outputs from step 1 to transform byte DNs of the image file into floating-point radiance values. This calibration step eliminates changes in DNs caused by fluctuations in the TIMS response over time.

3. ASTRTCH2 converts the floating-point values from step 2 back into byte DNs. This format change facilitates further image processing.

4. C130RECT corrects the image from step 3 for panoramic distortion. The results are six radiometrically calibrated and panoramically corrected byte files, one for each TIMS band.

5. Any three of the six files from step 4 (usually TIMS bands 1, 3, and 5) are then used as inputs to EIGEN, the program that produces the decorrelation (d) transformation (stretch) of the TIMS data.

Appendix E. Interactive Geological Interpretation System

The interactive geological interpretation system (IGIS) is a package of VICAR programs that enables a geologist to analyze interactively coregistered image and digital terrain data. IGIS is under development by the JPL Cartographic Applications Group. The current version provides a suite of general image-processing and display capabilities, as well as special functions to (1) trace geologic contacts on images using a graphics overlay plane, (2) solve three-point problems for determination of dip and strike of geologic features identified on images, (3) measure topographic slope, and (4) generate topographic profiles. Hard copies of IGIS results are provided by MIPL film recorders.

IGIS accepts up to four coregistered input data sets in byte format. The first three data sets (for example, 3 TM channels) provide a color composite image that is displayed on a color monitor for interactive photogeologic interpretation. The fourth is a DEM. Separate transfer functions are available for each of the three color display planes used for image data. A color table also is available for color display of a single image data set.

In either color (three channel) or pseudocolor (one channel) configuration, the internal transfer function may be output to a disk data set and used in batch operation.

Graphics overlay is accomplished by connecting an image memory plane to a graphics lookup table that consists of 3 tables (red, green, and blue). IGIS provides the user with up to 256 graphics colors. The graphics overlay plane may be selectively erased for editing.

Geologic applications are activated by keyword when prompted by IGIS. Input is by keyboard or interactive input/output device, normally a track-ball. Graphics colors are selected as required. After a task is finished, the geologist has the option to continue or to eliminate the graphics produced by the last task.

Additional IGIS routines are being developed that use capabilities of the Cartographic Applications Group's Intergraph CAD/CAM system. IGIS image, elevation, and graphics information will be transferred to the Intergraph using IBIS software. Data will be displayed on the Intergraph system using the Intergraph digital terrain modeling package and interactive graphics design software. These added capabilities, in conjunction with JPL-written Intergraph user commands, will be used to generate geologic cross sections, rotatable 3-D block diagrams, and stratigraphic columns.

A brief description of the four IGIS functions that were used to create Figure 11 is provided below. Keywords are listed in the same order as they might be used in a typical geologic interpretation session.

NORTH creates a north arrow and labels north with the letter N. A program-generated bar scale is automatically created. Both the north arrow and scale are interactively positioned on the image by the geologist.

ROCK enables the geologist to interactively draw geologic contacts that are traced on the image with the cursor controlled by the trackball.

DIP provides automatic dip-and-strike determination by solution of three-point problems. The program first queries the geologist as to what type of input will be used: (1) selecting 3 points manually, or (2) defining a box that encompasses a segment of a geologic contact delineated previously with the ROCK routine. If the second option is selected, the program automatically selects three optimal points on the encompassed contact segment for dip-and-strike calculation. After IGIS computes dip and strike, a standard dip/strike symbol is drawn on the graphics plane at a location selected by the geologist with the trackball. The geologist is then asked if the result is acceptable. If the reply is "N," the symbol is erased; if the reply is Y, IGIS returns to the DIP subroutine and the geologist is prompted for input to the next dip-and-strike determination.

GRAPH enables the geologist to create graphics manually. Options exist for drawing trackball-controlled lines and geometric figures such as circles and rectangles, and for writing text. Thus, geologic information such as field-sampling sites and borehole locations, and structural and lithologic designations can be labeled during interactive photogeologic interpretation.

REFERENCES CITED

Abrams, M. J., J. E. Conel, H. R. Lang, and H. N. Paley, 1985, The joint NASA/Geosat test case project final report, part 2: AAPG, 1292 p.

Bailey, G. B., and P. D. Anderson, 1982, Applications of Landsat imagery to problems of petroleum exploration in Qaidam basin, China: AAPG Bulletin, v. 66, p. 1348-1354.

—— J. R. Francica, J. L. Dwyer, and M. S. Feng, 1982, Extraction of geologic information from Landsat multispectral scanner and thematic mapper simulator data from the Uinta and Piceance basins, Utah and Colorado: Proceedings of the International Symposium on Remote Sensing of Environment, v. 1, p. 43-70.

Barker, J. L., ed., 1985, Landsat-4 science characterization early results: NASA Conference Publication 2355, 2114 p.

Bracken, P. A., 1983, Remote sensing software systems, in R. N. Colwell, ed., Manual of remote sensing, second ed.: American Society of Photogrammetry, p. 807-839.

Collins, W., S. H. Chang, and J. T. Kuo, 1981, Infrared airborne spectroradiometer survey results in western Nevada area: Columbia University Aldrich Laboratory of Applied Geophysics final report to NASA, contract JPL 955832, 61 p.

Colwell, R. N., ed., 1983, Manual of remote sensing, second ed.: American Society of Photogrammetry, 2440 p.

Compton, R. R., 1962, Manual of field geology: New York, John Wiley, 378 p.

Conel, J. E., H. R. Lang, E. D. Paylor, and R. E. Alley, 1985, Preliminary spectral and geologic analysis of Landsat-4 thematic mapper data, Wind River basin area, Wyoming: IEEE Transactions on Geoscience and Remote Sensing, v. GE-23, p. 562-573.

COSMIC, 1985, COSMIC software catalog 1985 ed.: Computer Software Management and Information Center, NASA-CR-174070, 367 p.

Elassal, A. A., and V. M. Caruso, 1983, Digital elevation models: USGS Circular 895-B, 40 p.

Gillespie, A. R., A. B. Kahle, and F. D. Palluconi, 1984, Mapping alluvial fans in Death Valley, California, using multichannel thermal infrared images: Geophysical Research Letters, v. 11, p. 1153-1156.

Goetz, A. F. H., and L. C. Rowan, 1981, Geologic remote sensing: Science, v. 211, p. 781-791.

—— J. B. Wellman, and W. L. Barnes, 1985, Optical remote sensing of the earth: Proceedings of the IEEE, v. 73, p. 950-969.

Grose, L. T., 1972, Tectonics, in W. W. Mallory, ed., Geologic atlas of the Rocky Mountain region: Rocky Mountain Association of Geologists, p. 34-44.

Halbouty, M. T., 1976, Application of LANDSAT imagery to petroleum and mineral exploration: AAPG Bulletin, v. 60, p. 745-793.

—— 1980, Geologic significance of Landsat data for 15 giant oil and gas fields: AAPG Bulletin, v. 64, p. 8-36.

Hoover, G. L., and A. B. Kahle, 1986, A portable spectrometer for use from 5 to 15 micrometers: Jet Propulsion Laboratory Publication 86-19, 48 p.

Hunt, G. R., 1977, Spectral signatures of particulate minerals in the visible and near infrared: Geophysics, v. 42, p. 501-513.

—— 1980, Electromagnetic radiation: the communication link in remote sensing, in B. S. Siegal and A. R. Gillespie, eds., Remote sensing in geology: New York, John Wiley, p. 5-45.

—— and J. W. Salisbury, 1975, Mid-infrared spectral behavior of sedimentary rocks: Air Force Cambridge Research Laboratory, Environmental Research Paper 520, 49 p.

Jet Propulsion Laboratory, 1985, Airborne imaging spectrometer science investigators guide to AIS data, appendix III: Jet Propulsion Laboratory Publication, 17 p.

Kahle, A. B., and L. C. Rowan, 1980, Evaluation of multispectral middle infrared aircraft images for lithologic mapping in the East Tintic Mountains, Utah: Geology, v. 8, p. 234-239.

Keefer, W. R., 1965a, Geologic history of Wind River basin, central Wyoming: AAPG Bulletin, v. 49, p. 1878-1892.

—— 1965b, Stratigraphy and geologic history of the uppermost Cretaceous, Paleocene, and lower Eocene rocks in the Wind River basin, Wyoming: USGS Professional Paper 495-A, 77 p.

Lang, H. R., ed., 1985, Report of the workshop on geologic applications of remote sensing to the study of sedimentary basins, Lakewood, Colorado, January 10-11: Jet Propulsion Laboratory Publication 85-44, 89 p.

Leith, W., and W. Alvarez, 1985, Structure of the Vakhsh fold-and-thrust belt, Tadjik SSR: geological mapping on a Landsat image base: GSA Bulletin, v. 96, p. 875-885.

Love, J. D., 1960, Cenozoic sedimentation and crustal movement in Wyoming: American Journal of Science, v. 258A, p. 204-214.

—— 1970, Cenozoic geology of the Granite Mountains area, central Wyoming: USGS Professional Paper 495C, 154 p.

—— and A. C. Christiansen, 1985, Geologic map of Wyoming: USGS Map, scale 1:500,000.

Marsh, S. E., and J. B. McKeon, 1983, Integrated analysis of high-resolution field and airborne spectroradiometer data for alteration mapping: Economic Geology, v. 78, p. 618-632.

Mazer, A. S., M. Martin, M. Lee, and J. E. Salomon, in press, Image processing software for imaging spectrometry data analysis: Remote Sensing of Environment.

McGookey, D. P., 1972, Cretaceous System, in W. W. Mallory, ed., Geologic atlas of the Rocky Mountains: Rocky Mountain Association of Geologists, p. 190-228.

Miall, A. P., 1984, Principles of sedimentary basin analysis: New York, Springer-Verlag, 490 p.

Miller, V. C., and C. F. Miller, 1961, Photogeology: New York, McGraw-Hill, 248 p.

Nomenclature Committee, 1956, Wyoming stratigraphy, part I: subsurface stratigraphy of the pre-Niobrara formations in Wyoming: Wyoming Geological Association, 98 p.

Palluconi, F. D., and G. R. Meeks, 1985, Thermal infrared multispectral scanner (TIMS): an investigator's guide to TIMS data: Jet Propulsion Laboratory Publication 85-32, 24 p.

Paylor, E. D., M. J. Abrams, J. E. Conel, A. B. Kahle, and H. R. Lang, 1985, Performance evaluation and geologic utility of Landsat-4 thematic mapper data: Jet Propulsion Laboratory Publication 85-66, 68 p.

Ragan, D. M., 1973, Structural geology: an introduction to geometrical techniques: New York, John Wiley, 208 p.

Salomonson, V. V., P. L. Smith, Jr., A. B. Park, W. C. Webb, and T. J. Lynch, 1980, An overview of progress in the design and implementation of Landsat-D systems: IEEE Transactions on Geoscience and Remote Sensing, v. GE-18, p. 137-145.

Soha, J. M., and A. A. Schwartz, 1978, Multispectral histogram normalization contrast enhancement: Proceedings of the Fifth Canadian Symposium on Remote Sensing, p. 86-93.

Vane, G., and A. F. H. Goetz, 1985, Proceedings of the airborne imaging spectrometer data analysis workshop: Jet Propulsion Laboratory Publication 85-41, 173 p.

Viljoen, R. P., M. J. Viljoen, J. Grootenboer, and T. G. Longshau, 1983, ERTS-1 imagery: an appraisal of applications in geology and mineral exploration, in K. Watson and R. D. Ragan, eds., Remote sensing: Society of Exploration Geophysicists Geophysics Reprint Series 3, p. 58-92.

Watson, K., 1985, Remote sensing—a geophysical perspective: Geophysics, v. 50, p. 2595-2610.

Williams, R. S., Jr., and W. D. Carter, eds., 1976, ERTS-1, a new window on our planet: USGS Professional Paper 929, 362 p.

Woodward, T. C., 1957, Geology of Deadman Butte area, Natrona County, Wyoming: AAPG Bulletin, v. 41, p. 212-262.

APPLICATION OF STRUCTURES MAPPED FROM LANDSAT IMAGERY TO EXPLORATION
FOR STRATIGRAPHIC TRAPS IN THE PARADOX BASIN*

by

Merin, I.S., and Michael, R.C.,
Earth Satellite Corporation
7222 47th Street
Chevy Chase, MD   20815

ABSTRACT

Significant quantities of petroleum occur in algal buildups
of Pennsylvanian age in the Paradox Basin. Isopach and lithofacies mapping by others suggest that low relief paleostructures
appear to have controlled Pennsylvanian seafloor topography and
thus the distribution of the buildups. Several workers have
reported that these paleostructures trend northwest and northeast. Therefore, the basin can be visualized as a mosaic of
fault blocks that were differentially active through geologic
time. The buildups are elongate northwest and their distribution and overall shape appear to be controlled by northwest-
trending paleostructures. Some larger buildups, i.e., Ismay,
show local northeast-trending thicks within a overall northwest-
trending buildup.

Examination of Landsat imagery revealed an extensive network of northwest and northeast-trending lineaments that parallel linear patterns apparent on aeromagnetic, gravity, and subsurface isopach data. Additionally, outcrops along selected
lineaments contain fractures that parallel these lineaments
which suggests that the lineaments are related to fundamental,
i.e., basement, fracture zones along which algal buildups may
have developed. Comparison of the fracture network to the
distribution of algal thickening reveals these buildups occur
predominantly along northwest-trending lineaments. Local disruptions within, and apparent terminations of, the buildups
correspond to cross-cutting northeast-trending linemanets. This
relationship provides guidance to locating prospective algal
buildups. Integration of these data with detailed subsurface
mapping can refine some leads into prospects. Several of these
features have been reported to have successfully been drilled.

INTRODUCTION:

Significant quantities of petroleum occur in algal buildups, i.e., carbonate mounds, of
Pennsylvanian age in the Paradox basin (Figure 1). Previous work has shown that the basin
can be visualized as a mosaic of basement fault blocks that were differentially active
through time. A number of workers have suggested that movement along these fault blocks
during Pennsylvanian time governed seafloor bathymetry and, thus, the distribution of the
carbonate mounds. Using an analysis of Landsat imagery, integrated with geophysical and
subsurface data, we have mapped structures that we believe mark recurrently active basement
fractures. Analysis of these data provides guidance for locating prospective algal
buildups.

---

* Presented at the International Symposium on Remote Sensing of Environment, Fourth Thematic
Conference: Remote Sensing for Exploration Geology, San Fancisco, April 1-4, 1985.

183

INTERPRETATION OF LANDSAT IMAGERY

Interpretation of Landsat imagery at a scale of 1:250,000 shows numerous folds, faults and lineaments to be present in this area (Figure 2). Although there are lineaments of a number of different orientations there appears to be two predominant trends: N38°-55°W and N23°-66°E and two minor sets at N0°-9°W and N85°-75°W (Figure 3). A number of workers have shown that numerous northeast and northwest-trending faults are present in this region and that these features have been reactivated through geologic time. These include: a northeast-trending Precambrian wrench-fault system (the Colorado lineament; Warner, 1978); numerous northeast-trending faults (Baars, 1966; Stevenson and Baars, 1977). Additionally, west to west-northwest to trending faults (e.g., the Verdue graben; Huff and Lesure, 1965) and north to north-northwest-trending faults are locally abundant in this region. The correlation of the lineament trends to the predominant trends of previously recognized faults encourages one to speculate that many of these lineaments are analogous structures and may be either extensions of previously mapped structures or newly discovered structures.

We selected about 35 lineaments mapped from Landsat imagery to examine in detail on the ground. All of these features consisted in part of linear segments of drainage, many of which were also aligned with linear scarps. We found no outcrop exposed along many of these features. Of those lineaments along which we did locate outcrop, in nearly all instances we observed measurable offset of strata on opposite sides of the lineament, or observed joints that parallel the lineament. This suggests that most of the lineaments we mapped are fracture zones (joints or faults).

Comparison of the lineament data with aeromagnetic and gravity data shows good correspondence. Many individual lineaments mapped on Landsat imagery correlate with individual alignments apparent on contoured aeromagnetic and gravity data. This implies that many of the lineaments mapped on Landsat imagery not only mark fractures at the surface, but also mark fundamental (i.e., basement) fractures.

Based on the above data we are confident that many of the lineaments mapped using Landsat imagery mark geologically significant zones of fractures that represent either extensions of previously mapped structures or mark previously unrecognized structures. To be used effectively in an exploration program, an analysis of Landsat imagery must be integrated with a good understanding of the tectonics and stratigraphy of a basin.

TECTONIC EVOLUTION OF THE PARADOX BASIN

The structures of the Paradox basin region are the result of intermittant tectonic activity extending from Precambrian time to the present. The late Paleozoic orogenic event that formed the Ancestral Rocky Mountains produced the Paradox basin and adjacent Uncompahgre Uplift. It appears that an ancient system of northeast and northwest-trending basement faults cut the Paradox basin into a series of panels or blocks and these faults actively influenced sedimentation patterns during the Paleozoic. Warner (1978) showed that a broad northeast-trending Precambrian wrench zone, i.e., the Colorado lineament, passes through this region. Hite (1975) suggested that this zone is composed of numerous faults some of which have indications of left-slip (Figure 4). Baars (1966), and Stevenson and Baars (1977) used isopach and facies maps to show that a series of fault blocks of small vertical relief were active from late Cambrian to Mississippian time. These features principally trend northwest and influenced seafloor bathymetry that in turn governed sedimentary facies and thickness patterns of Cambrian, Devonian and Mississippian rocks. Swabo and Wengerd (1977) extended the Mississippian data to show that a network of northwest-trending paleostructures extended across the region (Figure 5). Late Paleozoic orogenic activity reactivated these early Paleozoic structural zones, producing sedimentary facies and thickness patterns that mimic the orientation of these trends.

Pennsylvanian and early Permian deformation and subsidence in the Paradox basin region is part of the broader deformation of the Ancestral Rockies, an event related to plate interaction along the southern and eastern margins of North America (Kluth and Coney, 1981). Suturing of the southern margin of North America with other continental masses was the result of right-lateral coupling as South America moved westward relative to North America (Kluth and Coney, 1981; Dewey, 1982; Figure 6). The Ancestoral Rockies developed when that portion

184

of the craton located between the Cordilleran geosyncline and the Ouachita/Marathon collision zone, was wrenched and pushed northwestward as the collision progressed (Kluth and Coney, 1981). This produced foreland type deformation characterized by block faulted mountains (e.g., Uncompahgre Uplift, Ancestral Front Range) and rapidly subsiding fault bounded troughs (e.g., Paradox basin, Oquirrih basin). Apparently some of the foreland basins are the result of transtension, as data from the Oquirrih basin indicates (Jordan and Douglas, 1980). Data from the Paradox basin are sparse but, the large volume of marine sediments imply significant subsidence of the basin, which suggests transtension rather than tranpression.

Nature and Distribution of Carbonate Mounds

The carbonate mounds of the Paradox basin occur in the Paradox Formation (Figure 7), a complex cyclic unit consisting of carbonate rocks interbedded with black, dolomitic shale and gypsum that grades basinward to evaporite, interbedded with black, dolomitic shale. The carbonate mounds are lens-shaped buildups, principally composed of partially dolomitized carbonate mud with locally abundant oolitic and bioclastic debris (principally the phylloid algae Ivanovia). The typical reservoir consists of porous, partially dolomitized and leached algal and oolitic micrite, that may be grain supported in places. The mounds grade laterally and vertically into pellet rich micrite (Baars and Stevenson, 1982; Peterson and Hite, 1969; Peterson, 1966; Choquette, 1983).

Most of the individual mounds appear to be elongated in a northwest direction and many are abruptly terminated by northeast trends (Peterson and Hite, 1969; Ellias, 1963; Choquette, 1983). This is parallel to early Paleozoic structural trends, implying these trends were reactivated during the late Paleozoic and influenced the distribution of the mounds. Isopach and structure maps show that some mounds are located along northwest-trending paleostructures: areas of algal thickening correspond to structural highs and to thin areas of the overlying strata (Berghorn and Reid, 1981; Lehman, 1981). In places where drilling has been dense (e.g., the Ismay field, Cache field, Gothic Mesa field, Tocito Dome field), isopach data show that the individual thick "pods" within the field are aligned northeast, whereas the overall trend of the field is northwest (Figure 8: Ellias, 1963; Reid and Burghorn, 1981). This suggests that the northwest-trending paleostructures may govern the regional distribution and shape of the overall trend of the field whereas the northeast-trending structures play a more localized role, such as defining the shape of an individual mound.

Furthermore, it appears that bathymetry had a significant influence on the diagenetic processes that led to the development of reservoir porosity. The carbonate mounds were bathymetrically and structurally higher than the surrounding seafloor and were deposited in subtidal environments that underwent periodic episodes of intertidal to supertidal exposure (e.g., mounds contain carbonates marked by desiccation cracks; Martin, 1981; Krivanek, 1981). This resulted in the leaching of aragonite grains (e.g., oolites, Ivonovia debris) as well as the lithification and local dolomitization of the largely micritic mounds during periods of exposure when meteoric water migrated through the mounds. Moldic porosity developed in places where micrite (calcite) coated aragonite grains were leached by meteoric water. Local dolomitization of carbonate mud produced additional porosity in places. The result is a reservoir quality partially dolomitized, porous and permeable algal or oolitic micrite in which the porosity principally consists of molds of leached grains in a matrix of micrite. The overall pattern is that of deposition of a carbonate facies along structurally positive features in a restricted marine basin, followed by subaerial exposure that produced dolomitization and selective leaching to yield a highly porous and permeable reservoir rock.

Exploration Methodology

Comparisons of the fracture network derived from Landsat imagery (i.e., lineaments) to the distribution of algal buildups reveals that many of the buildups occur along regionally extensive northwest-trending lineaments. The new Tricentral discoveries (R25E, T37-38S, Utah) in addition to many other fields such as Ismay, Bug, Papoose Canyon and Gothic Mesa are all located along regionally extensive northwest-trending lineaments in places where these structures are cut by northeast-trending lineaments (Figure 9). Local disruptions within, and apparent terminations of, these mounds in many places correspond to cross-cutting, northeast-trending lineaments. For example, note the Ismay field lies along a zone

185

507

of regionally extensive northwest-trending lineaments and the individual algal buildups in the field parallel northeast-trending lineaments that cross the field (Figure 10). This implies that regionally extensive northwest-trending structural zones appear to govern the basinwide distribution of the mounds, whereas the northeast-trending zones appear to govern the shape and local limits of individual buildups.

Structures mapped from Landsat imagery are a valuable adjunct to subsurface or geophysical data in an exploration program for algal mounds in the Paradox basin. In areas of sparse subsurface control features mapped from imagery may be used to extrapolate trends of previously recognized structures. Similarly, structures that have gone unrecognized because of the paucity of geophyscial and subsurface data in an area may be recognized on the imagery as analogs to well documented features elsewhere. Field investigations in the Paradox basin have shown a consistent correlation of lineaments with zones of closely spaced joints and, in places, faults. The imagery shows fracture zones as areally discrete phenomena, unlike contoured subsurface data where the location of faults are inferred, except where a drill hole actually penetrated a fault. Our experience is that Landsat imagery reveals faults that, in many places, have offsets too small to be detected seismically. This ability to map faults with a small offset may be particularly valuable in the exploration for algal mounds because small, fault-related, changes in seafloor bathymetry can produce enough local relief to support development of algal bioherms.

ACKNOWLEDGEMENT

The manuscript benefited from thorough and critical reviews by J.R. Everett and O.R. Russell. We thank Dana Strouse, Heidi Schweikart and Dorothy McCabe for preparation of maps and illustrations and Earth Satellite Corporation for permission to publish this work.

REFERENCES

Baars, D.L., 1966, Pre-Pennsylvanian paleotectonics - key to basin evaluation and petroleum occurrences in Paradox basin, Utah and Colorado: American Association of Petroleum Geologists Bulletin, v. 50, p. 2080-2111.

Baars, D.L., and Stevenson, G.M., 1982, Subtle Stratigraphic traps in Paleozoic rocks of Paradox basin, in The Deliberate Search for the Subtle Trap, American Association of Petroleum Geologists Memoir 32 p. 131-158.

Choquette, P.W., 1983, Platy Algal reef mounds Paradox basin, in Scholle, P.A., et al., eds., Carbonate Depositional Environments, American Association of Petroleum Geologists Memoir 33, p. 454-462.

Dewey, J.F., 1982, Plate tectonics and the evolution of the British Isles: Journal of the Geological Society of London, v. 139, p. 371-412.

Elias, G.K., 1963, Habitat of Pennsylvanian algal bioherms, Four Corners Area, in Shelf carbonates of the Paradox basin: A Symposium Fourth Field Conference 1963, Four Corners Geological Society, p. 185-202.

Hite, R.J., 1975, An unusual northeast-trending fracture zone and its relations to basement wrench faulting in northern Paradox basin, Utah and Colorado: Canyonlands Country 8th Field Conference Guidebook, Four Corners Geological Society, p. 217-233.

Huff, L.C., and Lesure, F.G., 1965, Geology and Uranium deposits of Montezuma Canyon area, San Juan County: U.S. Geological Survey Bulletin 1190, 102 p.

Jordan, T.E., and Douglas, R.C., 1980, Paleogeography and structural development of the Late Pennsylvanian to Early Permian Oquirrh basin, northwestern Utah, in Fouch, T.D., and Magathan, E.R., eds., Paleozoic paleogeography of the west-central United States, Rocky Mountain Paleogeography Symposium 1: The Rocky Mountain Section Society Economic Paleontologists and Mineralogists, p. 217-238.

186

Kluth, C.F., and Coney, P.J., 1981, Plate tectonics of the Ancestral Rocky Mountains: Geology, v. 9, p. 10-15.

Krivanek, C.M., 1981, New fields and exploration drilling, Paradox basin, Utah and Colorado, in Wiegard, D.L., ed., Geology of the Paradox Basin: Rocky Mountain Association of Geologists, p. 77-83.

Lehman, D.D., 1981, Productive Pennsylvanian carbonate mounds southeast of the Aneth area, Utah, in Wiegand, D.L., ed., Geology of the Paradox basin: Rocky Mountain Association of Geologists, p. 83-88.

Martin, G.W., 1981, Patterson field San Juan County, Utah, in Wiegard, D.L., ed., Geology of the Paradox basin: Rocky Mountain Association of Geologists, p. 61-70.

Muehlberger, W.R., 1980, Texas lineament revisited, New Mexico Geological Society Guidebook, Trans-Pecos Region, p. 113-121.

Peterson, J.A., 1966, Stratigraphic versus structural controls on carbonate-mound hydrocarbon accumulation, Aneth area, Paradox basin: American Association of Petroleum Geologists Bulletin, v. 50, n. 10, p. 2068-2081.

Peterson, J.A., Hite, R.J., 1969, Pennsylvanian evaporite-carbonate cycles and their relation to petroleum occurrence, Southern Rocky Mountains: American Association of Petroleum Geologists Bulletin, v. 53, n. 4, p. 884-908.

Reid, F.S., and Berghorn, C.B., 1981, Facies recognition and hydrocarbon potential of the Pennsylvanian Paradox Formation, in Geology of the Paradox Basin, Wiegard, D.L., ed., Rocky Mountain Association of Geologists, p. 111-118.

Stevenson, G.M., and Baars, D.L., 1977, Pre-carboniferous paleotectonics of the San Juan Basin, in San Juan Basin III: New Mexico Geological Society, 28th Guidebook, p. 99-110.

Szabo, E., and Wengerd, S.A., 1975, Stratigraphy and tectogenesis of the Paradox basin, in Canyonlands Country 8th Field Conference Guidebook, Four Corners Geological Society, p. 193-210.

Warner, L.A., 1978, Colorado Lineament: A middle Precambrian wrench fault system: Geological Society of America Bulletin, v. 89, p. 161-171.

Figure 1

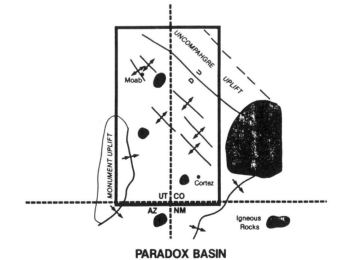

PARADOX BASIN

187

509

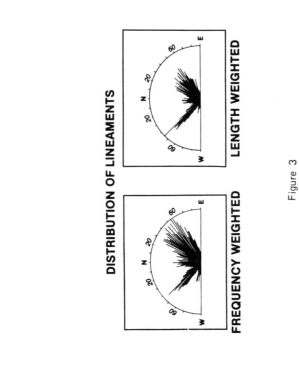

DISTRIBUTION OF LINEAMENTS

FREQUENCY WEIGHTED

LENGTH WEIGHTED

Figure 3

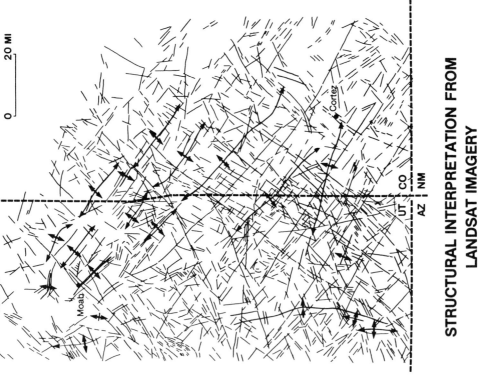

STRUCTURAL INTERPRETATION FROM
LANDSAT IMAGERY

Figure 2

20 MI

0

Moab

Cortez

UT   CO
AZ   NM

188

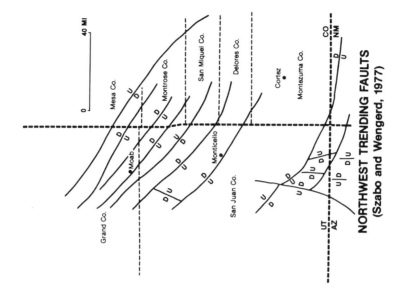

**NORTHWEST TRENDING FAULTS
(Szabo and Wengerd, 1977)**

Figure 5

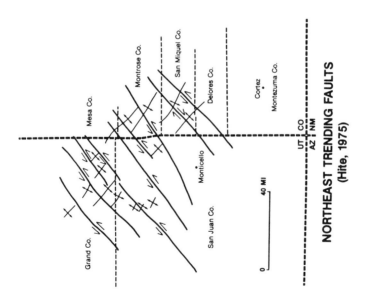

**NORTHEAST TRENDING FAULTS
(Hite, 1975)**

Figure 4

189

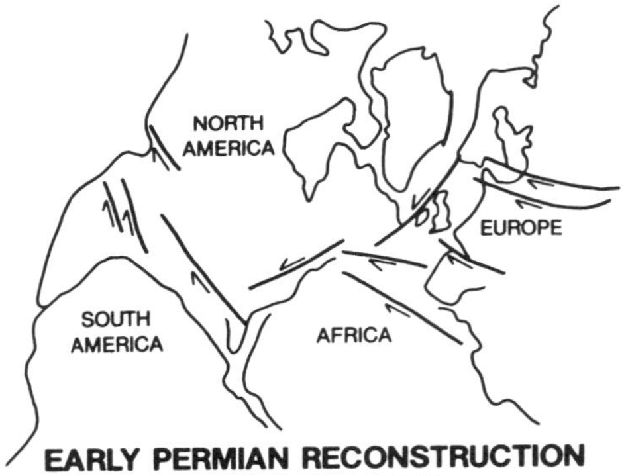

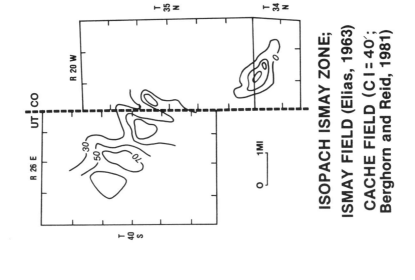

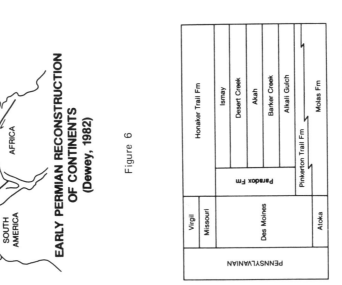

EARLY PERMIAN RECONSTRUCTION OF CONTINENTS (Dewey, 1982)

Figure 6

PENNSYLVANIAN	Virgil	Honaker Trail Fm	
	Missouri		
	Des Moines	Paradox Fm	Ismay
			Desert Creek
			Akah
			Barker Creek
			Alkali Gulch
		Pinkerton Trail Fm	
	Atoka	Molas Fm	

PENNSYLVANIAN STRATA OF PARADOX BASIN

Figure 7

ISOPACH ISMAY ZONE; ISMAY FIELD (Elias, 1963) CACHE FIELD (C I = 40′; Berghorn and Reid, 1981)

Figure 8a

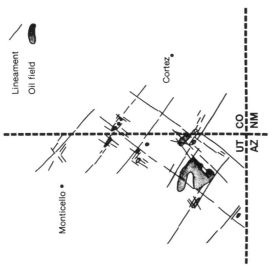

REGIONALLY EXTENSIVE LINEAMENTS AND OIL FIELDS

Figure 9

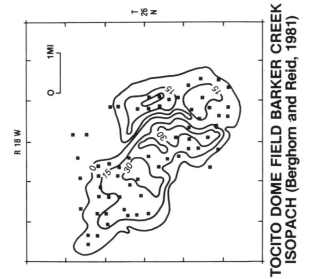

TOCITO DOME FIELD BARKER CREEK ISOPACH (Berghorn and Reid, 1981)

Figure 8b

191

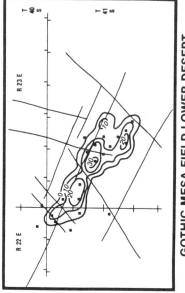

GOTHIC MESA FIELD LOWER DESERT CREEK ISOPACH (Berghorn and Reid, 1981) LINEAMENTS FROM LANDSAT IMAGERY

Figure 10b

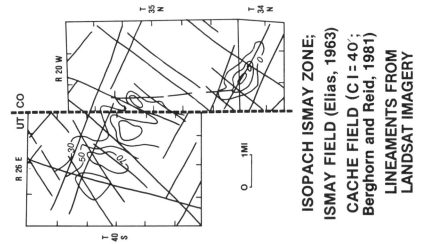

ISOPACH ISMAY ZONE; ISMAY FIELD (Elias, 1963)

CACHE FIELD (C I = 40'; Berghorn and Reid, 1981)

LINEAMENTS FROM LANDSAT IMAGERY

Figure 10a

The American Association of Petroleum Geologists
V. 68, No. 11 (November 1984), P. 1718-1731, 13 Figs.

Tectonic Framework of Powder River Basin, Wyoming and Montana, Interpreted from Landsat Imagery[1]

RONALD W. MARRS[2] and GARY L. RAINES[3]

ABSTRACT

Linear features in the Powder River basin, Wyoming and Montana, were interpreted from Landsat images and analyzed to define major lineaments. Lineaments identified include several that trend northwest and a prominent set that trends northeast. These lineaments represent broad (5-10 km or 3- 6 mi) linear zones where smaller, parallel and subparallel linear features are concentrated at the surface. The smaller linear features are interpreted as possible expressions of joints, fractures, folds, or lithologic boundaries produced by periodic readjustment along basement-block boundaries.

The lineaments discussed in this report were visually interpreted from maps of linear features and contour maps showing concentrations of linear features. Lineaments were subsequently compared to mapped structures, outcrop patterns, geophysical data, and isopach maps to assess their geologic significance. Correlations of these lineaments with mapped structures, geophysical gradients, or facies changes strongly support the interpretation that they represent the surface expression of boundaries of crustal blocks that have been periodically reactivated through time. The northeast and northwest patterns provide evidence that a systematic, rectilinear pattern of crustal blocks formed early in the earth's history and has largely controlled subsequent adjustments of the earth's crust.

These findings suggest the potential for depositional control of sedimentary units by structural adjustments between basement blocks and, thus, lead to the conclusion that lineaments may be used as guides in petroleum and mineral exploration if favorable source and host rocks are present.

INTRODUCTION

With the recent popularity of the plate-tectonics theory, continent-size structural elements and zones of dislocation in the deep crustal rocks have become increasingly important. The earth's crust has been deformed periodically through time, and major zones of crustal movement commonly are reactivated each time the crust is stressed (Weimer, 1980). These major zones of crustal weakness may even react to diurnal stresses (such as earth tides) and,

therefore, they may remain active during periods of tectonic quiescence. The exposed surface rocks, regardless of age, should reflect this continued adjustment in the relative abundance of faults, folds, fractures, and joints along these major zones of weakness.

The launch of Landsat-1 in 1972 provided a synoptic view of the earth's surface, resulting in new applications for geologic mapping and exploration. One of the first applications of the satellite images was the identification of linear features representing faults on other geologic trends (Hodgson et al, 1976). A major problem, however, in using the Landsat interpretations in structural or tectonic analysis is that the linear features that may be directly correlated with known geologic structure are of relatively little interest because they provide no new geologic information; linear features that cannot be related to known structures are viewed with suspicion. Interpreters generally are not able to distinguish geologically significant features (Hodgson et al, 1976).

The abundance of linear features that may be identified on each satellite image can make field checking or feature-by-feature correlation with other data an impossible task. A more efficient method for evaluating linear features is essential to make effective use of them in geologic studies. Techniques have been pioneered by Levandowski et al (1976) and Sawatzky and Raines (1977, 1981), and applied by Raines (1978), Raines et al (1978), and Turner et al (1982). Sawatzky and Raines' (1981) technique involves a statistical analysis that defines major trends and concentrations of linear features. These can be interpreted within a geologic framework without validating each individual linear feature. The use of this method and its potential for tectonic analysis of large regions are illustrated by earlier work in the Powder River basin, Wyoming and Montana (Raines et al, 1978), and in Sonora, Mexico (Raines, 1978; Turner et al, 1982). The critical concept derived from these analyses is that many deep-seated zones of tectonic adjustment correlate with concentrations of surface linear features that result either from recurrent movement of crustal blocks or from differential compaction along block boundaries. Statistical analysis of linear features interpreted from satellite images serves as an efficient way to detect such zones.

The objective of this study was to use the synoptic view of Landsat to detect linear features, and then to examine the features statistically to define major lineaments that may represent tectonic elements in the Powder River basin. This report presents an interpretation from imagery of the tectonic framework of the Powder River basin that may serve as an exploration guide for petroleum and minerals and as a model of the tectonic development of the Wyoming basins province.

---

[1]Manuscript received, August 17, 1983; accepted, April 3, 1984.
[2]University of Wyoming, Dept. of Geology and Geophysics, Laramie, Wyoming 82701.
[3]U.S. Geological Survey, P.O. Box 25046, Denver Federal Center, Denver, Colorado 80225.

GEOLOGIC SETTING

The Powder River basin, Wyoming and Montana (Figure 1), is a large asymmetric syncline with the west limb dipping steeply away from the Bighorn front. It is bounded on the west by the Bighorn Mountains and the Casper arch. The southern and southeastern boundaries of the Powder River basin are formed by the northern terminus of the Laramie Range (at Casper Mountain) and the Hartville uplift; the Black Hills and associated folds, extending northward, form the eastern boundary. The Powder River basin is bounded on the northeast by the Miles City arch, but the major northwest-trending structural low continues into the adjoining Ashland and Bull Mountain synclines.

Depositional patterns of gravels in the Eocene Kingsbury Conglomerate Member and in the Moncrief Member of the Wasatch Formation along the western flank of the Powder River basin (Sharp, 1948) suggest that the basin acquired much of its present structural configuration during Late Cretaceous and Tertiary orogenic episodes (Blackstone, 1949). Although the overall structural pattern of the Powder River basin appears simple relative to the smaller intermontane basins of Colorado, Wyoming, and Montana, complex structures are present along the flanks of bounding uplifts (Cloos and Cloos, 1934; Wilson, 1938; Hose, 1955; Mapel, 1959; Weimer, 1961; Hoppin and Jennings, 1971; Wyoming Geological Association, 1971; Hoppin et al, 1973; Blackstone, 1981). Structures of lesser magnitude are in evidence throughout the basin (Osterwald and Dean, 1961). These subtle structures have strongly influenced depositional patterns (Davis, 1969; Webb, 1969; Shurr, 1979; Weimer, 1980; Barlow and Doelgen, 1981; Flores, 1981; Slack, 1981); so we chose first to search for tectonic elements that might control the structures and then to relate these elements to the regional geology and depositional patterns.

IDENTIFICATION OF LINEAMENTS

Much confusion has arisen regarding the terms "linear," "linear feature," and "lineament." Hoppin (1974) provided a summary of usage and problems. He suggested that the term linear or linear feature be used to designate "... single rectilinear elements commonly, but not necessarily, of structural origin." He further stated that "Lineaments are generally rectilinear lines or zones of structural discordance of regional (100 km or longer) extent." We have attempted to adhere to this distinction, and we believe that the linear concentrations of linear features, which we will call lineaments, do indeed represent regional zones of structural discordance.

To define these lineaments, six Landsat scenes of the Powder River basin were computer-enhanced and interpreted for surface linear features (Figure 1). The linear features were then digitized for statistical analysis of orientations and trends. Contour maps were constructed to reveal major concentrations of linear features (Figures 2, 3). Trend intervals to be contoured were selected on the basis of the dominant orientations of linear features as revealed by the histogram of length-weighted azimuths

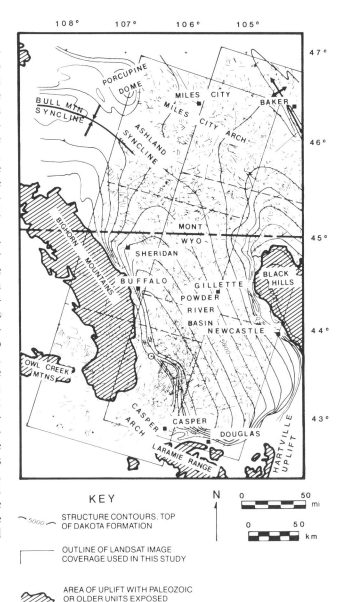

Figure 1—Structure contour map of Powder River basin, Wyoming and Montana (after Waring, 1976; Renfro and Feray, 1972), showing surrounding uplifts, area of Landsat coverage used in study, and linear features interpreted from six Landsat images.

and by statistical analysis of the trend/frequency data represented by the histogram. It is apparent that the trends of the linear features fall generally within two populations, one trending N12°-53°W (307°-348° azimuth), the other trending N30°-85°E. Contour maps constructed for these two intervals (Figures 2, 3) and the subsequent interpretation of contour trends reveal major northwest- and northeast-trending lineaments. The structural analysis is made by combining the interpretation of contour maps into a single map of lineaments and comparing the resulting pattern to known structures and geophysical data.

The contour maps were visually interpreted to identify contiguous highs and directional trends in the contour patterns. The criteria used in identifying lineaments on the contour maps are as follows.

517

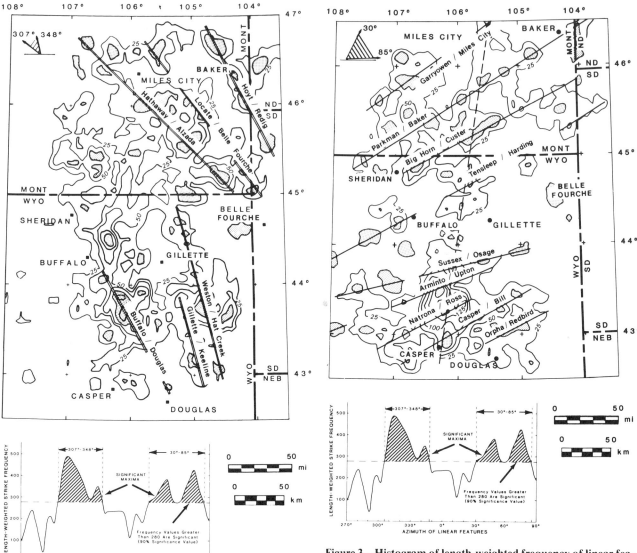

Figure 2—Histogram of length-weighted frequency of linear features in each 1° azimuth interval and contour map of linear features in azimuth interval 307°-348°. Potential lineaments in 307°-348° azimuth interval (solid lines) are interpreted from alignments of contour highs (stippled areas).

Figure 3—Histogram of length-weighted frequency of linear features in each 1° azimuth interval and contour map of linear features in azimuth interval 30°-85°. Potential lineaments (solid lines) in that same interval are interpreted from alignments of contour highs. Dashed trend connects strong contour highs (stippled areas); however, it is not within azimuth interval of features contour, so it is not designated a potential lineament. Lineaments are considered valid only if they fall within one of the intervals identified as statistically significant (cross-hatched areas) in previous stage of analysis.

1. Elongate contour highs and rows of highs paralleling the contoured azimuth trends were interpreted as prime indicators of potential lineaments. These were marked as potential lineaments on each contour map.

2. Only those potential lineaments having azimuth within the angular interval contoured were considered valid. For example, on the contour map of linear features with azimuth interval 307°-348°, only lineaments trending between 307° and 348° were retained in the final interpretation. This limitation is based on the consideration that adjustment along deep-seated zones of dislocation should produce some surface fractures and other structures that roughly parallel the deeper trend (Moody and Hill, 1956; Hobbs, 1976).

3. Lineaments were drawn as lines giving the best fit to the crests of contour highs and honoring, as much as possible, major features and trends observed on the linear-feature map (Figures 2, 3).

Each lineament was then evaluated as a possible tectonic element. It was recognized, also, that such lineaments might represent surface trends that reflect cultural patterns or changes in the physical character of the exposed rocks. The nature of exposed rock units strongly influences the number of detectable joints and fractures, and controls the formation of erosional patterns that may be produced by wind or other surface agents. Major lineaments were assigned names indicating their approximate locations with respect to settlements or distinctive physiographic features.

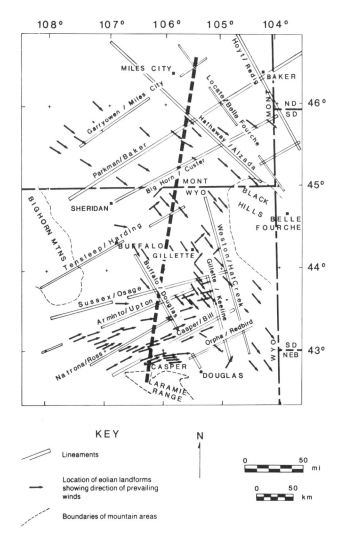

KEY   N

⬥ Lineaments

→ Location of eolian landforms
  showing direction of prevailing
  winds

––– Boundaries of mountain areas

0   50
         mi

0   50
         km

Figure 4—Map showing orientation and location of longitudinal eolian landforms and their relationship to potential lineaments. Eolian features were identified from high-altitude aerial photography (1:125,000 scale). Dominant wind flow parallels trend of these eolian features.

In making an evaluation of each major lineament, it is essential to consider the possible sources of bias or error produced by the processing or analysis procedure. For example, Figure 3 shows a prominent alignment of contour highs trending north-northeast across the central part of the basin (dashed line). Its trend is not within the statistically defined azimuth window, but it is a prominent feature on the contour map. By the interpretation criteria used, this dashed line would not be selected as a lineament. Its validity is further suspect because it corresponds generally to the zone of overlap between the two Landsat image strips from which the linear features were interpreted (Figure 1). The concentration of linear features represented by this zone is attributed to the fact that each image strip was interpreted separately and the two interpretations were combined. The zone of overlap received double attention, hence more features are identified in the overlap zone. This trend must, therefore, be considered an artifact produced by the analysis.

NORTHWEST-TRENDING LINEAMENTS

A pattern of six, somewhat regularly spaced, northwest-trending lineaments is apparent in Figure 2 (Hoyt/Redig, Locate/Belle Fourche, Hathaway/Alzada, Weston/Hat Creek, Gillette/Keeline, and Buffalo/Douglas). In the northern Powder River basin, these lineaments trend N30°-45°W, are approximately parallel, and are spaced about 50 km (30 mi) apart. In the southern part of the basin the lineaments trend N20°-30°W. Northwest-trending lineaments represent 35% of the length-weighted linear features and constitute the dominant trend as indicated by the histogram (Figure 2). However, in evaluating these lineaments, it may be particularly important to consider the possible effects of strong northwest winds and a pronounced northwest-trending drainage pattern evident in the northern part of the Powder River basin. In an effort to evaluate the possibility for wind erosion enhancing the northwest-trending linear features, a map of eolian landforms (Figure 4) was compiled, using both high-altitude aerial photographs, and techniques described by Marrs and Gaylord (1978). Direct comparison of the eolian landforms map with the maps of northwest-trending linear elements (Figure 2) reveals a striking parallelism between eolian landforms and lineaments in the northern Powder River basin. The trends of landforms and the prevailing wind-flow patterns are so nearly coincident with the trends of lineaments that it seems likely that wind erosion has accentuated the expression of these northwest-trending lineaments. However, regular spacing of the lineaments and the corresponding northwest trends in gravity data (Figure 5), magnetic data (Figure 6), petroleum production (Figure 8), and isopach data (Figures 9, 10, 11) in this same area cannot be explained solely as a function of surface topography and geomorphology.

NORTHEAST-TRENDING LINEAMENTS

Nine lineaments trend northeastward across the Powder River basin and are spaced roughly 55 km (35 mi) apart (Figure 3). On the histogram (Figure 3) these lineaments are represented by a broad (55°) interval of linear features totaling 34% of the mapped features. The lineaments were named as follows: (1) Garryowen/Miles City, (2) Parkman/Baker, (3) Big Horn/Custer, (4) Tensleep/Harding, (5) Sussex/Osage, (6) Arminto/Upton, (7) Natrona/Ross, (8) Casper/Bill, and (9) Orpha/Redbird. Although wind-direction indicators (Figure 4) show a strong east-northeast component (N65°-80°E) in the southern Powder River basin, their dominant direction is not exactly parallel to the lineaments, most of which trend approximately N60°-65°E. Northwesterly wind-direction indicators are dominant in the northern Powder River basin (Marrs and Gaylord, 1978); but both northwest- and northeast-trending lineaments are present in the northern reaches of the basin. The lineaments show no marked influence of a gradually changing wind-flow direction, so it appears that the wind plays only a secondary role in emphasizing the surface expression of lineaments in the Powder River basin.

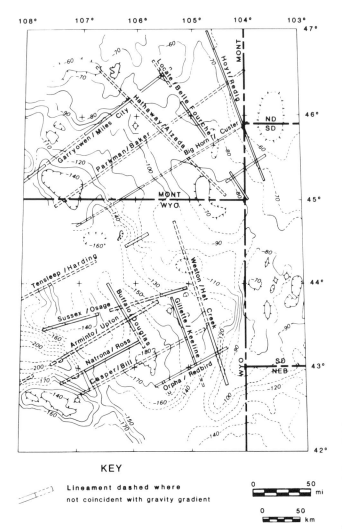

KEY

Lineament dashed where
not coincident with gravity gradient

0   50 mi

0   50 km

Figure 5—Simple Bouguer gravity map of Powder River basin, showing lineament trends (outlined with solid lines) that correlate with gravity gradients; lineaments that have no expression in the gravity data are outlined with dashed lines. Dashed gravity contours occur in areas of widely spaced gravity data. Hachures indicate closed gravity highs (outside hachures) and closed gravity lows (inside hachures).

CORRELATION OF LINEAMENTS WITH STRUCTURE AND GEOPHYSICAL DATA

If the lineaments defined through the analysis of Landsat images represent deep-seated structures, they should be in evidence in various aspects of the basin geology as depicted on structural and geophysical maps. They should also project into the surrounding uplifts where they might be expressed as faults, shear zones, or fold belts. If one assumes that aligned groups of linear features (lineaments) represent zones of Precambrian tectonic adjustment that formed in Precambrian time and that these zones have been periodically reactivated during later periods of crustal movement, then fractures should be expected where lineaments intersect the competent older rocks cropping out in the uplifts flanking the basin. Later intrusions might also take advantage of these zones of crustal weakness in making their way toward the surface. Conse-

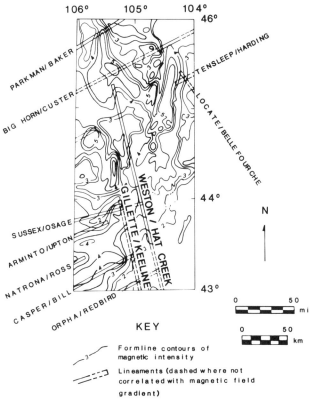

KEY

Formline contours of
magnetic intensity

Lineaments (dashed where not
correlated with magnetic field
gradient)

Figure 6—Formline map of total-intensity magnetic field, eastern Powder River basin, showing correlation of lineaments with regional magnetic field. Magnetic data are available only for eastern Powder River basin. Data are from U.S. Geological Survey unpublished maps.

quently, intrusive bodies might be concentrated along lineaments, particularly where they intersect other major structures. If these zones were periodically reactivated, such that they affected topography during periods of sedimentation, contrasts in depositional pattern should occur in the sedimentary section vertically stacked in a zone parallel to lineaments. The zones of the tectonic adjustment might also be indicated by folding, fracturing, or thinning of various sedimentary units (Tanner, 1962; Shurr, 1979). The evidence of contrast across each major lineament might be found in surface outcrop, in well data, and/or in geophysical data. The contrasts could also be reflected in the localization of petroleum or mineral deposits, the distribution of which should be strongly influenced by periodic readjustments occurring along each major tectonic zone (Weimer, 1980; Slack, 1981). Relevant models for these types of structures are reported by Tweto and Sims (1963) and Shoemaker et al (1978).

With these ideas in mind, each of the lineaments was systematically evaluated by correlation with bedrock and surficial geology maps (Figure 12), gravity and magnetic data (Figures 5, 6), isopach maps (Figures 9, 10, 11), maps of petroleum production (Figure 8), geologic structure maps (Figures 1, 7), and cross sections. In making such correlations, it is necessary to keep in mind the possibility that the major lineaments may be diffuse in nature (Tweto and Sims, 1963; Shoemaker et al, 1978). That is, a sharp break

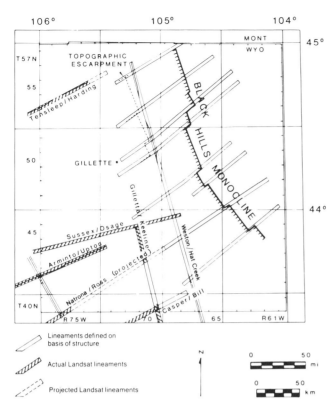

Figure 7—Map showing structurally-defined lineaments (Slack, 1981) and their relation to Landsat-defined lineaments along western flank of the Black Hills uplift.

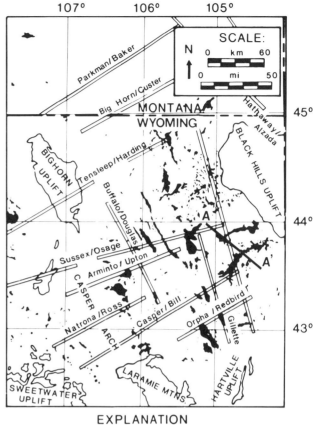

EXPLANATION

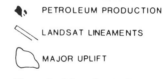

PETROLEUM PRODUCTION

LANDSAT LINEAMENTS

MAJOR UPLIFT

Figure 8—Map of petroleum production in Powder River basin, showing correlation of production trends with trends of Landsat-defined lineaments. Data on producing areas from unpublished maps, Wyoming Geological Survey.

in the Precambrian basement rocks could be represented by a rather broad zone (several kilometers wide) of structures and contrasting lithologic characteristics in the younger sediments and at the modern surface. If the basement features are broad, as described by Tweto and Sims (1963), then the structures at the surface could be even broader. In either case, the tectonic adjustment would probably diffuse upward into the overlying blanket of sedimentary rocks forming a broad zone that affects differing lithologic characteristics and structures in different ways (Tanner, 1962; Shoemaker et al, 1978) as the adjustment propagates upward. Consequently, the lineaments defined by this analysis, with widths on the order of 5-15 km (3-9 mi), must be considered zones rather than lines.

Evaluation of Northeast-Trending Lineaments

Northeast-trending lineaments commonly are associated with major tectonic elements throughout the Rocky Mountain area. Work in northern Mexico (Raines, 1978; Turner et al, 1982) related a strong pattern of northeast-trending lineaments to structure and copper mineralization in that area. The Colorado mineral belt trends northeastward and repeatedly shows the influence of northeast-trending tectonic elements (Lovering and Goddard, 1950). Hills and Houston (1979) related a northeast-trending lineament in south-central Wyoming to the Mullen Creek/Nash Fork shear zone, which is interpreted as a through-going structure of major proportions (7 km or 4.35 mi wide and several hundred kilometers long).

Thomas (1971) summarized the tectonics of southwest Wyoming as a set of northeast- and northwest-trending plates bounded by lineaments that respond to compressional forces by combined action of pure shear and plate coupling. Slack (1981) correlated both structure and petroleum-production trends in the eastern Powder River basin with northeast-trending linear features. Slack (1981, p. 730) commented further regarding the nature of the structural lineaments that he found trending northeastward in the area of the Belle Fourche arch:

Underlying basement zones of weakness are thought to be shear zones of Precambrian age analogous to the Mullen Creek/Nash Fork shear zone of southeast Wyoming. Stratigraphic evidence suggests that the structural lineaments which form the Belle Fourche arch have rejuvenated periodically throughout the Phanerozoic. Subtle movements along lineaments have affected depositional environments and hydrocarbon accumulation in virtually all significant reservoirs in the northern two-thirds of the basin.

Several known structures and major rock-domain boundaries are associated with the northeast-trending lineaments defined in this study (Figure 12). Chief among these are the following: the correlation of the Big Horn/Custer lineament with the trend of the Kearny Creek fault complex; the Tensleep/Harding lineament with the Little Missouri fault zone; the Arminto/Upton lineament with the juncture of the Casper arch and Bighorn Mountains and the trend of Tertiary intrusive bodies in the Black Hills; the Orpha/Redbird lineament with the Long Mountain fault zone; and the coincidence of the Natrona/Ross and Orpha/Redbird lineament with the boundaries of the deepest part of the Powder River basin. However, the northeast-trending lineaments are transverse to the main structural grain of the Powder River basin and the trend of most basin-flank structures. Strong evidence for the tectonic significance of the Powder River basin lineaments is also found to correlate with gradients in the geophysical data (Figures 5, 6) and in trends observed in petroleum-producing units in northeast Wyoming (Figure 8). Magnetic data are available only for the eastern Powder River basin, but comparison of the magnetic formline map (Figure 6) with the northeast-trending pattern of lineaments shows that the dominant trend of magnetic gradients in the southern part of the basin parallels the five northeast-trending lineaments in that region (Orpha/Redbird, Casper/Bill, Natrona/Ross, Arminto/Upton, and Sussex/Osage). The gravity map (Figure 5) reveals trends corresponding, in part, to lineaments in both the southern and northern parts of the basin. That one part of a lineament may parallel a strong gravity gradient, whereas another segment of the same lineament shows no expression in the gravity contours, suggests that the basement relief or lithologic character may vary considerably along the trend of the lineament. Consequently, many of the mapped lineaments parallel local changes in gradient and, at places, cut across steep gradients. Petroleum-production trends are evident on Figure 8 and were convincingly related to northeast-trending lineaments by Barlow and Doelger (1981) and Slack (1981) (Figure 7). Trends identified by Slack correspond closely to the Natrona/Ross, Arminto/Upton, and Tensleep/Harding lineaments identified from the Landsat image analysis. Slack's (1981) lineaments were defined on the basis of known structural offsets and correlated with petroleum-production and reservoir-rock patterns (Figures 7, 8). Although Slack showed evidence of additional lineaments not defined in this analysis, little doubt exists that these trends, defined on the basis of different data sets, are intimately related. Patterns shown in Figure 8 suggest that other lineaments mapped from Landsat (Buffalo/Douglas and Gillette/Keeline) may also correlate with petroleum distribution.

Evaluation of Northwest-Trending Lineaments

Northwest-trending lineaments (Figure 2) are spaced approximately 50 km (30 mi) apart and trend N20°-45°W. The possible influence of strong northwest winds (Figure

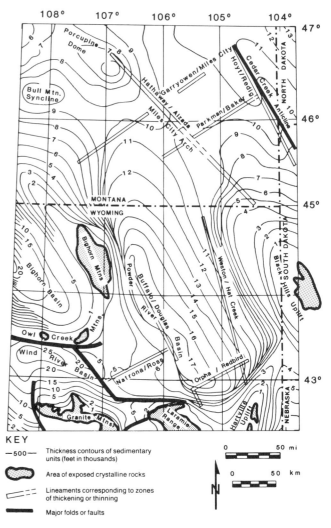

KEY

—500—   Thickness contours of sedimentary units (feet in thousands)

      Area of exposed crystalline rocks

‑ ‑ ‑   Lineaments corresponding to zones of thickening or thinning

━━━   Major folds or faults

0   50 mi

0   50 km

N

Figure 9—Isopach map showing total thickness of Phanerozoic rocks in Powder River basin and surrounding areas (after Rocky Mountain Association of Geologists, 1972, p. 56). Lineaments corresponding to zones of thickening or thinning are superimposed on isopach map.

4) on the formation of northwest-trending lineaments in the northern Powder River basin must be considered, yet correlation with structural data is also strong. For example, the Hoyt/Redig lineament parallels a set of northwest-trending folds associated with the Cedar Creek anticline (Figure 12). These folds extend northward into Montana from the Black Hills uplift and exhibit definite evidence of movement early in post-Ordovician time. The gravity data indicate that the basin falls off sharply to the southwest across this lineament. Other northwest-trending folds parallel the Locate/Belle Fourche and Hathaway/Alzada lineaments farther to the west (cf. Fig-

Figure 10—Isopach maps showing relationships between thickness of various sedimentary sequences and lineament defined from Landsat. Figure 10A shows Ordovician units thinning to south in vicinity of the Sussex/Osage and Tensleep lineaments. Figure 10B reveals a strong positive ridge of Devonian age along Hoyt Redig trend and southeastward thinning across Powder River basin. Figure 10C reveals a complex, blocky depositional pattern for Pennsylvanian sediments. Block boundaries commonly correlate with lineaments. Figure 10D shows a block-type depositional pattern for Permian units; greatest thickness of sediments is concentrated in a lineament-bounded block in southern Powder River basin. (Isopach data from Rocky Mountain Association of Geologists, 1972, p. 56, 81, 94, and 116.)

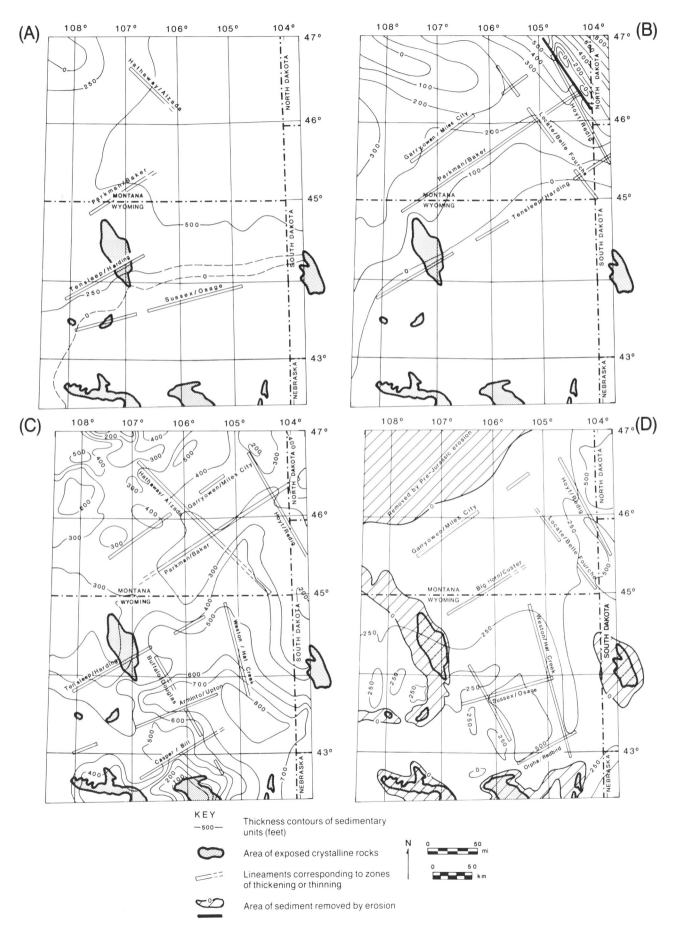

KEY

—500—   Thickness contours of sedimentary units (feet)

Area of exposed crystalline rocks

Lineaments corresponding to zones of thickening or thinning

Area of sediment removed by erosion

ures 2, 12). Drainage throughout the area parallels the Locate/Belle Fourche and Hathaway/Alzada lineaments and the mapped structures. Parallel gradients are detectable in the magnetic data (Figure 6), but lack of coincident gravity gradients indicates that the vertical expression in basement rocks is neither large nor persistent. The gravity map (Figure 5) shows shallow gradients in a northwest-trending pattern. Only the Hoyt/Redig trend correlates with a strong gravity gradient.

The Weston/Hat Creek, Gillette/Keeline, and Buffalo/Douglas lineaments trend N20°-25°W, paralleling the axis of the Powder River basin (Figure 12). The Buffalo/Douglas lineament is, in fact, coincident with the deep basin axis in the southern part of the basin. As in other lineaments, the geophysical data (Figures 5, 6) reveal basement trends paralleling these lineaments, but only a few segments of the lineaments are coincident with steep gradients. It is apparent from this relationship that the basement contrast across these lineaments is irregular and commonly insignificant. The most convincing evidence for the structural and stratigraphic significance of these lineaments is their notable correlation with the northwest trend of petroleum production in the south-central part of the Powder River basin (Figure 8). The striking parallelism of the lineaments and the production trends strongly suggests a genetic interrelationship. However, these lineaments are not expressed in mapped surface structures, and, of the three, only the Buffalo/Douglas lineament is correlated with known structure at depth. Its south end is coincident with the axis of the Powder River basin, and, where the basin axis shifts westward, the lineament continues northward along a structural trend described by Blackstone (1981) as the Buffalo Deep structure.

CORRELATION WITH PATTERNS OF SEDIMENTATION AND MINERALIZATION

If the lineaments represent the structural boundaries of crustal blocks that are periodically reactivated, they should exert a marked influence on rates of sedimentation and environments of deposition and erosion. Topographically high blocks would shed clastic sediment onto adjacent low blocks, with associated facies changes and sediment contrast roughly paralleling block boundaries. A facies change might not correspond precisely to the zone of adjustment. Rather, it could be shifted laterally to either side of the zone of maximum vertical displacement as a function of the amount and rate of vertical displacement, and rate of sedimentation; however, the resulting lithologic contrast should, at least, remain dominantly parallel to the zone of adjustment. Subsequently, the change in facies could influence the formation or localization of mineral deposits or petroleum accumulations.

Frequent minor adjustment may have taken place along these lineaments even during nonorogenic periods so we might expect to find the resulting depositional contrasts expressed as subtle intraformational facies changes or gradual changes in sand size, clay content, or abundance of organic material. Such changes might be considered insignificant in a general sense, but they are the kind of contrasts that can strongly affect the migration and accu-

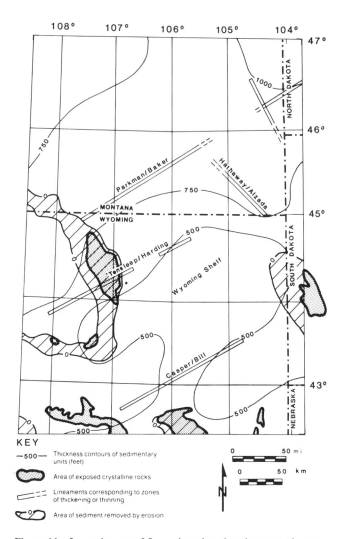

KEY

—500— Thickness contours of sedimentary units (feet)

Area of exposed crystalline rocks

Lineaments corresponding to zones of thickening or thinning

Area of sediment removed by erosion

Figure 11—Isopach map of Jurassic units, showing a northwest-trending shelf formed across Powder River basin area in the Jurassic (after Rocky Mountain Association of Geologists, 1972, p. 180; McKee et al, 1956). Shelf is flanked by Casper/Bill and Tensleep/Harding lineaments.

mulation of petroleum or deposition of uranium, not only through changing permeability, porosity, and chemical conditions, but also by differential loading and compaction that could lead to the formation of sympathetic folds and fractures. Ultimately, the contrasts in chemistry, lithology, and structure resulting from minor vertical adjustment between blocks might be the deciding factors in the localization of petroleum, coal, uranium, and other minerals. This influence may be apparent in the coincidence of mineralized trends or petroleum reservoirs with the trends of basement-block boundaries. Consequently, an essential part of this evaluation was the comparison of trends of lineaments with lithofacies and isopach maps (Figures 9, 10, 11) (McKee et al, 1956, 1959, 1967a, 1967b) of occurrences of petroleum and minerals (Wyoming Geological Survey, 1972, 1979) and other lineament maps (Hoppin, 1974; Hoppin and Jennings, 1971; Hoppin et al, 1973; Slack, 1981; Barlow and Doelger, 1981). The total thickness of Phanerozoic rocks in the Powder River basin ranges up to 18,000 ft (5,500 m). Regional isopach maps

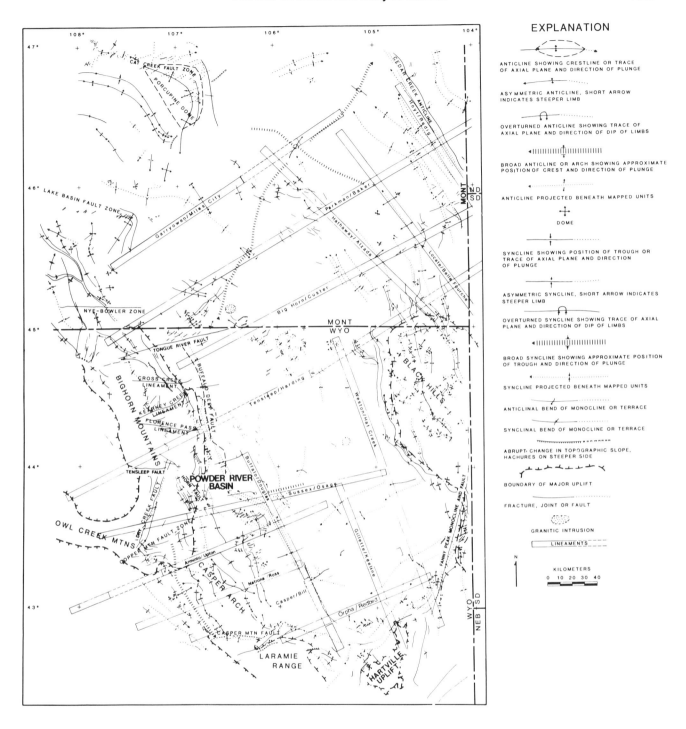

Figure 12—Composite structure map of Powder River basin, Wyoming and Montana. Data from Wilson (1938); Demorest (1941), Osterwald and Dean (1961); Hoppin and Jennings (1971), Shapiro (1971), Kiilsgaard et al (1972), Denson and Horn (1975), Colton (1978), Lisenbee (1978), Love et al (1977, 1978a, b, c), Blackstone (1981).

(Figure 9) (Rocky Mountain Association of Geologists, 1972, p. 56) show the total section thickening toward the deep basin axis that parallels the basin with a slightly positive central area trending northeast across the basin.

Deposition in the Powder River basin appears to have been fairly uniform during the Cambrian, with a clastic wedge shed from a craton on the east into an ocean basin to the west. The first hint of a major depositional contrast

in the Powder River basin is visible in the Ordovician sediments, some of which show an abrupt thinning south-southeastward across the basin (Figure 10A) (Rocky Mountain Association of Geologists, 1972, p. 81) along a linear trend roughly coincident with the Sussex/Osage lineament. This feature probably represents the northern boundary of a northeast-trending transcontinental arch that was present across northwest Colorado and southeast

525

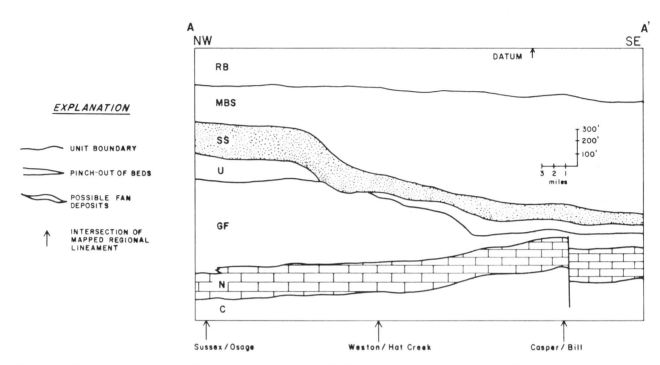

Figure 13—Diagrammatic representation adapted after Asquith's (1970) cross section, showing locations (arrows) where cross section intersects regional lineaments mapped in this study. Note that facies changes, penecontemporaneous faulting, and rapid changes in thickness of sedimentary units occur proximal to mapped regional lineaments. Line of cross section (A-A´) is shown on Figure 8. RB = Red Bird silty member; MBS = Mitten Black Shale member; SS = Sharon Springs member; U = unnamed; GF = Gammon Ferruginous member; N = Niobrara formation; C = Carlile shale.

Wyoming during much of the Ordovician. Isopach maps of Silurian rocks show the pronounced formation of a deep basin to the northeast in the area of the Williston basin (Rocky Mountain Association of Geologists, 1972, p. 87), but the northeast trend that was dominant in the Powder River basin during the Ordovician appears again in the Devonian, paralleling the zero isopach line of Devonian sedimentary rocks (Figure 10B) (Rocky Mountain Association of Geologists, 1972, p. 94). The dominance of the northeast structural trend weakens during the Mississippian, and throughout the Pennsylvanian. The depositional pattern in the Pennsylvanian is very complex, with isopachs indicating northeast- and north-trending structures (Figure 10C) (Rocky Mountain Association of Geologists, 1972, p. 116) and an obvious indication of a northwest-trending positive block in the area of the Cedar Creek anticline, southeast Montana. Northeast trends again dominated the regional depositional pattern of the Powder River basin during the Permian and Triassic Periods (McKee et al, 1959, 1967a, b; Rocky Mountain Association of Geologists, 1972). The depositional pattern suggests the formation of local basins and uplifts bounded by northeast- and northwest-trending structures (Figure 10D) (Rocky Mountain Association of Geologists, 1972, p. 154). The boundaries of these blocks are notably coincident with the trends of the lineaments. A northeast-trending topographic high (the Wyoming shelf) was a dominant feature in the southern Powder River basin in the Jurassic (Figure 11) (McKee et al, 1956), but this gave way to regional subsidence and the formation of a mid-continental sea during the Cretaceous Period. Periodic

fluctuations in sea level and the geographic extent of this sea produced a complex transgressive/regressive pattern of deposition. Horizontal and vertical changes in lithology are common and may be abrupt. The Upper Cretaceous rocks are of particular interest because they are the major petroleum-producing rocks in the Powder River basin. Lithofacies maps of the Cretaceous rocks in this area show both northwest and northeast trends of the shoreline and associated facies (Weimer, 1961, p. 25-26; Isabell et al, 1976, p. 167). The influence of such trends upon the localization of petroleum is apparent in maps showing the location and distribution of oil production in the Powder River basin (Figure 8). Northwest-trending reservoirs are dominant, particularly in the central and southwestern Powder River basin (Crews et al, 1976), and most of them produce from Upper Cretaceous units. However, northeast trends are displayed in the Clareton and Fiddler Creek fields near Newcastle, Wyoming, and in the Bell Creek field of southeast Montana, all of which produce from Lower Cretaceous rocks.

Final confirmation of the close relationship between the lineaments, as interpreted from Landsat imagery, and patterns of structure and sedimentation in the Powder River basin is seen in the comparison of the lineament maps with cross sections compiled from available well data. Figure 13 is an example showing a stratigraphic cross section compiled by Asquith (1970, his Figure 12) from 23 electric logs from wells in the eastern Powder River basin. Such comparisons show that structures and marked changes in thickness and character of sedimentary units are localized at lineaments. The cross sections also allow determination of the timing and magnitude of structural adjustments occurring along lineaments. It is apparent from such correlations that the rate of adjustment varies with time and that even the sense of movement on structures associated with the lineaments varies from one geologic period to another.

CONCLUSIONS

Concentrations of linear features identified from Landsat images reveal a pattern of intersecting northwest- and northeast-trending regional lineaments. Comparison of these lineaments with mapped folds, faults, facies changes, geophysical data, and other geologic phenomena indicates that they represent zones of adjustment along boundaries of crustal subplates or basement blocks. These block boundaries form a systematic, rectilinear pattern of structures. Mapped examples showing the character of these block boundaries are the Dewey and Long Mountain fault zones (Gott et al, 1974), the Buffalo Deep fault zone (Blackstone, 1981), and the Cedar Creek structural trend (Osterwald and Dean, 1961). Movement of blocks causes fracture and folds to form along the trend of the block boundaries, such as the Little Missouri fault zone (Lisenbee, 1978) and Cedar Creek anticline (Colton et al, 1978; Osterwald and Dean, 1961). Attendant topographic adjustments influence sedimentary deposition patterns such that facies changes, transport patterns, and contrasts in geologic environment also parallel the block boundaries. The resulting contrast in the sedimentary layers overlying the block boundaries enhances the formation of structures and chemical or lithologic gradients in these areas. The final result may be the concentration of petroleum and strata-bound minerals along the block boundaries or in trends paralleling the block boundaries.

Evidence suggests that the proposed basement-block boundaries in the Powder River basin have been periodically reactivated throughout the Phanerozoic, with minor adjustments occurring even during periods of little orogenic or tectonic activity. Similar patterns of recurrent fault-block movement have also been documented in Colorado and adjacent areas where they influence depositional patterns in sedimentary basins (Weimer, 1980).

Perhaps the most important implication of these continually active zones of weakness is that, once formed, they control structural patterns indefinitely. Other workers have cautioned against using fractures and linear elements exposed at the surface to infer forces and kinematics of deformation because a unique determination of stresses is impossible without information about the three-dimensional aspects of deformation (Bombolakis, 1979). Our work in the Powder River basin demonstrates that various stress fields would probably yield very nearly the same strain pattern regardless of the applied forces of deformation, simply because the previously formed zones of weakness along basement-block boundaries may respond to the stresses even if they are not ideally oriented with respect to the directions of applied stress.

Because these lineaments have been periodically reactivated, they controlled patterns of sedimentation through-

out the basin. This control is evident in observed trends of oil and gas reservoirs in Cretaceous and other sedimentary units. The potential for depositional control of favorable source and reservoir rocks by these lineaments leads to the conclusion that they may be used as guides in petroleum and mineral exploration.

The technique used in this study to identify lineaments and to interpret the tectonic framework of the basin is inexpensive and efficient. When evaluated relative to structural data, geophysics, sedimentary depositional patterns, surface morphology, and production patterns, the technique appears promising as a tool for basin analysis. It avoids many of the pitfalls of feature-by-feature structural interpretation of satellite imagery by statistically treating the interpretations so that the results are more objective and interpretable on a regional scale.

REFERENCES

Asquith, D. O., 1970, Depositional topography and major marine environments, Late Cretaceous, Wyoming: AAPG Bulletin, v. 54, p. 1184-1224.
Barlow, J., and M. Doelger, 1981, Powder River basin and relationship to tectonic activity (abs.), in J. R. Steidtmann, ed., Sedimentary tectonics—principles and applications: Wyoming Geological Association Meeting, Laramie, May 3-5, 1981, University of Wyoming unpublished report, p. 4.
Blackstone, D. L., 1949, Structural pattern of the Powder River basin: Wyoming Geological Association Guidebook, 4th Annual Field Conference, p. 35-36.
——— 1981, Compression as an agent in deformation of the east-central flank of the Bighorn Mountains, Sheridan and Johnson Counties, Wyoming: University of Wyoming, Contributions to Geology, v. 19, no. 2, p. 105-122.
Bombolakis, E. G., 1979, Some constraints and aides for analysis of fracture and fault development, in M. H. Podwysocki and J. L. Earle, eds., Proceedings of Second International Conference on Basement Tectonics, Newark, Delaware, 1976: Basement Tectonics Committee, Denver, Colorado, p. 289-305.
Cloos, E., and H. Cloos, 1934, Precambrian structure of the Beartooth, the Bighorn, and the Black Hills uplifts and its coincidence with Tertiary uplifting (abs.): GSA Proceedings, 1933, p. 56.
Colton, R. B., S. T. Whitaker, W. C. Ehler, J. Holligan, and C. G. Bowles, 1978, Preliminary geologic map of the Ekalaka 1° × 2° quadrangle, southeastern Montana and western North and South Dakota: U.S. Geological Survey Open-File Report 78-493, scale 1:250,000.
Crews, G. C., J. A. Barlow, Jr., and J. D. Haun, 1976, Upper Cretaceous Gammon, Shannon, and Sussex Sandstones, central Powder River basin, Wyoming, in Geology and energy resources of the Powder River basin: Wyoming Geological Association Guidebook, 28th Annual Field Conference, p. 9-20.
Davis, J. F., 1969, Uranium deposits in the Powder River basin: University of Wyoming, Contribution to Geology, v. 8, no. 2, p. 131-142.
Demorest, M. H., 1941, Critical structural features of the Bighorn Mountains Wyoming: GSA Bulletin, v. 52, p. 161-175.
Denson, N. M., and G. H. Horn, 1975, Geologic and structure map of the southern part of the Powder River basin, Converse, Niobrara, and Natrona Counties, Wyoming: U.S. Geological Survey Miscellaneous Investigations Map I-77, scale 1:125,000.
Gott, B. G., D. E. Wolcott, and C. B. Bowles, 1974, Stratigraphy of the Inyan Kara Group and localization of uranium deposits, southern Black Hills, South Dakota and Wyoming: U.S. Geological Survey Professional Paper 763, 57 p.
Hills, A. F., and R. S. Houston, 1979, Early Proterozoic tectonics of the Central Rocky Mountains, North America: University of Wyoming, Contribution to Geology, v. 17, no. 2, p. 89-109.
Hobbs, B. E., 1976, An outline of structural geology: New York, John Wiley and Sons, 571 p.
Hodgson, R. A., S. P. Gay, Jr., and J. Y. Benjamins, 1976, eds., Proceed-

ings of First International Conference on New Basement Tectonics, Salt Lake City, Utah, 1974: Utah Geological Association, 636 p.

Hoppin, R. A., 1974, Lineaments: their role in tectonics of central Rocky Mountains: AAPG Bulletin, v. 58, p. 2260-2273.

—— and T. V. Jennings, 1971, Tectonic elements, Bighorn Mountain region, Wyoming-Montana: Wyoming Geological Association Guidebook, 23rd Annual Field Conference, p. 39-48.

—— R. D. Manley, D. M. Tappmeyer, and N. E. Voldseth, 1973, Geological interpretation of ERTS-1 imagery, Bighorn Mountains: University of Wyoming, Contribution to Geology, v. 12, no. 2, p. 33-42.

Hose, R., 1955, Geology of the Crazy Woman area, Johnson County, Wyoming: U.S. Geological Survey Bulletin 1027-B, 118 p.

Isabell, E. G., C. W. Spencer, and T. Seitz, 1976, Petroleum geology of the Well Draw field, Converse County, Wyoming, in Geology and energy resources of the Powder River basin: Wyoming Geological Association Guidebook, 28th Annual Field Conference, p. 165-174.

Kiilsgaard, T. H., G. E. Erickson, L. L. Pattern, and C. L. Bieniewski, 1972, Mineral resources of the Cloud Peak primitive area, Wyoming: U.S. Geological Survey Bulletin 1371-C, 60 p.

Kleinkopf, M. D., and J. A. Redden, 1975, Bouguer gravity, aeromagnetic, and generalized geologic maps of part of the Black Hills of South Dakota and Wyoming: U.S. Geological Survey Geophysical Investigations Map GP-903, scale 1:250,000.

Levandowski, D. W., T. V. Jennings, and T. W. Lehman, 1976, Relations between ERTS lineaments, aeromagnetic anomalies, and geologic structures in north-central Nevada, in R. A. Hodgson, S. P. Gay, and J. Y. Benjamins, eds., Proceedings of First International Conference on New Basement Tectonics, Salt Lake City, Utah, 1974: Utah Geological Association, Publication 5, p. 106-117.

Lisenbee, A. L., 1978, Laramide structure of the Black Hills uplift, South Dakota-Wyoming-Montana, in Laramide folding associated with basement block faulting in the western United States: GSA Memoir 151, p. 165-196.

Love, J. D., 1978, Preliminary geologic map of the Gillette 1° × 2° quadrangle, northeastern Wyoming and western South Dakota: U.S. Geological Survey Open-File Report 78-343, scale 1:250,000.

—— A. C. Christiansen, and J. L. Earle, 1978a, Preliminary geologic map of the Sheridan 1° × 2° quadrangle, northern Wyoming: U.S. Geological Survey Open-File Report 78-456, scale 1:250,000.

—— and L. W. McGrew, 1977, Geologic map of the Newcastle 1° × 2° quadrangle, northeastern Wyoming and western South Dakota: U.S. Geological Survey Miscellaneous Field Studies Map MF-883, scale 1:250,000.

—— —— and C. K. Sever, 1978b, Preliminary geologic map of the Torrington 1° × 2° quadrangle, southeastern Wyoming and western Nebraska: U.S. Geological Survey Open-File Report 78-535, scale 1:250,000.

—— J. L. Weitz, and R. K. Hose, 1955, Geologic map of Wyoming: U.S. Geological Survey, scale 1:500,000.

—— A. C. Christiansen, J. L. Earle, and R. W. Jones, 1978c, Preliminary map of the Arminto 1° × 2° quadrangle, central Wyoming: U.S. Geological Survey Open-File Report 78-1089, scale 1:250,000.

Lovering, T. S., and E. N. Goddard, 1950, Geologic map of the Front Range mineral belt, Colorado: U.S. Geological Survey Professional Paper 223, plate 2.

Mapel, W. J., 1959, Geology and coal resources of the Buffalo-Lake De-Smet area, Johnson and Sheridan Counties Wyoming: U.S. Geological Survey Bulletin 1078, 148 p.

Marrs, R. W., and D. R. Gaylord, 1978, A guide to the interpretation of wind flow characteristics from eolian landforms: University of Wyoming, Department of Geology, 39 p.

McKee, E. D., S. S. Oriel, V. E. Swanson, M. E. MacLachlan, K. B. Ketner, J. W. Goldsmith, R. Y. Bell, and D. J. Jameson, 1956, Paleotectonic maps of the Jurassic system: U.S. Geological Survey Miscellaneous Investigations Map I-175, scale 1:5,000,000.

—— K. B. Ketner, M. E. MacLachlan, J. Goldsmith, J. C. MacLachlan, and M. R. Mudge, 1959, Paleotectonic maps of the Triassic system: U.S. Geological Survey Miscellaneous Investigations Map I-300, scale 1:5,000,000.

—— —— et al, 1967a, Paleotectonic investigations of the Permian System in the United States: U.S. Geological Survey Professional Paper 515, 271 p.

—— —— et al, 1967b, Paleotectonic maps of the Permian System: U.S. Geological Survey Miscellaneous Investigations Map I-450, scale 1:5,000,000.

Meuschke, J. L., R. W. Johnson, and J. R. Kirby, 1963, Aeromagnetic

map of the southwestern part of Custer County, South Dakota: U.S. Geological Survey Geophysical Investigations Map GP-362, scale 1:62,500.

Moody, J. D., and M. J. Hill, 1956, Wrench-fault tectonics: GSA Bulletin, v. 67, p. 1207-1246.

Osterwald, F. W., 1956, Relation of tectonic elements in Precambrian rocks to uranium deposits in the Cordilleran foreland of the western United States: U.S. Geological Survey Professional Paper 300, p. 329-335.

—— and B. G. Dean, 1961, Relation of uranium deposits to tectonic pattern of the central Cordilleran foreland: U.S. Geological Survey Bulletin 1087-1, p. 337-390.

Palmquist, J. C., 1978, Laramide structures and basement block faulting: two examples from the Big Horn Mountains, Wyoming: GSA Memoir 151, p. 125-138.

Raines, G. L., 1978, Porphyry copper exploration model for northern Sonora, Mexico: U.S. Geological Survey Journal of Research, v. 6, p. 51-58.

—— T. W. Offield, and E. S. Santos, 1978, Remote-sensing and subsurface definition of facies and structure related to uranium deposits, Powder River basin, Wyoming: Economic Geology, v. 73, p. 1706-1723.

Renfro, H. B., and D. E. Feray, 1972, Geological highway map of the northern Rocky Mountain region: AAPG Map 5, United States Geological Highway Map Series, scale 1:2,000,000.

Robinson, C. S., W. J. Mapel, and M. H. Bergendahl, 1964, Stratigraphy and structure of the northern and western flanks of the Black Hills uplift, Wyoming, Montana, and South Dakota: U.S. Geological Survey Professional Paper 404, 134 p.

Rocky Mountain Association of Geologists, 1972, Geologic atlas of the Rocky Mountain region: Denver, Hirshfeld Press, 331 p.

Russell, W. L., 1929, Drainage alignments in the western Great Plains: Journal of Geology, v. 37, p. 249.

Sawatzky, D. L., and G. L. Raines, 1977, Analysis of lineament trends and intersections (abs.): GSA Abstracts with Programs, v. 9, p. 759.

—— —— 1981, Geologic uses of linear-feature maps derived from small-scale images, in D. W. O'Leary and J. L. Earle, eds., Proceedings of Third International Conference on Basement Tectonics, Durango, Colorado, 1978: Basement Tectonics Committee, Denver, Colorado, p. 91-100.

Shapiro, L. H., 1971, Structural geology of the Fanny Peak lineament, Black Hills, Wyoming and South Dakota, in Symposium on Wyoming tectonics and their economic significance: Wyoming Geological Association Guidebook, 23rd Annual Field Conference, p. 61-64.

Sharp, R. P., 1948, Early Tertiary fanglomerate, Big Horn Mountains, Wyoming: Journal of Geology, v. 56, p. 1-15.

Shoemaker, E. M., R. L. Squires, and M. J. Abrams, 1978, Bright Angel and Mesa Butte fault systems of northern Arizona, in R. B. Smith and B. P. Eaton, eds., Cenozoic tectonics and regional geophysics of the western Cordillera: GSA Memoir 152, p. 341-367.

Shurr, G. W., 1979, Lineament control of sedimentary facies in the northern Great Plains, United States, in M. H. Podwysocki and J. L. Earle, eds., Proceedings of Second International Conference on Basement Tectonics, Newark, Delaware, 1976: Basement Tectonics Committee, Denver, Colorado, p. 413-422.

Slack, P. B., 1981, Paleotectonics and hydrocarbon accumulation, Powder River basin, Wyoming: AAPG Bulletin, v. 65, p. 730-743.

Tanner, W. F., 1962, Surface structural patterns obtained from strike-slip models: Journal of Geology, v. 70, p. 101-107.

Thomas, G. E., 1971, Continental plate tectonics, southwest Wyoming, in Symposium on Wyoming tectonics and their economic significance: Wyoming Geological Survey, 23rd Annual Field Conference Guidebook, p. 103-123.

Turner, R. L., G. L. Raines, M. D. Kleinkopf, and J. L. Lee-Moreno, 1982, Regional northeast-trending structural control of mineralization, northern Sonora, Mexico: Economic Geology, v. 77, p. 25-37.

Tweto, O., and P. K. Sims, 1963, Precambrian ancestry of the Colorado mineral belt: GSA Bulletin, v. 79, p. 991-1014.

Waring, J., 1976, Regional distribution of environments of the Muddy sandstone, in Geology and energy resources of the Powder River basin: Wyoming Geological Association, 28th Annual Field Conference Guidebook, p. 83-96.

Webb, M. D., 1969, Stratigraphic control of sandstone uranium deposits in Wyoming: University of Wyoming, Contribution to Geology, v. 8, no. 2, p. 121-130.

Weimer, R. J., 1961, Uppermost Cretaceous rocks in central and southern Wyoming, and northwest Colorado, *in* Symposium on Late Cretaceous rocks of Wyoming: Wyoming Geological Association Guidebook, 16th Annual Field Conference, p. 17-28.

———— 1980, Recurrent movements on basement faults, a tectonic style for Colorado and adjacent areas: Rocky Mountain Association of Geologists Symposium, Denver, Colorado, p. 23-35.

Wilson, C. W., Jr., 1938, The Tensleep fault, Johnson and Washakie Counties, Wyoming: Journal of Geology, v. 46, p.868-881.

Wyoming Geological Association, 1971, Wyoming tectonics symposium: Wyoming Geological Association Guidebook, 23rd Annual Field Conference, 185 p.

———— 1976, Powder River guidebook: Wyoming Geological Association, 28th Annual Field Conference, 328 p.

Wyoming Geological Survey, 1972, Wyoming energy resources map: scale 1:500,000.

———— 1979, Wyoming mines and minerals map: scale 1:500,000.

The American Association of Petroleum Geologists Bulletin
V. 70, No. 4 (April 1986), P. 453-455, 2 Figs.

Tectonic Framework of Powder River Basin, Wyoming and Montana, Interpreted from Landsat Imagery: Discussion[1]

RICHARD C. MICHAEL and IRA S. MERIN[2]

Marrs and Raines (1984) presented an approach for analyzing Landsat imagery that we believe is a valuable adjunct to more conventional approaches. We concur with many of their conclusions, particularly that some lineaments mapped from satellite images mark important structural features and that these features can be useful in the search for minerals and hydrocarbons. However, we object to defining lineaments solely on the basis of the type of statistical approach proposed by the authors. We do not believe, as Marrs and Raines appear to, that the most effective use for linear features is exclusively to define regionally extensive zones, called lineaments.

Marrs and Raines defined lineaments from contour maps of the density of linear features. Only linear features that belong to statistically defined dominant orientations are contoured, with each set of dominant trends contoured separately. Features that belong to minor orientations are excluded from the analysis. Lineaments are then selected by identifying those "contiguous highs and directional trends" in the contoured data that parallel the orientation interval that has been contoured.

We do not question the validity of these lineaments as regionally extensive zones of linear features. However, we do have three principal objections to this approach. First, the approach obscures the significance of individual linear features. We note a substantial body of literature demonstrating that specific linear features can be correlated with geologically significant features (Vernon, 1951; Lattman and Matzke, 1961; Bechtold et al, 1973; Merifield and Lamar, 1974; Wier et al, 1974; Podwysocki et al, 1982; Salgat, 1983; Merin and Moore, 1985; Merin and Michael, in press). We believe that individual linear features have geologic significance in the Powder River basin as well. Our analysis of the same region shows that in numerous places linear features correspond to specific subsurface features.

Second, the approach risks excluding potentially important features by restricting the analysis to those features that fall within significant trends. For example, a basement shear zone marked at the surface by gash fractures would not be identified using the Marrs and Raines approach because these fractures are typically oriented

about 30° to the shear zone responsible for their origin. A well-documented example of such a feature is the Lake Basin fault zone of south-central Montana, a west-northwest-trending left-slip basement fault marked at the surface by linear northeast-trending gash fractures. The Marrs and Raines approach would not identify such a zone, precisely because the surficial faults comprising the zone are oriented significantly different than the overall trend of the zone.

Finally, by restricting the statistical analysis to length-weighted frequency distributions of linear features, Marrs and Raines implied that longer features are inherently more significant than shorter features. We do not believe this assumption is necessarily correct.

Marrs and Raines (1984) stated that their technique "avoids many of the pitfalls of feature-by-feature structural interpretation . . . so that the results are more objective and interpretable on a regional scale." We agree that feature-by-feature interpretation has many drawbacks, and indeed, the approach presented by Marrs and Raines appears to cut through the clutter of Landsat imagery analyses by treating the data statistically. However, we believe this approach creates new pitfalls by statistically removing the interpreter from the original data source.

We disagree with the authors' use of the word "objective" as applied to their results. The objectivity of an interpretation resides in the original interpretation (i.e., Figure 1 of Marrs and Raines, 1984), which is inevitably subject to both data and interpreter biases. Finally, Marrs and Raines stated that their lineaments correlate well with production trends. For example, on page 1726, they discussed northwest-trending lineaments, stating that "the most convincing evidence for the structural and stratigraphic significance of these lineaments is their notable correlation with the northwest trend of petroleum production in the south-central part of the Powder River basin" (their Figure 8). We see no "notable" correlation.

The potential significance of individual linear features and their application to hydrocarbon exploration in the Powder River basin is illustrated in the two examples presented below.

Analysis of Landsat imagery in the vicinity of the Hilight oil field shows that northwest- and northeast-trending linear features dominate this area and locally correspond in position and trend to the isopach pattern of the high-resistivity (i.e., porosity) Muddy Sandstone present in this field (Figure 1). The upper Muddy reservoirs at the Hilight oil field contain high-porosity/permeability lenses that trend obliquely to the overall trend of the productive sands (Berg, 1976; Larberg, 1981). Berg (1976) suggested

[1]Manuscript received, May 30, 1985; accepted, October 31, 1985.
[2]Earth Satellite Corporation, 7222 47th Street, Chevy Chase, Maryland 20815.
We extend our sincere thanks to Donald B. Segal and John R. Everett of the Earth Satellite Corporation and Terri L. Purdy of the U.S. Geological Survey for their insightful comments as well as numerous reviews of this discussion.

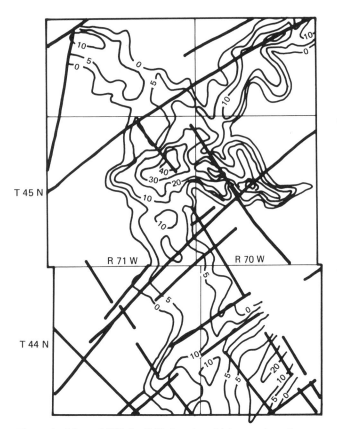

Figure 1—Map of Hilight field showing thickness of sandstones that exceed 25 ohm-meters resistivity for total Muddy section, with superimposed linear features mapped from Landsat imagery (contours from Berg, 1976). Scale: 1 in. = 3 mi. C.I. = 5 ohm-meters.

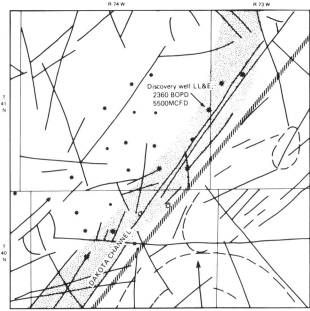

Figure 2—Interpretation of Landsat imagery for part of Powder River basin, Wyoming, near Buck Draw field, Campbell and Converse Counties. Hachured line represents a regional lineament. Stippled pattern represents the Dakota channel. Scale: 1 in. = 8,000 ft.

that this porosity trend resulted from sands reworked during a postdepositional transgression. Slack (1981) reported that northeast-trending structures produced fault-controlled topography that influenced deposition of Muddy sediments. Thus, the correlation of northeast-trending linear features with thick sands suggests that these zones are the result of deposition on fault-controlled topography. In addition, postdepositional fracturing along these trends may have increased reservoir porosity and permeability, thus contributing to the correlation of the high-resistivity sands with linear features.

Figure 2 is a Landsat interpretation of part of the southwestern Powder River basin in the vicinity of the recently developed Buck Draw oil field. The discovery well at Buck Draw was completed during the final phase of the Landsat interpretation. Development drilling within the field postdates the interpretation.

Reservoir rocks at the Buck Draw field consist of estuarine sandstones and associated lithologies of the Dakota Formation. A northeast-trending estuarine channel, mapped along the southwest margin of the field, coincides with northeast-trending linear features mapped using Landsat imagery. In turn, these features are associated with a northeast-trending regional lineament. Initial production rates and reserves from wells within the channel and, hence, along the zone of closely spaced northeast-trending linear features are considerably higher than from

wells outside the channel. The strong apparent correlation between channel development, Landsat linear features, and high production rates is explained as follows. First, syndepositional structural growth is thought to have locally influenced the distribution of Dakota estuarine channels and associated sands throughout the southern Powder River basin. In particular, northeast-trending fault zones, which have a well-documented history of recurrent growth (Shurr, 1979; Slack, 1981; Marrs and Raines, 1984), were active during Early Cretaceous subsidence of the Sevier foreland. Correlation between Dakota channels and linear features, such as those observed at Buck Draw, appears to be a manifestation of the relationship between structure and depositional environments. Second, and more speculatively, postdepositional fracturing along the same trend seemed to enhance the quality of the Dakota reservoir rock. Fracture enhancement of the reservoir most likely occurred during late Laramide (Eocene) tectonism when maximum compressive stress was oriented northeast, and preexisting northeast-trending fracture zones were subjected to tensional stresses. Postorogenic uplift may have caused additional movement along these fractures.

A statistical treatment of the data derived from interpreting the Landsat imagery obscures the specific correlation between the Dakota channel, production rates, and individual features mapped on Landsat imagery. Specific subsurface phenomena, such as the estuarine channel at Buck Draw, correlate with specific features visible on Landsat imagery—in this instance, the local linear features associated with a regionally extensive zone of linear features. Similarly, insights into the relationship between fractures and high production rates can be gained only by comparing individual linear features with production data from individual wells.

In general, the statistical approach outlined by Marrs and Raines (1984) undermines one of the major attributes of Landsat imagery—the ability to display structures as areally discrete features. This attribute complements the interpretation of subsurface data, where the ability to delineate a feature accurately depends on the well distribution in the area. Furthermore, the discrete nature of features derived from interpreting Landsat imagery, coupled with the synoptic perspective of the imagery, provides a framework for extrapolating features mapped in areas of good subsurface or seismic control into areas of poor control.

The ultimate goals of structural interpretation of Landsat imagery in basin analysis for petroleum exploration are: to refine previously recognized structures, to map previously unrecognized structures, and to determine to what degree these structures have influenced the accumulation of hydrocarbons in the region. Much of this interpretation must be done on "feature-by-feature" basis. The statistical approach that Marrs and Raines used augments—but does not replace—more traditional approaches to the structural interpretation of imagery. As Marrs and Raines pointed out, whenever possible, structural analysis of Landsat imagery should be supported by other data sets (i.e., fieldwork or geophysical and subsurface data) that help define the issues to be addressed and help us evaluate individual features mapped on Landsat imagery.

REFERENCES CITED

Bechtold, I. C., M. A. Liggett, and J. F. Childs, 1973, Regional tectonic control of Tertiary mineralization and recent faulting in the southern Basin-Range province—an application of ERTS-1 data, *in* Symposium on Significant Results Obtained from ERTS-1, v. 1: NASA Special Publication 327, p. 425-432.

Berg, R. R., 1976, Hilight Muddy field—Lower Cretaceous transgressive deposits in the Powder River basin, Wyoming: Mountain Geologist, v. 13, p. 33-45.

Larberg, G. M., 1981, Depositional environments and sand body morphologies of the Muddy sandstones at Kitty field, Powder River basin, Wyoming: Wyoming Geological Association Guidebook 31, p. 117-135.

Lattman, L. H., and R. H. Matzke, 1961, Geological significance of fracture traces: Photogrammetric Engineering, v. 27, p. 435-438.

Marrs, R. W., and G. L. Raines, 1984, Tectonic framework of Powder River basin, Wyoming and Montana, interpreted from Landsat imagery: AAPG Bulletin, v. 68, p. 1718-1731.

Merin, I. S., and R. C. Michael, in press, Application of structures mapped from Landsat imagery to exploration for stratigraphic traps in the Paradox basin: International Symposium on Remote Sensing of Environment, Fourth Thematic Conference: Remote Sensing for Exploration Geology, San Francisco, April 1-4.

———— and W. R. Moore, 1985, Application of Landsat imagery to hydrocarbon exploration in Niobrara Formation, Denver basin (abs.): AAPG Bulletin, v. 69, p. 288.

Merifield, P. M., and D. L. Lamar, 1974, Lineaments in a basement terrane of the Peninsular Ranges, southern California, *in* R. A. Hodgson, S. P. Gay, and J. Y. Benjamins, eds., Proceedings of the First International Conference on the New Basement Tectonics: Utah Geological Association Publication 5, p. 94-105.

Podwysocki, M. H., H. A. Pohn, J. D. Phillips, M. D. Krohn, T. L. Purdy, and I. S. Merin, 1982, Evaluation of remote sensing, geological, and geophysical data for south-central New York and northern Pennsylvania: USGS Open-File Report 82-319, 179 p.

Salgat, B., 1983, Scully field—Marion County, Kansas (abs.): AAPG Bulletin, v. 67, p. 1327.

Shurr, G. W., 1979, Lineament control of sedimentary facies in the northern Great Plains, United States, *in* M. H. Podwysocki and J. K. Earl, eds., Second International Conference on Basement Tectonics, Newark, Delaware 1976: Basement Tectonics Committee, Denver, Colorado, p. 413-422.

Slack, P., 1981, Paleotectonics and hydrocarbon accumulation, Powder River basin, Wyoming: AAPG Bulletin, v. 65, p. 730-743.

Vernon, R. O., 1951, Geology of Citrus and Levy Counties, Florida: Florida Geologic Survey Bulletin 33, p. 47-52.

Weimer, R. J., J. J. Emme, C. L. Farmer, L. O. Anna, L. D. Thomas, and R. L. Kidnev, 1982, Tectonic influences on sedimentation, Early Cretaceous, east flank Powder River basin, Wyoming and South Dakota: Colorado School of Mines Quarterly, v. 77, 59 p.

Wier, C. E., F. J. Wobber, O. R. Russell, R. V. Amato, and T. V. Leshendok, 1974, Application of ERTS-1 imagery to fracture related mine safety hazards in the coal mining industry: NASA, 136 p.

(No Reply Received)

The American Association of Petroleum Geologists Bulletin
V. 70, No. 4 (April 1986), P. 351-359, 15 Figs.

Application of Landsat Imagery to Oil Exploration in Niobrara Formation, Denver Basin, Wyoming[1]

I. S. MERIN[2] and W. R. MOORE[3]

ABSTRACT

The Niobrara Formation produces oil from fractures in several places in the Denver basin. The Niobrara is an oil-prone, mature source rock that entered the oil-generating window during the Laramide orogeny. The Laramide orogeny began with maximum compressive stress oriented east-northeast during the Late Cretaceous to Paleocene, and ended with maximum stress oriented to the northeast in the Eocene. We believe the Eocene phase activated northeast-trending extension fractures that may have acted as pathways for migration and loci for storage of oil, locally generated in the Niobrara. Theoretically, the fracture pressures related to oil generation in the Niobrara would preferentially open and fill this northeast-trending fracture system. Using Landsat imagery to map fractures in the northern part of the Denver basin, we have identified areas prospective for Niobrara oil production within an exploration fairway that is based on subsurface isopach and resistivity mapping. Support for this concept is the location of wells, reported to produce oil from the Niobrara, along a zone of northeast-trending lineaments.

INTRODUCTION

The Niobrara Formation produces oil from fractures in several places in the Denver basin (Figure 1). We used Landsat imagery to map the orientation and distribution of fractures in the northern part of this basin, and regional tectonics of the basin to predict the motion of a given fracture through time. Then integrating the Landsat analysis with well-log data, we predicted which fracture systems are the most favorable prospects for fracture production in the Niobrara Formation.

Oil production from fractures in the Niobrara Formation appears to occur in two areas of the Denver basin: (1) along the west flank of the basin within anticlines where the Niobrara is fractured (e.g., Loveland, Berthoud fields; Mallory, 1977), and (2) along the mildly deformed central part

of the basin in intensely fractured areas (e.g., Silo field; McCaslin and Williams, 1984). Although both types of accumulation are within fractured reservoirs, they may represent two different styles of fracture-controlled oil accumulations and may require different exploration methods. Open fractures tend to develop in folded strata in places of maximum curvature (e.g., where the rate of change in the dip or strike is greatest) and not necessarily at the structural crest (Stearns and Friedman, 1972). Therefore, the distribution of fracture porosity and permeability along the deformed flank of a basin (e.g., the western flank of the Denver basin) is governed largely by flexures within strata. In mildly to slightly deformed parts of a basin, the distribution of fracture-induced porosity and permeability is more significantly influenced by reactivation of previously existing fractures (faults) rather than by flexures, which play only a local role. Maps showing the rate of change in the dip of strata (Harnett, 1968) may reveal fracture distribution in areas where strata are folded; however, such maps provide little information on fractures in slightly deformed areas. Landsat imagery may be used to map flexures in strata, as well as faults and joints, and thus may be used to map fracture orientation and distribution along basin flanks as well as within the less deformed central part of a basin.

METHODS

We used Landsat imagery to identify geologic surface structures, and subsurface well-log data to map the structure, thickness, and resistivity patterns of the Niobrara Formation. These data were integrated into a data package that, along with a detailed knowledge of the local geology, formed the basis for identifying prospective areas for petroleum exploration in the northern Denver basin.

A digitally enhanced Landsat multispectral-scanner false-color composite of bands 4, 5, and 7 was produced at a scale of 1:125,000 and interpreted using standard photointerpretation techniques for folds, faults, and lineaments. The image is a late summer scene, acquired on August 15, 1973, and has a sun azimuth of N49°W and a sun elevation of 53°.

The lineaments mapped consist of scale-independent alignments of stream segments, tonal patterns, or geomorphic features that might mark geologic structures, such as joints or faults (see O'Leary et al, 1976). Mapping such features is useful because they may mark extensions of previously mapped geologic features or locations of previously unknown geologic features. Man-made linear features, such as changes in furrow patterns or section-line roads, were not mapped. To help identify the principal trends, we

©Copyright 1986. The American Association of Petroleum Geologists. All rights reserved.

[1]Manuscript received, May 24, 1985; accepted, January 10, 1986.

[2]Earth Satellite Corporation, 7222 47th Street, Chevy Chase, Maryland 20815.

[3]Suite 1212, 518 17th Street, Denver, Colorado 80202.

The manuscript benefited from thorough and critical reviews by J. R. Everett, O. R. Russell, J. P. Lockridge, and W. S. Kowalik, and from extensive discussion with R. F. Livaccari on Laramide tectonics. We thank Dana Strouse and Martha Williams for preparing maps and illustrations, Mike Ruth and Estella Nkwate for producing the lineament rose diagrams, and Earth Satellite Corporation for permission to publish this work.

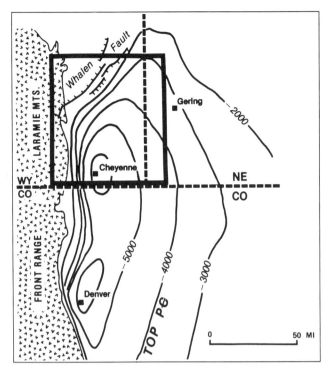

Figure 1—Project location shown on structure contour map of Precambrian rocks. C.I. = 1,000 ft.

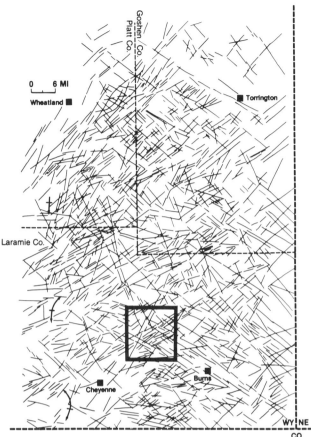

Figure 2—Lineaments, folds, and faults interpreted from 1:125,000-scale Landsat imagery. Box marks area depicted in Figure 15.

digitized all lineaments and produced length-weighted and frequency-weighted rose diagrams.

Available well logs were obtained for wells within the project area. From these data, we constructed isopach, resistivity, and structure contour maps of the Niobrara Formation. Core data for the Niobrara-Codell interval were obtained for four wells within the Silo field area (T15-16N, R64-65W, Laramie County, Wyoming; see Figure 15).

BASIN STRUCTURE

Landsat imagery of the northern Denver basin at a scale of 1:125,000 shows numerous folds, faults, and lineaments (Figure 2). Although lineaments are oriented in several directions, two directions appear to predominate: northeast (N41-65°E) and northwest (N42-58°W) (Figure 3). Several workers have shown that numerous northeast- and northwest-trending faults are present in this part of the basin and that these features have been reactivated through geologic time, including the following:

1. A northeast-trending Precambrian wrench-fault system—the Colorado lineament (Warner, 1978) (Figure 4).

2. A prominent northeast-trending linear isopach anomaly of Permian strata defining the southern edge of the Hartville uplift (Momper, 1963) (Figure 5).

3. The Whalen fault system—a system of northeast-trending, late Tertiary, high-angle normal faults (McGrew, 1961) that overlie early Tertiary thrusts (Droullard, 1963) and mark the surficial expression of the southern boundary of the Hartville uplift.

4. Northeast-trending basement faults interpreted (Weimer, 1978, 1980; Silverman, 1984) to have influenced Early and Late Cretaceous sedimentation patterns.

5. A prominent west-northwest–trending zone along the North Platte River interpreted (Thomas, 1971) to have undergone left-slip faulting during the Laramide orogeny.

Because lineaments mapped using Landsat imagery are predominantly parallel to known faults or tectonic zones, we are confident that many of these lineaments mark geologically significant fracture zones. The lineaments represent either extensions of previously mapped structural features or previously unrecognized structural features (see Bechtold et al, 1973; Wier et al, 1974; Podwysocki et al, 1982; Salgat, 1983; Merin and Michael, 1985).

We visualize the Denver basin as a mosaic of fault blocks, with fractures trending predominantly northeast and northwest defining the block edges. Based on knowledge of the tectonic history of this basin, we modeled the theoretical motion on the fractures defining these block edges.

STRATIGRAPHY AND HYDROCARBON POTENTIAL

Figure 6 shows part of the stratigraphic column for the Denver basin. The Niobrara Formation consists of chalky

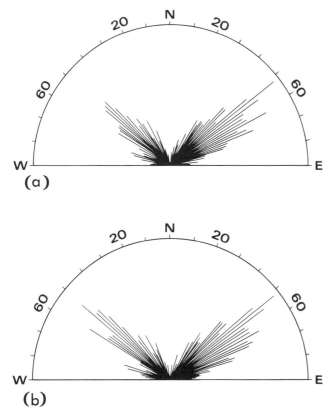

Figure 3—(a) Frequency-weighted rose diagram. Sun azimuth is N49°W. (b) Length-weighted rose diagram. Sun azimuth is N49°W.

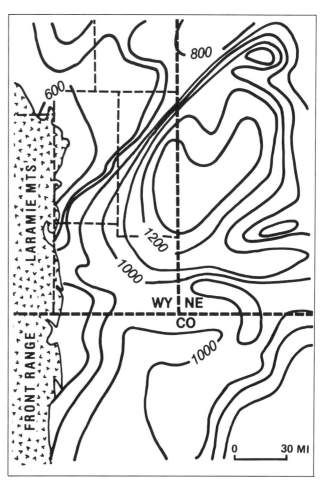

Figure 5—Isopach map of Permian rocks (Momper, 1963). C.I. = 100 ft.

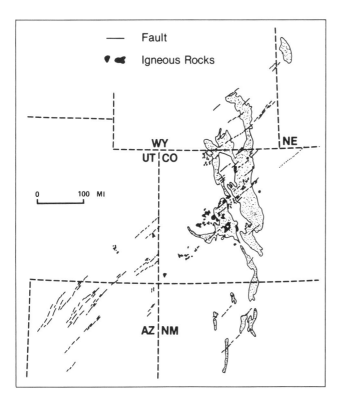

Figure 4—Colorado lineament (Warner, 1978).

limestones and interbedded calcareous shales; the upper part is the Smoky Hill Chalk Member, and the lower part is the Fort Hays Limestone Member.

Porosity and permeability of the Niobrara Chalk decrease with depth. Along the shallow east flank of the basin chalk, permeabilities exceed 0.5 md, and biogenic gas is produced from small structures (Lockridge and Scholle, 1978). Pollastro and Scholle (1984) reported that even at these shallow depths, faults and fractures are probably essential for development of commercially successful gas wells. Along the central and western parts of the basin where the Niobrara chalks occur at depths of 6,000-7,000 ft (1,800-2,100 m), such as in this project area, matrix permeability is less than 0.01 md, and porosity is less than 10% (Scholle, 1977). Reservoir conditions at these depths are poor, except in places where fractures provide additional pore space and enhance permeability.

Core data from wells in the Silo field (T15-16N, R64-65W, Laramie County, Wyoming, see Figure 15) reveal that fractures are important for reservoir porosity and permeability. Core analyses from the AMOCO Goertz B-1 well (Sec. 31, T16N, R64W) and the Champlin Lee 41-5 well (Sec. 5, T15N, R64W) show oil-saturated, vertical fractures in the Niobrara Formation (information from Wyoming

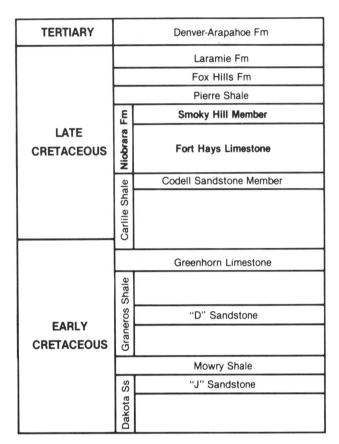

Figure 6—Cretaceous and Tertiary stratigraphic column for Denver basin (Clayton and Swetland, 1980).

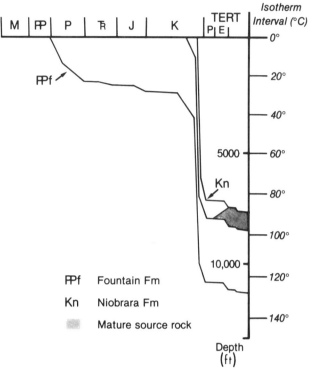

Figure 7—Burial curve of Denver basin (MacMillan, 1980).

State Oil and Gas Commission, November 1985). A core description of the AMOCO State of Wyoming X 1 well (Sec. 8, T15N, R64W) shows the Codell formation to have oil-saturated, vertical fractures (information from Petroleum Information, November 1985).

The Niobrara Formation is an oil-prone, mature source rock having as much as 3.4% total organic carbon (Clayton and Swetland, 1980; Pollastro and Martinez, 1985), and it has been in the oil-generating window since the early Eocene (MacMillan, 1980) (Figure 7). Therefore, oil generation from the Niobrara may be partly contemporaneous with the Laramide orogeny. Source rocks tend to fracture during oil generation, and the resultant fractures are parallel to the ambient maximum compressive stress. Thus, determining the stress field present during the Laramide orogeny will provide insight to the orientation of fractures generated, or reactivated, during oil generation and expulsion within the Niobrara Formation. Fractures parallel to the maximum compressive stress are prime candidates for oil migration and storage in this formation.

LARAMIDE OROGENY

The Laramide orogeny was a two-phase orogenic event that began with the maximum compressive stress oriented east-northeast during the Late Cretaceous to Paleocene (early Laramide), and ended with maximum compression oriented northeast during the Eocene (late Laramide) (Chapin and Cather, 1981; Gries, 1983). The principal sources of data supporting a two-phase orogeny are overprinted structures of Laramide age, and Laramide dikes of two primary orientations (Chapin and Cather, 1981; Rehrig and Heidrick, 1976). Based on the orientation of Laramide dikes and compressional structures, Chapin and Cather (1981) showed that the axis of compression changed from east-northeast in the early phase to northeast in the late phase of the Laramide orogeny. In general, north-south to northwest-trending compressional structures—e.g., folds, thrusts, and basement-cored uplifts such as the Front Range and the Wind River Mountains—formed during the early phase of the Laramide orogeny. Most northwest to east-west-trending compressional structures—e.g., folds, thrusts, and basement-cored uplifts such as the north flank of the Laramie, Uinta, and Owl Creek Mountains—formed during the late phase of the Laramide orogeny. Livaccari and Keith (1984, 1985) produced a series of paleotectonic maps (Figures 8, 9) showing that Engebretson et al's (1984) plate-motion data are compatible with the orientation of Laramide structures and dikes. These maps support Chapin and Cather's (1981) idea that the principal horizontal stress changed from east-northeast in the early phase to northeast in the late phase of Laramide orogeny.

FRACTURE BEHAVIOR THROUGH TIME

Based on the above model of the Laramide orogeny, we can predict the most likely motion on a particular fracture during the orogeny (Figures 10, 11). Strain theory predicts

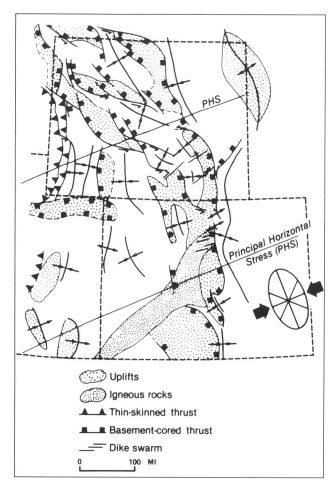

Figure 8—Paleotectonic map of Colorado and Wyoming showing orientation of principal horizontal stress (PHS), uplifts, and structures active during early Laramide orogeny, Maastrichtian to Paleocene (simplified from Livaccari and Keith, 1985).

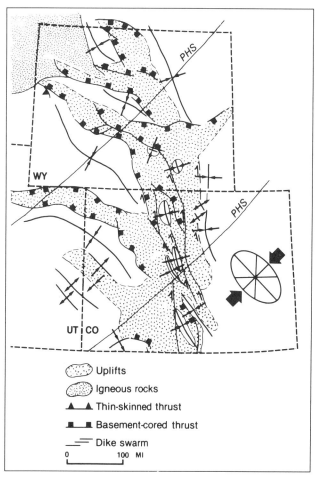

Figure 9—Paleotectonic map of Colorado and Wyoming showing orientation of principal horizontal stress (PHS), uplifts, and structures active during late Laramide orogeny, Eocene (simplified from Livaccari and Keith, 1985).

that an east-northeast–oriented principal horizontal compressive stress will cause fractures trending northeast and northwest to behave as right-slip faults and left-oblique thrust faults, respectively. This theoretical motion is similar to that of fractures reactivated, or produced, during the early phase of the Laramide orogeny in the Denver basin (Figure 10). If the principal horizontal compressive stress changed to a northeast orientation, as suggested for the late phase of the Laramide orogeny, then northeast-trending fractures (former zones of right-slip faulting) would be properly oriented for extension during the late Laramide orogeny. Additionally, extensional fractures formed during the late Laramide orogeny would theoretically trend northeast (Figure 11).

Because the Niobrara Formation was generating oil during the later phase of the Laramide orogeny (i.e., during the Eocene), we can use this model to predict the orientation of fractures produced, and reactivated, by the fracture pressures related to hydrocarbon generation in this formation. Theoretically, these fractures are extensional and parallel to the maximum compressive stress. The principal horizontal stress affecting the Denver basin during the late phase of the

Laramide orogeny was oriented northeast. Therefore, northeast-trending extensional fractures may have acted preferentially as channels for expulsion and migration of oil in response to the fracture pressures related to generation.

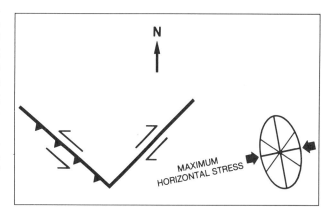

Figure 10—Theoretical motion of previously existing northeast- and northwest-trending faults subjected to an east-northeast-oriented maximum horizontal stress during early Laramide orogeny.

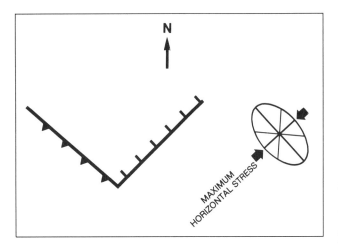

Figure 11—Theoretical motion of previously existing northeast- and northwest-trending faults subjected to a northeast-oriented maximum horizontal stress during late Laramide orogeny.

During the early phase of the Laramide orogeny, the principal horizontal stress was oriented east-northeast, so northeast-trending fractures probably were right-shear zones during the early Laramide orogeny and extensional zones during the late Laramide orogeny. Therefore, northeast-trending fractures that were reactivated, or generated, during the early Laramide orogeny would most likely have been held open during the late Laramide orogeny. Thus, northeast-trending fractures are primary candidates for the migration and storage of oil produced from the Niobrara Formation.

EXPLORATION TRENDS

The optimal location to test a fracture prospect is a naturally fractured area where fractures are interconnected, apparently filled with hydrocarbons, and sealed by interbedded shales. To be interconnected, fractures must intersect and have different azimuth or dip. Apparent hydrocarbon saturation can be indicated by mapping well-log resistivity values of the stratigraphic interval in question. Where shale is interbedded with chalk, such as in the Niobrara Formation, the shale beds can act as an effective seal because the shale behaves plastically, bending around the brittle, fractured chalk (Harnett, 1968; Mallory, 1977). Therefore, the optimum location to test fracture-controlled production in the Niobrara Formation probably would be marked by high well-log resistivity values and would have a dense population of intersecting fractures. Our model pre-

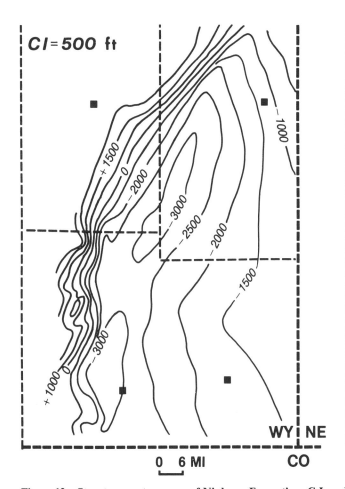

Figure 12—Structure contour map of Niobrara Formation. C.I. = 500 ft.

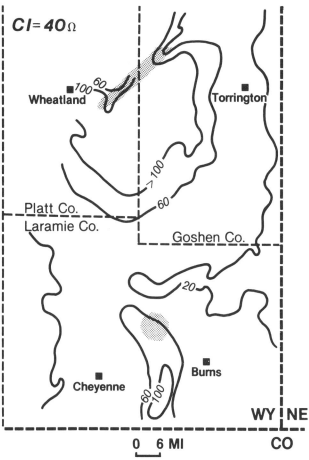

Figure 13—Resistivity map of Niobrara Formation. Shading indicates location of wells with oil shows in the Niobrara. C.I. = 40Ω.

dicts that the northeast-trending fractures acted as pathways for oil migration and as loci for storing oil generated in the Niobrara Formation because these fractures theoretically were sheared and then extended during oil generation. If this model is accurate, these fractures should be more important for fracture-induced porosity and permeability in the Niobrara than fractures of other azimuths. Therefore, the best places for open, oil-filled fractures to develop in the Niobrara Formation are: (1) areas with a high density of intersecting fractures where one fracture set trends northeast, and (2) areas with a high density of northeast-trending fractures.

Meissner (1978) showed that by mapping resistivity values, areas of apparent hydrocarbon saturation can be identified in a given formation (such as the Bakken shale of the Williston basin). A basinwide distribution of resistivity values should show increasing resistivity with increasing burial depth, with greater depth corresponding to greater maturation levels. Comparison of our resistivity map with our structural contour map of the Niobrara Formation shows that fair correlation exists in the northern part of our project area (near Goshen Hole); however, no correlation exists in the southern part, near Cheyenne (Figures 12, 13). Therefore, if resistivity indicates hydrocarbon saturation in the Niobrara Formation, then burial depth is not the only fac-

tor governing maturation in this area. Comparison of isopach and resistivity maps (Figures 13, 14) shows that depositionally thin areas of the Niobrara correlate with areas where Niobrara resistivity values are high, indicating possible facies control of organic carbon or paleostructural control of geothermal gradient. Isopach thins may indicate areas of increased organic carbon deposition or broad paleostructural highs that were associated with increased geothermal gradient. Whatever the explanation, we are confident that increasing resistivity values do reflect increasing hydrocarbon saturation in the Niobrara in this region because: (1) most reported oil shows in the Niobrara Formation are located within the 60-ohm resistivity value of this formation (Figure 13); and (2) Smagala et al (1984) demonstrated that vitrinite reflectance and resistivity values of the Niobrara Formation are directly proportional, meaning that resistivity indicates the maturation level.

Our lineament map corresponds well with Niobrara oil production data (Figure 15). Most of the wells reported to produce oil from the Niobrara in this part of the basin occur in R65-64W, T15-16N, Laramie County, Wyoming. These wells are located in an area of closely spaced, intersecting, northeast- and northwest-trending lineaments that is along a regionally extensive zone of northeast-trending lineaments. Of the 44 wells that produce oil in this area, 35 wells, or 79.5%, are located within 0.1 mi (0.16 km) of a lineament. To test whether this correlation could be a random occurrence, we produced four independent attempts of randomly plotted wells in this 12 × 12-mi (19 × 19-km) area using a grid of 0.1 mi (0.16 km). In these random plots,

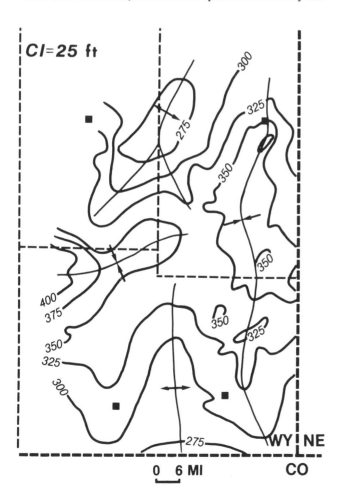

Figure 14—Isopach map of Niobrara Formation. C.I. = 25 ft.

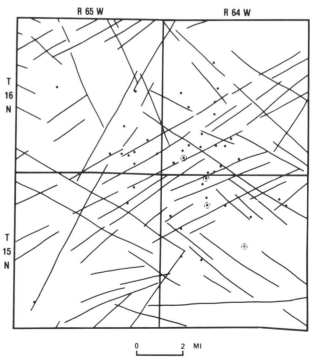

Figure 15—Distribution of lineaments and oil wells in Silo field area, Laramie County, Wyoming. Dots mark location of wells producing oil from Niobrara Formation. Wells for which core data were obtained are circled.

43%, 34%, 32%, and 45% of the wells were located within 0.1 mi (0.16 km) of a lineament. Additionally, as previously mentioned, core data for three wells producing oil from the Niobrara Formation show that oil-saturated vertical fractures are present. Two of these wells (Goertz and Lee) are located within 0.1 mi (0.16 km) of a northeast-trending lineament. However, the core description of the dry hole (Unioil 1 Hillsdale) did not report fractures in the Niobrara, and this well is located more than 0.5 mi (0.8 km) from any lineament. This report increases our confidence that oil production from these wells is strongly influenced by fractures and that Landsat imagery can be used effectively to map these fractures. Additionally, of the 35 wells located along lineaments, 24 wells, or about two-thirds, are located along northeast-trending lineaments, which supports our idea that northeast-trending fractures might be pathways for oil migration and loci for storage in the Niobrara Formation.

CONCLUSIONS

We used the areal distribution of resistivity values in the Niobrara Formation to identify an exploration fairway for Niobrara fracture production. Once such a fairway is identified, optimal areas for exploration may be identified using fracture patterns and density determined from Landsat imagery. Based on this approach, numerous places in the northern Denver basin appear prospective for Niobrara fracture production.

This approach may be applied to explore for oil in fractured reservoirs elsewhere. It requires: (1) defining and mapping a basement fracture network, (2) understanding the tectonic history of an area, (3) understanding the timing of source rock maturation, and (4) mapping the geographic extent of mature source rock. However, to apply this approach, one set of basement fractures must parallel the principal horizontal compressive stress during an orogenic event that is time contemporaneous with oil generation, which is probably the norm, rather than the exception, in many places in the Rocky Mountains.

To use Landsat imagery effectively, the Landsat analysis must be integrated with subsurface data, geophysical data, and a detailed understanding of the geologic history of the area. Analysis of Landsat imagery should be used as one tool of an integrated data package that can yield critical information to help identify the best places to drill for oil. A similar approach also may be used to explore for oil in fractured reservoirs elsewhere.

REFERENCES CITED

Bechtold, I. C., M. A. Liggett, and J. F. Childs, 1973, Regional tectonic control of Tertiary mineralization and recent faulting in the southern Basin-Range province: an application of ERTS-1 data: Symposium on Significant Results Obtained from ERTS-1, NASA, v. 1, p. 425-432.

Chapin, C. E., and S. M. Cather, 1981, Eocene tectonics and sedimentation in the Colorado Plateau, in B. D. Dickenson and D. P. Payne, eds., Relations of tectonics to ore deposits in the Southern Cordillera: Arizona Geological Society Digest, v. 14, p. 173-197.

Clayton, J. L., and P. J. Swetland, 1980, Petroleum generation and migration in Denver basin: AAPG Bulletin, v. 64, p. 1613-1633.

Droullard, E. K., 1963, Tectonics of the southeast flank of the Hartville uplift, Wyoming, in D. W. Bolyard and P. J. Katich, eds., Guidebook to the geology of the northern Denver basin and adjacent uplifts: Rocky Mountain Association of Geologists 14th Annual Conference, p. 176-178.

Engebretson, D. C., A. Cox, and G. A. Thompson, 1984, Correlation of plate motions with continental tectonics: Laramide to Basin-Range: Tectonics, v. 3, p. 115-120.

Gries, R., 1983, North-south compression of Rocky Mountain foreland structures, in J. D. Lowell, ed., Rocky Mountain foreland basins and uplifts: Rocky Mountain Association of Geologists, p. 9-32.

Harnett, R. A., 1968, Niobrara oil potential: Wyoming Geological Association Earth Science Bulletin, v. 1, p. 37-48.

Livaccari, R. F., and S. B. Keith, 1984, Tectonic evolution of Sevier-Laramide foreland structures from latest Jurassic through the Eocene (abs.): AAPG Bulletin, v. 68, p. 501.

——— ——— 1985, Paleotectonic maps: Sevier to Laramide orogeny: Earth Satellite Corporation and Magmachen Exploration.

Lockridge, J. P., and P. A. Scholle, 1978, Niobrara gas in eastern Colorado and northwestern Kansas, in J. D. Pruit and P. E. Coffin, eds., Energy resources of the Denver basin: Rocky Mountain Association of Geologist Guidebook, p. 35-49.

MacMillan, L., 1980, Oil and gas of Colorado: a conceptual view, in H. C. Kent and K. W. Porter, eds., Symposium on the Denver basin: Rocky Mountain Association of Geologists, p. 191-197.

Mallory, W. W., 1977, Oil and gas from fractured shale reservoirs in Colorado and northwest New Mexico: Rocky Mountain Association of Geologists Special Publication 1, 38 p.

McCaslin, J., and B. Williams, 1984, U.S. drilling spotty, heavy in some areas: Oil & Gas Journal, v. 82 (October 22), p. 145-152.

McGrew, L. W., 1961, Geology of the Fort Laramie area, Platte and Goshen Counties, Wyoming: USGS Bulletin 1141-F, p. F1-F38.

Meissner, F. F., 1978, Petroleum geology of the Bakken Formation, Williston basin, North Dakota and Montana, in The economic geology of the Williston basin: Montana Geological Association 24th Annual Conference, Williston Basin Symposium, p. 207-227.

Merin, I. S., and R. C. Michael, 1985, Application of structures mapped from Landsat imagery to exploration for stratigraphic traps in Paradox basin (abs.): AAPG Bulletin, v. 69, p. 287-288.

Momper, J. A., 1963, Nomenclature, lithofacies, and genesis of Permo-Pennsylvanian rocks—northern Denver basin, in D. W. Bolyard, and P. J. Katich, eds., Guidebook to the geology of the northern Denver basin and adjacent uplifts: Rocky Mountain Association of Geologists 14th Field Conference, p. 41-67.

O'Leary, D. W., J. D. Friedman, and H. A. Pohn, 1976, Lineament, linear, lineation: some proposed new standards for old terms: GSA Bulletin, v. 87, p. 1463-1469.

Podwysocki, M. H., H. A. Pohn, J. D. Phillips, M. D. Krohn, T. L. Purdy, and I. S. Merin, 1982, Evaluation of remote sensing, geological, and geophysical data for south-central New York and northern Pennsylvania: USGS Open-File Report 82-319, 59 p.

Pollastro, R. M., and C. J. Martinez, 1985, Mineral, chemical and textural relationships in rhythmic-bedded hydrocarbon productive chalk of the Niobrara Formation, Denver basin, Colorado: The Mountain Geologist, v. 22, no. 2, p. 55-63.

——— and P. A. Scholle, 1984, Hydrocarbons exploration, development from low permeability chalks, Upper Cretaceous Niobrara Formation, Rocky Mountain region: Oil & Gas Journal, v. 82 (April 23), p. 140-145.

Rehrig, W. A., and T. L. Heidrick, 1976, Regional tectonic stress during the Laramide and late Tertiary intrusive periods, Basin and Range province, Arizona: tectonics of Arizona: Arizona Geological Society Digest, v. 10, p. 205-228.

Salgat, B., 1983, Scully field—Marion County, Kansas (abs.): AAPG Bulletin, v. 67, p. 1327.

Scholle, P. A., 1977, Chalk diagenesis and its relation to petroleum exploration: oil from chalks, a modern miracle?: AAPG Bulletin, v. 61, p. 982-1009.

Silverman, M. R., 1984, Petroleum geology and exploration of Scotts Bluff trend, northeastern Denver basin, Nebraska: AAPG Bulletin, v. 68, p. 527-528.

Smagala, T. M., C. A. Brown, and G. L. Nydegger, 1984, Log-derived indicator of thermal maturity, Niobrara Formation, Denver basin, Colorado, Nebraska, Wyoming, *in* J. Woodward et al, eds., Symposium on hydrocarbon source rocks of the greater Rocky Mountain region: Rocky Mountain Association of Geologists, p. 355-364.

Stearns, D. W., and M. Friedman, 1972, Reservoirs in fractured rock, *in* Stratigraphic oil and gas fields—classification, exploration methods, and case histories: AAPG Memoir 16, p. 82-106.

Thomas, G. E., 1971, Continental plate tectonics: southwest Wyoming, *in* A. B. Renfro, ed., Symposium on Wyoming tectonics and their economic significance: Wyoming Geological Association 23rd Annual Field Conference on Wyoming Tectonics, p. 103-124.

Warner, L. A., 1978, Colorado lineament: a middle Precambrian wrench fault system: GSA Bulletin, v. 89, p. 161-171.

Weimer, R. J., 1978, Influence of transcontinental arch on Cretaceous marine sedimentation: a preliminary report, *in* J. D. Pruit and P. E. Coffin, eds., Symposium on energy resources of Denver basin: Rocky Mountain Association of Geologists, p. 211-221.

———— 1980, Recurrent movement on basement faults, a tectonic style for Colorado and adjacent area, *in* H. C. Kent and K. W. Porter, eds., Symposium on the Denver basin: Rocky Mountain Association of Geologists, p. 23-35.

Wier, C. E., F. J. Wobber, O. R. Russell, R. V. Amato, and T. V. Leshendok, 1974, Application of ERTS-1 imagery to fracture related mine safety hazards in the coal mining industry: ERTS Program Office, NASA, 136 p.

Reprinted by permission of the Environmental
Research Institute of Michigan from *Proceedings of
the Seventh Thematic Conference on Remote Sensing
for Exploration Geology*, 1989, v. 2, p. 1111-1123.

RECOGNIZING THRUST FAULTS, AND EXPLORATION IMPLICATIONS*

Gary L. Prost
Amoco Production Company
Houston, Texas U.S.A.

ABSTRACT

Thrusts as a class of faults are the most difficult
to recognize because the low angle of the fault causes
the trace to follow topography rather than form a lineament. Clues to recognizing thrusting are illustrated in
the Sulaiman Range of Pakistan. They include repeated
section, an irregular thrust trace that is overall convex
in the direction of transport, abrupt changes in strike,
dip, and structural style, overlapping folds with common
vergence toward the foreland, folds with amplitude
greater than one quarter of the wavelength, and folds
with axial length several times wavelength. Tear faults,
imbricate fans, fold geometry and vergence may provide
clues to subthrust structure.

Anticlines carried on thrusts can provide hydrocarbon traps. A correct remote sensing interpretation
will lead to recognizing fold asymmetry and subsequent
offset of the structure at depth. One can also infer a
sequence of fold development, depth of erosion and
breaching, and hanging wall thickness. This knowledge
can contribute to a successful exploration program.

1.   INTRODUCTION

An example of world-class production from a fold-thrust province can
be seen in the Wyoming overthrust. By 1981 seventeen fields with an estimated recoverable 6.7 billion barrels of oil and 58.4 Tcf of gas had been
found (P.I., 1981). Attention is directed to other thrust belts around the
world because structural complexity and remoteness hinder exploration,
creating the potential for large new discoveries. Remote sensing has been
used in the early mapping and evaluation of poorly studied regions (e.g.,
Banks and Warburton, 1986; Berry and Nishidai, 1988). These maps help
establish models of tectonic development, depth of burial, timing of generation and expulsion, and timing of structural growth. Thus, the
exploration implications of a proper interpretation of thrust belts are
such that geologists can, before visiting an area, determine the location
of structural traps, which traps are worth testing, which part of the
structure to test, and how deep to drill.

The geomorphic evidence for thrusting is examined and compared to the
expression of other faults and unconformities. No single feature identifies thrusting, but rather the coincidence of several factors builds a case

---

*Presented at the Seventh Thematic Conference on Remote Sensing for
Exploration Geology, Calgary, Alberta, Canada, October 2-6, 1989.

1111

for a thrust interpretation. An example from Pakistan illustrates the geomorphic criteria in an area with excellent exposures and classic thrust features. The structure interpretation is supported by construction of cross sections, and together these are used to suggest areas of exploration interest.

2. INTERPRETATION CRITERIA

Most faults are recognized by abrupt changes in strike, dip, or rock type, often associated with an escarpment. The trace of a high-angle fault tends to be linear, hence the term "lineament". The trace of a thrust is generally irregular, following topographic contours, and making identification of these faults difficult. There are, however, both direct and indirect criteria for recognizing thrusts.

Direct indications of thrusting include abrupt changes in strike or dip, representing different structures in the hanging wall and footwall (Fig. 1). The hanging wall is often topographically higher than the footwall, and the fault trace runs along a break in slope at the base of the upper plate. Where there is no topographic relief, the hanging wall can be recognized because the strike of bedding is usually parallel to the thrust front. In many cases the fault trace in plan view is convex in the direction of transport (Fig. 2). Part of the irregular fault trace may be a result of erosional outliers, or klippen, in front of the main sheet. Leading edge anticlines characterize thrusts that terminate by ramping up-section; otherwise the hanging wall units generally dip the same direction as the thrust fault. Imbricate fans, where thrust splays cut the hanging wall into stacked slivers, appear as zones of parallel ridges, and are recognized as thrusts by observing repeated section. Folds carried on a thrust tend to have a common asymmetry, verging toward the foreland. These often form imbricated, or stacked anticlines or synclines, depending on the level of erosion (Fig. 3). Detachment folds in the hanging wall tend to be concentric, formed by flexural slip, and have amplitudes greater than a quarter wavelength. The axial length of these folds is generally several times the wavelength, and these folds are frequently sinuous (Fig. 4). Abrupt changes in structural style, as from tight to open folding, or imbricate ridges to folds, often occur across thrust faults (Fig. 5).

Indirect clues to thrusting include tear faults, lateral ramp anticlines, monoclines, and relaxation faults. Tear faults tend to be linear and have characteristics of strike-slip faults, but can also have normal offset, or have monoclines developed along trend. They generally end at the thrust front, either abruptly or by curving into the thrust fault (Fig. 6). Folds on either side of a tear fault may have opposite vergence, or be at different stages of structural development. Monoclines form in the hanging wall where a tear fault lies above a faulted footwall. In other cases lateral ramp anticlines form over a footwall fault (Fig. 7). Relaxation faults develop on the backlimb of hanging wall folds or where the thrust ramps upsection. These listric normal faults are scoop-shaped, concave toward the hinterland. Thrusts terminate along strike by transferring their displacement to folds or overlapping thrust faults. Thus folds that merge into faults along strike may suggest thrusting.

Angular unconformities have some of the same characteristics as thrusts. They are characterized by abrupt changes in strike and/or dip. Folding above the unconformity will be seen below the unconformity, unlike many hanging wall folds. Bedding above an unconformity is, by definition, less deformed than below, whereas the upper plate of a thrust may be more deformed than the lower. Thrusts repeat the section; unconformities do not.

Reverse faults superimpose younger rocks on older, as do thrusts. The high angle of the fault plane causes a linear fault trace, distinguishing this class of faults from thrusts.

Subthrust plays are receiving increasing attention, and are still more difficult to recognize. Folding that occurred post-thrusting will be expressed at the surface and will extend below the hanging wall, but the age of folding generally cannot be determined with certainty from remote sensing data alone. Subthrust highs can act as a buttress, causing deeper erosion into the sheet and tear faults along the margins of the uplift. Subthrust structure may be implied by changes in thickness of the hanging wall. The thickness of the thrust plate can be estimated on the basis of the geometry of detached folds. A series of curves developed by Jamison (1987) relate fold interlimb angle and backlimb dip to the ratio of fold amplitude (from elevation data) divided by thrust sheet thickness. Hanging wall thickness changes can be hung from surface elevations to reveal the shape of the thrust surface. Fold vergence can also be used to speculate whether the thrust sheet is riding up onto a structure (hinterland verging) or moving over a subthrust structure (foreland verging; Dunne and Ferrill, 1988). Imbricate fans develop over footwall faults or where a thrust ramps upsection. These criteria are illustrated in the following example.

3.   SULAIMAN RANGE, PAKISTAN

The Sulaiman Ranges consist of imbricate packages of Mississippian through Neogene age units thrust south-southeast by the collision of India with Asia (Asrarullah and Abbas, 1979; Kazmi and Rana, 1982). The main episode of thrusting began about 80 million years ago and continued to 53 million years (Powell, 1979). Some shortening continues. The Sulaiman Range is a thrust that slid southward between the Jacobabad high to the west and the Sahiwal-Horunabad high to the east (Farah, et al., 1977). Piggyback anticlines formed along thrusts that splay upward from the main detachment, probably in Eocambrian evaporites (not exposed). The Landsat interpretation suggests that the youngest thrusting is in the south, and involves the youngest units (Paleocene-Eocene). This is revealed by the pattern of folding seen in the southwest part of the range (Figs. 8,9). The anticlines at Sui and Uch, as well as Zindapir, are recent features, as indicated by upturned gravel along their flanks. They are probably carried on thrusts whose leading edges are buried beneath alluvium of the Indus plain (Fig. 10). This is suggested because the fold axes are parallel to but in front of the exposed thrusts. Large faults with strike-slip displacement die out in folds or bedding, and are considered tear faults. The displacement along these faults, marked by monoclines along part of their extent, is right-lateral in the west and left-lateral in the east, allowing southward translation of the Sulaiman thrusts. Production to date is limited to the fringes of the thrust belt (Dhodak, Sui, Uch, and seeps at Tadri, Dunagan, Sarpusht, and Spintangi). In part this is due to poor access to the interior, but it is also evident from the progression of units that folds are breached to increasingly deeper levels as one proceeds into the thrust belt. Interior folds may have breached reservoirs, or source rock maturity may be at peak generation around the thrust margins, as indicated by seeps and fields, but in the gas window or past peak generation toward the interior.

A Landsat structural interpretation shows the principal thrust features of this region (Fig. 11). Note the curved thrust front, the imbricate anticlines, the consistent fold asymmetry, and the tear faults. An attempt to interpret subthrust structure is based on fold vergence patterns and an analysis of fold geometry, suggesting subthrust surface dip and thrust sheet thickness, respectively (Fig. 12). Topographic maps (Tactical Pilotage Charts, 1:500,000) were used to obtain fold amplitudes. Dips are estimated from Landsat imagery to the nearest five degrees.

A good prospect combines several favorable factors, such as large size, unbreached reservoirs, and relatively simple deformation. These criteria can be observed on imagery and are used to rank prospects. Other factors, such as proximity to mature source rock or timing of generation with respect to structural growth require other information sources. A cross section through the thrust requires that the image interpreter think in a third dimension and constrains the map to geometrically plausible structures. Depth information thus provided can help lay out a seismic survey and perhaps even guide the drill. Bearing in mind that the structural high shifts at depth in thrusted folds, one can locate structures such as Tadri where drilling the backlimb will test both the surface fold and a footwall high. A seep on the west plunge only increases confidence in the existence of nearby mature source rock.

CONCLUSIONS

The convergence of several lines of evidence leads to the recognition of thrusts, among the most difficult structures to interpret using remote sensing. Among those features most characteristic of thrusts are repeated section, irregular fault trace, asymmetric folding with common vergence, discordant strikes and dips, imbricate ridges, stacked folds, and folds with amplitude greater than quarter wavelength and axial length many times the wavelength.

Subthrust prospects may also be surmised. Fold geometry and vergence can be used to speculate on the depth and shape of the thrust surface, revealing possible subthrust plays. Imbricate fans and anticlines develop over thrust ramps.

Unconformities can be confused with thrust faults on imagery. Folding affects bedding above and below the unconformity; hanging wall folds often have no counterpart below the decollament. Thrusts repeat the section, unlike unconformities. Reverse faults are most like thrusts, but dip at angles greater than 45 degrees. Reverse faults are more likely to be expressed as lineaments.

Remote sensing interpretation in an exploration program is helpful prior to leasing, prior to geophysical surveys, and prior to detailed field mapping. The proper interpretation of thrusting can provide the location and distribution of anticlines, establish fold asymmetry, give insights as to the depth of erosion, provide a relative age of folding across the thrust belt, and perhaps reveal the location of subthrust folds.

REFERENCES

Asrarulla, Z. Ahmad, and S. G. Abbas, 1979, Ophiolites in Pakistan; an introduction: in A. Farah and K. A. De Jong (eds.), Geodynamics of Pakistan, Geologic Survey of Pakistan, p. 181-185.

Banks, C. J., and J. Warburton, 1986, 'Passive-roof' duplex geometry in the frontal structures of the Kirthar and Sulaiman mountain belts, Pakistan: Journal Structural Geology, v. 8, p. 229-237.

1114

Berry, J. L., and T. Nishidai, 1988, Hydrocarbon potential of part of the margin of the Tarim Basin from Landsat: a case history: Proc. of 6th Thematic Conf. on Remote Sensing for Exploration Geology, Houston, Texas, May 1988, p. 49-63.

Dunne, W. M., and D. A. Ferrill, 1988, Blind thrust systems: Geology, v. 16, p. 33-36.

Farah, A., M. A. Mirza, M. A. Ahmad, M. H. Butt, 1977, Gravity field of the buried shield in the Punjab plain, Pakistan: Geol. Soc. Am. Bull., v. 88, p. 1147-1155.

Jamison, W. R., 1987, Geometric analysis of fold development in overthrust terranes: Journal Structural Geology, v. 9, p. 207-219.

Kazmi, A. H., and R. A. Rana, 1982, Tectonic map of Pakistan: Geologic Survey of Pakistan, 1:2,000,000.

P.I., 1981, The Overthrust belt-1981: Petroleum Information, p. 15-50.

Powell, C. McA., 1979, A speculative tectonic history of Pakistan and surroundings: some constraints from the Indian Ocean: in A. Farah and K. A. DeJong (eds.), Geodynamics of Pakistan, Geologic Survey of Pakistan, p. 5-24.

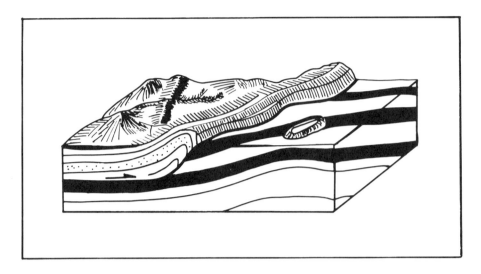

FIGURE 1. Diagram of sinuous leading-edge anticline illustrating
irregular fault trace, erosional outlier (klippe), and
discordant strikes/dips across the fault. Hanging wall
beds strike parallel to the thrust trace and dip parallel to
the fault.

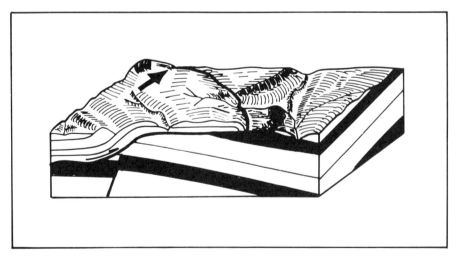

FIGURE 2. Thrusting is suggested by a break in slope, thrust trace
convex in the direction of transport (arrow).

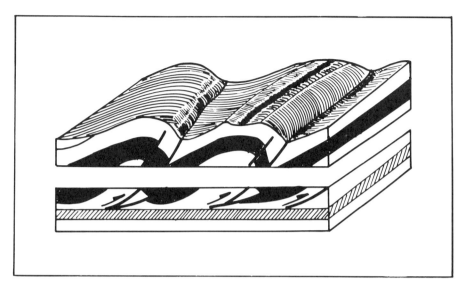

FIGURE 3. Stacked anticlines (or synclines, depending on level of erosion) imply thrust splays.

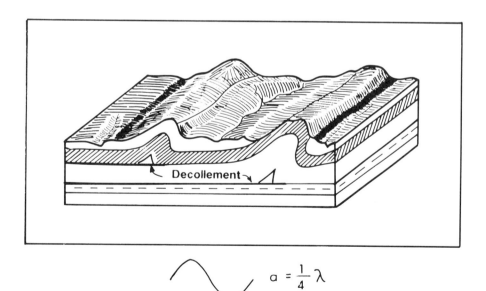

$$a = \frac{1}{4}\lambda$$

FIGURE 4. Folds with amplitudes greater than a quarter wavelength, and axial length several times wavelength, imply detached folding.

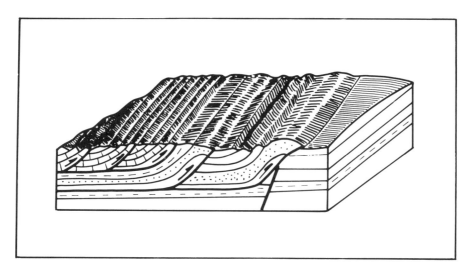

FIGURE 5. Abrupt changes in structural style, as from imbricate sheets to leading edge folds, suggest thrusting.

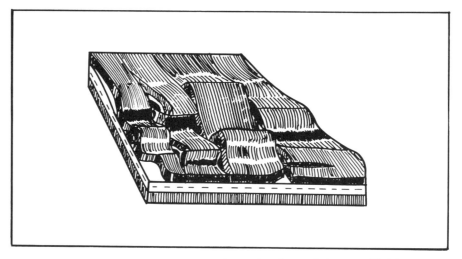

FIGURE 6. Tear faults can curve into the frontal thrust. Folds may have opposite vergence on either side of a tear fault. Displacement can appear normal or strike-slip. The thrust transfers displacement to folding along strike.

1118

552

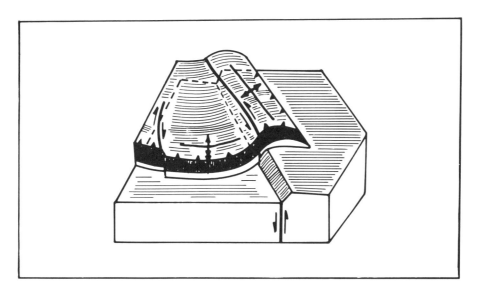

FIGURE 7. Lateral ramp anticlines form over footwall faults.

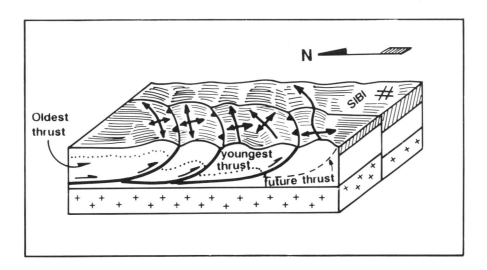

FIGURE 8. Sequence of thrusting northeast of Sibi.
Pattern of overlapping folds and progressively
younger units southward suggest that
thrusts are youngest to the South.

1119

Figure 9. Landsat image of the southern Sulaiman Range showing repeated section exposed in stacked anticlines (A, B, C), tear fault/monocline (D), and tear fault curving into a thrust (E).

1120

Figure 10. Landsat image of the southern Sulaiman Range showing young folds at Sui (A) and Uch (B), curved thrust front (C), and older, more deeply-eroded folds such as Tadri (D).

1121

FIGURE 11. LANDSAT STRUCTURAL INTERPRETATION OF PART OF THE SULAIMAN RANGE.

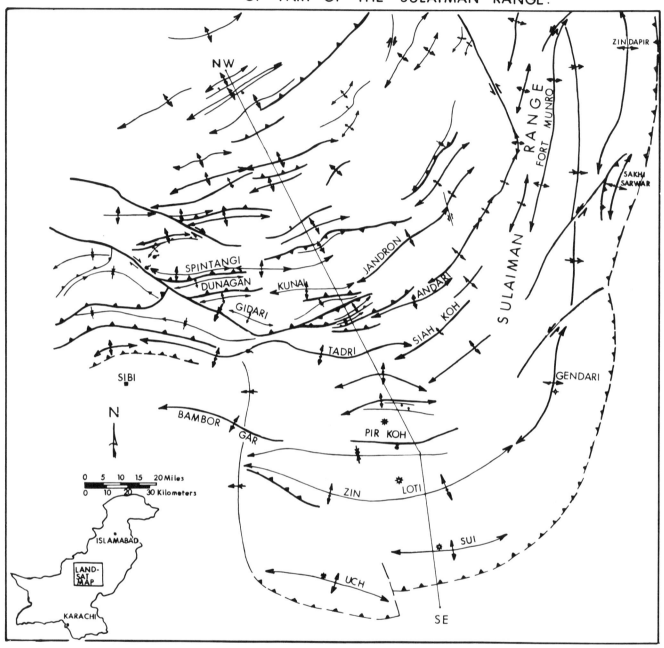

FIGURE 12. CROSS SECTION FROM NORTHWEST TO SOUTHEAST THROUGH THE SULAIMAN RANGE (SEE FIGURE 11 FOR LOCATION).

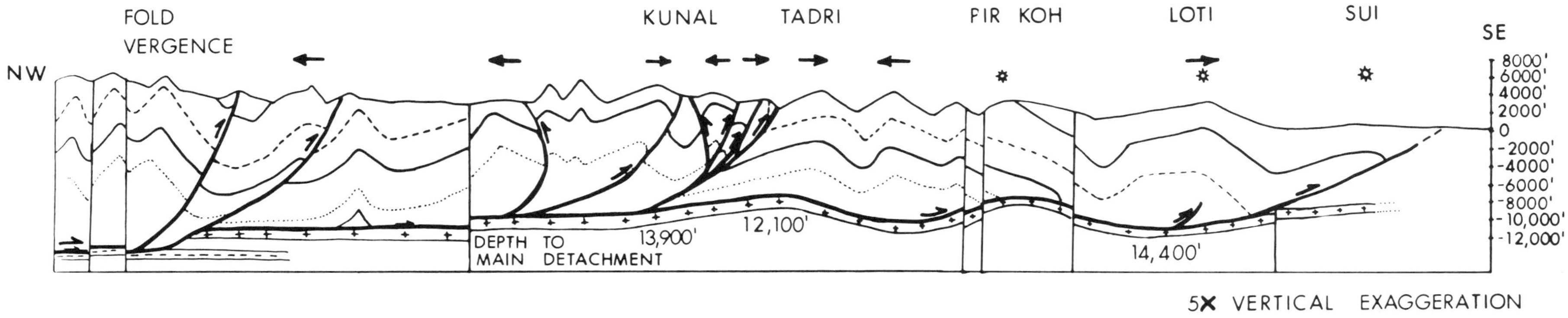

SPECTRAL REMOTE SENSING INVESTIGATION OF LISBON VALLEY, UTAH*

Donald B. Segal, Michael DeC. Ruth and Ira S. Merin
Earth Satellite Corporation
Chevy Chase, MD., USA

Hiroshi Wantanabe and Koichi Soda
Japex Geoscience Institute, Inc.
Tokyo, Japan

Osamu Takano and Masaharu Sano
Earth Resource Satellite Data Analysis Center
Tokyo, Japan

ABSTRACT

Specific diagenetic mineral assemblages within the Wingate Formation are closely associated with hydrocarbon production at the Lisbon Valley Anticline. These minerals can be identified with remotely sensed data acquired in the visible and near-infrared portions of the spectrum. The Wingate Formation, exposed along the southwestern flank of the structure, has a relatively uniform composition and appearance over the entire Colorado Plateau, except at isolated localities such as Lisbon Valley, where it is locally bleached. Previous workers have suggested that hydrocarbon microseepage may account for the association of bleached Wingate sandstone with uranium mineralization and oil and gas production at Lisbon Valley. Landsat Multispectral Scanner (MSS) and airborne Thematic Mapper Simulator (TMS) data are used to map the bleached Facies on the basis of albedo, lack of ferric-iron, and the abundance of clay minerals. Two spatially distinct occurrences of the bleached facies exist: a) rocks overlying the reservoir at the Lisbon Field, and b) the Three Step Hill area, located downdip from the Little Valley Field.

Preliminary analysis using TMS data suggests that two types of bleached rock can be distinguished: a clay rich rock exposed at the Lisbon Field and a moderately clay rich rock exposed at Three Step Hill. The broad widths and positions of the spectral bands preclude making unique mineralogic determinations with these data. Airborne high-resolution spectroradiometric data, thin sections, and XRD data show that an abundance of clay minerals are mappable within the bleached Wingate facies. The XRD analysis suggests that both the bleached and unbleached rocks contain more kaolinite than illite. The airborne spectral response of the bleached rocks exposed over the Lisbon Field is characteristic of kaolinite. The Three Step Hill area has an airborne spectral response characteristic of a muscovite-like clay.

The results of this study suggest that a correlation may exist between the abundance of clay minerals, particularly kaolinite, and apparent hydrocarbon microseepage. Because one of the principal differences between the bleached and unbleached rocks is in the relative abundance of clay minerals, areas of potential hydrocarbon induced diagenetic alteration may be mappable using broad band sensors.

* Presented at the International Symposium on Remote Sensing of Environment, Third Thematic Conference, Remote Sensing for Exploration Geology, Colorado Springs, Colorado, April 16-19, 1984.

273

SITE DESCRIPTION

Lisbon Valley is a northwest-trending doubly plunging anticline located in San Juan County, Utah (Figure 1). Lisbon Valley is ten miles long and about seven miles wide. Physiographic features associated with Lisbon Valley and present within the study area include Big Indian Valley and Three Step Hill to the southwest.

The Lisbon Valley anticline is a non-diapiric salt-cored fold which has been breached by erosion and undergone subsequent collapse. The physiography of the study area is controlled by the asymmetric geometry of the anticline and associated faults developed during the phases of salt growth and collapse in the anticlinal core. Cuestaform ridges and hogback structures are well developed on the southwest limb of the anticline. Steeply incised drainages, are common in the southern part of the study area.

Stratigraphy

The stratigraphic sequence exposed in the Lisbon Valley study area includes units from the Pennsylvanian Hermosa Group through the Cretaceous Dakota Sandstone, with three major unconformities. Pre-Paradox formations of Mississippian, Devonian, and Cambrian age underlie the exposed sequence and lie unconformably on a Precambrian basement. Tertiary and Quaternary alluvial deposits cover bedrock exposures in much of the study area. Mississippian, Devonian, Pennsylvanian, and Permian rocks produce petroleum at Lisbon Valley in three fields. The Mesozoic Wingate Formation is the most distinctively bleached unit at Lisbon Valley. It is overlain by the Kayenta Formation and the Navajo Formation.

The Late Triassic Wingate Formation is easily recognized as a thick-bedded erosion-resistant, cliff-forming sandstone prominently displayed in the southwest dipping cuesta of the west flank of the Lisbon Valley Anticline. Most of the sandstone is orange-red to gray-orange, with black streaks of desert varnish. Steep to vertical cliffs display pervasive vertical joints, tangential cross-bedding on a grand scale, and may show the effects of spalling of huge blocks of sandstone along vertical joint faces.

The mineralogical composition of the Wingate Sandstone is consistent both locally and regionally. Based on petrographic analysis, an average distribution of the mineral constituents is 70% quart, 12% feldspar, 11% calcite cement, 5% siliceous cement and quartz overgrowths, and less than one percent dark minerals such as biotite, muscovite, and ferromagnesians (Cater and Craig, 1970). The mineral grains are well-sorted and well-rounded. Thin shale laminae are found at the base of many of the cross-beds.

In specific locations on the southwest flank of Lisbon Valley the Wingate Formation is bleached to gray and displays anomalous mineralogical and weathering patterns. Specific effects include anomalous carbonate and limonite concentrations and the transformation of the Wingate Formation from an erosionally resistant sandstone to a friable unit which weathers easily into small irregular mounds where it is bleached. In many respects, including overall color, sorting, and weathering aspect, the bleached Wingate appears similar to outcrops of the Jurassic Navajo Formation and are easily distinguished by superposition relations to the underlying units.

Petroleum Geology

Three oil and gas fields currently produce in the Lisbon Valley study area. These include the Lisbon Field, the Little Valley Field, and the Big

274

Indian Field. Of these, the Lisbon Field is the largest, having estimated ultimate reserves of 42,850,000 barrels of oil and 250 MMCFG. Estimated reserves of the Little Valley Field are 150,000 BO and 18 MMCFG and at the Big Indian Field are 115,000 BO and 16 MMCFG from its Mississippian trap (Clark, 1978).

The Lisbon Field produces from a faulted anticlinal trap in the Devonian McCracken Sandstone and the Mississippian Redwall (Leadville) Limestone. Clastic breaks in the Pennsylvanian Paradox formation have shows of oil, but have never been produced. The gross thicknesses of the producing formations are 113 feet in the McCracken and 328 to 534 feet in the Redwall Formation (Clark, 1978).

The crest of the Mississippian structural culmination is located approximately four miles west of the axis of the Lisbon Valley salt anticline. The disparity of the pre- and post-Pennsylvanian structures suggests that field production is related to basement structure which predates the salt deposition (Clark, 1978). Accumulation in the Mississippian is controlled by a fault on the northeast side of the structure which exhibits a throw of 1,200 to 1,600 feet, and by an oil-water contact downdip. Closure of the structure is estimated at 1,900 feet, supporting an oil column 1,800 feet high (Clark, 1978). Thus, the hydrocarbon reservoir is filled almost to the spill point, making it an ideal location for study of the effects of vertical leakage.

Conel and Neisen (1981) have documented the close spatial correlation between commercially recoverable uranium deposits, exposures of bleached Wingate Sandstone, and the spatial distribution of oil and gas production at Lisbon Valley (Figure 2). Sour gas leaking from the hydrocarbon reservoir is believed to have created a reducing and acidic environment that mobilizes and redeposits previously stable ferric-iron minerals as sulfides, contributes to uranium mineralization, and geochemically controls many of the other mineralogical anomalies expressed in the rocks overlying the production.

HYDROCARBON MICROSEEPAGE

Oil and gas seeps have long been used in exploration for hydrocarbons. Indirect evidence of hydrocarbon microseepage may include a suite of closely interrelated geomorphic, chemical, and mineralogical anomalies, and magnetic, radiation, or related geophysical anomalies. Additionally, since plant physiology is strongly affected by soil chemistry, hydrocarbon induced alteration of the soil can have geobotanical consequences. Any one or combination of these anomalies may be detected by remotely sensed data in the appropriate exploration environment. Research into direct detection of reliable signatures caused by alteration phenomena is in the fore-front of geologic remote sensing.

Although there are a variety of diverse manifestations of hydrocarbon induced alteration, many anomalies are directly attributed to an essential premise. Leaking hydrocarbons and associated fluids such as hydrogen sulfide produce a reducing environment, thereby initiating diagenetic Eh/pH controlled reactions in the rocks and soil above hydrocarbon accumulations. These chemical reactions cause mobilization or precipitation of minerals in the stratigraphic column above the leaking reservoir. Mineralogic changes may, in turn, promote geomorphic or physical alteration of the surface and near-surface strata above the hydrocarbon reservoir.

The central zone of alteration attributed to hydrocarbon seepage is commonly known as the "alteration chimney." In many places, the alteration chimney is surrounded by a "halo" of geochemical influences that also are genetically related to the alteration and leakage phenomenon (Duchscherer, 1982). A distinction is made between the chimney and halo, because they may

275

represent dual expressions of hydrocarbon induced alteration that are seemingly contradictory (such as a radiation low surrounded by a radiation high). The chimney and halo together comprise the alteration "aureole," as shown in Figure 3. The geometric configuration of the alteration aureole, however, may be complicated by faults, the geometry of the trapping mechanism, and the chemical nature of the rocks that comprise the overlying strata.

Bleaching of Redbeds

Anomalous coloration of exposed rocks in hydrocarbon provinces has been noted by many authors in a variety of environments (Donovan, 1974; Ferguson, 1979, Conel and Niesen, 1981). In unaltered rocks, the red color is caused by the presence of ferric-iron in hematite, goethite, and other limonitic minerals. Bleaching of redbeds presumably results from the acidification or reduction of iron within the alteration aureole. Thus, ferric-iron adsorbed onto grain surfaces of the host rock is converted to the ferrous state and either re-precipitated as ferrous compounds or dissolved, leaving the formation with a bleached coloration. Because of the association with limonitic minerals, this bleaching can be detected using broad-band visible wavelength spectral remote sensing systems such as the MSS.

Radioactive Anomaly

In some areas uranium tends to associate with petroleum accumulations leading to anomalous gamma radiation concentrations that may have prospecting applications in the appropriate geological and geochemical setting. Radiation anomalies are hydrocarbon induced alteration phenomena, generally referred to as "radiation halos" (Armstrong and Heemstra, 1973; Roberts et al., 1978).

As in the mobilization of iron, the uranium mineralization phenomenon is controlled primarily by redox conditions in the subsurface. The geochemical behavior of uranium is essentially opposite that of iron because uranium is insoluble in its reduced state and soluble when oxidized. Reducing conditions, such as those caused by the local presence of hydrocarbons or hydrogen sulfide, favor the precipitation of uranium minerals. Oxidizing conditions, such as those encountered in shallow circulating meteoric waters, tend to mobilize uranium (Armstrong and Heemstra, 1973).

Clay Mineralization

The Eh-pH environment of the reduced chimney apparently induces diagenetic alterations that modify the population of clay minerals in the strata overlying a leaking reservoir. The leaking fluids may contain quantities of hydrogen sulfide and carbon dioxide sufficient to be slightly acidic. Acidic solutions may cause feldspars to be replaced by clay minerals such as kaolinite, illite, and chlorite. This appears to be true at the Cement Oil Field where Lilburn and Al-Shaieb (1983) have documented the precipitation of late diagenetic kaolinite and illite-smectite clays in the surface rocks overlying the petroleum accumulation. The leaking fluids have presumably caused enhanced weathering of feldspars to produce suites of clay minerals that are stable in the chemically anomalous setting. Many of the clay minerals can be detected and uniquely identified on the basis of diagnostic spectral absorption features.

Tonal and Textural Anomalies

Donovan (1981), Marrs and Kaminsky (1977), Saunders (1980), Collins et al. (1974), and others claim that many oil and gas fields are associated with tonal anomalies attributed to lithologic or soil alterations. Olmstead et al. (1976), Ferguson (1979), and Donovan (1974, 1981) report that many tonal

276

562

anomalies consist of erosionally altered redbeds overlying oil fields. The erosional variation is attributed to a change in cementing material from clay and hematite to calcite and dolomite. This change in cement results largely from Eh-pH controlled reactions involving oxidation of hydrocarbons occurring in strata overlying a leaking reservoir.

SPECTRAL DISCRIMINATION

The visible and near-infrared (0.4 - 2.5 micrometre) spectral reflectance properties of rocks and soils are dominated by electronic processes and vibrational processes that take place in transitional metal anions and in anion groups, respectively (Hunt, 1977). These processes give rise to absorption bands, which can be used to discriminate some geologic materials in multispectral images and to provide constraints on the mineralogical composition (Figure 4). Identification in the classic petrologic sense is rarely possible, however.

Ferric and ferrous iron are the most common sources of electronic absorption bands in terrestial rocks. The most intense ferric absorption bands occur near 0.7 and 0.9 micrometres; subordinate bands are centered at about 0.40, 0.45, and 0.49 micrometres (Hunt and Salisbury, 1970). The most intense vibrational absorption bands occur near 1.4, 1.9, 2.2, and 2.35 micrometres. Although the 1.4 and 1.9 micrometre bands are useful for analyzing laboratory spectra, atmospheric absorption by water generally precludes recording useful spectra or images in these spectral regions.

Although a variety of materials can be distinguished using MSS data, identification of specific lithologies on the basis of spectral response is rarely possible. Notably, ferric-iron bearing (limonitic) minerals, such as hematite, goethite, and jarosite, are readily distinguished from other lithologies in the MSS wavelength region. These minerals can be uniquely identified using high-resolution radiometric data owing to the presence of diagnostic spectral absorption features (Figure 5).

Recently developed technology permits discrimination of a variety of lithologies on the basis of the presence of narrow and well-defined absorption bands located in the near-infrared (2.0-2.5 micrometre) portion of the spectrum. The most common sources of absorption bands in this wavelength region are overtones of fundamental hydroxyl (OH) bond stretching and bending vibrations that occur near 2.77 micrometres (Hunt and Salisbury, 1970). The exact position and relative depths of absorption bands in the 2.0-2.5 micrometre spectral region are dependent upon the coordination of the OH radical with cations in the mineral lattice structure. Minerals that can be uniquely identified in this wavelength region include a variety of clays, micas, and carbonates (Figure 6). Carbonates exhibit distinct absorption features in this spectral range owing to overtone vibrations of the CO_3 ion (Hunt and Salisbury, 1971).

METHODS

Four levels of remotely sensed data were examined in this study, including Landsat multispectral scanner (MSS), airborne Thematic Mapper Simulator (TMS), aerial photography, and high-resolution airborne spectroradiometrics. The interpretive process involved a progression from large-scale, coarse resolution data to high-resolution data that covered an increasingly smaller area in greater detail.

Initially, Landsat MSS false-color composite imagery was produced using a decorrelation procedure based on the inverted principal component contrast enhancement (Soha and Schwartz, 1978). Because of the broad widths and positions of the MSS bands, only indications of overall surface reflectivity (albedo) were interpretable from this image. Determination of specific

277

minerals responsible for the observed albedo differences was not possible. Landsat MSS principal component and color-ratio composite imagery were processed in order to enhance the spectral contrasts diagnostic of limonitic rocks and vegetation (Rowan et al., 1974; Rowan and Abrams, 1978; Segal, 1983). In both image products, limonitic exposures were defined on the basis of large spectral contrasts between MSS bands 4 and 5 and MSS bands 6 and 7 (Figure 5). Vegetated areas were readily delineated on the basis of diagnostic contrasts between MSS bands 4 and 6 and MSS bands 6 and 7 (Figure 4, No. 6).

The Lisbon Valley, Utah, airborne TMS data were acquired by the Jet Propulsion Laboratory on August 28, 1979, using the NS001 scanner system (Mission 407, Site 379). Natural-color and false-color composite images were constructed using TMS bands 1, 2, 3 and TMS bands 2, 3, 4, respectively. Although these image products provided a familiar data base, as with the MSS false-color composite image, specific mineralogic determinations were not possible. Thus, standard band ratioing techniques were employed to enhance the spectral contrasts diagnostic of limonitic rocks, vegetation, and clay minerals. The TMS ratio 4/3 was selected in order to define vegetation, because of the presence of chlorophyll absorption in the TMS band 3 (0.65 micrometre) region and the lack of absorption in the TMS band 4 (0.82 micrometre) wavelength range (Figure 4, No. 6). The TMS 1/3 ratio was selected to differentiate limonitic versus non-limonitic rocks based on the falloff towards shorter wavelengths diagnostic of ferric-iron bearing minerals (compare curves 1 and 4, Figure 4). Clay minerals containing hydroxyl (OH) within their lattices and vegetation containing molecular water were defined by absorption in the TMS band 6 (2.2 micrometre) region relative to the higher reflectance in the TMS band 5 (1.6 micrometre) wavelength range (Figure 4, No. 2; Figure 6).

U.S. Geological Survey quadrangle-centered black-and-white aerial photography covering the Lisbon Valley area (Mission GS-VDXU, 1975) were acquired at a scale of 1:80,000. The photography was examined in stereo and mosaicked at 1:80,000 scale. Photomosaics, enlarged to scales of 1:48,000 and 1:24,000 were used to plot the positions of flight lines for the airborne spectrometer and also served as base maps for geologic field work. Copies of the photography were mounted on plexiglass and used for navigation during the airborne data acquisition.

Based on the results of reconnaissance field work and the image interpretations, flight lines for the airborne spectroradiometer were selected and plotted on a 1:24,000 scale photomosaic (Figure 7). Approximately 40 line-miles of data were collected. Visible (0.40 to 1.0 micrometre) and near-infrared (2.0 to 2.5 micrometre) airborne data were acquired on June 28, 1983, between 11:00 A.M. and 1:00 P.M. The Collins instrument provides 512 (2 nanometre wide) bands in the visible and 64 (8 nanometre wide) bands in the near-infrared. The spectroradiometer was drape flown at approximately 2,000 feet above the terrain in a twin engine Piper Aztec at an average speed of 110 miles per hour. The resulting spatial resolution is estimated at 20 metres. The data were collected by Geophysical Environmental Research, Inc.

Data processing consisted of waveform analysis and band-ratioing techniques. The waveform processing is based upon a best fit of Chebyshev polynomials to the spectral data and utilizes a mathematical pattern recognition approach. The Chebyshev processing was provided by the subcontractor and is discussed in Collins et al. (1983).

Band-ratioing was employed in order to enhance subtle spectral contrasts diagnostic of specific minerals. Band-ratios were constructed in order to separate limonitic versus non-limonitic rocks and vegetation, clay minerals, and carbonates. Using the visible data, a compound ratio of (.55/.65)-

278

(.75/.9)) was used to provide maximum separability between limonitic rocks and vegetation (Segal, 1983). The data were degraded by averaging spectral channels: band widths of 21 nanometres were used in order to reduce high-frequency variation. In the near-infrared data, ratios of the 2.1/2.15, 2.05/2.2, and 2.1/2.35 micrometre wavelength regions were used to differentiate kaolinite-rich rocks from background clay signatures and carbonates, respectively. Band widths of 17 nanometres were used for the near-infrared data ratios. Maps of the ratio values were produced by plotting the along-flight line variation in ratio value on a digitized data base. All ratio maps were produced by plotting the variation in ratio value as vertical distance from the trace of the flight line. The flight line trace was used as a base line, representing the average ratio value. Plots of unfiltered and spatially filtered data were examined. Spatial filtering consisted of plotting a moving average of every seven resolution cells. This served to reduce high-frequency variation due to small-scale topographic relief and instrument noise.

A total of 20 lithologic samples of the Wingate Formation were collected and subjected to whole rock chemical analysis and laboratory spectral reflectance measurements. Five of these were analyzed using x-ray diffraction. Four thin-sections were examined, including three bleached and one unbleached sample. Rock and soil were sampled in proportion to their relative surface exposure in order to approximate the field of view of the airborne instrument. Whole rock chemical analysis of 18 elements were obtained using the Barringer Lasertrace™ analytic technique. Laboratory reflectance spectra were obtained using the Barringer Portable Reflectance Spectrometer (REFSPEC™II). Laboratory reflectance data were collected in the 0.45 to 1.0 and 1.0 to 2.5 micrometre regions, yielding spectral resolution on the order of 1.5 and 3.5 nanometres, respectively. Mineralogic determinations, using x-ray diffraction, were performed on both whole rock and size fractionated samples. Both the greater than 2 micron residue and less than 2 micron "clay" size fractions were analyzed from 4° to 35° 2-theta. The "clay" sized fraction was glycolated and run through a restricted range from 4° to 14° 2-theta to refine the estimate of clay minerals present.

RESULTS

REMOTE SENSING DATA

Examination of Landsat MSS and airborne TMS false-color composite images reveals that the bleached Wingate Sandstone can be detected on the basis of overall albedo by using bands located primarily in the visible portion of the spectrum. Two spatially distinct occurrences of bleached Wingate are mappable, namely the Lisbon Field and Three Step Hill areas (Figure 7). The ability to delineate the bleaching in these images is not specifically related to the presence or absence of particular minerals, because the spectral band widths and their positions preclude unique mineralogical determinations. Examination of aerial photography and TMS imagery shows that topography in the bleached Wingate Sandstone is more rounded than in unbleached portions of the same formation.

Use of carefully selected spectral bands and specific digital enhancement techniques allows the detection of surface bleaching which can be directly attributed to specific mineralogies. Examination of the MSS color-ratio composite (CRC) and principal component images reveals that the bleached Wingate exposures are mappable primarily on the basis of a lack of limonite. Vegetation cover is minimal and consists primarily of Pinyon, Juniper and Sage. The distribution of vegetative cover appears similar on both the bleached and unbleached exposures. Analysis of the TMS CRC image confirms the absence of limonite in the bleached Wingate and also shows that these two areas are associated with an abundance of clay minerals. Thus,

279

although both the MSS and TMS data provide information diagnostic of limonite and vegetation, the TMS data allows further mineralogical determinations to be made owing to the diagnostic near-infrared spectral signature of OH-bearing minerals. The broad width of 2.2 micrometre TMS band precludes determination of specific clay mineralogies.

Both visible and near-infrared high-resolution airborne spectro-radiometric data were examined. The high spectral contrast of vegetation in the visible wavelengths (Figure 4, No. 6), precluded detection of the various lithologies where vegetation cover exceeds 30%. This is especially true for non-limonitic exposures, because of their generally flat visible spectral response (compare Nos. 1, 4, and 6, Figure 4). However, because of spectral contrast in the 2.0 to 2.5 micrometre near-infrared region, lithologies containing appreciable quantities of clays and/or carbonates that exhibit a large degree of spectral contrast were generally detectable under vegetation covers of 65-70%. Thus, in areas where vegetation cover exceeds 30% we see differentiation of underlying rock or soil types in the near-infrared data, whereas diagnostic signatures in the visible spectral range tend to be dominated (masked) by the vegetation.

Analysis of the visible spectroradiometric data generally confirms the results obtained with the MSS data. However, using the high-resolution data, the overwhelming effect of vegetation on the visible spectral response of the surface in conjunction with the small spot size of the airborne system largely limits the effective utility of the data for mapping surface bleaching on the basis of ferric-iron content. Pinyon and/or Juniper cover exceeding approximately 25% generally precluded detection of the iron anomalies with the visible data.

Figure 8 is a map showing the results of the visible spectroradiometric survey. Using the (.55/.65)/(.75/.9) compound ratio, low, medium, and high values are representative of limonitic rocks, non-limonitic rocks, and vegetation, respectively. Note the moderately high values associated with many of the Wingate exposures in the Lisbon Field and Three Step Hill areas. These values are largely a result of vegetation cover; however, several small exposures of bleached Wingate formation are apparent. These areas are displayed as medium ratio values and are plotted directly along the flight lines. A transition to relatively low compound ratio values, representing limonitic exposures, is apparent over the remainder of the Wingate Formation outcrops. High ratio values are observed where the flight lines cross small alluvial valleys and the dip-slope of the Hermosa Formation (NE of Big Indian Valley) due to the abundance of vegetation cover. Conversely, the abundance of ferric-iron in the maroon and red-colored Cutler Formation (exposed in Big Indian Valley) is associated with very low compound ratio values.

Because of the high spectral resolution in the 2.2 micrometre region the airborne near-infrared spectroradiometric data confirm and supplement the results interpreted from the broad-band TMS data. Unlike the visible portion of the spectrum, the high-frequency near-infrared clay mineral absorption bands dominate the spectral response of the surface such that the underlying lithology can be resolved even with approximately 70% vegetative cover.

The results interpreted from 2.05/2.2 band ratio map (not shown) are very similar to those observed with the 2.1/2.15 ratio. The 2.05/2.2 ratio serves as an indicator of overall clay content, although kaolinite may display slightly larger contrasts than other clay minerals. The 2.05/2.2 ratio values for the Lisbon Field area are very high, whereas the bleached exposures in the Three Step Hill area show moderate values. The unbleached Wingate exposures also show moderate to low 2.05/2.2 ratio values. This suggests that there is an overall abundance of clay minerals that is uniquely associated with the bleached Wingate exposures in the Lisbon Field area.

280

The results interpreted from the 2.1/2.15 ratio map reveal an anomalous geographic distribution of kaolinite within the Wingate Formation that is spatially associated with hydrocarbon production at the Lisbon Field. Figure 9 is a map showing the results of the airborne near-infrared spectro-radiometric survey. The 2.1/2.15 band ratio provided an effective means for defining the presence of kaolinite clay due to the large contrast between these two spectral regions that is characteristic of the kaolinite signature (Figure 6). Other clay minerals do not show as strong a contrast. For example, montmorillonite is essentially spectrally flat within this wavelength range.

Note the very large 2.1/2.15 ratio values that are associated with the bleached Wingate exposures in the Lisbon Field area. The spatial pattern of kaolinite occurrence defined with the spectroradiometric data corresponds closely with the bleaching mapped using the MSS and TMS color-ratio composite images.

The Three Step Hill area, located on the southern flank of the Lisbon Anticline, also appears bleached and clay-rich on the MSS and TMS imagery. However, the airborne near-infrared spectral response in this area is diagnostic of a muscovite-like clay rather than kaolinite. Notably, there is a small amount of gas production, known as the Little Valley Field (T30S, R25E NW 1/4 Sec. 27), located two miles updip from the Three Step Hill area.

LABORATORY SPECTRA

Laboratory spectra were examined for a total of 15 samples obtained from the Wingate Formation. Without exception, all spectra exhibit resolvable absorption features in the 0.48-, 1.4-, 1.9-, and 2.2-micrometre regions. These features indicate the presence of ferric iron, molecular water, and clay, respectively. Although the Wingate Formation is a relatively clean sandstone, clay mineral absorption features are readily apparent.

The spectra examined can be separated into two groups on the basis of overall albedo, depth of the 0.48 micrometre ferric-iron band, and hydroxyl absorption features located in the 2.2 micrometre region. The latter of these criteria appears to provide the most consistent and reliable basis for separation. The distinction of the two groups on the basis of albedo, and visible and near-infrared spectral characteristics corresponds closely with the "bleached" versus "unbleached" nomenclature determined on the basis of their appearance in the field.

Relative to the unbleached exposures, the bleached facies of the Wingate Formation exhibit a higher overall albedo, shallow 0.48 micrometre ferric-iron bands, and 2.2 micrometre bands that are dominated by the presence of kaolinite. Alternatively, the unbleached facies tend to show a relatively lower albedo, high contrast ferric iron features and 2.2 micrometre bands that are indicative of muscovite-like clays. For comparison, Figure 10 shows a portion of the near-infrared spectra for a bleached sample (L-10) and an unbleached sample (LV-2) of the Wingate Formation. Note the well developed doublet centered at 2.16 and 2.2 micrometres and the broad, shallow absorption feature in the 2.38 micrometre region that is characteristic of kaolinite (sample L-10). The near-infrared signature for the unbleached sample (LV-2) shows a minor shoulder in the 2.25 micrometre region, a sharp minimum at 2.2 micrometres, and a very wide wing rising toward the longer wavelengths. The relative sharpness of the 2.2 micrometre features and the broadness of the wings approaching the band are diagnostic of a muscovite-like clay material.

All bleached samples collected in the area of the Lisbon Field were classified as kaolinitic. Conversely, all unbleached samples appear to exhibit spectral features characteristic of muscovite. Of particular

281

interest is the observation that the only bleached sample that does not appear kaolinitic is sample L-15. This sample was obtained from the area just west of Three Step Hill (T30S, R25E, SW 1/4 Sec. 32). This area appears analogous to the Lisbon Valley Oil Field in both Landsat MSS and airborne TMS images. These data indicate that the exposed Wingate Formation at the Lisbon Valley Oil Field and at the Three Step Hill are both ferric-iron poor and clay-rich. However, the limited laboratory spectral data suggest that there is a difference in clay mineralogy between these two areas.

GEOCHEMISTRY

The mineralogy of the Wingate Formation was determined using X-ray diffraction (XRD) and the petrographic microscope (Table 1). Examples of XRD traces representing bleached and unbleached Wingate samples are given in Figure 10. The principal differences between bleached and unbleached facies of the Wingate Formation are the presence of plagioclase, the abundance of potassium feldspars, and a relative lack of clay minerals in the unbleached Wingate. Bleached samples contain an abundance of clay minerals, particularly kaolinite, and lack plagioclase feldspar. Additionally, unbleached Wingate samples contain small quantities of hematite and lack siderite, whereas bleached Wingate samples contain trace quantities of siderite and lack hematite. These differences in ferric versus ferrous iron (hematite versus siderite) content enable differentiation of bleached and unbleached Wingate exposures using broad-band remote sensing instruments that are sensitive to the visible portion of the spectrum. Differences in the total quantity of clay minerals between the bleached and unbleached Wingate permit mapping the distribution of bleached Wingate using broad-band near-infrared remote sensing instruments, such as the TMS.

Twenty Wingate samples were subjected to whole rock geochemical analysis to determine the concentrations of certain elements in the bleached and unbleached rocks. Results of this evaluation suggest that the principal chemical difference between the bleached and unbleached samples is the higher degree of variation in elemental concentration in bleached samples, which typically display coefficients of variation on the order of 100 to 1,000 times those of the unbleached samples. The disparity in variation is particularly apparent in the concentrations of the elements calcium, iron, carbon, potassium, and silicon. The nonuniform distribution of these elements in the bleached portion of the formation probably reflects incomplete chemical (mineralogical) reactions related to (or occurring as a result of) the bleaching process.

The principal cause for the red color of redbeds is grain surface coatings consisting of ferric oxide minerals or poorly crystalline ferric oxide mineral mixtures with clay minerals (Walker, 1967). It is reasonable to assume that bleaching of redbeds will be correlated with the quantity of ferric oxide minerals in the rock. This appears to be true for the Wingate Formation: thin-section analysis reveals that hematite occurs principally as coatings on grains and that nearly all of the grains of the unbleached Wingate are coated, whereas most of the grains composing the bleached Wingate lack coatings. Additionally, even though the XRD analysis shows no evidence of hematite, it does reveal that some of the bleached Wingate samples contain trace quantities of siderite. The unbleached Wingate does not contain siderite. That ferric oxide coatings may be poorly crystalline (Walker, 1967) could explain why the XRD analysis of these samples of Wingate did not reveal even trace amounts of hematite.

Although the bleached Wingate facies lacks disseminated hematite as grain-surface coatings, it does contain localized accumulations of ferric oxide nodules. Individual nodules may be as large as four inches in diameter and tend to occur in aggregates. Whole rock chemical analysis of some of the bleached samples containing ferric oxide nodules reveals that these rocks

282

568

are iron rich, containing substantially more total iron then the mean value
for the unbleached Wingate. Thin-section analysis reveals that these ferric
oxide nodules consist of areas in which the pore spaces between the sand
grains are completely occluded by hematite.

The presence of locally iron-rich, bleached Wingate, containing more
iron than that present in the unbleached Wingate, suggests that a process of
local enrichment of iron has occurred in the altered rock. Because the iron
rich bleached Wingate samples lack significant quantities of disseminated
ferric oxides, these rocks probably contain significant quantities of
ferrous-iron bearing minerals (e.g., sulfides or carbonates) and/or local
pockets of ferric oxides. None of the bleached samples analyzed by XRD
contained localized pockets of ferric oxides; however, such occurrences
appear to be common in places where the Wingate is locally bleached.

It is clear that the dominant clay mineral present in the Wingate
Formation is kaolinite and that the quantity of kaolinite appears to be
enriched in the bleached portions of the Wingate. Examination of thin-
sections of samples of bleached and unbleached Wingate reveals that kaolinite
occurs in a "book-like", apparently pore-filling fashion and that the
bleached Wingate contains significantly more kaolinite than the unbleached
Wingate. Additionally, some of the pore-filling kaolinite "books" present in
the bleached Wingate appear to have locally etched grains of quartz. Such
occurrences of kaolinite are clearly diagenetic rather than detrital (Wilson
and Pittman, 1977). One of the principal mechanisms of kaolinite formation
is by the alteration of feldspars (Keller, 1978). The presence of smaller
quantities of potassium feldspar, no plagioclase feldspar and larger
quantities of kaolinite in the bleached Wingate samples compared to the
unbleached samples implies that the feldspars were altered to kaolinite in
this instance. Moreover, the lack of any trace quantity of plagioclase
feldspar (as revealed by XRD) is strong evidence that the environment
governing the formation of kaolinite in the bleached Wingate was one in which
feldspars were unstable with respect to kaolinite. This implies the
existence of an acidic environment (Garrels and Christ, 1965).

SUMMARY AND CONCLUSIONS

Table 2 summarizes the results interpreted from the MSS, TMS, airborne
spectroradiometric, laboratory spectral and XRD analyses. Bleached Wingate
exposures were readily identified with the MSS and TMS data as a result of
ferric-iron content. In addition, bleached exposures were mappable with the
TMS data on the basis of relative abundance of clay minerals. The airborne
near-infrared spectroradiometric and laboratory spectral results show a
strong correspondence. These high-resolution data not only provide a basis
for differentiation of bleached and unbleached exposures but also enable
discrimination of the bleached Wingate exposed at the Lisbon Field from that
exposed in the Three Step Hill area.

Comparison of XRD and spectral data of Wingate samples shows that
spectral differentiation of bleached and unbleached Wingate can be made on
the basis of total clay content and clay type. The XRD analysis shows that
both unbleached and bleached rocks contain substantially more kaolinite than
illite; however, bleached rocks contain significantly more total clay than
unbleached rocks. Rocks having a spectral response characteristic of
kaolinite are those in which XRD analysis reveals greater than 10% kaolinite
clay. The single unbleached sample that was X-rayed contains only 1% illite
and 9% kaolinite, and has a spectral response characteristic of a muscovite-
like clay rather than kaolinite.

Although the amount of XRD data is limited, some inferences can be made
regarding the expected spectral signature of a sample containing various
mixtures of illite and kaolinite. It is clear that the ratio of kaolinite to

283

illite does not control the spectral response of a sample. Bleached samples exhibit kaolinite/illite ratios ranging between 4:1 and 8:1, whereas the unbleached sample has a 9:1 ratio. The spectral difference between bleached and unbleached Wingate is presumably a function of total clay content, particularly the abundance of kaolinite. Apparently, there is a critical threshold that must be reached before the diagnostic kaolinite clay spectral signatures are resolvable. This value appears to be about 10% even if the ratio of kaolinite to illite is as large as 9:1. For example, a sample containing 12% kaolinite and 3% illite appears to have a spectral response characteristic of kaolinite. Thus, there appears to be a threshold quantity of kaolinite (approximately 10%) that must be exceeded for a sample rich in kaolinite to have a spectral response characteristic of kaolinite.

Differences in the clay minerology between unbleached exposures and bleached exposures at the Lisbon Field and Three Step Hill areas were detected using airborne near-infrared data that was degraded both spectrally and spatially. Ratios were calculated using 17 nanometre band widths and were plotted using a seven-cell spatial filter, resulting in a simulated spatial resolution of approximately 20 x 140 metres. This suggests that broad-band near-infrared sensors, such as the TMS or TM systems, may provide a basis for distinguishing these three areas based strictly on differences in the abundance of clay minerals. In fact, preliminary examination of the TMS 1.6/2.2 band ratio shows that the bleached Wingate exposed in the Lisbon Field area can be differentiated from the Three Step Hill area and from unbleached Wingate exposures on the basis of this broad-banded spectral ratio.

In conclusion, anomalous surface mineralogy first detected through analysis of Landsat MSS and airborne TMS data has been correlated with known hydrocarbon production. Airborne spectroradiometric and laboratory data confirm and supplement the results by enabling specific mineralogic determinations to be made. The results of the study reveal an anomalous concentration of clay minerals within the Wingate Formation that is spatially associated with known hydrocarbon production at Lisbon Valley. The XRD analysis confirms that the bleached and unbleached rocks can be distinguished on the basis of relative abundance of clay minerals. The Lisbon Field area is bleached and rich in kaolinite. The Three Step Hill area also appears bleached and clay-rich on the MSS and TMS imagery, however, this area displays an airborne near-infrared spectral response diagnostic of a muscovite-like clay rather than kaolinite. Even though no known hydrocarbon production is directly associated with this specific portion of the structure, hydrocarbons are produced updip at the Little Valley Field. These data suggest that a correlation may exist between the abundance of clays, particularly kaolinite, and apparent hydrocarbon microseepage. laboratory spectral reflectance measurements, XRD, and thin-section analysis support the results obtained using the remotely sensed data and suggest that the effects of hydrocarbon microseepage may be mappable with broad-band sensors.

284

REFERENCES

Armstrong, F.E., and Heemstra, R.J., 1973, Radiation halos and hydrocarbon reservoirs; A Review: U.S. Bureau of Mines Information Circular 8579, 51 p.

Cater, F.W., and Craig, L.C., 1970, Geology of the salt anticline region in southwestern Colorado: U.S.G.S. Professional Paper 637, 80 p., 2 maps, 1:62,500 scale.

Clark, C.R., 1978, Lisbon, in Fassett, J.E., ed., Oil and Gas of the Four Corners Area: v. II, p. 662-665.

Collins, R.J., Petzel, G.H., and Everett, J.R., 1974, An evaluation of the suitability of ERTS data for petroleum exploration; Type III final report under contract NASA-21735, NTIS Report no. E74-10704, 149 p.

Collins, W., Chang, S.H., Raines, G., Canney, F., and Ashley, R., 1983, Airborne biogeophysical mapping of hidden mineral deposits: Economic Geology, V. 78, p. 737-749.

Conel, J.E., and Niesen, P.L., 1981, Remote sensing and uranium exploration at Lisbon Valley, Utah: IEEE, International Geoscience and Remote Sensing Symposium, v. 1, p. 318-324.

Donovan, T.J., 1974, Petroleum microseepage at Cement, Oklahoma - Evidence and mechanism: Am. Assoc. Petrol. Geol. Bull., v. 58, p. 429-446.

Donovan, T.J., 1981, Geochemical prospecting for oil and gas from orbital and suborbital altitudes, in Gottlieb, B.M., ed., Unconventional methods in exploration for petroleum and natural gas: Symposium II, Southern Methodist University Press, Dallas, p. 96-115.

Ferguson, J.D., 1979, The subsurface alteration and mineralization of Permian redbeds overlying several oil fields in southern Oklahoma, (Part 1): Shale Shaker, v. 29, n. 8, p. 172-208.

Ferguson, J.D., 1979, The subsurface alteration and mineralization of Permian redbeds overlying several oil fields in southern Oklahoma, (Part 2): Shale Shaker, v. 29, n. 9, p. 200-208.

Garrels, R.M., and Christ, C.L., 1965, Solutions, minerals, and equilibria: Harper and Row, New York, 450 p.

Hunt, G.R., 1977, Spectral signatures of particulate minerals in the visible and near-infrared: Geophysics, v. 42, n. 3, p. 501-513.

Hunt, G.R., and Salisbury, J.W., 1970, Visible and near-infrared spectra of minerals and rocks, I. Silicate Minerals: Modern Geology, v. 1, p. 283-300.

Hunt, G.R., and Salisbury, J.W., 1971, Visible and near-infrared spectra of minerals and rocks: II. Carbonates: Modern Geology, v. 2, p. 23-30.

Keller, W.D., 1978, Kaolinization of feldspar as displayed in scanning electron micrographs: Geology, v. 6, p. 184-188.

285

Lilburn, R.A., and Al-Shaieb, Z., 1983, Geochemistry and isotopic composition of hydrocarbon-induced diagenetic aureole (HIDA), Cement Field, Oklahoma (Part 1): Shale Shaker, December 1983, p. 40-56.

Marrs, R.W., and Kaminsky, B., 1977, Detection of petroleum-related soil anomalies from Landsat, in 29th Annual Field Conference: Wyoming Geology Association Guidebook, p. 353-365.

Molenaar, C.M., 1981, Mesozoic stratigraphy of the Paradox Basin an overview, in Geology of the Paradox Basin: Rocky Mountain Assoc. of Geologists, Denver, p. 119-128.

Olmstead, R.W., Houson, R.E., May, R.T., and Owens, R.T., 1976, Summary of the stratigraphy, sedimentology and mineralogy of Pennsylvanian and Permian rocks of Oklahoma in relation to uranium-resource potential: Oklahoma State University Contract At. (05-1)-1641.

Roberts, A.A., Donovan, T.J., Dalziel, M.C., and Forgey, R.L., 1978, Application of helium surveys to petroleum exploration [abs.]: U.S. Geol. Survey Prof. Paper, n. 1100, p. 24.

Rowan, L.C., and Abrams, M.J., 1978, Evaluation of Landsat multispectral scanner images for mapping altered rocks in the East Tintic Mountains, Utah: U.S. Geological Survey Open-File Report 78-736, 73 p.

Rowan, L.C., Wetlaufer, P.H., Goetz, A.F.H., Billingsley, F.C., and Stewart, J.H., 1974, Discrimination of rock types and detection of hydrothermally altered areas in south central Nevada by the use of computer enhanced ERTS images: U.S.G.S. Professional Paper 883, 35 p.

Saunders, D.F., 1980, Use of Landsat geomorphic and tonal anomalies in petroleum prospecting: Unconventional Methods in Exploration for Petroleum and Natural Gas, II, Southern Methodist University Press, Dallas, 1980.

Segal, D.B., 1983, Use of Landsat multispectral scanner data for the definition of limonitic exposures in heavily vegetated areas: Economic Geology, v. 78, p. 711-722.

Soha, J.M., and Schwartz, A.A., 1978, Multispectral histogram normalization contrast enhancement, Presented to the 5th Canadian Symposium on Remote Sensing, Victoria, August 1978.

Walker, T.R., 1967, Formation of redbeds in modern and ancient deserts: Geol. Soc. Amer. Bull., v. 78, p. 353-368.

Weir, G.W., Puffett, W.P., and Dodson, C.L., 1961, Preliminary geologic map of the Mount Peale 4 NW quadrangle, San Juan County, Utah: U.S. Geological Survey Mineral Investigations Field Studies Map MF-151, scale 1:24,000.

Wilson, M.D., and Pittman, E.D., 1977, Authigenic clays in sandstones: recognition and influence on reservoir properties and paleoenvironmental analysis: Jour. Sed. Pet., v. 47, p. 3-31.

286

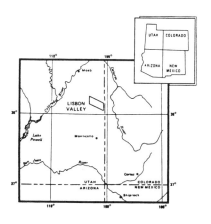

Figure 1. Location map of the Lisbon Valley Study Area

287

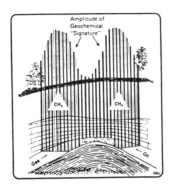

Figure 3. Schematic diagram of a Geochemical Aureole (from Duchscherer, 1982)

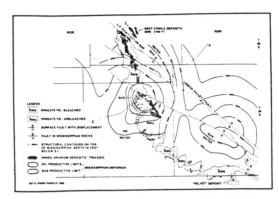

Figure 2. Distribution of bleached Wingate outcrops, Chinle ore deposits and Mississippian oil and gas accumulations on Lisbon Valley anticline. Solid lines around ore bodies represent approximate limits of mineralization (after Conel and Niesen, 1981).

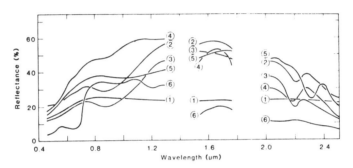

Figure 4. Field-acquired reflectance spectra: 1. unaltered tuff fragments and soil; 2. argillized andesite fragments; 3. silicified dacite; 4. opaline tuff; 5. tan marble; 6. ponderosa pine. The gaps at 1.4 and 1.9 μm are the result of atmospheric water absorption (from Goetz and Rowan, 1981).

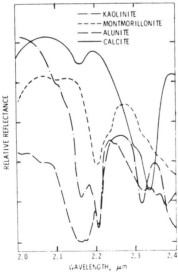

Figure 6. Laboratory spectral reflectance curves for kaolinite, montmorillonite, alunite and calcite (from Hunt and Ashley, 1979).

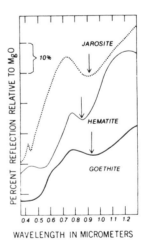

Figure 5. Laboratory spectral reflectance curves for jarosite, hematite, and goethite (from Hunt and Ashley, 1979).

288

574

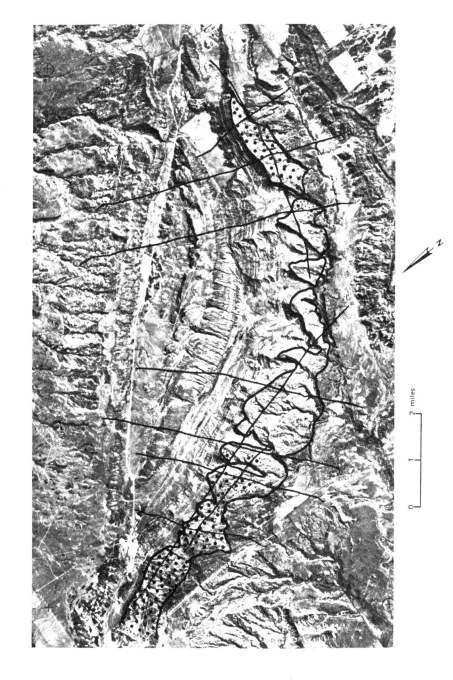

Figure 7.   Photomosaic of Lisbon Valley, showing the location of airborne spectroradiometric flight lines and exposures of the Wingate Formation.

Bleached exposures in the Lisbon Field area (left) and Three Step Hill area (right) are shown by the stippled pattern.

289

575

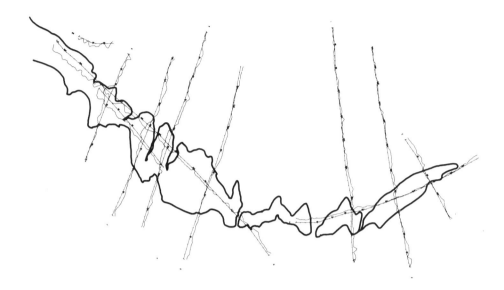

Figure 9. Map showing the variation in the 2.1/2.15 band ratio. Note the very high values associated with the bleached Wingate exposed in the Lisbon Field area.

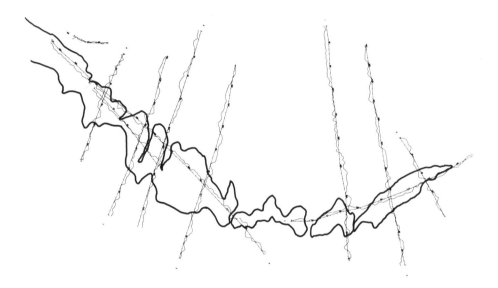

Figure 8. Map showing the variation in the (.55/.65)/(.75/.9) compound ratio. Note the relatively high values associated with the bleached exposures and the transition to lower ratio values associated with the unbleached Wingate.

290

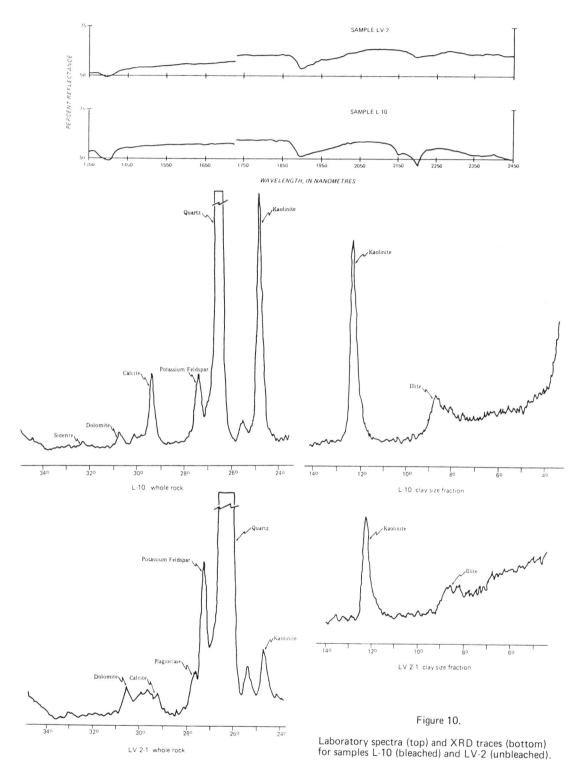

Figure 10.

Laboratory spectra (top) and XRD traces (bottom) for samples L-10 (bleached) and LV-2 (unbleached).

291

577

TABLE 1: XRD AND THIN SECTION ANALYSIS
XRD ANALYSIS

	Quartz	Potassium Feldspar	Plagioclase	Calcite	Dolomite	Siderite	Hematite	Illite	Kaolinite
Bleached									
L-5	45	19	0	3	3	<0.5	0	4	27
L-10	53	14	0	3	<0.5	<0.5	0	4	25
L-26	43	14	0	6	1	<0.3	0	4	32
SLV14	62	20	0	1	<1	<1	0	3	12
Unbleached									
LV-2	64	23	2	<1	1	0	0	1	9
* L-7	64	25	3	1	4	0	0	?	?

THIN-SECTION ANALYSIS

	Quartz	Orthoclase	Microline	Plagioclase	Calcite or Dolomite	Hematite	Kaolinite
Bleached							
LV4-T	>15	<15?	<5	0	<5	<1	<15
LV3-T	>15	>15?	<5	0	<5	<1	<15
Unbleached							
LV2-T	>15	<15?	<5	<5	<10	<10	<1

* Run from 20° 2-theta to 35° 2-theta. Percentage values were
  calculated based on the relative quantity of minerals detectable
  in this range.

TABLE 2: SUMMARY OF RESULTS

FACIES	SAMPLE	MSS	TMS	AIR	LAB	CARBONATE	XRD KAOLINITE	ILLITE
BLEACHED								
	L-5	NL	NL,CR	mk	K	5	27	4
	L-9	NL	NL,CR	B	K			
	L-10	NL	NL,CR	sK	K	3	25	4
	L-11	NL	NL,CR	mK	K			
	* L-15	NL	NL,CR	mC	M			
	L-23	NL	NL,CR	sK	K			
	L-24	NL	NL,CR	mK	Km			
	L-26	NL	NL,CR	mK	K	7	32	4
	SLV-14	NL	NL,CR		K	1	12	3
	LV-4	NL	NL,CR	mK				
	LV-40	NL	NL,CR	B	K			
UNBLEACHED								
	L-6	L	L,CP	mC	M			
	L-7	L	L,CP	B	mK			
	L-13	L	L,CP	mM	M			
	LV-2	L	L,CP	B	M	<1	9	1

* Three Step Hill area

LEGEND

MSS, TMS

L – limonite
NL – non-limonite

TMS

CP – clay-poor
CR – clay-rich

AIR

B – background
sK – strong kaolinite
mK – moderate kaolinite
mM – moderate muscovite
mC – moderate carbonate

LAB

K – kaolinite
M – muscovite
Km – kaol./musc. mixture
     kaolinite dominant
mK – musc./kaol. mixture
     muscovite dominant

292

Reprinted by permission of the Environmental
Research Institute of Michigan from *Proceedings of
the Third Thematic Conference on Remote Sensing for
Exploration Geology*, 1984, v. 1, p. 229-247.

RECONNAISSANCE HYDROCARBON PROSPECT GENERATION WITH INTEGRATED

POTENTIAL FIELD AND REMOTELY SENSED DATA

SOUTHERN ILLINOIS BASIN*

Darcy L. Vixo and J. Gregory Bryan
Aero Service Division
Western Geophysical Company of America
Houston, Texas, U.S.A.

INTRODUCTION

The purpose of this project was to research techniques for defining probable areas of hydrocarbon occurrence in the southern Illinois Basin using the low cost, reconnaissance tools, LANDSAT, Defense Mapping Agency (DMA) Bouguer gravity, and aeromagnetic data.

The fundamental geological premise on which this research was based is that basement features, shallow, subsurface structural controls, and surface lineaments, can be recognized and correlated by the integrated interpretation of gravity/magnetics, second vertical derivative gravity and LANDSAT imagery.

The general method involved an initial comparison of the gravity and magnetic data with interpreted Landsat imagery and with the distribution of oil and gas fields illustrated on the "Oil and Gas Industry in Illinois" map (Meents, 1977). General relationships were observed and subsequent cross-correlations were made. Specific geophysical criteria were ultimately derived which defined regions of rather small areal extent that were characterized by a significant level of hydrocarbon occurrence. (To date, no gas fields are known to exist in the study area.)

The study area lies within the southern portion of the Illinois Basin between Townships 7-4 south and Ranges 2-5 east (see Figure 1). Known major geologic features bound the area. The deepest part of the Illinois Basin, the Fairfield Basin, is located to the east of the study area. The Cottage Grove Fault System, which appears to be a complex wrench fault zone, exhibits dominantly right lateral slip, (Nelson et al, 1981) and trends east-southeast bounding the study area to the south. On the west, the study area encompasses a portion of the generally north-trending DuQuoin Monocline which dips steeply to the east into the Fairfield Basin.

Approximately 2,700 to 4,000 meters of Cambrian through Pennsylvanian aged sediments overlie Precambrian crystalline basement in the study area. Flat-lying Pennsylvanian rocks, largely covered by Illinoian glacial drift and Quaternary alluvial sediments, are present on the surface. Hydrocarbon reserves in the south-central portion of the Illinois Basin are almost equally distributed between structural and stratigraphic traps, and are usually less than 1,200 meters deep. The hydrocarbons are found in Mississippian sandstones and bioclastic carbonate rocks, reefs and local structural traps developed in Silurian carbonate rocks, fracture-controlled dolomitized streaks in Ordovician rocks, porous dolomites associated with an unconformity at the top of the Lower Ordovician, and sandstones of Cambrian age. The total area for individual, proven hydrocarbon reservoirs in the south-central Illinois Basin is reported to vary between 0.04 and 73.0 square kilometers; however,

---

*Presented at the International Symposium on Remote Sensing of Environment, Third Thematic Conference, Remote Sensing for Exploration Geology, Colorado Springs, Colorado, April 16-19, 1984.

229

most of the hydrocarbon reservoirs appear to occupy areas varying between 0.5 and 6 square kilometers.

DATA

LANDSAT color additive composites (CAC) of the Carbondale, Illinois, April 10, 1976, path 024 row 034 (I.D. 154930) MSS scene and the Paducah, Kentucky, October 18, 1982, path 022 row 034 (I.D. 4009415574) Thematic Mapper (TM) scene were processed using Aero Service's GeoImages® software package and Interactive Digital Image Manipulation System (IDIMS). In GeoImages® and IDIMS processing, a 5 x 5 pixel spatial filter is used to enhance the interpretability of subtle linear and curvilinear features. Selection of the optimum filter size is critical in order not to over- or under-enhance the image. The 512 x 512 pixel subscenes of the LANDSAT data were subjected to this filtering to obtain the optimum edge enhancement for the structure and drainage in the study area.

An interpretation of lineaments and curvilineaments was performed, without reference to known producing structures in the area, on both the Carbondale MSS and Paducah TM scenes (see Figures 2 and 3). In the present study, lineaments of two kilometers or more in length were mapped. Lineaments may be defined by tonal changes; alignments of topographic features such as depressions, hills, stream or valley segments, wind or water gaps or shoreline segments; faults; fractures; alignments of vegetation, soil or different rock types. Curvilineaments often reflect intrusive bodies, impact craters, faults or reefs.

Subsequently, the LANDSAT-interpreted lineaments were discriminantly digitized with respect to hydrocarbon producing and nonproducing areas, then analyzed using Aero Service's Digital Lineament Analysis Package (DLAP). Lineaments in each cover type area were analyzed as individual groups and the lineaments were separated according to their direction into 18 sets of 10 degrees each and plotted on separate overlays. The curvilineaments and the total lineaments were also plotted on individual overlays. The reasons for separation and plotting these sets were to permit:

1) rapid and comprehensive identification of individual lineaments of a given directional set,

2) identification of trends,

3) recognition of presence or absence of a set in a given cover type or types, and

4) identification and location of intersection points of two or more sets of lineaments, whether in the area covered by the whole scene or a particular cover type.

Rose diagrams were also generated in the DLAP processing to aid in the statistical analysis of the interpreted linear features (see Figure 4). Plotting of the rose diagram for the total count of all lineaments or any given cover type is achieved by drawing a circle with a radius equal to the maximum percentage of events falling in a given 10 degree section of the diagram. This circle is divided into eighteen sections of 10 degrees each. The maximum percentage is then taken as a standard and percentages are normalized and plotted accordingly. The rose diagrams thus produced, show the

---

GeoImages® -- A registered servicemark of Aero Service Division, Western Geophysical Company of America.

230

normalized percentages of the number of lineaments within each increment for either given cover type.

The LANDSAT-interpreted lineaments were also subjected to a lineament density analysis. The data were digitized, gridded and contoured. In the current survey, a square grid mesh of 3.81 kilometers was used for the Carbondale lineaments and a square grid mesh of 2.54 kilometers was used for the Paducah lineaments. Once gridded, calculations were made to determine the total length of interpreted linear features per square kilometer (see Figure 5).

Variations in lineament density may be indicative of a number of geologically related phenomena. Where the density of the lineaments is high, for example, more faulting or fracturing might be expected, thus, possibly defining fault zones and possibly a higher hydrocarbon trapping potential. In another instance, areas where sediments are affected by basement structures, zones of high lineament density may be indicative of shallow basement, whereas zones of low lineament density may show where basement is deeper.

The Bouguer gravity data utilized were a subset of the DMA regional gravity library. The state of Illinois has an unusually high density of gravity stations (approximately 10% of the total library) which, for the most part, were collected and compiled by L. D. McGinnis at Northern Illinois University for the State Geological Survey. Approximately five hundred gravity stations occur within the 1,600 square kilometer study area, at an average station spacing of approximately 3 kilometers. The gravity coverage is generally uniform, though some areas have a greater density of gravity station control. Comparisons between the known oil fields and the gravity station locations established that gravity data were obtained over nearly all known oil fields, or were located within 500 meters or less from the mapped boundary of any fields. Gravity station data within the study area were gridded in a 500 meter square mesh, smoothed using a 1,500 meter radius moving second order polynomial fit filter, and contoured at a one milligal interval (see Figure 6).

Second vertical derivative and band-pass filtered gravity maps were generated from the gridded Bouguer data in order to aid in the resolution of more shallow structures. Second vertical derivative algorithms accentuate the high frequency noise components within the original data; therefore, if an optimal second vertical derivative gravity map is to be generated, it is necessary to use the proper high-cut frequency filter to eliminate noise and enhance geologically related high frequency gravity anomalies.

The high-cut frequency filter selection was based on both systematic and empirical evaluations. The evaluation techniques included power spectral analysis of the Bouguer gravity data, which generally aids in the separation of signal from noise. The signal/noise separation in the area was determined to be at a frequency between 0.144 and 0.250 cycles per kilometer (see Figure 7). To further define the correct filter frequency, several separate second vertical derivative gravity maps were created using various high-cut filter applications between the above-stated limits. The resultant maps were then compared to the original Bouguer data to determine which maps most accurately reflected the Bouguer anomalies. The technique of multiple filter evaluation indicated that the 0.20 cycles per kilometer filter appeared to produce the best second vertical derivative gravity map (see Figure 8).

To further evaluate the gravity data, the gridded Bouguer and second vertical derivative data were entered on the IDIMS. The principle advantage of using the IDIMS for this project was real time computer, color contouring which allowed rapid color assignment of the gravity image values in a number of different ways to enhance various subtle aspects of the data sets. This, for example, helped emphasize subtle linear trends previously unobserved in

231

the Bouguer and second vertical derivative gravity anomaly maps, and subtle relationships between the Bouguer gravity data and the positions of known oil fields.

In addition to the DMA gravity and LANDSAT data, aeromagnetic data from the Illinois State Geological Survey (see Figure 9) were qualitatively evaluated. The Oil and Gas Industry in Illinois map (Meents, 1977) was used as a comparative base for establishing relationships between areas of hydrocarbon production and the geophysical trends observed in the gravity, magnetic and LANDSAT imagery (see Figure 10).

PROSPECT GENERATION CRITERIA

In the current investigation, geophysical and remote sensing correlations were only attempted for structurally related oil fields. To define prospect areas, each data set was initially compared individually with the oil field map and correlation criteria were established. Areas in the combined data sources that confirmed positive correlation were outlined as areas of more probable hydrocarbon occurrence and labeled "Prospect Zones."

Bristol (1975) suggests that faulting plays an important role in the formation of traps in this area. In an effort to incorporate surface structural constraints, lineament or fracture density was considered in light of the other geophysical criteria. Approximately 67% of the known oil fields fell within zones of relatively high lineament or fracture density.

Observations made from the LANDSAT lineaments, aided by the digital analysis, suggest that there may be a significant correlation that can be made between the occurrence of northwest- and east-trending lineaments and the locations of known hydrocarbon reservoirs. Comparisons of these overlays indicated that approximately 80% of these intersecting or near intersecting pairs were associated with oil fields; however, the limited number of lineaments displaying these trend directions may be too few to yield statistically valid conclusions.

Comparisons made with the Bouguer data indicate that approximately 78% of the structurally related oil fields occur on major east- or north-trending positive gravity anomalies. The amplitude and wavelength of these anomalies suggest that they may be basement related, with the east-west and north-south trending features possibly being associated with the generations of deformation evidenced by the Cottage Grove Fault System and the DuQuoin Monocline, respectively.

Interestingly, 78% of the structurally related oil fields showed a similar relationship to the anomalies in the magnetic data. Specifically, oil fields were located on, or just north of, the high of major east- or north-trending magnetic anomalies. Comparison of the correlated gravity and magnetic anomalies suggests the possibility of the same or similar basement sources.

Geologic evidence supports at least two possible explanations for basement related magnetic and gravity anomalies. The first possibility is a source such as ultrabasic intrusive bodies, which display a higher density and magnetic susceptibility than surrounding rock types. Alternatively, the magnetic and gravity anomalies represent structural relief associated with the Precambrian basement surface. If the magnetic and gravity anomaly sources are strictly lithologic contrasts, their apparent relationship to known hydrocarbon reserves is obscure. Perhaps, in this case, the sources represent erosionally remnant ridges (monadnocks) on the Precambrian surface. Then, with burial, differential compaction of the surrounding sediments possibly produced slopes allowing hydrocarbon migration toward the ridge axes.

232

From drilling records, it is known that most of the hydrocarbon production occurs at 1,200 meters or less. Therefore, in an attempt to classify the more shallow structural controls on the formation of hydrocarbon traps, second vertical derivative gravity data were analyzed. Approximately 67% of the known structurally related oil fields are associated with positive second vertical derivative gravity anomalies.

RESULTS

In an effort to eliminate areas of lower prospect probability, the above-described criteria were plotted in map format in the following categories:

I.   positive lineament density anomalies,

II. positive lineament density and Bouguer gravity anomalies,

III. positive lineament density, Bouguer gravity and magnetic total intensity anomalies,

IV. positive lineament density, Bouguer gravity, magnetic total field and second vertical derivative gravity anomalies (see Figure 11).

Table I illustrates the areas of production versus total prospect in the total study area and in each of the four categories.

TABLE I

CATEGORY	PRODUCTIVE AREA (KM^2)	PROSPECT AREA (KM^2)	PERCENTAGE (PROD./PROS. AREA)	APPROX. CHANCES FOR STRIKE
TOTAL	69	1600	4.3	1:23
I	34	671	5.1	1:20
II	17	270	6.3	1:16
III	16	156	10.3	1:10
IV	16	112	14.3	1:7

Figure 12 graphically illustrates the results in Table I. These observations strongly suggest that the application of the above-described criteria in this area, though in restricting the areas of production, actually seem to increase the probability of oil field occurrence within a prospect zone. In the current investigation, it was noted that all but three of the defined prospect zones had oil fields associated with them. To further check the integrity of this method of reconnaissance prospect generation, another test was performed. Comparison of the prospect map with the most recent (1983) "Oil and Gas Fields in Illinois" (after prospect selection was finalized) revealed that two of those three previously mentioned prospect zones also had subsequent hydrocarbon production.

SUMMARY, CONCLUSIONS, AND RECOMMENDATIONS

The purpose of this project was to research techniques for defining probable areas of hydrocarbon occurrence in the southern Illinois Basin using the low cost, reconnaissance tools of LANDSAT, DMA Bouguer gravity and aeromagnetic data.

The MSS and TM LANDSAT data were interpreted for linear and curvilinear features. The interpretations were digitized and analyzed using Aero Service's Digital Lineament Analysis Package. Additionally, the digitized interpretations were used to generate lineament density contour maps.

233

The gravity maps utilized were a Bouguer anomaly map contoured at 1 milligal and a second vertical derivative gravity anomaly map contoured at 0.1 milligal/kilometer^2.

The developed data sets were then individually compared to the 1977 "Oil and Gas Industry in Illinois" map to define relationships. Once correlations were identified, the geophysical data sets were coregistered and areas at overlapping positive correlation were outlined as prospect zones.

Trend analysis of the interpreted LANDSAT lineaments suggests that certain trend directions (i.e., northwest and east) can be attributed to the areas labeled as productive zones. Further, it was noted that approximately 80% of these lineament pairs displayed close proximal relationships to known oil fields.

The lineament density analysis revealed that 67% of the lineament density highs were related to known oil field positions. These observations, though perhaps somewhat influenced by the presence of the extensive glacial cover within the study area, nevertheless suggest the possibility of significant trend and lineament or fracture density relationships to producing oil fields as evidenced by the DLAP and lineament density results.

The Bouguer gravity and aeromagnetic data display strong correlations with known hydrocarbon occurrence. Additionally, comparisons of the Bouguer and aeromagnetic anomaly maps suggest the source or sources for the more broad wavelength anomalies are similar or the same. The geologic sources for these anomalies are probably basement related structures associated with generations of deformation evidenced by the Cottage Grove Fault System and the DuQuoin Monocline.

The second vertical derivative gravity anomaly map displays good correlation with the oil map especially when confined by the positive correlation area of the Bouguer gravity and aeromagnetic maps. In this zone of overlapping criteria of magnetic and Bouguer data, nearly all the second derivative anomalies are associated with known oil fields and appear to be reliable indicators for identifying areas of higher probability of hydrocarbon occurrence.

The results reported here might be improved by subsequent follow-up work over a larger study area to ensure statistically more valid conclusions; utilization of higher quality potential field data, particularly high sensitivity aeromagnetic data and more closely spaced gravity data; and more detailed characterization of specific structural trapping mechanisms by means of seismic data, well log correlations and geophysical modeling.

Reconnaissance prospect generation utilizing geologically well-constrained geophysical and remote sensing criteria has the potential of being developed into a valuable exploration technique, allowing rapid, low cost evaluation of well-developed or frontier hydrocarbon provinces.

ACKNOWLEDGEMENTS

The authors wish to thank Aero Service Division for sponsoring this project and to express particular gratitude to Glen T. Penfield, Dr. Allen Feder, John Kelley, Rick Prucha, Richard Ray, Peter Emmet, and LuAnn Smith.

234

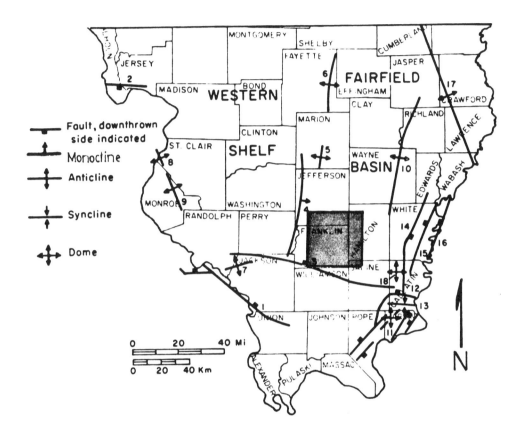

Figure 1. - LOCATION OF STUDY AREA (SHADED).
(after Heigold, 1976,fig.5). Structural features
in and around the southwestern Illinois study area
1-Ste. Genevieve Fault Zone; 2-Cap au Gres Flexure;
3-Cottage Grove Fault System; 4-Du Quoin Monocline;
5-Salem Anticline; 6-Louden Anticline; 7-Ava-Campbell
Hill Anticline; 8-Dupo Anticline; 9-Waterloo Anticline;
10-Clay City Anticlinal Belt; 11-Hicks Dome; 12-Shawneetown
Fault Zone; 13-Eagle Valley Syncline; 14-Ridgway Fault;
15-Herald-Phillipstown Fault; 16-Maunie Fault; 17-La Salle
Anticlinal Belt; 18-Omaha Dome.

235

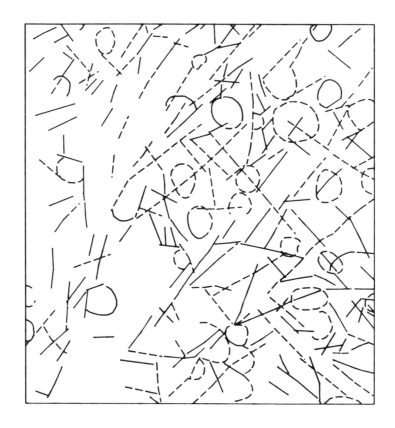

Figure 2. - INTERPRETED LINEAMENTS AND CURVILINEAMENTS
FROM CARBONDALE MSS SCENE. SCALE APPROX. 1 : 300,000.

236

Figure 3. – INTERPRETED LINEAMENTS AND CURVILINEAMENTS
FROM PADUCAH TM SCENE. SCALE APPROX. 1 : 300,000.

237

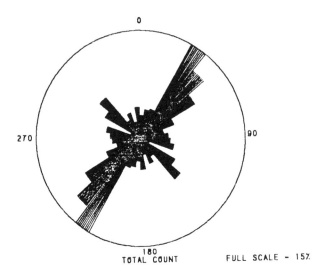

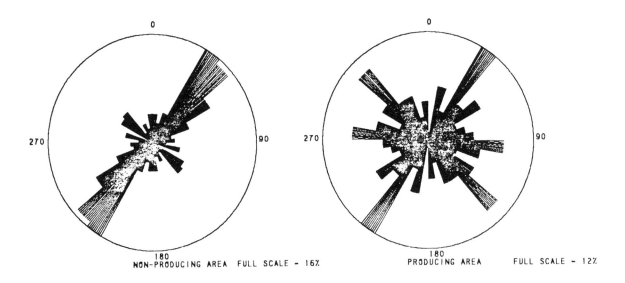

Figure 4. – ROSE DIAGRAMS, CARBONDALE MSS INTERPRETATION.
Observed trend directions for Total, Producing and Non-producing areas.

238

Figure 5. - LINEAMENT DENSITY MAP, CARBONDALE MSS INTERPRETATION.
Grid Mesh = 3.81 x 3.81 km. Contour Interval = 0.2 km. per sq. km.

239

Figure 6. – DMA BOUGUER GRAVITY. Contour Interval = 1 Milligal.

240

590

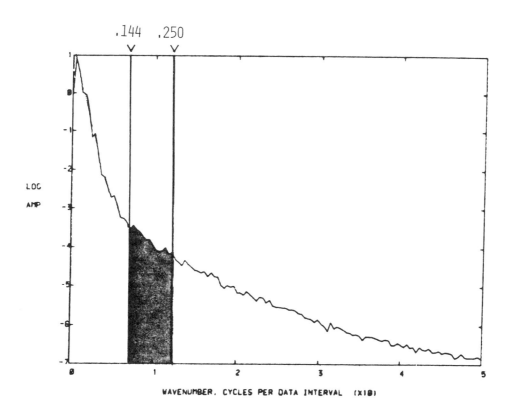

Figure 7. - AMPLITUDE SPECTRUM. Selected frequency cuttoffs: 0.144,0.250.

241

Figure 8. - SECOND VERTICAL DERIVATIVE GRAVITY MAP.
Contour Interval = 0.1 milligal per sq. km.

242

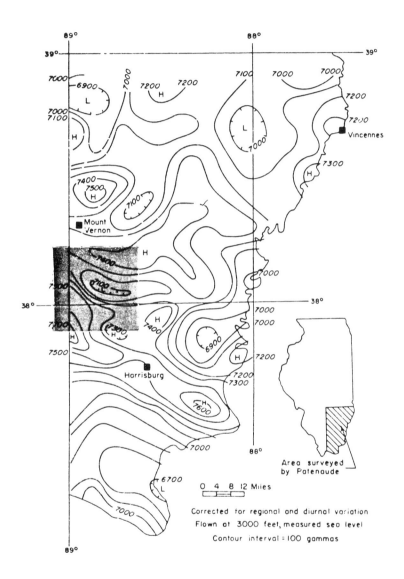

Figure 9. – TOTAL FIELD AEROMAGNETIC MAP OF SOUTHEASTERN
ILLINOIS (after Patenaude, 1964). Study area is shaded.

243

593

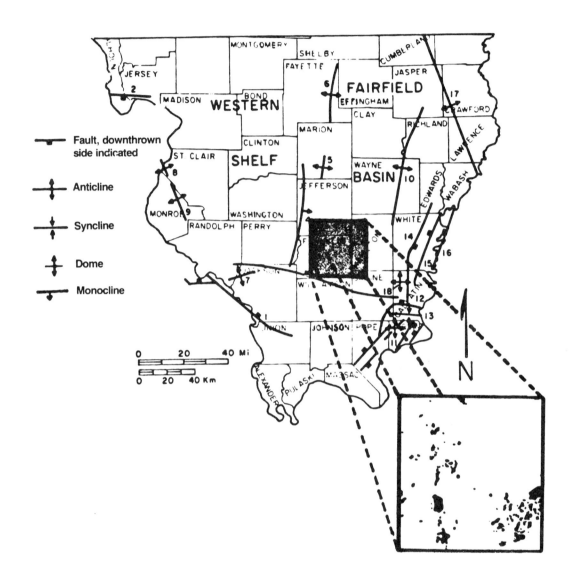

Oilfield Location Diagram

Figure 10. – Structurally related oil fields (after Meents, 1977).

244

Figure 11. - FINAL PROSPECT ZONES (HACHURED), where all
four prospecting criteria are met.

245

595

Figure 12. - Bar Chart of productive areas at each level
of prospect generation. I) Positive lineament
density; II) Positive lineament density and
Bouguer gravity anomalies; III) Positive
lineament density, Bouguer gravity and magnetic
anomalies; IV) Positive lineament density,
Bouguer gravity, magnetic and second vertical
derivative gravity anomalies.

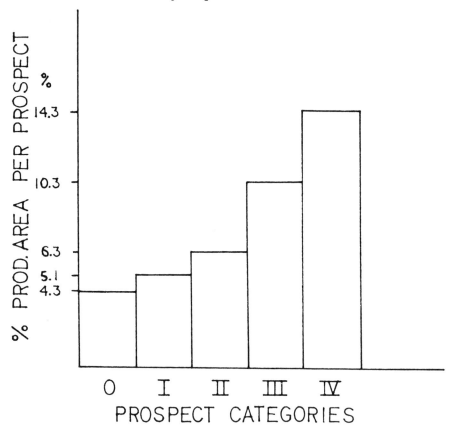

SELECTED REFERENCES

Bond, D. C., Atherton, E., Bristol, H. M., Buschbach, T. C., Stevenson, D. L., Becker, L. E., Dawson, T. A., Fernalld, E. C., Schwalb, H., Wilson, E. N., Statler, A. T., Stearns, E. G., and Buchner, J. H., 1971, Possible future petroleum potential of region 9 - Illinois Basin, Cincinnati Arch, and Northern Mississippi Embayment, in I. H. Cram ed., Future petroleum provinces of the United States--their geology and potential: AAPG Memoir 15, pp. 1165-1218.

Bristol, H. M., 1975, Structural geology and oil production of Northern Gallatin County and Southernmost White County, Illinois: Illinois State Geological Survey, Illinois Petroleum 105, p. 20.

Damon, Norman, ed., 1970, International oil and gas development, part 2, U. S., Canada, and Foreign: International Oil Scouts Association, v. XL, pp. 67-87.

Harding, T. P., 1974, Petroleum traps associated with wrench faults: American Association of Petroleum Geologist Bulletin, v. 58, no. 7, pp. 1290-1304.

Heigold, P. C., 1976, An aeromagnetic survey of Southwestern Illinois: Illinois State Geological Survey, circular 495, scale 1:250,000.

Kolata, D. R., Treworgy, J. D., and Masters, J. M., 1981, Structural framework of the Mississippi embayment of Southern Illinois: Illinois State Geological Survey, circular 516, 38 p.

McGinnis, L. D., 1970, Tectonics and the gravity field in the continental interior: Journal of Geophysical Research, v. 75, no. 2, pp. 317-31.

McGinnis, L. D., Ervin, C. P., 1974, Earthquakes and block tectonics in the Illinois Basin: Geology, v. 2, no. 10, pp. 517-19.

Meents, W. F., 1977, Oil and gas industry in Illinois: Illinois State Geological Survey, scale 1:500,000.

Nelson, W. J., Krausse, H. F., and Bristol, H. M., 1981, The Cottage Grove Fault System in southern Illinois: Illinois State Geological Survey, circular 522, 65 p.

Oil and gas fields in Illinois, 1983, Illinois State Geological Survey, scale 1:360,000.

O'Leary, D. W., Friedman, J. D., and Pohn, H. A., 1976, Lineaments, linear, lineations: Some proposed new standards for old terms: Geological Society of America Bulletin, v. 87, pp. 1463-1469.

Paternaude, R. W., 1964, Results of regional aeromagnetic surveys of eastern upper Michigan, central lower Michigan, and southeastern Illinois: University of Wisconsin, Geophysics and Polar Research Center Rpt. 64-2, 51 p.

Treworgy, J. D., 1981, Structural features in Illinois - a compendium: Illinois State Geological Survey, circular 519, 22 p., scale 1:680,000.

Vanderberg, J. and Elyn, J. R., 1983, Petroleum industry in Illinois, 1981: Illinois State Geological Survey, Illinois Petroleum 124, 136 p.

Willman, J. B. et al, compilers, 1967, Geologic Map of Illinois: Illinois State Geological Survey, scale 1:500,000.

247

Reprinted by permission of the Environmental
Research Institute of Michigan from *Proceedings of
the Fourth Thematic Conference on Remote Sensing
for Exploration Geology*, 1985, v. 1, p. 381-389.

REMOTE SENSING AND SURFACE GEOCHEMICAL STUDY OF RAILROAD VALLEY NYE COUNTY, NEVADA*

V. T. Jones and S. G. Burtell
Exploration Technologies, Inc., 3911 Fondren, Houston, TX 77063, U.S.A.

R. A. Hodgson
Geological Consulting Services, 403 Liberty St. Jamestown,
PA 16134, U.S.A.

Tom Whelan and Charlie Milan
Woodward Clyde Oceaneering, 7330 Westview, Houston, TX 77055, U.S.A

Takeshi Ando, Kinya Okada, and Takashi Agatsuma
Japex Geoscience Institute, Inc., Akasaka Twin Tower, 2-17-22
Akasaka, Minato-Ku, Tokyo 107, Japan

Osamu Takano
Earth Resources Satellite Data Analysis Center (ERS-DAC),
No. 39 Mori Bldg., 4-5 Azabudai, 2 Chome, Minato-Ku, Tokyo 106, Japan

ABSTRACT

A remote sensing and soil gas geochemical survey has been completed in Railroad Valley, Nevada for Japex, Tokyo Japan. The initial survey was designed around an existing stuctural model published by Foster and Dolly (1979). This study provides a test of this model with remote sensing and geochemical information. Combined interpretation, using SAR, TM and TDCS images has suggested a series of major fracture lines which define regional fault and fracture systems. Many fracture lines cross the valley graben and may reflect structural subdivisions within the graben blocks, some of these confirm some of the structural divisions identified by Foster and Dolly.

There appears to be a reasonable correlation between the mapped fracture systems, the geochemical anomalies, and the existing oil fields, although any one set of data alone does not delineate the fields. A comparison of the geochemical data with the oil fields and remote sensing interpretations shows good correlations with the identified fields. Light hydrocarbon magnitude and compositional anomalies appear to reflect preferential migration along certain fracture systems on the flanks of the valley updip from the oil reservoirs. Although direct soil gas anomalies do not occur over the reservoirs, adjacent valley boundary faults do exhibit a significant flux of hydrocarbons updip from all the presently known accumulations.

---

*Presented at the Fourth Thematic Conference: "Remote Sensing for Exploration Geology", San Francisco, California, April 1-4, 1985.

INTRODUCTION

The objective of this study is to determine whether or not geochemical anomalies occur at locations that can be predicted by any of the proposed structural models of Railroad Valley reservoirs, and to establish the relationship of structures to interpretations from remote sensing data in Railroad Valley. The four principal elements in this project are: 1) the analysis and interpretation of remote sensing data; 2) geological field checking of the interpretations, 3) the field collection and analysis of geochemical data, and 4) the integration of the geological, remote sensing, and geochemical data.

GENERAL GEOLOGY

Although the Basin and Range Province has been sparsely explored for oil and gas, surface macroseeps are known, and shows of oil and gas have been encountered in many exploratory wells. Railroad Valley has been the most actively explored valley within the Great Basin since the discovery of the Eagle Springs Field in 1954. It currently has four producing fields: Eagle Springs, Trap Spring, Bacon Flat and Grant Canyon, from which over 10 million barrels of oil have been produced. The Trap Spring and Eagle Springs fields are combination stratigraphic, truncation subcrop fault traps, which also may be true of the subeconomic Currant discovery. Production occurs from matrix and fracture porosity in reservoirs in the lacustrine Sheep Pass Formation (Cretaceous and Eocene) and the Garrett Ranch volcanic group (Oligocene), (Dolly 1979). The most unique feature of these fields is that production occurs from the highest position on the lowermost fault block at the basin margin. On the adjacent higher fault blocks the reservoir beds were removed or altered by erosion during Basin and Range deformation (Foster 1979). The Grant Canyon and Bacon Flat fields are also located in structurally low, intermediate fault blocks, with production reported to be from the Middle Devonian Simonson dolomite (McCaslin, 1984). It has not been reported whether production is confined to fractured zones or to primary porosity in the dolomite or whether the Sheep Pass Formation and Garrett Volcanic Ranch Group are present and/or productive in this area. The sources of the oils in Railroad Valley are probably a combination of the Late Cretaceous-Eocene Sheep Pass Formation and Mississippian Chainman shale with the additional possibility of a varying degree of mixing from these sources within any particular reservoir (Picard, 1960).

REMOTE SENSING

The remote sensing study of Railroad Valley was based on the examination of Synthetic Aperture Radar (SAR), Thematic Mapper Landsat Imagery (TM), and TDCS-processed TM data.

SAR - Synthetic Aperture Radar

The airborne radar covering Railroad Valley which was interpreted for this project was synthetic aperture radar purchased from Aero Service. This data comprises a portion of their non-exclusive proprietary survey, 81-5 and primarily delineates the topographic aspects of the terrain. Topographic elements result from the effect of climatic events acting over time on developing regional and local geologic structures and areal lithologic differences at the surface. Also of importance is whether the environment is degradational (as in the mountains) or aggradational (as in the main valley floor). Topographic features obviously are of greatest magnitude in degradational environments. Much more subtly expressed but equally extensive topographic features (as regards structural significance) are found in aggradational environments. Some of these are recognized through drainage alignments, extensive low-relief scarps or ridges, or abrupt distinctive changes in the fabric of the

382

topography across a line. The superior spatial resolution of the SAR allows the definition of such low relief features in the central part of Railroad Valley including several in the playa area of the valley. The linear radar features mapped, whether rectilinear, curvilinear, oval or circular, are believed to correspond to zones of a higher level of structural activity. Most commonly this is an increased density of fracturing along a line and predominantly in the direction of the line, whether rectilinear of curvilinear. Less commonly they may correspond to faults or a series of faults of varying displacements depending on other structural factors. Least commonly the lines or segments of lines may correspond to narrow flexures which may or may not be faults at some other stratigraphic level (Figure 1).

LANDSAT

Landsat imagery used in the mapping and interpretation of structural and morphological features of interest in Railroad Valley was provided by Japex Geoscience Institute, Inc., sponsor of the project. Images used for interpretation were Thematic Mapper (TM) images covering the main (central) region of Railroad Valley, provided as paper Polaroid prints of CRT displays. Black and white images were also provided in each of the TM spectral bands 1-5 and 7 with the exception of Data from band 6 (Thermal IR Emission) which was too poor a quality for interpretation and two composite color images of combined spectral bands 2, 3 and 4.

Much more information than had been initially anticipated was derived from the several images studied. The data from all sources appear totally compatible with respect to azimuthal distribution and, to a large extent, with respect to areal distribution and location of specific features. It was found, however, that each image provided a different aspect of the area and in each case, additional data. As a result there is not necessarily a direct correspondence of features mapped from one image to another. The pattern of linear features is fairly complex but not random. Elements of the pattern fall generally into well-defined azimuthal groupings of about N45E , N65E , E-W , N60W , N40W , N25W , and N-S (these values are + 5°). Within the NE and NW trending sets there is also a tendency for a grouping into sub-equally spaced belts within the set. These require study to determine if they correspond to the larger Landsat lineaments apparent on Landsat regional mosaics.

Three sets of CRT color composite images were generated by Japex to further enhance Landsat (TM) features in areas of low topographic relief and high (and narrow range) reflectance values such as the pediments, bajadas and the playa of Railroad Valley. Of the three sets of data generated, the TDCS set yielded a maximum amount of useful structural information with respect to visual inspection. Within the TDCS data set the discrimination of lithologic and soil differences of significance was excellent and allowed the recognition and mapping of a number of previously unrecognized morpho-structural elements in the playa and sand dune areas of the central part of the valley. It is apparent from the TDCS images that the primary trend of the main graben of the valley is about NNE and not N-S, although there are important N-S structural elements which appear in the northern part of the valley, as well as N-W. These data, in conjunction with those from the regional mosaics suggests that the central (and probably deepest) part of the Railroad Valley graben underlies the playa of the valley (Figure 2).

Active Structural Lines: In mapping the several color and black and white images, it was noted that there were several well-defined regional lines or zones which appear to be comprised of a series of closely-spaced, small-order features. They are particularly apparent in the playa and bajada areas but are also readily defined as continuing features in the pediments and mountains east and west of the valley. They follow generally the azimuthal lines displayed by the smaller order features but, those striking NW, NE and in the south about E-W clearly cross mountains and valleys without consideration for the local structural geometrics. The fact that they are so

383

well displayed in the imagery suggests recent structural activity along their lines.

Composite Structural Lines: In studying the maps showing the linear features derived from the radar and Thematic Mapper imagery it became apparent that many of the shorter lines lie in narrow belts which can be readily defined visually. An overlay was placed on each of the three interpretive maps, and the boundaries of the belts drawn. The three sets of data were then compiled on a single sheet and are shown on Fig. 3.

The features of this map show what may be considered the larger-scale composite TM lineaments of the mapped area. This map is of particular interest in that it shows that although a number of the composite lineaments are fully defined on a single image, several are composed of segments each
segment of which is defined on a single set of data. In particular these are lines crossing the valley in the NE-SW direction. In this example it is clearly illustrated how a single linear structural element may have aspects which are not totally defined by a single type of remote sensor.

A comparison of the remote sensing features mapped show a number of coincidences of ASL's and linear belts. Where this occurs it provides an added dimension to the structural inferences of the lineament. Taken together the linear features compiled from this unique combination of remote sensing data outline the primary structural framework of the mapped area.

GEOCHEMISTRY AND REMOTE SENSING INTERPRETATION

The geochemical 4' probes technique used for this study is the same as that described by Matthews, Jones and Richers (1984) at the 1984 Symposium held in Colorado Springs, Co. The composition of Railroad Valley C2/C3 data as compared to 4 foot probe surveys at Patrick Draw, Wyoming and Lost River, West Virginia show a close similarity to Patrick Draw predicting an oil potential for this area (Figure 4). Contoured geochemical data reveals a number of well defined magnitude anomalies which seem to be indirectly related to the petroleum production in Railroad Valley (Figure 5). The anomalies for the most part lie on the edges of the graben blocks and along the valley boundary faults. These anomalies reflect up-dip migration of light hydrocarbons along fault and fracture planes which are an active component of continuing basin formation. The hydrocarbons which originate at depth in the basin where source formations are undergoing maturation are channeled along existing fracture systems and contained in permeable sand and gravel valley fill formations and migrate toward the valley margins. When the gas comes in contact with fractures related to basin and range normal faulting it is communicated to the surface along such pathways (Figure 6). This relationship is apparent in the vicinity of the Eagle Springs and Grant Canyon, Bacon Flat oil fields. The Trap Springs field has a direct propane anomaly over the northern half of the field which continues onto the adjacent uplifted block. This direct geochemical anomaly may be related to direct vertical migration along the mapped fault from this fractured reservoir.

High magnitude geochemical anomalies have also been identified over shallow buried pediment surfaces and along the edges of structural blocks as identified by Foster (1979),(Figure 7). These structures are bound by the active structural lines as identified by remote sensing data. The fractures and faults of this type control both the size and shape of the graben blocks. An examination of the strike of these linear features reveals that the majority of the high magnitude geochemical sites lie along these features which also bind the limits of groups of high magnitude sites. The intersections of one or more of these features also show high magnitudes in some cases. The linear features do not show anomalous values in all areas and seem to be dependent on an active source of light hydrocarbons. The northeast part of the geochemical area, in the vicinity of Currant is very closely correlated to divisions as defined by the lineaments. Geochemical anomalies in this area are sectioned off

384

and bounded by these active structural lines.

High magnitude geochemical sites are also located on pediment surfaces far removed from the valley bounding faults that control up dip migration of gases. The anomalies are over Paleozoic blocks which may not have a high hydrocarbon potential such as sites at the south eastern limit of the survey area. The sites are located over Ordivocian and possibly even Cambrian formations from which there is little source rock potential. The source of the measured hydrocarbons is probably lateral migration along regional fracture systems that extend across the basin and into the adjacent ranges. Hydrocarbons from mature subsurface formations migrate along the fractures which are probably also closely related to the localization of oil in reservoirs. This type of migration could extend into the adjacent pediment areas. An examination of the location of these anomalies versus compositestructural lines from all the remote sensing data shows that the high magnitude samples were collected near linear systems which are probably a result of local or regional fracturing.

CONCLUSIONS

Remote sensing imagery is able to identify major fault and fracture systems which are related to the structural development of Railroad Valley. These systems define individual structural blocks which are confirmed by near surface geochemical data. Geochemical magnitude anomalies are located along fractures and faults updip from the oil fields and other prospective areas of the valley. Anomalies over pediment surfaces may reflect fracture migrated gas which probably originates in the graben structures. A combination of remote sensing and soil gas sampling can identify major near surface gas migration pathways in this complexly faulted valley. A model for interpretation is suggested by this study.

REFERENCES

Dolly, E.D., 1979. Geological Techniques used in the discovery of Trap Spring Field. p. 455-467 in RMAG-UGA: 1979 Basin and Range Symposium.

Duey, H.D., 1978. Trap Spring Oil Field, Nye County, Nevada. p. 469-476 in RMAG-UGA: 1979 Basin and Range Symposium.

Foster, N.H., 1979. Geomorphic exploration utilized in the discovery of Trap Spring Field, Nye County Nevada. p. 477-486 in RMAG-UGA: 1979 Basin and Range Symposium.

Jones, V.T., and Drozd, R.J., 1983. Predictions of oil and gas potential by near surface geochemistry: Am. Assoc. Petroleum Geologists Bulletin, v.67, no. 6, p. 932-952.

Matthews, M.D., Jones, V.T., and Richers, B.M., 1984. Remote Sensing and surface hydrocarbon leakage. Presentation at the International Symposium on Remote Sensing of Environment. Third Thematic Conference, Remote Sensing for Exploration Geology, Colorado Springs Colorado, April 16-19, 1984.

McCaslin, 1984, AAPG looks at Utah, Nevada basins: Oiland Gas Journal, 8 Oct. 1984, p. 107-108.

Picard, 1960, On the origin of oil, Eagle Springs Field, Nye County Nevada. p. 237-244 in Guidebook to the geology of east-central Nevada, Intermtn. Assoc. Petroleum Geologists and Eastern Nevada Geol. Soc., 11th Ann. Field Conf., Salt Lake City Utah.

385

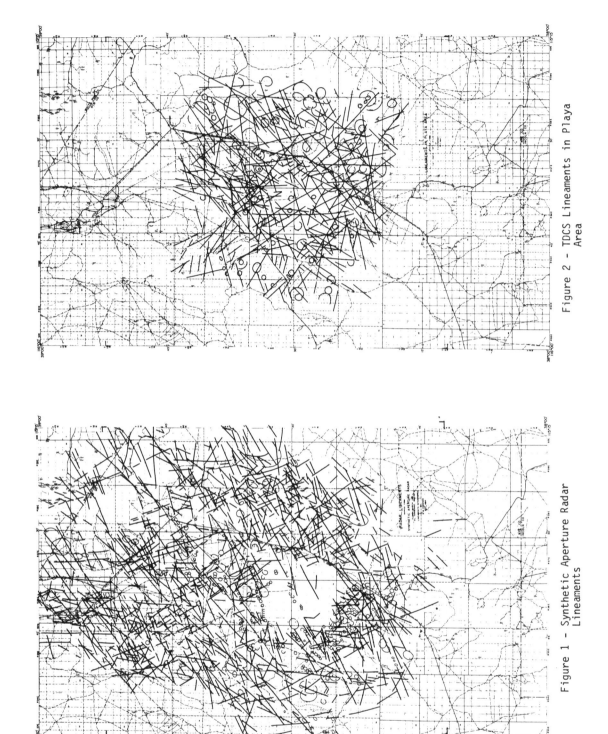

Figure 2 - TDCS Lineaments in Playa
Area

Figure 1 - Synthetic Aperture Radar
Lineaments

386

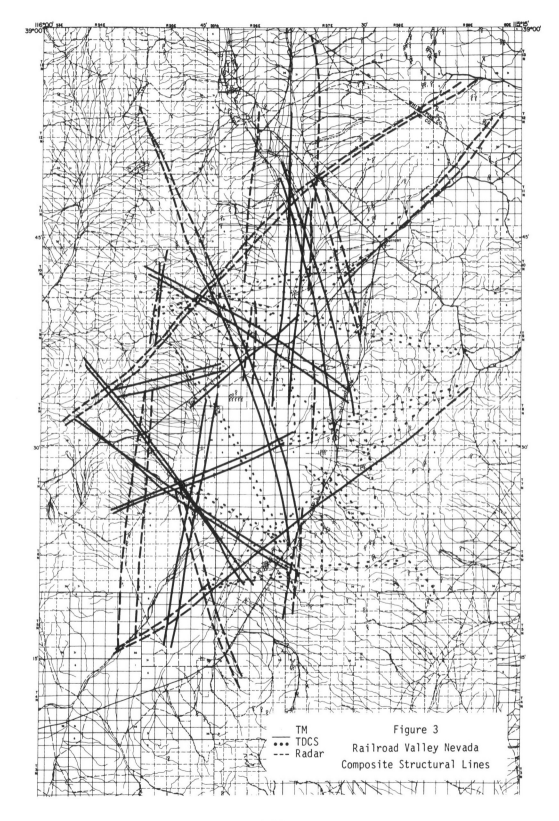

	TM	Figure 3
	TDCS	Railroad Valley Nevada
	Radar	Composite Structural Lines

387

605

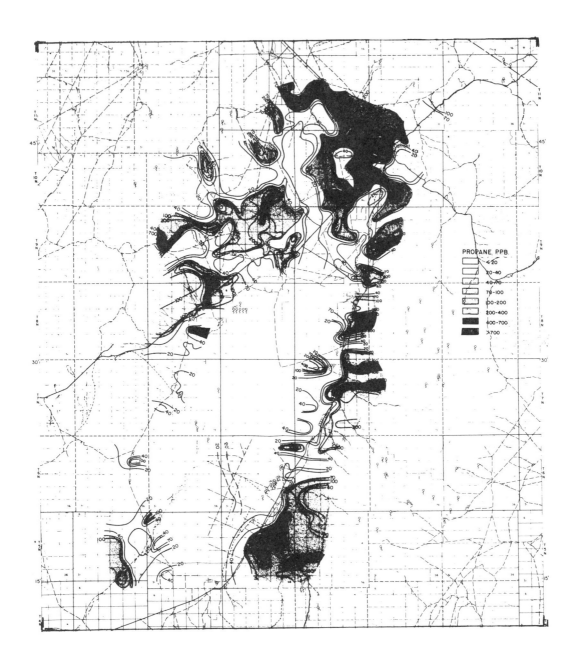

Figure 5
Contoured Propane Magnitudes

388

606

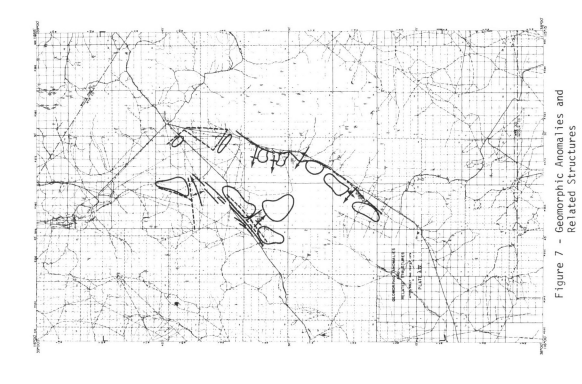

Figure 7 - Geomorphic Anomalies and
Related Structures

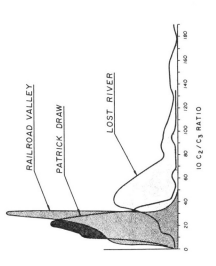

Figure 4. Ethane to Propane Ratio (C2/C3)
Probe Data from Railroad Valley, Nevada;
Patrick Draw; Sweetwater County, Wyoming
and Lost River in Hardy County, West
Virginia

RAILROAD VALLEY

PATRICK DRAW

LOST RIVER

10 C₂/C₃ RATIO

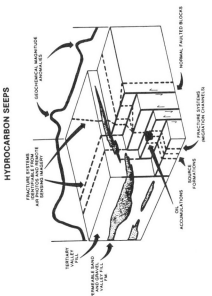

Figure 6 - Hydrocarbon Seeps

HYDROCARBON SEEPS

389

607